# Non-linear Finite Element Analysis of Solids and Structures

VOLUME 1: ESSENTIALS

# Non-linear Finite Element Analysis of Solids and Structures

## VOLUME 1: ESSENTIALS

**M. A. Crisfield**
*FEA Professor of Computational Mechanics*
*Department of Aeronautics*
*Imperial College of Science, Technology and Medicine*
*London, UK*

JOHN WILEY & SONS
Chichester · New York · Brisbane · Toronto · Singapore

Telephone (+44) 1243 779777

Email (for orders and customer service enquiries): cs-books@wiley.co.uk
Visit our Home Page on www.wileyeurope.com or www.wiley.co.uk

Reprinted April 2000, August 2001, September 2003



***Other Wiley Editorial Offices***

John Wiley & Sons Inc., 111 River Street, Hoboken, NJ 07030, USA

Jossey-Bass, 989 Market Street, San Francisco, CA 94103-1741, USA

Wiley-VCH Verlag GmbH, Boschstr. 12, D-69469 Weinheim, Germany

John Wiley & Sons Australia Ltd, 33 Park Road, Milton, Queensland 4064, Australia

John Wiley & Sons (Asia) Pte Ltd, 2 Clementi Loop #02-01, Jin Xing Distripark, Singapore 129809

John Wiley & Sons Canada Ltd, 22 Worcester Road, Etobicoke, Ontario, Canada M9W 1L1

***British Library Cataloguing in Publication Data***

A catalogue record for this book is available from the British Library

ISBN 0471 97059 X

Typeset by Thomson Press (India) Ltd., New Delhi. India
Printed and bound in Great Britain by TJI Digital, Padstow, Cornwall
This book is printed on acid-free paper responsibly manufactured from sustainable forestry in which at least two trees are planted for each one used for paper production.

# Contents

# Preface

This book was originally intended as a sequal to my book *Finite Elements and Solution Procedures for Structural Analysis, Vol 1—Linear Analysis*, Pineridge Press, Swansea, 1986. However, as the writing progressed, it became clear that the range of contents was becoming much wider and that it would be more appropriate to start a totally new book. Indeed, in the later stages of writing, it became clear that this book should itself be divided into two volumes; the present one on 'essentials' and a future one on 'advanced topics'. The latter is now largely drafted so there should be no further changes in plan!

Some years back, I discussed the idea of writing a book on non-linear finite elements with a colleague who was much better qualified than I to write such a book. He argued that it was too formidable a task and asked relevant but esoteric questions such as 'What framework would one use for non-conservative systems?' Perhaps foolishly, I ignored his warnings, but I am, nonetheless, very aware of the daunting task of writing a 'definitive work' on non-linear analysis and have not even attempted such a project.

Instead, the books are attempts to bring together some concepts behind the various strands of work on non-linear finite elements with which I have been involved. This involvement has been on both the engineering and research sides with an emphasis on the production of practical solutions. Consequently, the book has an engineering rather than a mathematical bias and the developments are closely wedded to computer applications. Indeed, many of the ideas are illustrated with a simple non-linear finite element computer program for which Fortran listings, data and solutions are included (floppy disks with the Fortran source and data files are obtainable from the publisher by use of the enclosed card). Because some readers will not wish to get actively involved in computer programming, these computer programs and subroutines are also represented by flowcharts so that the logic can be followed without the finer detail.

Before describing the contents of the books, one should ask 'Why further books on non-linear finite elements and for whom are they aimed?' An answer to the first question is that, although there are many good books on linear finite elements, there are relatively few which concentrate on non-linear analysis (other books are discussed in Section 1.1). A further reason is provided by the rapidly increasing computer power and increasingly user-friendly computer packages that have brought the potential advantages of non-linear analysis to many engineers. One such advantage is the ability to make important savings in comparison with linear elastic analysis by allowing, for example, for plastic redistribution. Another is the ability to directly simulate the collapse behaviour of a structure, thereby reducing (but not eliminating) the heavy cost of physical experiments.

While these advantages are there for the taking, in comparison with linear analysis, there is an even greater danger of the 'black-box syndrome'. To avoid the potential dangers, an engineer using, for example, a non-linear finite element computer program to compute the collapse strength of a thin-plated steel structure should be aware of the main subject areas associated with the response. These include structural mechanics, plasticity and stability theory. In addition, he should be aware of how such topics are handled in a computer program and what are the potential limitations. Textbooks are, of course, available on most of these topics and the potential user of a non-linear finite element computer program should study such books. However, specialist texts do not often cover their topics with a specific view to their potential use in a numerical computer program. It is this emphasis that the present books hope to bring to areas such as plasticity and stability theory.

Potential users of non-linear finite element programs can be found in the aircraft, automobile, offshore and power industries as well as in general manufacturing, and it is hoped that engineers in such industries will be interested in these books. In addition, it should be relevant to engineering research workers and software developers. The present volume is aimed to cover the area between work appropriate to final-year undergraduates, and more advanced work, involving some of the latest research. The second volume will concentrate further on the latter.

It has already been indicated that the intention is to adopt an engineering approach and, to this end, the book starts with three chapters on truss elements. This might seem excessive! However, these simple elements can be used, as in Chapter 1, to introduce the main ideas of geometric non-linearity and, as in Chapter 2, to provide a framework for a non-linear finite element computer program that displays most of the main features of more sophisticated programs. In Chapter 3, these same truss elements have been used to introduce the idea of 'different strain measures' and also concepts such as 'total Lagrangian', 'up-dated Lagrangian' and 'corotational' procedures. Chapters 4 and 5 extend these ideas to continua, which Chapter 4 being devoted to 'continuum mechanics' and Chapter 5 to the finite element discretisation.

I originally intended to avoid all use of tensor notation but, as work progressed, realised that this was almost impossible. Hence from Chapter 4 onwards some use is made of tensor notation but often in conjunction with an alternative 'matrix and vector' form.

Chapter 6 is devoted to 'plasticity' with an emphasis on $J_2$, metal plasticity (von Mises) and 'isotropic hardening'. New concepts such as the 'consistent tangent' are fully covered. Chapter 7 is concerned with beams and rods in a two-dimensional framework. It starts with a shallow-arch formulation and leads on to 'deep-formulations' using a number of different methods including a degenerate-continuum approach with the total Lagrangian procedure and various 'corotational' formulations. Chapter 8 extends some of these ideas (the shallow and degenerate-continuum, total Lagrangian formulations) to shells.

Finally, Chapter 9 discusses some of the more advanced solution procedures for non-linear analysis such as 'line searches', quasi-Newton and acceleration techniques, arc-length methods, automatic increments and re-starts. These techniques are introduce into the simple computer program developed in Chapters 2 and 3 and are then applied to a range of problems using truss elements to illustrate such responses as limit points, bifurcations, 'snap-throughs' and 'snap-backs'.

It is intended that Volume 2 should continue straight on from Volume 1 with, for example, Chapter 10 being devoted to 'more continuum mechanics'. Among the subjects to be covered in this volume are the following: hyper-elasticity, rubber, large strains with and without plasticity, kinematic hardening, yield criteria with volume effects, large rotations, three-dimensional beams and rods, more on shells, stability theory and more on solution procedures.

## REFERENCES

At the end of each chapter, we will include a section giving the references for that chapter. Within the text, the reference will be cited using, for example, [B3] which refers to the third reference with the first author having a name starting with the letter 'B'. If, in a subsequent chapter, the same paper is referred to again, it would be referred to using, for example, [B3.4] which means that it can be found in the References at the end of Chapter 4.

## NOTATION

We will here give the main notation used in the book. Near the end of each chapter (just prior to the References) we will give the notation specific to that particular chapter.

### General note on matrix/vector and/or tensor notation

For much of the work in this book, we will adopt basic matrix and vector notation where a matrix or vector will be written in bold. It should be obvious, from the context, which is a matrix and which is a vector.

In Chapters 4–6 and 8, tensor notation will also be used sometimes although, throughout the book, all work will be referred to rectangular cartesian coordinate systems (so that there are no differences between the co- and contravariant components of a tensor). Chapter 4 gives references to basic work on tensors.

A vector is a first-order tensor and a matrix is a second-order tensor. If we use the direct tensor (or dyadic) notation, we can use the same convention as for matrices and vectors and use bold symbols. In some instances, we will adopt the suffix notation whereby we use suffixes to refer to the components of the tensor (or matrix or vector). For clarity, we will sometimes use a suffix on the (bold) tensor to indicate its order. These concepts are explained in more detail in Chapter 4, with the aid of examples.

### Scalars

$E$ = Young's modulus
$e$ = error

$f$ = yield function
$g$ = out-of-balance force or gradient of potential energy
$G$ = shear modulus
$I$ = 2nd moment of area
$J$ = det(**F**)
$k$ = bulk modulus
$K_t$ = tangent stiffness
$t$ = thickness
$u, v, w$ = displacements corresponding to coordinates $x, y, z$
$V$ = volume: $dV$ = increment of volume; also
$V$ = virtual work
$V_i$ = internal virtual work
$V_e$ = external virtual work
$x, y, z$ = rectangular coordinates
$\gamma$ = shear strain
$\varepsilon$ = strain
$\mu$ = shear modulus
$\lambda$ = load-level parameter
$\nu$ = Poisson's ratio
$\xi, \eta, \zeta$; = non-dimensional (natural) coordinates
$\sigma$ = stress
$\tau$ = shear stress
$\phi$ = total potential energy
$\chi$ = curvature

## Subscripts

2 = second order
4 = fourth order
cr = 'critical' (in relation to buckling)
e = external
ef = external (fixed)
g = global
i = internal
n = new
o = old
t = tangential
v = virtual

## Superscripts and special symbols

. = rate or time-derivative
T = transpose
: = contraction (see equation (4.6))
$\otimes$ = tensor product (see equation (4.31))
tr = trace (= sum of diagonal elements)

## Vectors

$\mathbf{b}$ = strain/nodal-displacement vector
$\mathbf{d}$ = displacements
$\mathbf{e}_i$ = unit base vectors
$\mathbf{g}$ = out-of-balance forces (or gradient of total potential energy)
$\mathbf{h}$ = shape functions
$\mathbf{p}$ = nodal (generalised) displacement variables
$\mathbf{q}$ = nodal (generalised) force variables corresponding to $\mathbf{p}$
$\boldsymbol{\varepsilon}$ = strain (also, sometimes, a tensor—see below)
$\boldsymbol{\sigma}$ = stress (also, sometimes, a tensor—see below)

## Matrices or tensors

(A subscript 2 is sometimes added for a second-order tensor (matrix) with a subscript 4 for a fourth-order tensor.)

$\mathbf{1}$ = Unit second-order tensor (or identity matrix)
$\mathbf{B}$ = strain/nodal-displacement matrix
$\mathbf{C}$ = constitutive matrices or tensors (with stress/strain moduli)
$\mathbf{D}$ = diagonal matrix in $\mathbf{LDL}^{\mathrm{T}}$
$\mathbf{H}$ = shape function matrix
$\mathbf{I}$ = identity matrix or sometimes fourth-order unit tensor
$\mathbf{K}$ = tangent stiffness matrix
$\mathbf{K}_{t\sigma}$ = initial stress or geometric stiffness matrix
$\mathbf{K}_o$ = linear stiffness matrix
$\mathbf{L}$ = lower triangular matrix in $\mathbf{LDL}^{\mathrm{T}}$ factorisation
$\delta_{ij}$ = Kronecker delta ($=1, i=j;\ =0, i \neq j$)
$\boldsymbol{\varepsilon}$ = strain

## Special symbols with vectors or tensors

$\delta$ = small change (often iterative or virtual) so that $\delta\mathbf{p}$ = iterative change in $\mathbf{p}$ or iterative nodal 'displacements', $\delta\mathbf{p}_v$ = virtual change in $\mathbf{p}$
$\Delta$ = large change (often incremental—from last converged equilibrium state) so that $\Delta\mathbf{p}$ = incremental change in $\mathbf{p}$ or incremental nodal 'displacements'

# 1 General introduction, brief history and introduction to geometric non-linearity

## 1.1 GENERAL INTRODUCTION AND A BRIEF HISTORY

At the end of the present chapter (Section 1.5), we include a list of books either fully devoted to non-linear finite elements or else containing significant sections on the subject. Of these books, probably the only one intended as an introduction is the book edited by Hinton and commissioned by the Non-linear Working Group of NAFEMS (The National Agency of Finite Elements). The present book is aimed to start as an introduction but to move on to provide the level of detail that will generally not be found in the latter book.

Later in this section, we will give a brief history of the early work on non-linear finite elements with a selection of early references being provided at the end of the chapter. References to more recent work will be given at the end of the appropriate chapters.

Following the brief history, we introduce the basic concepts of non-linear finite element analysis. One could introduce these concepts either via material non-linearity (say, using springs with non-linear properties) or via geometric non-linearity. I have decided to opt for the latter. Hence, in this chapter, we will move from a simple truss system with one degree of freedom to a system with two degrees of freedom. To simplify the equations, the 'shallowness assumption' is adopted. These two simple systems allow the introduction of the basic concepts such as the out-of-balance force vector and the tangent stiffness matrix. They also allow the introduction of the basic solution procedures such as the incremental approach and iterative techniques based on the Newton–Raphson method. These procedures are introduced firstly via the equations of equilibrium and compatibility and later via virtual work. The latter will provide the basis for most of the work on non-linear finite elements.

### 1.1.1 A brief history

The earliest paper on non-linear finite elements appears to be that by Turner *et al.* [T2] which dates from 1960 and, significantly, stems from the aircraft industry. The present review will cover material published within the next twelve years (up to and including 1972).

Most of the other early work on geometric non-linearity related primarily to the linear buckling problem and was undertaken by amongst others [H3, K1], Gallagher *et al.* [G1, G2]. For genuine geometric non-linearity, 'incremental' procedures were originally adopted (by Turner *et al.* [T2] and Argyris [A2, A3]) using the 'geometric stiffness matrix' in conjunction with an up-dating of coordinates and, possibly, an initial displacement matrix [D1, M1, M3]. A similar approach was adopted with material non-linearity [Z2, M6]. In particular, for plasticity, the structural tangent stiffness matrix (relating increment of load to increments of displacement) incorporated a tangential modular matrix [P1, M4, Y1, Z1, Z2] which related the increments of stress to the increments of strain.

Unfortunately, the incremental (or forward-Euler) approach can lead to an unquantifiable build-up of error and, to counter this problem, Newton–Raphson iteration was used by, amongst others, Mallet and Marcal [M1] and Oden [O1]. Direct energy search [S2,M2] methods were also adopted. A modified Newton–Raphson procedure was also recommended by Oden [O2], Haisler *et al.* [H1] and Zienkiewicz [Z2]. In contrast to the full Newton–Raphson method, the stiffness matrix would not be continuously updated. A special form using the very initial, elastic stiffness matrix was referred to as the 'initial stress' method [Z1] and much used with material non-linearity. Acceleration procedures were also considered [N2]. The concept of combining incremental (predictor) and iterative (corrector) methods was introduced by Brebbia and Connor [B2] and Murray and Wilson [M8, M9] who thereby adopted a form of 'continuation method'.

Early work on non-linear material analysis of plates and shells used simplified methods with sudden plastification [A1,B1]. Armen *et al.* [A4] traced the elasto-plastic interface while layered or numerically integrated procedures were adopted by, amongst others, Marcal *et al.* [M5, M7] and Whang [W1] combined material and geometric non-linearity for plates initially involved 'perfect elasto-plastic buckling' [T1, H2]. One of the earliest fully combinations employed an approximate approach and was due to Murray and Wilson [M10]. A more rigorous 'layered approach' was applied to plates and shells by Marcal [M3, M5], Gerdeen *et al.* [G3] and Striklin *et al.* [S4]. Various procedures were used for integrating through the depth from a 'centroidal approach' with fixed thickness layers [P2] to trapezoidal [M7] and Simpson's rule [S4]. To increase accuracy, 'sub-increments' were introduced for plasticity by Nayak and Zienkiewicz [N1]. Early work involving 'limit points' and 'snap-through' was due to Sharifi and Popov [S3] and Sabir and Lock [S1].

## 1.2 A SIMPLE EXAMPLE FOR GEOMETRIC NON-LINEARITY WITH ONE DEGREE OF FREEDOM

Figure 1.1(a) shows a bar of area $A$ and Young's modulus $E$ that is subject to a load $W$ so that it moves a distance $w$. From vertical equilibrium,

$$W = N \sin\theta = \frac{N(z+w)}{l''} \simeq \frac{N(z+w)}{l} \tag{1.1}$$

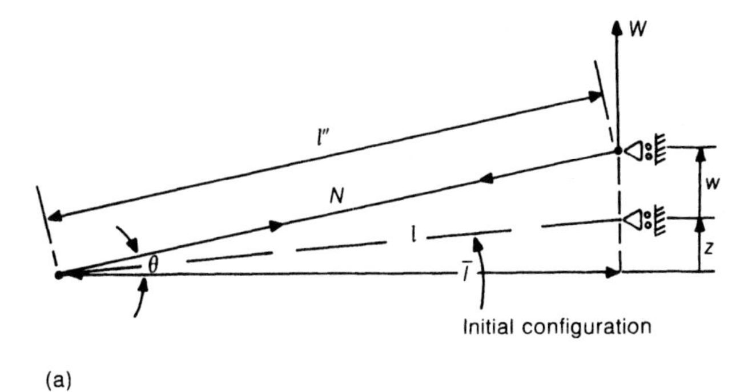


(a)

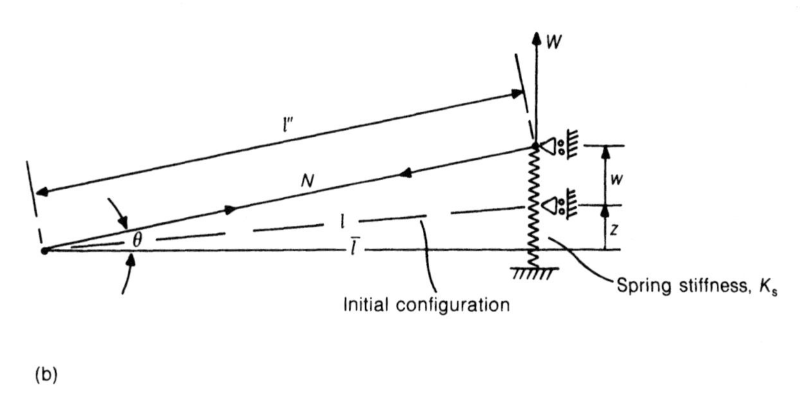


(b)

**Figure 1.1** Simple problem with one degree of freedom: (a) bar a lone (b) bar with spring.

where $N$ is the axial force in the bar and it has been assumed that $\theta$ is small. By Pythagoras's theorem, the strain in the bar is

$$\varepsilon = \frac{((z+w)^2 + \bar{l}^2)^{1/2} - (z^2 + \bar{l}^2)^{1/2}}{(z^2 + \bar{l}^2)^{1/2}} \tag{1.2}$$

$$= \left(1 + \left(\frac{z+w}{\bar{l}}\right)^2\right)^{1/2} \left(1 + \left(\frac{z}{\bar{l}}\right)^2\right)^{-1/2} - 1 \tag{1.3}$$

$$\simeq \left(1 + \frac{1}{2}\left(\frac{z+w}{\bar{l}}\right)^2\right)\left(1 - \frac{1}{2}\left(\frac{z}{\bar{l}}\right)^2\right) - 1 \tag{1.4}$$

$$= \left(\frac{z}{\bar{l}}\right)\left(\frac{w}{\bar{l}}\right) + \frac{1}{2}\left(\frac{w}{\bar{l}}\right)^2 \simeq \left(\frac{z}{l}\right)\left(\frac{w}{l}\right) + \frac{1}{2}\left(\frac{w}{l}\right)^2. \tag{1.5}$$

Although (1.5) is approximate, it can be used to illustrate non-linear solution procedures that are valid in relation to a 'shallow truss theory'. From (1.5), the force in the bar is given by

$$N = EA\varepsilon = EA\left(\left(\frac{z}{l}\right)\left(\frac{w}{l}\right) + \frac{1}{2}\left(\frac{w}{l}\right)^2\right) \tag{1.6}$$

and, from (1.1), the relationship between the load $W$ and the displacement, $w$ is given by

$$W = \frac{EA}{l^3}(z^2 w + \tfrac{3}{2}zw^2 + \tfrac{1}{2}w^3). \tag{1.7}$$

This relationship is plotted in Figure 1.2(a). If the bar is loaded with increasing $-W$, at point A (Figure 1.2(a)), it will suddenly snap to the new equilibrium state at point C. Dynamic effects would be involved so that there would be some oscillation about the latter point.

Standard finite element procedures would allow the non-linear equilibrium path to be traced until a point A′ just before point A, but at this stage the iterations would probably fail (although in some cases it may be possible to move directly to point C—see Chapter 9). Methods for overcoming this problem will be discussed in Chapter 9. For the present, we will consider the basic techniques that can be used for the equilibrium curve, OA′.

For non-linear analysis, the *tangent* stiffness matrix takes over the role of the stiffness matrix in linear analysis but now relates small changes in load to small changes in displacement. For the present example, this matrix degenerates to a scalar $\mathrm{d}W/\mathrm{d}w$ and, from (1.1), this quantity is given by

$$K_t = \frac{\mathrm{d}W}{\mathrm{d}w} = \frac{(z+w)}{l}\frac{\mathrm{d}N}{\mathrm{d}w} + \frac{N}{l} \tag{1.8}$$

$$= \frac{EA}{l}\left(\frac{z+w}{l}\right)^2 + \frac{N}{l} \tag{1.9}$$

$$= \frac{EA}{l}\left(\frac{z}{l}\right)^2 + \frac{EA}{l}\left(\frac{2zw + w^2}{l^2}\right) + \frac{N}{l}. \tag{1.10}$$

Equation (1.6) can be substituted into (1.10) so that $K_t$ becomes a direct function of the initial geometry and the displacement $w$. However, there are advantages in maintaining the form of (1.10) (or (1.9)), which is consistent with standard finite element formulations. If we forget that there is only one variable and refer to the constituent terms in (1.10) as 'matrices', then conventional finite element terminology would describe the first term as the linear stiffness matrix because it is only a function of the initial geometry. The second term would be called the 'initial-displacement' or 'initial-slope matrix' while the last term would be called the 'geometric' or 'initial-stress matrix'. The 'initial-displacement' terms may be removed from the tangent stiffness matrix by introducing an 'updated coordinate system' so that $z' = z + w$. In these circumstances, equation (1.9) will only contain a 'linear' term involving $z'$ as well as the 'initial stress' term.

The most obvious solution strategy for obtaining the load–deflection response OA′ of Figure 1.2(a) is to adopt 'displacement control' and, with the aid of (1.7) (or (1.6) and (1.1)), directly obtain $W$ for a given $w$. Clearly this strategy will have no difficulty with the 'local limit point' at A (Figure 1.2(a)) and would trace the complete equilibrium path OABCD. For systems with many degrees of freedom, displacement control is not so trivial. The method will be discussed further in Section 2.2.5. For the present we will consider load control so that the problem involves the computation of $w$ for a given $W$.

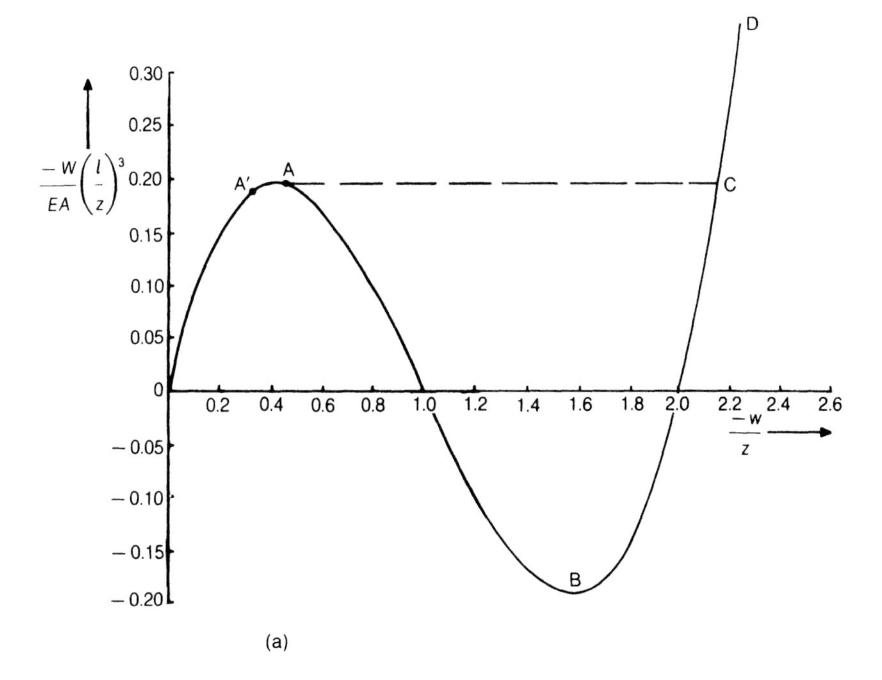


(a)

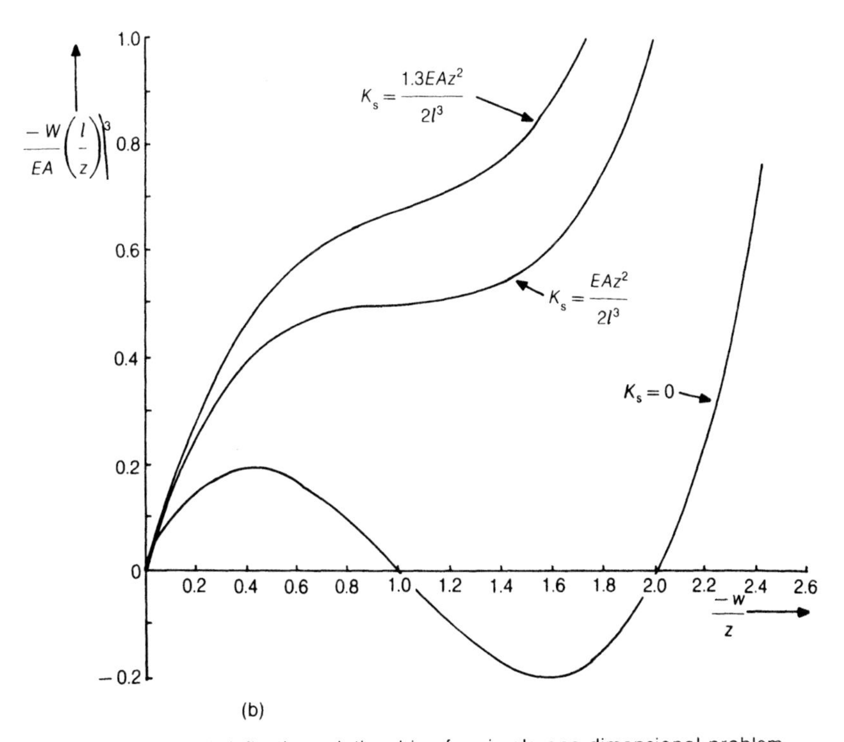


(b)

**Figure 1.2** Load/deflection relationships for simple one-dimensional problem.
(a) Response for bar alone.
(b) Set of responses for bar–spring system.

Before discussing a few basic solution strategies, some dimensions and properties will be given for the example of Figure 1.1(b) so that these solution strategies can be illustrated with numbers. The spring in Figure 1.1(b) has been added so that, if the stiffness $K_S$ is large enough, the limit point A of Figure 1.2(a) can be removed and the response modified to that shown in Figure 1.2(b). The response of the bar is then governed by

$$W = \frac{EA}{l^3}(z^2 w + \tfrac{3}{2} z w^2 + \tfrac{1}{2} w^3) + K_S w \tag{1.11}$$

which replaces equation (1.7). For the numerical examples, the following dimensions and properties have been chosen:

$$EA = 5 \times 10^7\,\mathrm{N}, \quad z = 25\,\mathrm{mm}, \quad l = 2500\,\mathrm{mm}, \quad K_S = 1.35\,\mathrm{N/mm}, \qquad \Delta W = -7\,\mathrm{N} \tag{1.12}$$

where $\Delta W$ is the incremental load. For brevity, the 'units' have been omitted from the following computations.

### 1.2.1 An incremental solution

An incremental (or Euler) solution scheme involves (Figures 1.2(a) and 1.3) repeated application of

$$\Delta w = \left(\frac{\mathrm{d}W}{\mathrm{d}w}\right)^{-1} \Delta W = K_t^{-1} \Delta W. \tag{1.13}$$

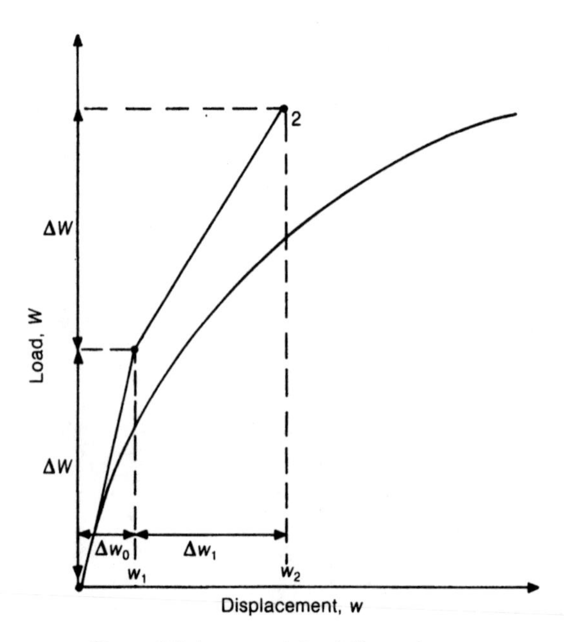


**Figure 1.3** Incremental solution scheme.

For the first step, $w_0$ and $N_0$ are set to zero so that, from (1.10):

$$K_0 = \frac{EA}{l}\left(\frac{z}{l}\right)^2 + K_S = 3.35 \tag{1.14}$$

and hence

$$w_1 = \Delta w_0 = K_0^{-1}\Delta W = \frac{l^3 \Delta W}{EAz^2 + K_S l^3} = -7/3.35 = -2.0896 \tag{1.15}$$

where $\Delta W$ $(-7)$ is the applied incremental load. From (1.6), the corresponding axial force is given by

$$N_1 = EA\left\{\left(\frac{z}{l}\right)\left(\frac{w_1}{l}\right) + \frac{1}{2}\left(\frac{w_1}{l}\right)^2\right\} = -400.45. \tag{1.16}$$

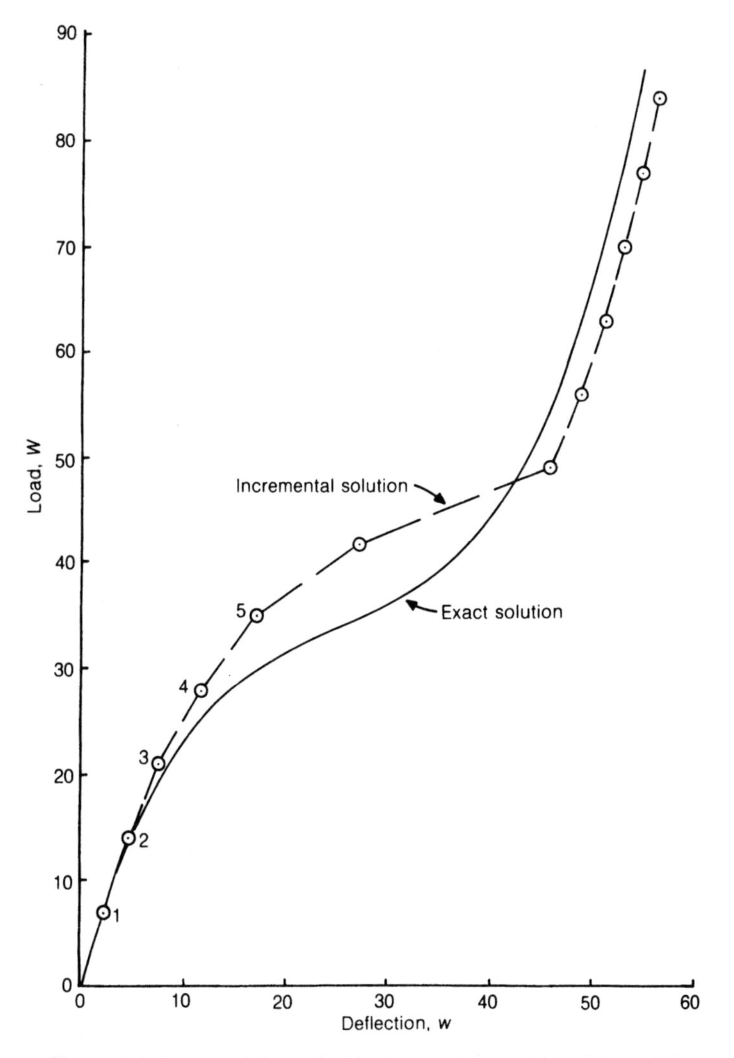


**Figure 1.4** Incremental solution for bar–spring problem ($K_s = 1.35$).

The second increment of load is now applied using (see (1.10))

$$K_1(w_1) = \frac{EA}{l}\left(\frac{z}{l}\right)^2 + \frac{EA}{l^3}(2zw_1 + w_1^2) + \frac{N_1}{l} + K_S = 2.8695 \qquad (1.17)$$

to give

$$\Delta w_1 = K_1^{-1}\Delta W = -7/2.8695 = -2.4394 \qquad (1.18)$$

so that

$$w_2 = w_1 + \Delta w_1 = -2.0896 - 2.4394 = -4.5290 \qquad (1.19)$$

and $N_2$ is computed from

$$N_2 = EA\left\{\left(\frac{z}{l}\right)\left(\frac{w_2}{l}\right) + \frac{1}{2}\left(\frac{w_2}{l}\right)^2\right\} = -823.76. \qquad (1.20)$$

Inevitably (Figures 1.3 and 1.4), the solution will drift from the true equilibrium curve. The lack of equilibrium is easily demonstrated by substituting the displacement $w_1$ of (1.15) and the force $N_1$ of (1.16) into the equilibrium relationship of (1.1). Once allowance is made for the spring stiffness $K_S$, this provides

$$W_1 = N_1\frac{(z + w_1)}{l} + K_S w_1 = \Delta W\left(1 + \frac{\Delta W}{2EA}\left(\frac{z}{l}\right)^3\right)\left(1 + \left(\frac{l}{z}\right)^3\frac{\Delta W}{EA}\right) + K_S\frac{\Delta W l}{EA}\left(\frac{l}{2}\right)^2 \qquad (1.21)$$

$$= -3.6698 - 2.8210 = -6.4908 \qquad (1.22)$$

which is only approximately equal to the applied load $\Delta W$ $(-7)$.

### 1.2.2 An iterative solution (the Newton–Raphson method)

A second solution strategy uses the well-known Newton–Raphson iterative technique to solve (1.7) to obtain $w$ for a given load $W$. To this end, (1.7) can be re-written as

$$g = \frac{EA}{l^3}(z^2 w + \tfrac{3}{2}zw^2 + \tfrac{1}{2}w^3) - W = 0. \qquad (1.23)$$

The iterative procedure is obtained from a truncated Taylor expansion

$$g_n \simeq g_o + \frac{dg_o}{dw}\delta w + \left(\frac{1}{2}\frac{d^2 g_o}{dw^2}(\delta w)^2\right) \qquad (1.24)$$

where terms such as $dg_o/dw$ imply $dg/dw$ computed at position 'o'. Hence, given an initial estimate $w_o$ for which $g_o(w_o) \neq 0$, a better approximation is obtained by neglecting the bracketed and higher-order terms in (1.24) and setting $g_n = 0$. As a result (Figure 1.5)

$$\delta w_o = -\left(\frac{dg_o}{dw}\right)^{-1} g_o(w_o) \qquad (1.25)$$

and a new estimate for $w$ is

$$w_1 = w_0 + \delta w_0. \qquad (1.26)$$

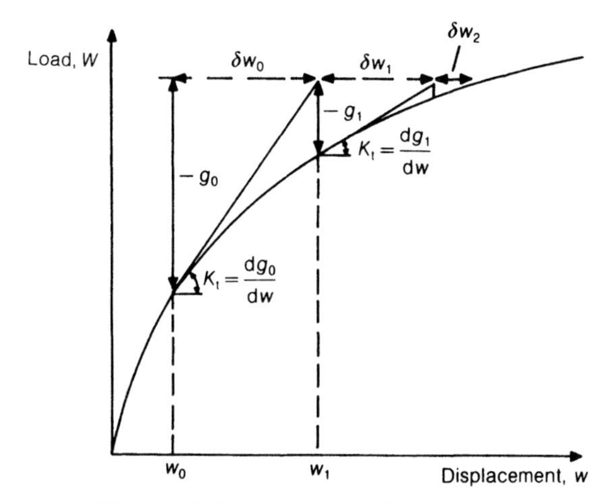


**Figure 1.5** The Newton–Raphson method.

Substitution of (1.25) into (1.24) with the bracketed term included shows that $g_n$ is proportional to $g_0^2$. Hence the iterative procedure possess 'quadratic convergence'.

Following (1.26), the iterative process continues with

$$\delta w_1 = -\left(\frac{\mathrm{d}g_1}{\mathrm{d}w}\right)^{-1} g_1(w_1). \tag{1.27}$$

In contrast to the previous incremental solutions, the $\delta w$s in (1.24)–(1.27) are *iterative* changes at the same fixed load level (Figure 1.5).

Equations (1.25) and (1.27) require the derivative, $\mathrm{d}g/\mathrm{d}w$, of the residual or out-of-balance force, $g$. But (1.23) was derived from (1.7) which, in turn, came from (1.1) so that an alternative expression for $g$, based on (1.1), is

$$g = \frac{N(z+w)}{l} - W \tag{1.28}$$

where $W$ is the fixed *external* loading. Consequently:

$$\frac{\mathrm{d}g}{\mathrm{d}w} = \frac{(z+w)}{l}\frac{\mathrm{d}N}{\mathrm{d}w} + \frac{N}{l} = K_t \tag{1.29}$$

which coincides with (1.8) so that $\mathrm{d}g/\mathrm{d}w$ is the tangent stiffness term previously derived in (1.8).

However, although $\mathrm{d}g/\mathrm{d}w$ will be referred to as $K_t$ and, indeed, involves the same formulae ((1.8)–(1.10)), there is an important distinction between (1.8), which is a genuine tangent to the *equilibrium path* ($W - w$), and $\mathrm{d}g/\mathrm{d}w$, which is to be used with an iterative procedure such as the Newton–Raphson technique. In the latter instance, $K_t = \mathrm{d}g/\mathrm{d}w$ does not necessarily relate to an equilibrium state since $g$ relates to some trial $w$ and is not zero until convergence has been achieved. Consequently, for equilibrium states relating to a stable point on the equilibrium path, such as points on the solid parts of the curve on Figure 1.6, $K_t = \mathrm{d}W/\mathrm{d}w$ will always be positive although $K_t = \mathrm{d}g/\mathrm{d}w$, as used in an iterative procedure, may possibly be zero or

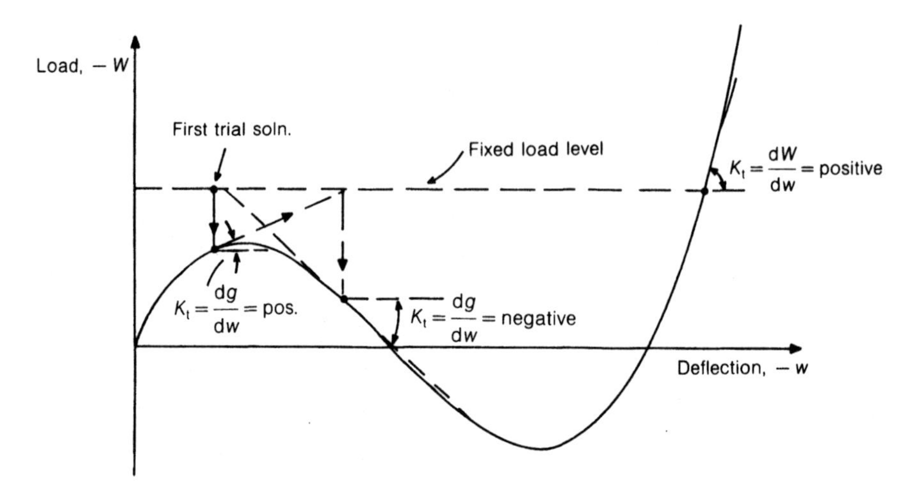


**Figure 1.6** Positive and negative tangent stiffnesses.

negative. This is illustrated for the Newton–Raphson method in Figure 1.6. Once the problems are extended beyond one variable, the statement '$K_t$ will always be positive' becomes '$K_t$ will always be positive definite' while '$K_t$ may possibly be zero or negative' becomes '$K_t$ may possibly be singular or indefinite'.

### 1.2.3 Combined incremental/iterative solutions (full or modified Newton–Raphson or the initial-stress method)

The iterative technique on its own can only provide a single 'point solution'. In practice, we will often prefer to trace the complete load/deflection response (equilibrium path). To this end, it is useful to combine the incremental and iterative solution procedures. The 'tangential incremental solution' can then be used as a 'predictor' which provides the starting solution, $w_0$, for the iterative procedure. A good starting point can significantly improve the convergence of iterative procedures. Indeed it can lead to convergence where otherwise divergence would occur.

Figure 1.7 illustrates the combination of an incremental predictor with Newton–Raphson iterations for a one-dimensional problem. A numerical example will now be given which relates to the dimensions and properties of (1.12) and starts from the converged, 'exact', equilibrium point for $W = -7$ (point 1 in Figure 1.4). This point is given by

$$w_1 = -2.2683, \quad N_1 = -433.08. \tag{1.30}$$

As a consequence of the inclusion of the linear spring, the out-of-balance force term, $g$, is given by

$$g = W_i(\text{bar}) + W_i(\text{spring}) - W_e = g(1.28) + K_s w. \tag{1.31}$$

The term $g(1.28)$ in (1.31) refers to equation (1.28). (Equation (1.23) could be used

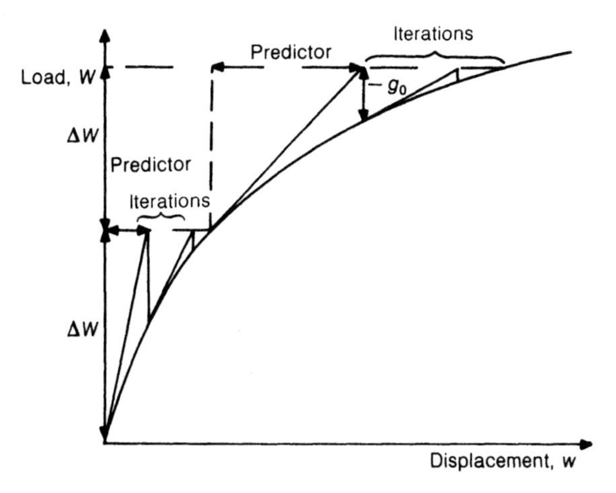


**Figure 1.7** A combination of incremental predictors with Newton–Raphson iterations.

instead.) At the starting point, (1.30), the tangent stiffness is given by

$$K_t = \frac{dg}{dw}(1.9) + K_s = 1.4803 + 1.35 \tag{1.32}$$

so that the incremental (tangential predictor-solution) would give

$$w = w_1 + \Delta w_1 = w_1 + K_t^{-1}\Delta W = w_1 - 7/2.8306 = -4.7415 \tag{1.33}$$

with (from (1.6))

$$N = -858.37. \tag{1.34}$$

Equation (1.31) now provides the out-of-balance force, $g$, as

$$g = g(1.28) + K_s w = -6.9557 + 14.0 - 6.4010 = 0.6432 \tag{1.35}$$

while the tangent stiffness is given by

$$K_t = \frac{dg}{dw}(1.9) + K_s = 0.97 + 1.35 = 2.32 \tag{1.36}$$

and the first iterative solution is, from (1.25)

$$\delta w = -0.6432/2.320 = -0.2773 \tag{1.37}$$

so that the total deflection is

$$w = -4.7415 - 0.2773 = -5.0188 \tag{1.38}$$

with (from (1.6)):

$$N = -903.0. \tag{1.39}$$

In order to apply a further iteration (1.31) gives

$$g = g(1.28) + K_s w = -7.2172 + 14.0 - 6.7754 = -0.0074 \tag{1.40}$$

and, from (1.27) and (1.40),

$$\delta w = -K_t^{-1} g = \left(\frac{\mathrm{d}g}{\mathrm{d}w}\right)^{-1} g = -0.0074/2.2664 = -0.0032 \tag{1.41}$$

and the total deflection is

$$w = -5.0188 - 0.0032 = -5.0220. \tag{1.42}$$

To four decimal places, this solution is exact and the next iterative change (which is probably affected by numerical round-off) is $-0.28 \times 10^{-6}$. From (1.33), the initial error is

$$e_0 = 4.7415 - 5.0220 = -0.2805 \tag{1.43}$$

while from (1.38)

$$e_1 = 5.0188 - 5.0220 = -0.0032 \tag{1.44}$$

and the next error is $e_2 = -0.28 \times 10^{-6}$. Hence

$$\frac{e_1}{e_0^2} = 0.04 \simeq \frac{e_2}{e_1^2} = 0.027 \tag{1.45}$$

which illustrates the 'quadratic convergence' of the Newton–Raphson method.

An obvious modification to this solution procedure involves the retention of the original (factorised) tangent stiffness. If the resulting 'modified Newton–Raphson' (or mN–R) iterations [O2, H1, Z2] are combined with an incremental procedure, the technique takes the form illustrated in Figure 1.8. Alternatively, one may only update $\mathbf{K}_t$ periodically [H1, Z2]. For example, the so-called $\mathbf{K}_t^1$ (or KT1) method would involve an update after one iteration [Z2].

Assuming the starting point of (1.30), the tangential solution would involve (1.32)–(1.34) as before. The resulting out-of-balance force vector would be given by (1.35) but (1.36) would no longer be computed to form $K_t$. Instead, the $K_t$ of (1.32)

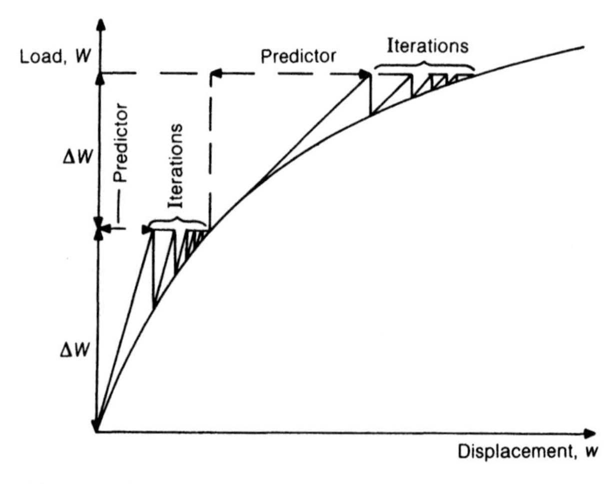


**Figure 1.8** A combination of incremental predictors with modified Newton–Raphson iterations.

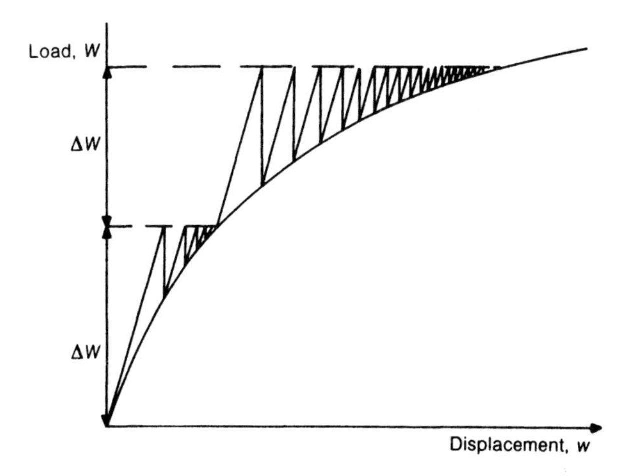


**Figure 1.9** The 'initial stress method' combined with an incremental solution.

would be re-used so that

$$\delta w = -0.6432/2.8303 = -0.2273, \qquad w = -4.9688. \tag{1.46}$$

Thereafter

$$g = -0.1210, \qquad \delta w = -0.1210/2.8303 = -0.04273, \qquad w = -5.0115 \tag{1.47}$$

$$g = -0.0239, \qquad \delta w = -0.0239/2.8303 = -0.00844, \qquad w = -5.0200 \tag{1.48}$$

etc. In contrast to (1.45),

$$\frac{e_1}{e_0} = 0.190 \simeq \frac{e_2}{e_1} = 0.198 \simeq \frac{e_3}{e_2} = 0.190 \tag{1.49}$$

which indicates the slower 'linear convergence' of the modified Newton–Raphson method. However, in contrast to the full N–R method, the modified technique requires less work at each iteration. In particular, the tangent stiffness matrix, $K_t$, is neither re-formed nor re-factorised.

The 'initial stress' method of solution [Z1] (no relation to the 'initial-stress matrix') takes the procedure one stage further and only uses the stiffness matrix from the very first incremental solution. The technique is illustrated in Figure 1.9.

## 1.3 A SIMPLE EXAMPLE WITH TWO VARIABLES

Figure 1.10 shows a system with two variables $u$ and $w$ which will be collectively referred to as

$$\mathbf{p}^{\mathrm{T}} = (u, w). \tag{1.50}$$

For this system, the strain of (1.5) is replaced by

$$\varepsilon = -\frac{u}{l} + \left(\frac{z}{l}\right)\left(\frac{w}{l}\right) + \frac{1}{2}\left(\frac{w}{l}\right)^2. \tag{1.51}$$

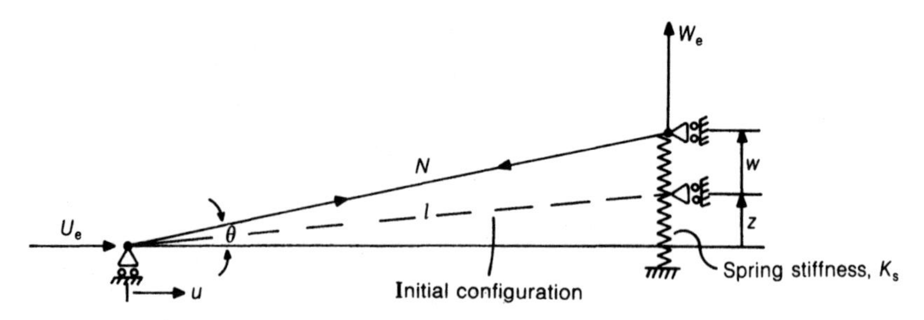


**Figure 1.10** Simple problem with two degrees of freedom.

(The term $(u/l)^2$ can be considered as negligible.) Resolving horizontally,

$$U_e + N\cos\theta \simeq U_e + N = 0 \tag{1.52}$$

while, resolving vertically,

$$W_e = N\sin\theta + K_s w \simeq \frac{N(z+w)}{l} + K_s w. \tag{1.53}$$

These equations can be re-written as

$$\mathbf{g} = \mathbf{q}_i - \mathbf{q}_e = \frac{N}{l}\begin{pmatrix} -l \\ z+w \end{pmatrix} + \begin{pmatrix} 0 \\ K_s w \end{pmatrix} - \begin{pmatrix} U_e \\ W_e \end{pmatrix} = \begin{pmatrix} U_i \\ W_i \end{pmatrix} - \begin{pmatrix} U_e \\ W_e \end{pmatrix} = \begin{pmatrix} 0 \\ 0 \end{pmatrix} \tag{1.54}$$

where $\mathbf{g}$ is an 'out-of-balance force vector', $\mathbf{q}_i$ an internal force vector and $\mathbf{q}_e$ the external force vector. The axial force, $N$, in (1.54) is simply given by

$$N = EA\varepsilon \text{ (equation (1.51)).} \tag{1.55}$$

In order to produce an incremental solution procedure, the internal force, $\mathbf{q}_i$, corresponding to the displacement, $\mathbf{p}$, can be expanded by means of a truncated Taylor series, so that

$$\mathbf{q}_i(\mathbf{p} + \Delta\mathbf{p}) = \mathbf{q}_i(\mathbf{p}) + \frac{\partial \mathbf{p}_i(\mathbf{p})}{\partial \mathbf{p}}\Delta\mathbf{p}. \tag{1.56}$$

Assuming perfect equilibrium at both the initial configuration $\mathbf{p}$ and the final configuration, $\mathbf{p} + \Delta\mathbf{p}$, equation (1.56) gives

$$\mathbf{q}_i(\mathbf{p} + \Delta\mathbf{p}) - \mathbf{q}_i(\mathbf{p}) = \mathbf{q}_e(\mathbf{p} + \Delta\mathbf{p}) - \mathbf{q}_e(\mathbf{p}) = \Delta\mathbf{q}_e = \frac{\partial \mathbf{q}_i}{\partial \mathbf{p}}\Delta\mathbf{p} = \mathbf{K}_t\Delta\mathbf{p} \tag{1.57}$$

or, in relation to the two variables $u$ and $w$,

$$\begin{pmatrix} \Delta U_e \\ \Delta W_e \end{pmatrix} = \mathbf{K}_t \begin{pmatrix} \Delta u \\ \Delta w \end{pmatrix} \tag{1.58}$$

where from (1.51), (1.54) and (1.55),

$$\mathbf{K}_t = \begin{bmatrix} \dfrac{\partial q_{i1}}{\partial p_1} & \dfrac{\partial q_{i1}}{\partial p_2} \\ \dfrac{\partial q_{i2}}{\partial p_1} & \dfrac{\partial q_{i2}}{\partial p_2} \end{bmatrix} = \begin{bmatrix} \dfrac{\partial U_i}{\partial u} & \dfrac{\partial U_i}{\partial w} \\ \dfrac{\partial W_i}{\partial u} & \dfrac{\partial W_i}{\partial w} \end{bmatrix} = \frac{EA}{l}\begin{bmatrix} 1 & -\beta \\ -\beta & \beta^2 + K_s l/EA \end{bmatrix} + \begin{bmatrix} 0 & 0 \\ 0 & N/l \end{bmatrix} \tag{1.59}$$

with

$$\beta = \frac{z + w}{l}. \tag{1.60}$$

The final matrix in (1.59) is the 'initial-stress matrix'. Clearly, the incremental procedure of Section 1.2.1 can be applied to this two-dimensional system using the general form

$$\Delta \mathbf{p} = \mathbf{K}_t^{-1} \Delta \mathbf{q}_e. \tag{1.61}$$

Alternatively, the tangent stiffness matrix of (1.59) can also be related to the Newton–Raphson iterative procedure and can be derived from a truncated Taylor series as in (1.24). For two dimensions this gives

$$\mathbf{g}_n = \begin{pmatrix} g_1 \\ g_2 \end{pmatrix}_n = \begin{pmatrix} g_1 \\ g_2 \end{pmatrix}_o + \begin{bmatrix} \dfrac{\partial g_1}{\partial p_1} & \dfrac{\partial g_1}{\partial p_2} \\ \dfrac{\partial g_2}{\partial p_1} & \dfrac{\partial g_2}{\partial p_2} \end{bmatrix}_o \begin{pmatrix} \delta p_1 \\ \delta p_2 \end{pmatrix} = \mathbf{g}_o + \frac{\partial \mathbf{g}}{\partial \mathbf{p}} \delta \mathbf{p} = \mathbf{g}_o + \mathbf{K}_t \delta \mathbf{p} \tag{1.62}$$

where $\mathbf{K}_t$ is again given by (1.59). The Newton–Raphson solution procedure now involves

$$\delta \mathbf{p} = -\left(\frac{\partial \mathbf{g}}{\partial \mathbf{p}}\right)^{-1} \mathbf{g}_o = -\mathbf{K}_t^{-1} \mathbf{g}_o. \tag{1.63}$$

We will firstly solve the 'perfect' system, for which $z$ (Figure 1.10) is zero. The applied load, $W_e$, will also be set to zero. In these circumstances, (1.58) and (1.59) give

$$\begin{pmatrix} \Delta U_e \\ 0 \end{pmatrix} = \begin{bmatrix} \dfrac{EA}{l} & 0 \\ 0 & K_s + \dfrac{N}{l} \end{bmatrix} \begin{pmatrix} \Delta u \\ \Delta w \end{pmatrix}. \tag{1.64}$$

The solution is

$$\Delta u = \frac{l}{AE} \Delta U_e, \qquad \Delta w = 0 \tag{1.65}$$

so that

$$u = \frac{l}{AE} U, \qquad w = 0. \tag{1.66}$$

These solutions remain valid while $(K_s + N/l)$ is positive and the matrix $\mathbf{K}_t$ is 'positive definite'. However, when

$$N = N_{cr} = -lK_s \tag{1.67}$$

the load $U$ reaches a 'critical value',

$$U_{cr} = -N_{cr} = lK_s = \frac{AE}{l} u_{cr} \tag{1.68}$$

at which $\mathbf{K}_t$ becomes singular, $\Delta u$ and $\Delta w$ are indeterminate and the system 'buckles'.

This example illustrates one particular use of the 'initial-stress matrix'. In general, for a perfect system (when the pre-buckled path is linear or 'effectively linear'), we can write

$$\mathbf{K}_t = \mathbf{K}_0 + \lambda \mathbf{K}_{t\sigma} \tag{1.69}$$

where $\mathbf{K}_0$ is the standard 'linear stiffness matrix' and $\mathbf{K}_{t\sigma}$ is the initial-stress matrix when computed for a 'unit membrane stress field' (in the present case, $N = 1$). The term $\lambda$ in (1.69) is the load factor that amplifies this initial stress field. As a consequence of (1.69), the buckling criterion becomes

$$\det(\mathbf{K}_o + \lambda \mathbf{K}_{t\sigma}) = 0 \tag{1.70}$$

which is an eigenvalue problem. Numerical solutions for the imperfect system (with $z$ (Figure 1.10) $\neq 0$) will be given in Chapter 2. For the present, we will derive a set of 'exact solutions'.

### 1.3.1 'Exact' solutions

The governing equations (1.54) have solutions

$$w = \left(\frac{U}{U_{cr} - U}\right) z \tag{1.71}$$

or

$$w + z = \frac{z}{(1 - U/U_{cr})} \tag{1.72}$$

as well as

$$\frac{u}{u_{cr}} = \frac{U}{U_{cr}} + \beta(\alpha + \tfrac{1}{2}\alpha^2) \tag{1.73}$$

where

$$\alpha = \frac{U}{U_{cr} - U}, \qquad \beta = \frac{EA}{U_{cr}}\left(\frac{z}{l}\right)^2 = \frac{EA}{lK_s}\left(\frac{z}{l}\right)^2 \tag{1.74}$$

and the 'buckling load', $U_{cr}$, and equivalent displacement, $u_{cr}$, have been defined in (1.68).

Equations (1.72) and (1.74) have been plotted in Figure 1.11 where the 'perfect solutions' relate to the system of Figure 1.10 with $z$ set to zero. The non-dimensionalising factor, $z_a$, in Figure 1.10(a) is the initial offset, $z$, for the imperfect system and any non-zero value for the perfect system. In plotting equation (1.73) in Figure 1.11(a), the factor $\beta$ of (1.74) has been set to 0.5 (i.e. as if using (1.12) but with $K_s = 4$).

The perfect solutions are stable up to point A from which the path AC (or A'C' in Figure 1.11(a)) is the post-buckling path. If the offset, $z$, in Figure 1.10 is non-zero, either the imperfect path EF or the equivalent path E'F' in Figure 1.11(a) will be followed, depending on the sign of $z$. At the same time, the 'load/shortening relationship' will follow OD in Figure 1.11(b). While these paths are fairly obvious, the solutions GH (or G'H') in Figure 1.11(a) and GH in Figure 1.11(b) are less obvious and could not be reached by a simple monotonic loading. Nonetheless they do

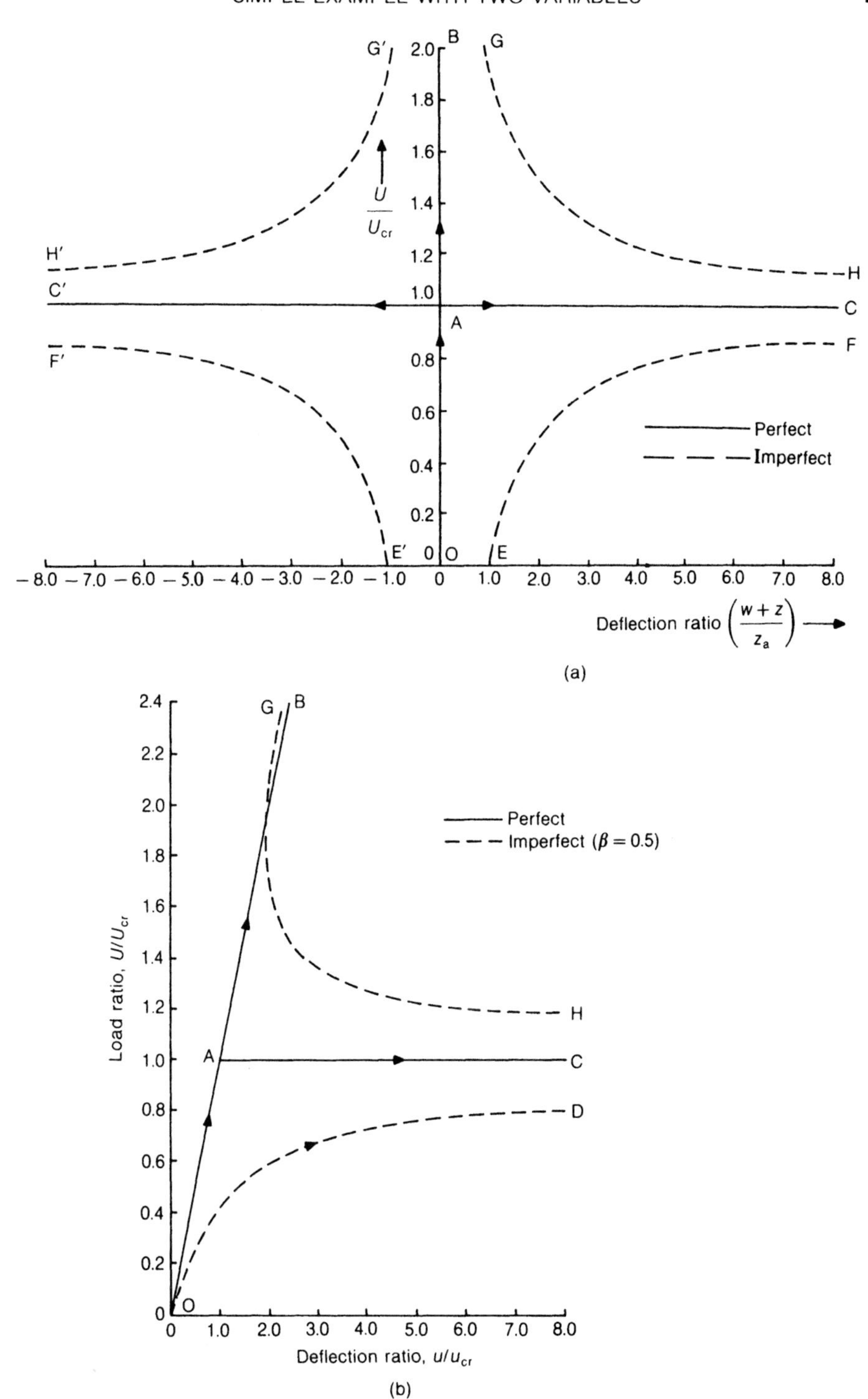


**Figure 1.11** Load/deflection relationships for two-variable bar–spring problem: (a) transverse deflection; (b) shortening deflection.

represent equilibrium states and their presence can cause difficulties with the numerical solution procedures. This will be demonstrated in Chapter 2 where it will be shown that it is even possible to accidentally converage on the 'spurious upper equilibrium states'.

Before leaving this section, we should note the inverted commas surrounding the word 'exact' in the title of this section. The solutions *are* exact solutions to the governing equations (1.54). However, the ltter were derived on the assumption of a small angle $\theta$ in Figure 1.10. Clearly, this assumption will be violated as the deflection ratios in Figure 1.11 increase, even if it is valid when $w$ is small.

### 1.3.2 The use of virtual work

In Section 1.2, the governing equations were derived directly from equilibrium. With a view to later work with the finite element method, we will now derive the out-of-balance force vector, $\mathbf{g}$ using virtual work instead. To this end, with the help of differentiation, the change in (1.51) can be expressed as

$$\delta\varepsilon = -\frac{\delta u}{l} + \left(\frac{z+w}{l}\right)\left(\frac{\delta w}{l}\right) + \left[\frac{1}{2}\left(\frac{\delta w}{l}\right)^2\right]_{\mathrm{h}}. \tag{1.75}$$

For really small virtual changes, the last, higher-order, square-bracketed term in (1.75) is negligible and

$$\delta\varepsilon_{\mathrm{v}} = -\frac{\delta u_{\mathrm{v}}}{l} + \left(\frac{z+w}{l}\right)\left(\frac{\delta w_{\mathrm{v}}}{l}\right) \tag{1.76}$$

where the subscript v means 'virtual'.

The virtual work undertaken by the internal and external forces can now be expressed as

$$V = \int \sigma\delta\varepsilon_{\mathrm{v}}\,\mathrm{d}V + K_{\mathrm{s}}w\delta w_{\mathrm{v}} - U_{\mathrm{e}}\delta u_{\mathrm{v}} - W_{\mathrm{e}}\delta w_{\mathrm{v}} = Nl\delta\varepsilon_{\mathrm{v}} + K_{\mathrm{s}}w\delta w_{\mathrm{v}} - U_{\mathrm{e}}\delta u_{\mathrm{v}} - W_{\mathrm{e}}\delta w_{\mathrm{v}}. \tag{1.77}$$

Substituting from (1.76) into (1.77) gives

$$V = \mathbf{g}^{\mathrm{T}}\delta\mathbf{p}_{\mathrm{v}} \tag{1.78}$$

where $\delta\mathbf{p}_{\mathrm{v}}^{\mathrm{T}} = (\delta u_{\mathrm{v}}, \delta w_{\mathrm{v}})$ and the vector $\mathbf{g}$ is of the form previously derived directly from equilibrium in (1.54). The principle of virtual work specifies that $V$ should be zero for any arbitrary small virtual displacements, $\delta\mathbf{p}_{\mathrm{v}}$. Hence (1.78) leads directly to the equilibrium equations of (1.54). Clearly, the tangent stiffness matrix, $\mathbf{K}_{\mathrm{t}}$, can be obtained, as before, by differentiating $\mathbf{g}$. With a view to future developments, we will also relate the latter to the variation of the virtual work. In general, (1.78) can be expressed as

$$V = \int \sigma^{\mathrm{T}}\delta\varepsilon_{\mathrm{v}}\,\mathrm{d}V - \mathbf{q}_{\mathrm{e}}^{\mathrm{T}}\delta\mathbf{p}_{\mathrm{v}} = (\mathbf{q}_{\mathrm{i}} - \mathbf{q}_{\mathrm{e}})^{\mathrm{T}}\delta\mathbf{p}_{\mathrm{v}} = \mathbf{g}^{\mathrm{T}}\delta\mathbf{p}_{\mathrm{v}} \tag{1.79}$$

from which

$$\delta V = \delta \mathbf{p}_v^T \delta \mathbf{g} = \delta \mathbf{p}_v^T \frac{\partial \mathbf{g}}{\partial \mathbf{p}} \delta \mathbf{p} = \delta \mathbf{p}_v^T \mathbf{K}_t \delta \mathbf{p}. \tag{1.80}$$

### 1.3.3 An energy basis

The previous developments can be related to the total potential energy. For the current problem, the latter is given by

$$\phi = \int \tfrac{1}{2} \mathbf{E} \varepsilon^2 \, \mathrm{d}V + \tfrac{1}{2} K_s w^2 - \mathbf{q}_v^T \mathbf{p} \tag{1.81}$$

or

$$\phi = \frac{1}{2} K_s w^2 + \frac{1}{2} EAl \left[ -\frac{u}{l} + \left(\frac{z}{l}\right)\left(\frac{w}{l}\right) + \frac{1}{2}\left(\frac{w}{l}\right)^2 \right]^2 - U_e u - W_e w. \tag{1.82}$$

If the loads $U_e$ and $W_e$ are held fixed, and the displacements $u$ and $w$ are subjected to small changes, $\delta u$ and $\delta w$ (collectively $\delta \mathbf{p}$), the energy moves from $\phi_o$ to $\phi_n$, where

$$\phi_n = \phi_o + \left(\frac{\partial \phi}{\partial \mathbf{p}}\right) \delta \mathbf{p} = \phi_o + \mathbf{g}^T \delta \mathbf{p} \tag{1.83}$$

or

$$\phi_n = \phi_o + K_s w \delta w + \frac{EA}{l}\left[\frac{u}{l} + \left(\frac{z}{l}\right)\left(\frac{w}{l}\right) + \frac{1}{2}\left(\frac{w}{l}\right)^2\right]\begin{pmatrix} -l \\ z + w \end{pmatrix}^T \begin{pmatrix} \delta u \\ \delta w \end{pmatrix} - \begin{pmatrix} U_e \\ W_e \end{pmatrix}^T \begin{pmatrix} \delta u \\ \delta w \end{pmatrix}. \tag{1.84}$$

The principle of stationary potential energy dictates that, for equilibrium, the change of energy, $\phi_n - \phi_o$, should be zero for arbitrary $\delta \mathbf{p}$ ($\delta u$ and $\delta w$). Hence equation (1.84) leads directly to the equilibrium equations of (1.54) (with $N$ from (1.55)). Equation (1.83) shows that the 'out-of-balance force vector' $\mathbf{g}$, is the gradient of the potential energy. Hence the symbol $\mathbf{g}$. The matrix $\mathbf{K}_t = \partial \mathbf{g}/\partial \mathbf{p}$ is the second differential of $\phi$ and is known in the 'mathematical-programming literature' (see Chapter 9) as the Jacobian of $\mathbf{g}$ or the Hessian of $\phi$.

## 1.4 SPECIAL NOTATION

$A$ = area of bar
$e$ = error
$K_s$ = spring stiffness
$N$ = axial force in bar
$u$ = axial displacement at end of bar
$U$ = force corresponding to $u$
$w$ = vertical displacement at end of bar
$W$ = force corresponding to $w$
$z$ = initial vertical offset of bar

$l$ = initial length of bar
$\beta$ = geometric factor (equation (1.60))
$\varepsilon$ = axial strain in bar
$\theta$ = final angular inclination of bar

## 1.5 LIST OF BOOKS ON (OR RELATED TO) NON-LINEAR FINITE ELEMENTS

Bathe, K. J., *Finite Element Procedures in Engineering Analysis*, Prentice Hall (1981).
Kleiber, M., *Incremental Finite Element Modelling in Non-linear Solid Mechanics*, Ellis Horwood, English edition (1989)
Hinton, E. (ed.), *Introduction to Non-linear Finite Elements*, National Agency for Finite Elements (NAFEMS) (1990)
Oden, J. T., *Finite Elements of Nonlinear Continua*, McGraw-Hill (1972)
Owen, D. R. J. & Hinton, E., *Finite Elements in Plasticity—Theory and Practise*, Pineridge Press, Swansea (1980)
Simo, J. C. & Hughes, T. J. R., *Elastoplasticity and Viscoplasticity*, Computational aspects, Springer (to be published).
Zienkiewicz, O. C., *The Finite Element Method*, McGraw-Hill, 3rd edition (1977) and with R. L. Taylor, 4th edition, Volume 2, to be published.

## 1.6 REFERENCES TO EARLY WORK ON NON-LINEAR FINITE ELEMENTS

[A1] Ang, A. H. S. & Lopez, L. A., Discrete model analysis of elastic-plastic plates, *Proc. ASCE*, **94**, EM1, 271–293 (1968).
[A2] Argyris, J. H., *Recent Advances in Matrix Methods of Structural Analysis*, Pergamon Press (1964)
[A3] Argyris, J. H., Continua and discontinua, *Proc. Conf. Matrix Methods in Struct. Mech.*, Air Force Inst. of Tech., Wright Patterson Air Force Base, Ohio (October 1965).
[A4] Armen, H., Pifko, A. B., Levine, H. S. & Isakson, G., Plasticity, *Finite Element Techniques in Structural Mechanics*, ed. H. Tottenham *et al.*, Southampton University Press (1970).
[B1] Belytschoko, T. & Velebit, M., Finite element method for elastic plastic plates, *Proc. ASCE, J. of Engng. Mech. Div.*, EM1, 227–242 (1972).
[B2] Brebbia, C. & Connor, J., Geometrically non-linear finite element analysis, *Proc. ASCE, J. Eng. Mech. Div., Proc.* paper 6516 (1969).
[D1] Dupius, G. A., Hibbit, H. D., McNamara, S. F. & Marcal, P. V., Non-linear material and geometric behaviour of shell structures, *Comp. & Struct.*, **1**, 223–239 (1971).
[G1] Gallagher, R. J. & Padlog, J., Discrete element approach to structural stability, *Am. Inst. Aero. & Astro. J.*, **1** (6), 1437–1439.
[G2] Gallagher, R. J., Gellatly, R. A., Padlog, J. & Mallet, R. H., A discrete element procedure for thin shell instability analysis, *Am. Inst. Aero. & Astro. J.*, **5** (1), 138–145 (1967).
[G3] Gerdeen, J. C., Simonen, F. A. & Hunter, D. T., Large deflection analysis of elastic–plastic shells of revolution, *AIAA/ASME 11th Structures, Structural Dynamics & Materials Conf.*, Denver, Colorado, 239–49 (1979).
[H1] Haisler, W. E., Stricklin, J. E. & Stebbins, F. J., Development and evaluation of solution procedures for geometrically non-linear structural analysis by the discrete stiffness method, *AIAA/ASME 12th Structure, Structural Dynamics & Materials Conf.*, Anaheim, California (April 1971).

[H2] Harris, H. G. & Pifko, A. B., Elasto-plastic buckling of stiffened rectangular plates, *Proc. Symp. on Appl. of Finite Element Meth. in Civil Engng.*, Vanderbilt Univ., ASCE, 207–253 (1969).

[H3] Holand, I. & Moan, T., The finite element in plate buckling, *Finite Element Meth. in Stress Analysis*, ed. I. Holand *et al.*, Tapir (1969).

[K1] Kapur, W. W. & Hartz, B. J., Stability of plates using the finite element method, Proc. ASCE, *J. Engng. Mech.*, **92**, EM2, 177–195 (1966).

[M1] Mallet, R. H. & Marcal, P. V., Finite element analysis of non-linear *structures, Proc. ASCE, J. of Struct. Div.*, **94**, ST9, 2081–2105 (1968).

[M2] Mallet, R. H. & Schmidt, L. A., Non-linear structural analysis by energy search, Proc. ASCE, *J. Struct. Div.*, **93**, ST3, 221–234 (1967).

[M3] Marcal, P. V. Finite element analysis of combined problems of non-linear material and geometric behaviour, *Proc. Am. Soc. Mech. Conf. on Comp. Approaches in Appl. Mech.*, (June 1969).

[M4] Marcal, P. V. & King, I. P., Elastic–plastic analysis of two-dimensional stress systems by the finite element method, *int. J. Mech. Sci.*, **9** (3), 143–155 (1967).

[M5] Marcal, P. V., Large deflection analysis of elastic–plastic shells of revolution, *Am. Inst. Aero. & Astro. J.*, **8**, 1627–1634 (1970).

[M6] Marcal, P. V., Finite element analysis with material non-linearities—theory and practise, *Recent Advances in Matrix Methods of Structural Analysis & Design*, ed. R. H. Gallagher *et al.*, The University of Alabama Press, pp. 257–282 (1971).

[M7] Marcal, P. V. & Pilgrim, W. R., A stiffness method for elasto-plastic shells of revolution, *J. Strain Analysis*, **1** (4), 227–242 (1966).

[M8] Murray, D. W. & Wilson, E. L., Finite element postbuckling analysis of thin elastic plates, Proc. ASCE, *J. Engine. Mech. Div.*, **95**, EM1, 143–165 (1969).

[M9] Murray, D. W. & Wilson, E. L., Finite element postbuckling analysis of thin elastic plates, *Am. Inst. of Aero. & Astro. J.*, **7**, 1915–1930 (1969).

[M10] Murray, D. W. & Wilson, E. L., An approximate non-linear analysis of thin-plates, *Proc. Air Force 2nd Conf. on Matrix Meth. in Struct. Mech.*, Wright-Patterson Air Force Base, Ohio (October 1968).

[N1] Nayak, G. C. & Zienkiewicz, O. C., Elasto-plastic stress analysis. A generalisation for various constitutive relationships including strain softening, *Int. J. Num. Meth. in Engng.*, **5**, 113–135 (1972).

[N2] Nayak, G. C. & Zienkiewicz, O. C., Note on the 'alpha-constant' stiffness method of the analysis of non-linear problems, *Int. J. Num. Meth. in Engng.*, **4**, 579–582 (1972).

[O1] Oden, J. T., Numerical formulation of non-linear elasticity problems, Proc. ASCE, *J. Struct. Div.*, **93**, ST3, paper 5290 (1967).

[O2] Oden, J. T., Finite element applications in non-linear structural analysis, *Proc. Conf. on Finite Element Meth.*, Vanderbilt University Tennessee (November 1969).

[P1] Pope, G., A discrete element method for analysis of plane elastic-plastic stress problems, Royal Aircraft Estab. TR SM65-10 (1965).

[P2] Popov, E. P., Khojasteh Baht, M. & Yaghmai, S., Bending of circular plates of hardening materials, *Int. J. Solids & Structs.*, **3**, 975–987 (1967).

[S1] Sabir, A. B. & Lock, A. C., The application of finite elements to the large-deflection geometrically non-linear behaviour of cylindrical shells, *Proc. Int. Conf. on Var. Meth. in Engng.*, Southampton Univ., Session VII, 67–76 (September 1972).

[S2] Schmidt, F. K., Bognor, F. K. & Fox, R. L., Finite deflection structural analysis using plate and shell discrete elements, *Am. Inst. Aero. & Astro. J.*, **6**(5), 781–791 (1968).

[S3] Sharifi, P. & Popov, E. P., Nonlinear buckling analysis of sandwich arches, Proc. ASCE, *J. Engng. Mech. Div.*, **97**, 1397–1411 (1971).

[S4] Stricklin, J. A., Haisler, W. E. & Von Riseseman, W. A., Computation and solution procedures for non-linear analysis by combined finite element–finite difference methods, *Computers & Structures*, **2**, 955–974 (1972).

[T1] Terazawa, K., Ueda, Y. & Matsuishi, M., Elasto-plastic buckling of plates by finite element method, *ASCE Annual Meeting and Nat. Meeting on Water Res. Engng.*, New Orleans, La., paper 845 (February 1969).

[T2] Turner, M. J., Dill, E. H., Martin, H. C. & Melosh, R. J., Large deflection of structures subject to heating and external load, *J. Aero. Sci.*, **27**, 97–106 (1960).
[W1] Whange, B., Elasto-plastic orthotropic plates and shells, *Proc. Symp. on Appl. of Finite Element Methods in Civil Engng.*, Vandebilt University, ASCE 481–516 (1969).
[Y1] Yamada, Y., Yoshimura, N., & Sakurai, T., Plastic stress-strain matrix and its application for the solution of elasto-plastic problems by the finite element method, *Int. J. Mech. Sci.*, **10**, 343–354 (1968).
[Z1] Zienkiewicz, O. C., Valliapan, S. & King, I. P., Elasto-platic solutions of engineering problems. Initial stress, finite element approach, *Int. J. Num. Meth. Engng.*, **1**, 75–100 (1969).
[Z2] Zienkiewicz, O. C., *The Finite Element in Engineering Science*, McGraw-Hill, London (1971).

# 2 A shallow truss element with Fortran computer program

In Sections 1.2.1–3, we obtained numerical solutions for the simple bar/spring problem with one degree of freedom that is illustrated in Figure 1.1. We also proposed, in Figure 1.10, a simple example with two degrees of freedom. However, no numerical solutions were obtained for the latter problem. Once the number of variables is increased beyond one, it becomes tedious to obtain numerical solutions manually, and a simple computer program is more appropriate.

Such a program will be of more use if its is written in a 'finite element context', so that different boundary conditions can be applied. So far, only indirect reference has been made to the finite element method. In this chapter, we will use the 'shallow truss theory' of Section 1.2 to derive the finite element equations for a shallow truss element. We will then provide a set of Fortran subroutines which allows this element to be incorporated in a simple non-linear finite element program. Flowcharts are given for an 'incremental formulation', a 'Newton–Raphson iterative procedure' and, finally, a combined 'incremental/iterative technique' that uses either the full or modified Newton–Raphson methods. Fortran programs, which incorporate the earlier subroutines, are then constructed around these flowcharts. Finally, the computer program is used to analyse a range of problems.

## 2.1 A SHALLOW TRUSS ELEMENT

We will now use the 'shallow truss theory' of Chapter 1 to derive the finite element equations for the shallow truss of Figure 2.1. The derivation will be closely related to the virtual work procedure of Section 1.3.2. Short-cuts could be used in the derivation but we will follow fairly conventional finite element procedures so that this example provides an introduction to the more complex finite element formulations that will follow. The element (Figure 2.1) has four degrees of freedom $u_1 = p_1$, $u_2 = p_2$, $w_1 = p_3$ and $w_2 = p_4$. Both the geometry and the displacements are defined with the aid of simple linear shape functions involving the non-dimensional coordinate, $\xi$, so

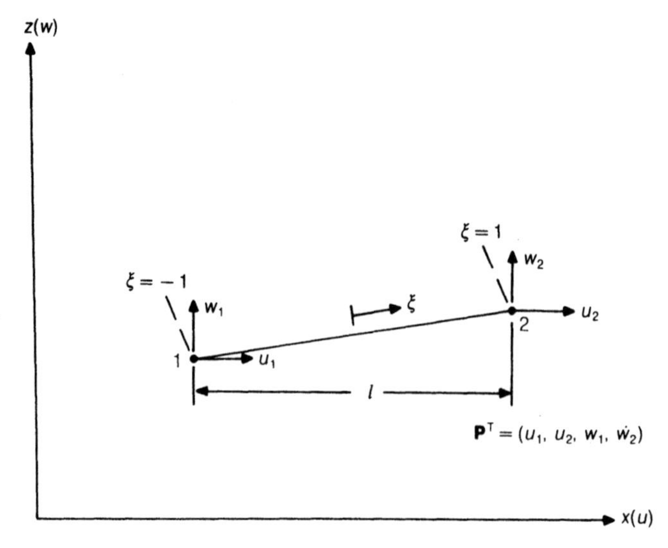


**Figure 2.1** A shallow truss element.

that

$$x = \frac{1}{2}\begin{pmatrix} 1-\xi \\ 1+\xi \end{pmatrix}^{\mathrm{T}} \begin{pmatrix} x_1 \\ x_2 \end{pmatrix}, \qquad z = \frac{1}{2}\begin{pmatrix} 1-\xi \\ 1+\xi \end{pmatrix}^{\mathrm{T}} \begin{pmatrix} z_1 \\ z_2 \end{pmatrix}, \tag{2.1}$$

$$u = \frac{1}{2}\begin{pmatrix} 1-\xi \\ 1+\xi \end{pmatrix}^{\mathrm{T}} \begin{pmatrix} u_1 \\ u_2 \end{pmatrix}, \qquad w = \frac{1}{2}\begin{pmatrix} 1-\xi \\ 1+\xi \end{pmatrix}^{\mathrm{T}} \begin{pmatrix} w_1 \\ w_2 \end{pmatrix}. \tag{2.2}$$

Following from (1.51), the strain in the bar is

$$\varepsilon = \frac{\mathrm{d}u}{\mathrm{d}x} + \left(\frac{\mathrm{d}z}{\mathrm{d}x}\right)\left(\frac{\mathrm{d}w}{\mathrm{d}x}\right) + \frac{1}{2}\left(\frac{\mathrm{d}w}{\mathrm{d}x}\right)^2. \tag{2.3}$$

From (2.1),

$$\frac{\mathrm{d}x}{\mathrm{d}\xi} = (x_2 - x_1)/2 = l/2 \tag{2.4}$$

while from (2.2),

$$\frac{\mathrm{d}u}{\mathrm{d}x} = \frac{\mathrm{d}u}{\mathrm{d}\xi}\frac{\mathrm{d}\xi}{\mathrm{d}x} = (u_2 - u_1)/l = u_{21}/l \tag{2.5}$$

where the shorthand $u_{21}$ has been used for $u_2 - u_1$. In a similar fashion,

$$\frac{\mathrm{d}w}{\mathrm{d}x} = w_{21}/l \qquad \frac{\mathrm{d}z}{\mathrm{d}x} = z_{21}/l. \tag{2.6}$$

Hence, from (2.3),

$$\varepsilon = \frac{u_{21}}{l} + \left(\frac{z_{21}}{l}\right)\left(\frac{w_{21}}{l}\right) + \frac{1}{2}\left(\frac{w_{21}}{l}\right)^2 \tag{2.7}$$

and the axial force in the bar is

$$N = EA\varepsilon \text{ (equation (2.7))}. \tag{2.8}$$

From (2.3), a change of strain, $\Delta\varepsilon$, corresponding to displacement changes $\Delta u$ and $\Delta w$ is given by

$$\Delta\varepsilon = \frac{\mathrm{d}\Delta u}{\mathrm{d}x} + \left(\frac{\mathrm{d}z}{\mathrm{d}x} + \frac{\mathrm{d}w}{\mathrm{d}x}\right)\frac{\mathrm{d}\Delta w}{\mathrm{d}x} + \frac{1}{2}\left(\frac{\mathrm{d}\Delta w}{\mathrm{d}x}\right)^2 \tag{2.9}$$

where the final (higher-order) term in (2.9) becomes negligible as $\Delta w$ gets very small. Using (2.5) and (2.6):

$$\Delta\varepsilon = \frac{\Delta u_{21}}{l} + \frac{1}{l^2}(z_{21} + w_{21})\Delta w_{21} + \frac{1}{2l^2}\Delta w_{21}^2. \tag{2.10}$$

If a set of virtual nodal displacements†,

$$\delta \mathbf{p}_v^T = (\delta u_{v1}, \delta u_{v2}, \delta w_{v1}, \delta w_{v2}) \tag{2.11}$$

are applied, the resulting strain is, from (2.9),

$$\delta\varepsilon_v = \frac{1}{l}\,\delta u_{v21} + \frac{1}{l^2}(z_{21} + w_{21})\,\delta w_{v21} = \mathbf{b}^T\delta\mathbf{p}_v \tag{2.12}$$

where

$$\mathbf{b}^T = \frac{1}{l}(-1, 1, -\beta, \beta) \tag{2.13}$$

with

$$\beta = \frac{z_{21} + w_{21}}{l}. \tag{2.14}$$

In deriving (2.12) from (2.10), the quadratic terms involving $\delta w_{v21}^2$ have been considered negligible.

The virtual work equation can (see (1.77)) be expressed as

$$V = \int \sigma\delta\varepsilon_V\,\mathrm{d}V - \mathbf{q}_e^T\delta\mathbf{p}_v = 0 \tag{2.15}$$

where $\mathbf{q}_e$ are the external nodal forces corresponding to the nodal displacements, $\delta\mathbf{p}_v$. Because $\delta\varepsilon_v$ can be expressed, via (2.12), in terms of $\delta\mathbf{p}_v$, equation (2.15) can be re-written as

$$V = \delta\mathbf{p}_v^T\mathbf{g} = \delta\mathbf{p}_v^T(\mathbf{q}_i - \mathbf{q}_e) = \delta\mathbf{p}_v^T\left(\int \sigma\mathbf{b}\,\mathrm{dV} - \mathbf{q}_e\right) = 0 \tag{2.16}$$

where $\mathbf{q}_i$ is the internal force vector, given by

$$\mathbf{q}_i = \int \sigma\mathbf{b}\,\mathrm{d}V = Nl\mathbf{b}. \tag{2.17}$$

†This ordering would not be the most convenient for element assembly, but the ordering could easily be altered prior to such assembly.

For equilibrium, (2.16) should be satisfied for any virtual displacements. $\delta\mathbf{p}_v$. Hence

$$\mathbf{g} = \mathbf{q}_i - \mathbf{q}_e = 0 \tag{2.18}$$

where $\mathbf{g}$ is the out-of-balance force vector.

From (1.80), $\mathbf{K}_t = \partial\mathbf{g}/\partial\mathbf{p}$ and a truncated Taylor expansion of $\mathbf{g}$, about an 'old' configuration, $\mathbf{g}_o$ gives

$$\mathbf{g}_n = \mathbf{g}_o + \frac{\partial\mathbf{g}}{\partial\mathbf{p}}\delta\mathbf{p} = \mathbf{g}_o + \mathbf{K}_t\delta\mathbf{p}. \tag{2.19}$$

Hence, from equations (2.17)–(2.19),

$$\mathbf{K}_t = \frac{\partial\mathbf{g}}{\partial\mathbf{p}} = \frac{\partial\mathbf{q}_i}{\partial\mathbf{p}} = l\mathbf{b}\frac{\mathrm{d}N}{\mathrm{d}\mathbf{p}} + lN\frac{\partial\mathbf{p}}{\partial\mathbf{p}}. \tag{2.20}$$

From (2.8) and (2.12):

$$\frac{\mathrm{d}N}{\mathrm{d}\mathbf{p}} = \frac{\mathrm{d}N}{\mathrm{d}\varepsilon}\frac{\partial\varepsilon}{\partial\mathbf{p}} = EA\mathbf{b}^{\mathrm{T}}. \tag{2.21}$$

Hence,

$$\mathbf{K}_t = EAl\mathbf{b}\mathbf{b}^{\mathrm{T}} + lN\frac{\partial\mathbf{b}}{\partial\mathbf{p}} \tag{2.22}$$

or

$$\mathbf{K}_t = \frac{EA}{l}\begin{bmatrix} 1 & -1 & \beta & -\beta \\ -1 & 1 & -\beta & \beta \\ \beta & -\beta & \beta^2 & -\beta^2 \\ -\beta & \beta & -\beta^2 & \beta^2 \end{bmatrix} + \frac{N}{l}\begin{bmatrix} 0 & 0 & 0 & 0 \\ 0 & 0 & 0 & 0 \\ 0 & 0 & 1 & -1 \\ 0 & 0 & -1 & 1 \end{bmatrix}. \tag{2.23}$$

This is the matrix equivalent of (1.9) with the second matrix being the 'initial stress' matrix. Equation (1.9) can be recovered by setting

$$u_1 = w_1 = z_1 = u_2 = 0, \qquad w_2 = w \qquad \text{and} \qquad z_2 = z. \tag{2.24}$$

## 2.2 A SET OF FORTRAN SUBROUTINES

We will now provide a set of Fortran subroutines to enable the solution of the simple bar–spring problems of Chapter 1 and others to be discussed in Section 2.6 and Chapters 3 and 9. In its most general form, the adopted bar–spring system is that shown in Figure 2.2, with the bar element of Figure 2.1 being surrounded by up to four linear 'earthed springs' and one horizontal linear spring, connecting variables 1 and 5. For many of the problems the linear spring $K_{s5}$ will be omitted; however, the potential to include this spring allows the solution of the complete range of NAFEMS bar–spring problems [C1, D1] which include both 'snap-throughs' and 'snap-backs' (Chapter 9). In relation to Figure 2.2, for the shallow truss elements, there is assumed to be no effective difference between $x_{21}$ and $l$.

The following subroutines are not designed for maximum computer efficiency but rather to illustrate the basic concepts. To a considerable extent, the subroutines are self-explanatory. However, a brief synopsis will be given above each routine. The

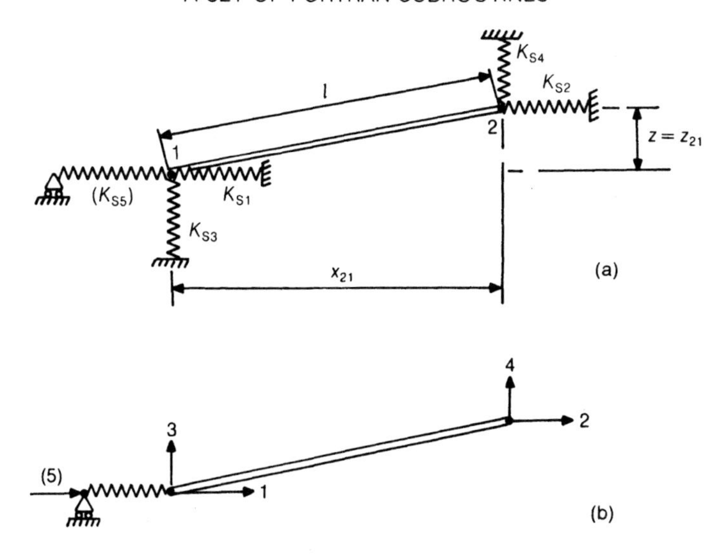


**Figure 2.2** Bar–spring system. (a) Bar element with springs; (b) variables.

computer programs of Sections 2.3–5 will incorporate these subroutines to produce programs for the analysis of shallow trusses. These computer programs are also designed to incorporate the deep-truss elements of Chapter 3 and hence the present subroutines will include some 'dummy variables' that will only be used for the deep trusses.

### 2.2.1 Subroutine ELEMENT

This subroutine forms the internal force vector FI(4) (using equation (2.17)) and/or the element tangent stiffness matrix AKT(4, 4) (using equation (2.23)) for the shallow truss element. Only the upper triangle is formed but the full 4 × 4 structure is used.

```
      SUBROUTINE ELEMENT (FI,AKT,AN,X,Z,P,E,ARA,AL,IWRIT,IWR,IMOD,
     1    IDUM,ADUM1,ADUM2)
C     ARGUMENTS IN LINE ABOVE AND ARRAY X NOT USED FOR SHALLOW TRUSS
C
C     FOR SHALLOW TRUSS ELEMENT
C     IMOD=1 COMPUTES INT. LD. VECT. FI
C     IMOD=2 COMPUTES TAN. STIFF. AKT
C     IMOD=3 COMPUTES BOTH
C
C     AN=INPUT TOTAL FORCE IN BAR
C     Z=INPUT=Z COORD VECTOR
C     P=INPUT=TOTAL DISP. VECTOR
C     AL=INPUT=LENGTH OF ELEMENT
C     EA=INPUT=YOUNGS MOD AND ARA (INPUT)=AREA OF ELEMENT
C
C     IF IWRIT.NE.0 WRITES OUT FI AND/OR AKT ON CHANNEL IWR
C
      IMPLICIT DOUBLE PRECISION (A-H,O-Z)
      DIMENSION AKT(4,4),FI(4),Z(2),P(4),X(2)
```

```
C
      EA=E*ARA
      EAL=EA/AL
      Z21=Z(2)-Z(1)
      W21=P(4)-P(3)
      BET=(Z21+W21)/AL
C
      IF (IMOD.NE.2) THEN
C     COMPUTES INT. FORCE VECT. (SEE (2.17))
      FI(1)=-1.D0
      FI(2)=-1.D0
      FI(3)=-BET
      FI(4)=BET
      DO 1 I=1,4
    1 FI(I)=AN*FI(I)
      IF (IWRIT.NE.0) THEN
      WRITE (IWR,1000) (FI(I),I=1,4)
 1000 FORMAT(/,1X,'INT. FORCE VECT. FOR TRUSS EL IS',/,1X,4G13.5)
      ENDIF
C
      ENDIF
C
      IF (IMOD.NE.1) THEN
C     COMPUTES TAN STIFF. MATRIX (UPPER TRIANGLE) (SEE (2.23))
      AKT(1,1)=1.D0
      AKT(1,2)=-1.D0
      AKT(1,3)=BET
      AKT(1,4)=-BET
      AKT(2,2)=1.D0
      AKT(2,3)=-BET
      AKT(2,4)=BET
      AKT(3,3)=BET*BET
      AKT(3,4)=-AKT(3,3)
      AKT(4,4)=BET*BET
      DO 12 I=1,4
      DO 12 J=1,4
   12 AKT(I,J)=EAL*AKT(I,J)
C
C     NOW ADD GEOM. OR INIT STRESS MATRIX (SEE (2.23))
C
      ANL=AN/AL
      AKT(3,3)=AKT(3,3)+ANL
      AKT(3,4)=AKT(3,4)-ANL
      AKT(4,4)=AKT(4,4)+ANL
      IF (IWRIT.NE.0) THEN
      WRIT (IWR,1001)
 1001 FORMAT(/,1X, 'TAN. STIFF. MATRIX FOR TRUSS EL. IS',/)
      DO 14 I=1,4
   14 WRITE (IWR,67) (AKT(I,J),J=1,4)
   67 FORMAT (1X,7G13.5)
      ENDIF
C
      ENDIF
C
      RETURN
      END
```

### 2.2.2 Subroutine INPUT

This subroutine inputs the data for the geometry, properties, boundary conditions and loading (for sample data see Section 2.6.1.1). The first variable, NV, defines the number of variables (4 or 5). If NV is four the spring $K_{s5}$ of Figure 2.2 is omitted. The subroutine also inputs EA, and AL, the length of the bar. An initial force in the bar (usually zero), $N_0$, is also input. For consistency with some of the work in Chapter 3 which includes large strains, the variable EA is subdivided into E and A (the variable ARA although for the current work this subdivision is unnecessary.

The subroutine requires the $z$ coordinates of nodes 1 and 2 (Figures 2.1 and 2.2). In addition, a fixed external load (or displacement) vector, $\mathbf{q}_e$ = QFI(NV) and a boundary condition counter, IBC(NV) are input. The constituent terms of these vectors relate to the four or five degrees of freedom (Figure 2.2).

Considering firstly standard load control, QFI(I) will contain a load (possibly zero) if the variable, I, is free. It will be zero if the variable, I, is constrained to zero. Simultaneously, IBC(I) will be set to zero if the variable is free or to unity if the variable is constrained to zero. To apply displacement control, QFI(I) is set to the magnitude of the fixed prescribed displacement (to be incremented), while IBC(I) is set to −1 (see Section 2.2.5 and the example in Section 2.6.4.4).

The routine inputs the number of linear 'earthed springs' (NDSP—up to four) followed by the degree-of-freedom numbers (IDSP(I), I = 1,NDSP) and the equivalent spring stiffnesses (AKSP(I),I = 1,NDSP). Finally, if NV is five, the subroutine inputs the stiffness of spring $K_{s5}$ (AK15) which connects variables 1 and 5 (Figure 2.2).

```
      SUBROUTINE INPUT (E,ARA,AL,QFI,X,Z,ANIT,IBC,IRE,IWR,AK14S,ID14S,
     1                  NDSP,NV,AK15,
     2                  ADUM1,IDUM)
C     ARGUMENTS IN LINE ABOVE AND ARRAY X NOT USED FOR SHALLOW TRUSS
C
C     READS INPUT FOR TRUSS ELEMENT
C
      IMPLICIT DOUBLE PRECISION (A-H, O-Z)
      DIMENSION X(2),Z(2),QFI(NV),IBC(NV),AK14S(4),ID14S(4)
C
      READ (IRE,*) NV,EA,AL,ANIT
      E=EA
      ARA=1.D0
      WRITE (IWR, 1000) NV,EA,AL,ANIT
 1000 FORMAT(/,1X,'NV=NO. OF VARBLS.=',I5,/,1X,
     1  'EA=',G13.5,/,1X,
     2  'AL=EL. LENGTH=',G13.5,1X,
     3  'ANIT=INIT. FORCE=',G13.5)
      IF (NV.NE.4.AND.NV.NE.5) STOP 'INPUT 1000'
      READ (IRE,*) Z(1),Z(2)
      WRITE (IWR,1001) Z(1),Z(2)
 1001 FORMAT(/,1X,'Z CO-ORD OF NODE 1=',G13.5,1X,
     1  'Z CO-ORD OF NODE 2=',G13.5)
      READ (IRE,*) (QFI(I),I=1,NV)
      WRITE (IWR,1002) (QFI(I),I=1,NV)
```

```
 1002 FORMAT(/,1X, 'FIXED LOAD OR DISP. VECTOR,QFI=',/,1X,5G13.5)
      WRITE (IWR, 1008)
 1008 FORMAT(/,1X, 'IF IBC(I)—SEE BELOW— =0, VARIABLE=A LOAD',/,1X,
     2    'IF IBC(I)—SEE BELOW—= -1, VARIABLE=A DISP.')
      READ (IRE,*) (IBC(I),I=1,NV)
      WRITE (IWR,1003) (IBC(I),I=1,NV)
 1003 FORMAT(/,1X,'BOUND. COND. COUNTER, IBC',/,1X,
     1    '=0, FREE: =1, REST. TO ZERO:= -1 REST. TO NON-ZERO',/,1X,
     2    5I5)
      READ (IRE,*) NDSP
      IF (NDSP.NE.0) THEN
      READ (IRE,*) (ID14S(I),I=1,NDSP)
      READ (IRE,*) (AK14S(I),I=1,NDSP)
      DO 40 I=1,NDSP
      WRITE (IWR,1004) AK14S(I), ID14S(I)
 1004 FORMAT(/,1X, 'LINEAR SPRING OF STIFFNESS',G13.5,/,1X,
     1    'ADDED AT VAR. NO. ',I5)
   40 CONTINUE
      ENDIF
C
      IF (NV.EQ.5) THEN
      READ (IRE,*) AK15
      WRITE (IWR,1005) AK15
 1005 FORMAT(/,1X, 'LINEAR SPRING BETWEEN VARBLS. 1 and 5 OF STIFF ',
     1      G13.5)
      ENDIF
C
      RETURN
      END
```

### 2.2.3 Subroutine FORCE

This subroutine computes the axial force $N$ in the bar using equations (2.7) and (2.8).

```
      SUBROUTINE FORCE(AN,ANIT,E,ARA,AL,X,Z,P,IWRIT,IWR,
     1      IDUM, ADUM1,ADUM2,ADUM3)
C     ARGUMENTS IN LINE ABOVE AND ARRAY X NOT USED FOR SHALLOW TRUSS
C
C     COMPUTES INTERNAL. FORCE IN AN SHALLOW TRUSS ELEMENT
C     USING (2.7) AND (2.8)
      IMPLICIT DOUBLE PRECISION (A-H, P-Z)
      DIMENSION Z(2),P(4),X(2)
C
      EA=E*ARA
      EAL=EA/AL
      U21=P(2)-P(1)
      W21=P(4)-P(3)
      Z21=Z(2)-Z(1)
      AN=U21+(Z21*W21/AL)+0.5D0*(W21*W21*AL)
      AN=EAL*AN+ANIT
      IF (IWRIT.NE.0) WRITE (IWR,1000) AN
```

```
 1000 FORMAT(/,1X,'AXIAL FORCE AN=  ',G13.5)
      RETURN
      END
```

### 2.2.4 Subroutine ELSTRUC

This subroutine puts the element stiffness matrix AKTE(4,4) into the structure stiffness matrix AKTS(NV,NV) (with NV=4 or 5), adds in the 'earthed springs' (if NDSP>0) and, if NV=5, the linear spring between variables 1 and 5. Depending on the input mode parameter, IMOD, the subroutine may alternatively, or also, apply similar operations to the internal forces FI.

```
      SUBROUTINE ELASTRUC(AKTE,AKTS, NV, AK15,ID14S,AK14S,NDSP,FI,PT,
     1                IMOD,IWRIT,IWR)
C
C     FOR IMOD=2 OR 3
C     PUTS EL-STIFF MATRIX AKTE(4,4) INTO STRUCT. STIFF AKTS(NV,NV)
C     IF NV=5, ALSO ADDS IN LINEAR SPRING AK15 BETWEEN VARBLS. 1&5
C     ALSO ADDS IN NDSP EARTHED LINEAR SPRINGS FOR VARBLS. 1-4
C     USING PROPERTIES IN AK14S(4) AND DEGS. OF F. IN IDSPS(4)
C     THROUGHOUT ONLY WORKS WITH UPPER TRIANGLE
C     FOR IMOD=1 OR 3
C     MODIFIES INTERNAL FORCE VECT., FI, TO INCLUDE EFFECTS FROM
C     VARIOUS LINEAR SPRINGS USING TOTAL DISPS., PT.
C
      IMPLICIT DOUBLE PRECISION (A-H,O-Z)
      DIMENSION AKTE(4,4),AKTS(NV,NV),ID14S(4),AK14S(4)
      DIMENSION FI(NV,)PT(NV)
C
      IF (IMOD.NE.2) THEN
C     MODIFY FORCES
      IF (INDSP.NE.0) THEN
C     FOR EARTHED SPRINGS
      DO 40 I=1,NDSP
      IDS=ID14S(I)
      FI(IDS)=FI(IDS)+AK14S(I)*PT(IDS)
   40 CONTINUE
      ENDIF
C
      IF (NV.EQ.5) THEN
C     MODIFY FOR SPRING BETWEEN VARBLS. 1 AND 5
      FI(1)=FI(1)+AK15*(PT(1)-PT(5))
      FI(5)=AK15*(-PT(1)+PT(5))
      ENDIF
C
      IF (IWRIT.NE.0) WRITE (IWR,1002) FI
 1002 FORMAT(/,1X,'STR. INT. FORCE VECT IS'/,1X,5G13.4)
C
      ENDIF
```

```
C
      IF (IMOD.NE.1) THEN
C     WORK ON STIFFNESS MATRIX; CLEAR STRUCT. STIFFNESS MATRIX
      DO 10 I=1,NV
      DO 11 J=1,NV
   11 AKTS(I,J)=0.D0
   10 CONTINUE
C
C     INSERT EL. STIFFNESS MATRIX
      DO 20 I=1,4
      DO 21 J=I,4
   21 AKTS(I,J) = AKTE(I,J)
   20 CONTINUE
C
C     SPRING BETWEEN VARBLS. 1&5
      IF (NV.EQ.5) THEN
      AKTS(1,1)=AKTS(1,1)+AK15
      AKTS(1,5)=AKTS(1,5)-AK15
      AKTS(5,5)=AKTS(5,5)+AK15
      ENDIF
C
C     EARTHED SPRINGS FOR VARBLS. 1-4.
      IF (NDSP.NE.0) THEN
      DO 30 I=1,NDSP
      IDS=ID14S(I)
      AKTS(IDS,IDS)=AKTS(IDS,IDS)+AK14S(I)
   30 CONTINUE
      ENDIF
C
      IF (IWRIT.NE.0) THEN
      WRITE (IWR,1001)
 1001 FORMAT(/,1X, 'FULL STRUCT. TAN. STIFF. IS',/)
      DO 50 I=1,NV
   50 WRITE (IWR,67) (AKTS(I,J),J=1,NV)
   67 FORMAT(1X,7G13.5)
      ENDIF
C
      ENDIF
C
      RETURN
      END
```

### 2.2.5 Subroutine BCON and details on displacement control

The tangent stiffness equations can be assumed to be of the form

$$\mathbf{q} = \begin{pmatrix} \mathbf{q}_f \\ \mathbf{q}_p \end{pmatrix} = \mathbf{K}\mathbf{p} = \begin{bmatrix} \mathbf{K}_{ff} & \mathbf{K}_{fp} \\ \mathbf{K}_{pf} & \mathbf{K}_{pp} \end{bmatrix} \begin{pmatrix} \mathbf{p}_f \\ \mathbf{p}_p \end{pmatrix} \tag{2.25}$$

where subscript f means 'free' and subscript p means 'prescribed'. (In practice, the ordering need not be of the form of (2.25).) Considering, firstly the case where the displacements $\mathbf{p}_p$ are constrained to zero, subroutine BCON effectively alters (2.25) to become

$$\mathbf{q} = \begin{pmatrix} \mathbf{q}_f \\ \mathbf{0} \end{pmatrix} = \mathbf{K}\mathbf{p} = \begin{bmatrix} \mathbf{K}_{ff} & \mathbf{0} \\ \mathbf{0} & \mathbf{I} \end{bmatrix} \begin{pmatrix} \mathbf{p}_f \\ \mathbf{p}_p \end{pmatrix}. \tag{2.26a}$$

This is achieved in subroutine BCON by setting the leading diagonal term of $\mathbf{K}$ = AKT(J,J) to unity if the boundary condition counter IBC(J) is unity (see Section 2.2.2). In addition, the remainder of the Jth row and column are set to zero. Hence a 'dummy equation' is introduced (similar, in concept, to the second row of (2.26a)). (It is assumed that no loads are applied at constrained nodes.)

If the displacements, $\mathbf{p}_p$, are to be constrained to non-zero values, instead of (2.26a), (2.25) is conceptually altered to

$$\mathbf{q} = \begin{pmatrix} \mathbf{q}_f - \mathbf{K}_{fp}\mathbf{p}_p \\ \mathbf{p}_p \end{pmatrix} = \mathbf{K}\mathbf{p} = \begin{bmatrix} \mathbf{K}_{ff} & \mathbf{0} \\ \mathbf{0} & \mathbf{I} \end{bmatrix} \begin{pmatrix} \mathbf{p}_f \\ \mathbf{p}_p \end{pmatrix}. \tag{2.26b}$$

These modifications are applied within subroutine BCON although, as previously discussed, the variables are not necessarily partitioned into the ordering of $\mathbf{p}^T = (\mathbf{p}_f^T, \mathbf{p}_p^T)$. In practice, the procedure of (2.26b) will cover the procedure of (2.26a) where $\mathbf{p}_p = \mathbf{0}$ and there is no need to distinguished between 'restrained variables' (with $\mathbf{p}_p = \mathbf{0}$) and 'prescribed variables' (with $\mathbf{p}_p \neq \mathbf{0}$). Hence the same procedure is applied whether the boundary condition counter, IBC(J) is 1 or $-1$. The distinction is introduced for clarity in the input and for use with some of the more advanced solution procedures of Chapter 9.

```
      SUBROUTINE BCON(AK,IBC,N,F,IWRIT,IWR)
C     APPLIES B. CONDS. TO MATRIX AK AS WELL AS
C     ALTERING 'LOAD VECTOR', F FOR PRESC. DISPS.
C     BY SETTING DIAG=1. AND ROW AND COL TO ZERO IN REST.
C     USES COUNTER IBC WHICH IS 0 IF FREE, 1 IF REST. TO ZERO,
C     -1 IF REST. TO NON-ZERO VALUE.
C     ON ENTRY F HAS LOADS FOR FREE VARIABLES AND DISPS. FOR
C     REST. (POSSIBLY ZERO) VARIABLES.
C     ON EXIT THE LATTER ARE UNCHANGED BUT LOADS ARE ALTERED
C
      IMPLICT DOUBLE PRECISION (A-H, O-Z)
      DIMENSIONS IBC(N)
      DIMENSION AK(N,N),F(N)
C
      IPRS=0
      DO 10 I=1,N
      II=IBC(I)
      IF (II.LT.0) IPRS=1
      IF (II.NE.0) AK(I,I)=1.D0
      IF (I.EQ.N) GO TO 10
      DO 20 J=1+1,N
      JJ=IBC(J)
      IF (II.EQ.0.AND.JJ.EQ.0) GO TO 20
C     ABOVE BOTH FREE, BELOW BOTH REST.
      IF (II.NE.0.AND.JJ.NE.0) GO TO 25
```

```
C       BELOW I REST OR PRESC.
        IF (II.NE.0) THEN
        F(J)=F(J)-AK(I,J)*F(I)
C       BELOW J REST OR PRESC
        ELSE
        F(I)=F(I)-AK(I,J)*F(J)
        ENDIF
     25 AK(I,J)=0.D0
     20 CONTINUE
     10 CONTINUE
C
        IF (IWRIT.NE.0) THEN
        WRITE (IWR,1000)
   1000 FORMAT(/,1X, 'STIFF. MAT. AFTER B. CONDS. IS',/)
        DO 30 I=1,N
     30 WRITE (IWR, 67) (AK(I,J),J=1,N)
        IF (IPRS.EQ.1) WRITE (IWR,1001) F
   1001 FORMAT(/,1X, 'MODIFIED LOAD VECTOR AFTER B CONDS. IS ',/,1X,5G13.4)
     67 FORMAT(1X,7G12.5)
        ENDIF
C
        RETURN
        END
```

### 2.2.6 **Subroutine** CROUT

Conceptually, subroutine CROUT applies the Crout factorisation [C2] to the tangent stiffness matrix, $\mathbf{K}_t$. Here, $\mathbf{L}$ is a lower triangular matrix with unit terms on the leading diagonal and $\mathbf{D}$ is a diagonal matrix containing the 'pivots'. In practice, the routine is entered with AK(N,N) containing the upper triangle of $\mathbf{K}_t$ and exits with AK(N,N) containing $\mathbf{L}^T$ while a vector $\mathbf{D}$(N) exists with the diagonal pivots.

```
        SUBROUTINE CROUT(AK,D,N,IWRIT,IWR)
C
C       INPUTS AK(N,N); OUTPUTS UPPER TRIANGLE IN AK AND DIAG
C       PIVOTS IN D(N)
C
        IMPLICIT DOUBLE PRECISION (A-H,O-Z)
        DIMENSION AK(N,N),D(N)
C
        D(1)=AK(1,1)
        DO 1 J=2,N
        DO 2 I=1,J-1
        A=AK(I,J)
        IF (I.EQ.1) GO TO 2
        DO 3 L=1, I-1
      3 A=A-AK(L,J)*AK(L,I)
      2 AK(I,J)=A
        DO 4 I=1,J-1
      4 AK(I,J)=AK(I,J)/AK(I,I)
        DO 5 L=1,J-1
```

```
    5 AK(J,J)=AK(J,J)-AK(L,J)*AK(L,J)*AK(L,J)
    1 D(J)=AK(J,J)
C
      IF (IWRIT.NE.0) THEN
      WRITE (IWR,1000)
 1000 FORMAT(/,1X, 'FACTORISED MATRIX IS',/)
      DO 10 I=1,N
   10 WRITE (IWR,67) (AK(I,J),J=1,N)
   67 FORMAT(1X,7G12.5)
      WRITE (IWR,1001)
 1001 FORMAT(/,1X, 'DIAG. PIVOTS ARE',/)
      WRITE (IWR,67) (D(I),I=1,N)
      ENDIF
C
      RETURN
      END
```

### 2.2.7 Subroutine SOLVCR

This subroutine applies forward and backward substitution on the vector **Q** using the previously obtained $\mathbf{LDL}^T$ factors (in AK and D–see Section 2.2.6). Hence **Q** enters as a load vector and exits as a displacement vector so that the routine obtains:

$$\mathbf{q} = \mathbf{K}_t^{-1}\mathbf{q} \tag{2.28}$$

```
      SUBROUTINE SOLVCR(AK,D,Q,N,IWRIT,IWR)
C
C     APPLIES FORWARD AND BACK CROUT SUBS ON Q
C
      IMPLICIT DOUBLE PRECISION (A-H,O-Z)
      DIMENSION AK(N,N),D(N),Q(N)
C
C     FORWARD SUBS.
      DO 1 J=2,N
      DO 2 L=1,J-1
    2 Q(J) = Q(J) - AK(L,J)*Q(L)
    1 CONTINUE
      IF (IWRIT.NE.0) THEN
      WRITE (IWR,1000) (Q(I),I=1,N)
 1000 FORMAT(/,1X,'DISP.INCS AFTER FORWARD SUBS. ARE',/,
     1       1X,7G12.5)
      ENDIF
C
C     BACK SUBS.
      DO 3 I=1,N
    3 Q(I) = Q(I)/D(I)
C
      DO 4 JJ=2,N
      J=N+2-JJ
      DO 5 L=1,J-1
    5 Q(L) = Q(L) - AK(L,J)?Q(J)
    4 CONTINUE
```

```
C
         IF (IWRIT.NE.0) THEN
         WRITE (IWR,1001) (Q(I),I=1,N)
   1001  FORMAT(/,1X,'DISP INCS. AFTER BACKWARD SUBS. ARE',/,
      1         1X,7G12.5)
         ENDIF
C
         RETURN
         END
```

## 2.3 A FLOWCHART AND COMPUTER PROGRAM FOR AN INCREMENTAL (EULER) SOLUTION

A computer program for an incremental solution can be generated by extending the concepts of Sections 1.2.1 and 1.3. In particular, solutions will be obtained for a load

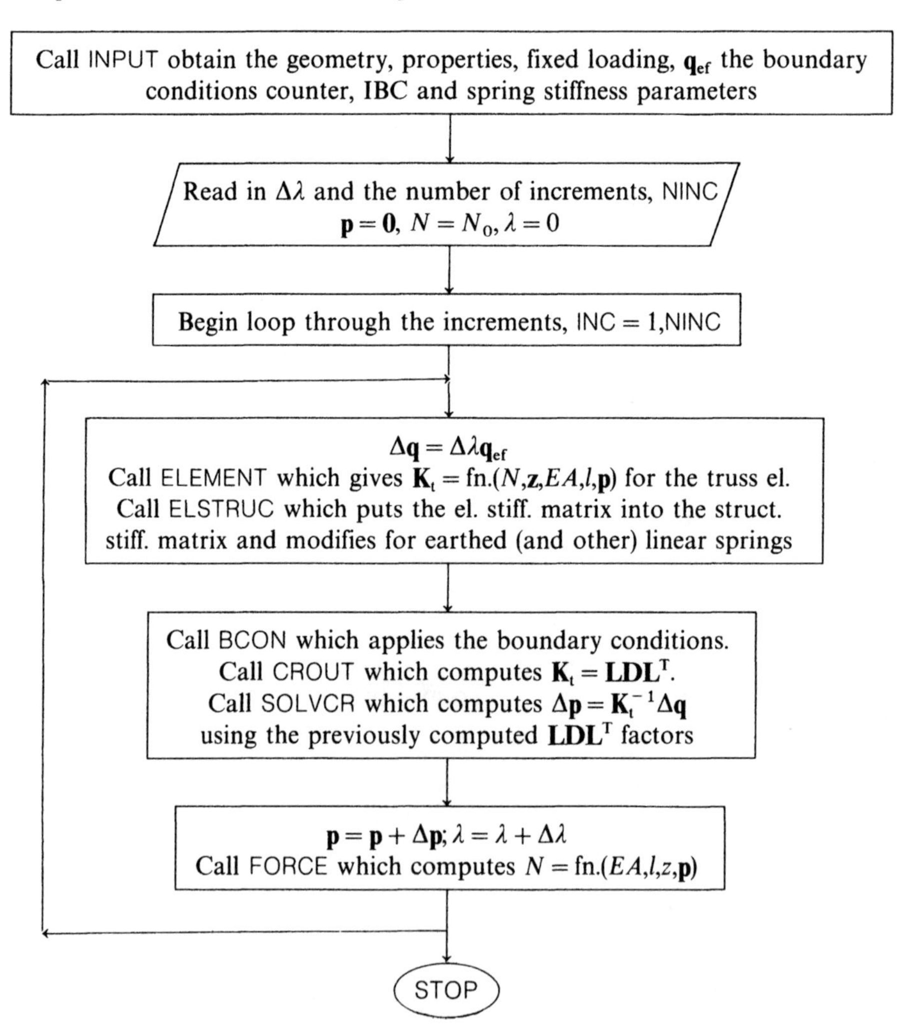


**Figure 2.3** Flowchart for an incremental solution (program NONLTA).

vector

$$\mathbf{q}_e = \lambda \mathbf{q}_{ef} \tag{2.29}$$

where $\mathbf{q}_{ef}$ is a fixed external load vector that is input in subroutine INPUT and $\lambda$ is the load-level parameter. For the present, we will work with fixed increments so that, in addition to $\mathbf{q}_{ef}$, the program inputs the load-increment factor, $\Delta\lambda$.

A flowchart for the incremental solution is given in Figure 2.3 and a computer program (NONLTA) follows.

### 2.3.1 **Program** NONLTA

```
      PROGRAM NONLTA
C
C     PERFORMS NON-LIN. INCREMENTAL SOLN. FOR TRUSS
C     NV=NO. OF VARIABLES (4 OR 5)
C     QFI=FIXED LOAD VECTOR
C     IBC=B. COND. COUNTER (0=FREE, 1=FIXED)
C     Z=C COORDS OF NODES
C     QINC=INC. LOAD VECTOR
C     PT=TOTAL DISP. VECTOR
C     AKTS=STRUCT. TAN. STIFF. MATRIX
C     AKTE=ELEMENT TAN. STIFF. MATRIX
C     FI (NOT USED HERE)=INTERNAL FORCES
C     D=DIAGONAL PIVOTS FROM LDL (TRAN) FACTORISATION
C     ID14S=VAR. NOS. (1-4) AT WHICH LIN EARTHED SPRINGS
C     AK14S=EQUIV. LINEAR SPRING STIFFNESSES
C     AK15=LINEAR SPRING STIFF BETWEEN VARBLS. 1 and 5 (IF NV=5)
C
      IMPLICIT DOUBLE PRECISION (A-H,O-Z)
      DIMENSION QFI(5),IBC(5),Z(2),QINC(5),PT(5),AKTE(4,4)
      DIMENSION FI(5),D(5),AK14S(4),ID14S(4),AKTS(25),X(2)
C     ARRAY X ABOVE NOT USED FOR SHALLOW TRUSS
C
      IRE=5
      IWR=6
      OPEN (UNIT=5,FILE=' ')
      OPEN (UNIT=6,FILE=' ')
C
      CALL INPUT(E,ARA,AL,QFI,X,Z,ANIT,IBC,IRE,IWR,AK14S,ID14S,NDSP,
     1           NV,AK15,
     2           POISS,ITYEL)
C     ARGUMENTS IN LINE ABOVE NOT USED FOR SHALLOW TRUSS
C     BELOW RELEVANT TO DEEP TRUSS BUT LEAVE FOR SHALLOW TRUSS
      ALN=AL
      ARN=ARA
C
      READ (IRE,*) FACI,NINC,IWRIT
      WRITE (IWR,1000) FACI, NINC,IWRIT
```

```
 1000 FORMAT(/,1X, 'INCREMENTAL LOAD FACTOR=',G13.5,/,1X,
     1       'NO. OF INC. (NINC)=  ',I5,/,1X,
     2       'WRITE CONTROL (IWRIT)=  ',I5,/,3X,
     3       '0=LIMITED ; 1=FULL')
C
      AN=ANIT
      FACT=0.D0
      DO 5 I=1,NV
    5 PT(I) =0.D0
C
C
      DO 100 INC=1,NINC
      FACT=FACT+FACI
      WRITE (IWR,1001) INC,FACT
 1001 FORMAT(//,1X,'INC=  ',I5,'LD. FACTOR=  ',G13.5)
      DO 10 I=1,NV
   10 QINC(I)=FACI*QFI(I)
C
C     BELOW FORMS EL. TAN. STIFF MATRIX AKT
      CALL ELEMENT (FI,AKTE,AN,X,Z,PT,E,ARA,AL,IWRIT,IWR,2,
     1      ITYEL,ALN,ARN)
C     ARGUMENTS IN LINE ABOVE NOT USED FOR SHALLOW TRUSS
C
      CALL ELSTRUC(AKTE,AKTS,NV,AK15,ID14S,AK14S,NDSP,FI,PT,
     1             2,IWRIT,IWR)
C     ABOVE PUTS EL.STIFF AKTE IN STRUC STIFF AKTS AND
C     ADDS EFFECTS OF VARIOUS LINEAR SPRINGS
C
      CALL BCON(AKTS,IBC,NV,QINC,IWRIT,IWR)
C     ABOVE APPLIES B. CONDITIONS
      CALL CROUT(AKTS,D,NV,IWRIT,IWR)
C     ABOVE FORMS LDL(TRAN) FACTORISATION INTO AKT AND D
      CALL SOLVCR(AKTS,D,QINC,NV,IWRIT,IWR)
C     ABOVE SOLVES EQNS. AND GETS INC. DISPS IN QIN
C
      DO 20 I=1,NV
   20 PT(I)=PT(I)+QINC(I)
C     ABOVE UPDATES TOTAL DISPS.
C
      WRITE (6,1002) (PT(I),I=1,NV)
 1002 FORMAT(/,1X, 'TOTAL DISPS. ARE',/,1X,5G13.5)
C
C     BELOW FORMS TOTAL FORCE IN BAR
      CALL FORCE(AN,ANIT,E,ARA,AL,X,Z,PT,IWRIT,IWR,
     1      ITYEL,ARN,ALN,POISS)
C     ABOVE ARGUMENTS NOT USED FOR SHALLOW TRUSS
  100 CONTINUE
C
      STOP 'NONLTA'
      END
```

## 2.4 A FLOWCHART AND COMPUTER PROGRAM FOR AN ITERATIVE SOLUTION USING THE NEWTON–RAPHSON METHOD

A computer program for an iterative solution can be generated by extending the concepts of Sections 1.2.2 and 1.3. Because an iterative procedure is being adopted, a convergence criterion must be introduced. A detailed discussion on convergence criteria will be given in Section 9.5.4. In the meantime, we will adopt a simple 'force criterion' whereby,

$$\begin{aligned} \|\mathbf{g}\| &= (\mathbf{g}^T\mathbf{g})^{1/2} < \beta \|\mathbf{q}_e\| \qquad \text{load control (a)} \\ \|\mathbf{g}\| &< \beta \|\mathbf{r}\| \qquad \text{displacement control (b)} \end{aligned} \tag{2.30}$$

where $\mathbf{g}$ is the out-of-balance force vector and $\mathbf{q}_e$ is the current total external load vector while $\mathbf{r}$ is the reaction vector (with non-zero values where the displacements are constrained). The introduction of the $\mathbf{r}$ terms in (2.30) allows the procedure to work when displacement control is adopted and the external load vector, $\mathbf{q}_e$, is zero. (In very rare circumstances, under displacement control, $\|\mathbf{r}\|$ may be zero; hence the alternative convergence criterion included in subroutine ITER (Section 2.4.2) and its flowchart (Figure 2.4, Section 2.4.2).) Typical values for $\beta$ lie between 0.001 and 0.01 although a tighter tolerance might be required for displacement control because $\mathbf{r}$ will generally have terms for each component so that, with the same $\beta$, (2.30b) will be less severe than (2.30a).

The reactions, $\mathbf{r}$, are simply computed as being equal to the internal forces, $\mathbf{q}_i$ (see (2.17)), at the constrained variables. In practice, in subroutine ITER (see Section 2.4.2), $\mathbf{r}$ is set to $\mathbf{q}_i$ for all variables. Hence, for the early iterations, $\mathbf{r}$ will include the out-of-balance forces at the free variables but, as the solution procedure converges, these terms will tend to zero.

For the master segment of the computer program (NONLTB—see below), we simply read in, as before, the geometric data and properties as well as the fixed external load vector, $\mathbf{q}_{ef}$, via subroutine INPUT. In addition, we input a starting 'trial vector', $\mathbf{p}_o$ = PT(NV) (possibly zero), and the required convergence tolerance, $\beta$ (see (2.30)). Finally, the program calls a subroutine ITER (see Section 2.4.2) which performs the Newton–Raphson iterations until convergence is achieved. A flowchart for this subroutine is given in Figure 2.4. The subroutine is designed to operate with the general incremental/iterative strategy of the next section. Hence it allows either full or modified N–R iterations. However, only the former may be used with the program NONLTB.

### 2.4.1 Program NONLTB

```
      PROGRAM NONLTB
C
C     PERFORMS NEWTON-RAPHSON ITERATION FROM STARTING PREDICTOR, PT
C     NV=NO. OF VARIABLES (4 OR 5)
```

```
C         IBC=B. COND. COUNTER (0=FREE, 1=FIXED)
C         Z=Z COORDS OF NODES
C         PT=TOTAL DISP. VECTOR
C         ID14S=VAR. NOS. (1-4) AT WHICH LINEAR EARTHED SPRINGS
C         AK14S=EQUIV. LINEAR SPRING STIFFNESSES
C         QFI=TOTAL LOAD VECTOR
C         AKTS=STRUCT. STIFF. MATRIX
C         AK15=LIN SPRING STIFF. BETWEEN VARBLS. 1 AND 5 (IF NV=5)
C         FI=INTERNAL FORCE VECTOR
C         GM=OUT-OF-BALANCE FORCE VECTOR
C         REAC=REACTIONS
C         X=X COORDS
C         ARGUMENTS IN COMMON/DAT2/ AND ARRAY X NOT USED FOR SHALLOW TRUSS
C
          IMPLICIT DOUBLE PRECISION (A-H,O-Z)
          COMMON /DAT/ X(2),Z(2),E,ARA,AL,ID14S(4),AK14S(4),NDSP,ANIT,AK15
          COMMON /DAT2/ ARN,POISS,ALN,ITYEL
          DIMENSION QFI(5),IBC(5),PT(5),AKTS(25),D(5),GM(5),FI(5)
          DIMENSION REAC(5)
C
          IRE=5
          IWR=6
          OPEN (UNIT=5,FILE=' ')
          OPEN (UNIT=6,FILE=' ')
C
          CALL INPUT(E,ARA,AL,QFI,X,Z,ANIT,IBC,IRE,IWR,AK14S,ID14S,NDSP,
         1         NV,AK15
         2         POISS,ITYEL)
C         ARGUMENTS IN LINE ABOVE NOT USED FOR SHALLOW TRUSS
C         BELOW RELEVANT TO DEEP TRUSS BUT LEAVE FOR SHALLOW TRUSS
          ALN=AL
          ARN=ARA
C
          READ (IRE*) (PT(I),I=1,NV)
          WRITE (IWR,2000) (PT(I),I=1,NV)
     2000 FORMAT(/,1X,2STARTING PREDICTOR DISPS ARE',/,1X,6G12.5)
          READ (IRE*) BETOK,IWRIT
          WRITE (IWR,2001) BETOK,IWRIT
     2001 FORMAT(/,1X,'CONV. TOL FACTOR, BETOK=  ',G12.5,/,1X,
         2      'DIAGNOSTIC WRITE CONTROL(IWRIT) =  ',15,/,3X,
         3      '0=NO ; 1=YES')
C         SET TO NEWTON-RAPHSON ITERATIONS
          ITERTY=1
C
          CALL ITER(PT,AN,BETOK,QFI,IBC,IWRIT,IWR,AKTS,D,ITERTY,NV,
         1         GM,FI,REAC)
C
          WRITE (IWR,1004), (PT(Ï,I=1,NV)
     1004 FORMAT(/,1X,'FINAL TOTAL DISPLACEMENTS ARE',/,1X,5G12.5)
          WRITE (IWR,1006) (REAC(I),I=1,NV)
```

```
1006 FORMAT(/,1X,'FINAL REACTIONS ARE',/,1X,5G12.5)
     WRITE (IWR,1005) AN
1005 FORMAT(/,1X, 'AXIAL FORCE IN BAR IS ',G12.5)
     STP 'NONLTB'
     END
```

### 2.4.2 Flowchart and computer listing for subroutine ITER

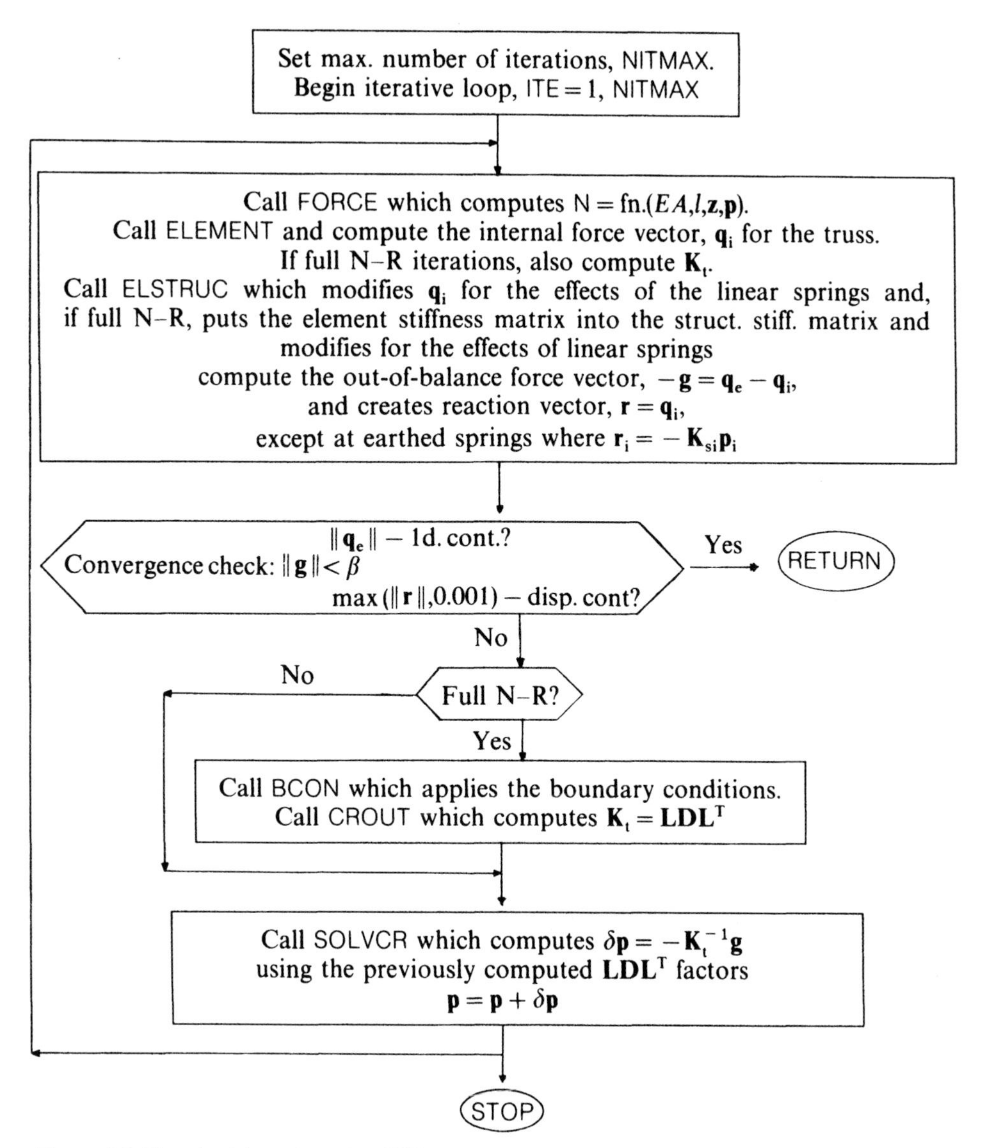


**Figure 2.4** Flowchart for subroutine ITER which performs equilibrium iterations and can be used with program NONLTB (N–R iteration) or NONLTC (combined incremental/iterative) solution.

```
      SUBROUTINE ITER(PT,AN,BETOK,QEX,IBC,IWRIT,IWR,AKTS,D,ITERY,NV,
     1                GM,FI,REAC)
C
C     INPUTS PREDICTOR DISPS. PT(NV) AND EXT. FORCE VECTOR QEX(NV)
C     ALSO BETOK=CONV. TOL, IBC=B. CON COUNTER
C     ITERATES TO EQUILIBRIUM: OUTPUTS NEW PT AND FORCE IN BAR,AN
C     IF ITERTY (INPUT)=1 USES FULL N-R, =2 USES MOD N-R
C     IN LATTER CASE, AKTS AND D INPUT AS CROUT FACTORS (D=PIVOTS)
C     LOCAL ARRAY IS AKTE=EL. STIFF. MATRIX
C     ARGUMENTS IN COMMON/DAT2/ AND ARRAY X NOT USED FOR SHALLOW TRUSS
C
      IMPLICIT DOUBLE PRECISION (A-H,O-Z)
      COMMON /DAT/ X(2),Z(2),E,ARA,AL,ID14S(4),AK14S(4),NDSP,ANIT,AK15
      COMMON /DAT2/ ARN,POISS,ALN,ITYEL
      DIMENSION PT(NV),QEX(NV),IBC(NV),REAC(NV)
      DIMENSION FI(NV),GM(NV),AKTS(NV,NV),D(NV),AKTE(4,4)
C
      SMALL=0.1D-2
      NITMAX=16
      IMOD=1
      IF (ITERTY.EQ.1) IMOD = 3
C
      DO 100 ITE=1,NITMAX
C
      IF (IWRIT.EQ.1) WRITE (IWR,1005) ITE
 1005 FORMAT(/,IX,'ITERATIVE LOOP WITH ITE=',I5)
C     BELOW CALCS FORCE IN BAR (AN)
      CALL FORCE(AN,ANIT,E,ARA,AL,X,Z,PT,IWRIT,IWR,
     1    ITYEL,ARN,ALN,POISS)
C     ABOVE ARGUMENTS NOT USED FOR SHALLOW-TRUSS
C
C     ABOVE CALCS FORCE IN BAR, AN: BELOW TAN STIFF AKT
C     (IF NR) AND INT. FORCE VECT. FI
      CALL ELEMENT(FI,AKTE,AN,X,Z,PT,E,ARA,AL,IWRIT,IWR,IMOD,
     1    ITYEL,ALN,ARN)
C     ABOVE ARGUMENTS NOT USED FOR SHALLOW TRUSS
C
C     BELOW PUTS EL. STIFF. MAT., AKTE, IN STR. STIFF., AKTS AND
C     ADDS IN EFFECTS OF VARIOUS LINEAR SPRINGS (IF NR)
C     ALSO MODIFIES INT. FORCE VECT. FI FOR SPRING EFFECTS
      CALL ELSTRUC(AKTE,AKTS,NV,AK15,ID14S,AK14S,NDSP,FI,PT,
     1      IMOD,IWRIT,IWR)
C
C     BELOW FORMS GM=OUT-OF-BALANCE FORCE VECTOR
C     AND REACTION VECTOR
      DO 10 I=1,NV
      GM(I)=0.D0
      REAC(I)=FI(I)
      IF (IBC(I).EQ.0) THEN
      GM(I)=QEX(I)-FI(I)
      ENDIF
```

```
   10 CONTINUE
   67 FORMAT(6G13.5)
   47 FORMAT(5I5)
C
C     OVERWRITE SPRING REACTION TERMS
      IF (NDSP.NE.0) THEN
      DO 50 I=1,NDSP
   50 REAC(ID14S(I)) = -AK14S(I)*PT(ID14S(I))
      ENDIF
C
C     BELOW CHECKS CONVERGENCE
      FNORM=0.D0
      GNORM=0.D0
      RNORM=0.D0
      IDSP=0
      DO 20 I=1,NV
      IF (IBC(I).EQ.0) FNORM=FNORM+QEX(I)*QEX(I)
      IF (IBC(I).EQ.-1) IDSP=1
      RNORM=RNORM+REAC(I)*REAC(I)
   20 GNORM=GNORM+GM(I)*GM(I)
      FNORM=DSQRT(FNORM)
      GNORM=DSQRT(GNORM)
      RNORM=DSQRT(RNORM)
      BAS=MAX(FNORM,SMALL)
C     BELOW DISP. CONTROL
      IF (IDSP.EQ.1) BAS = MAX(RNORM,SMALL)
      BET=GNORM/BAS
      ITEM=ITE-1
      WRITE (IWR,1001) ITEM,BET
 1001 FORMAT/,1X,'ITERN. NO.=  ',I5,' CONV. FAC.=  ',G13.5)
      IF (IWRIT.EQ.1) WRITE (IWR,1003) (GM(I),I=1,NV)
 1003 FORMAT(/,1X,'OUT-OF-BAL. FORCE VECTOR=  ',/,1X,4G13.5)
      IF (BET.LE.BETOK) GO TO 200
C
      IF (ITERTY.EQ.1) THEN
      CALL BCON(AKTS,IBC,NV,GM,IWRIT,IWR)
C     ABOVE APPLIES B. CONDITIONS
      CALL CROUT(AKTS,D,NV,IWRIT,IWR)
C     ABOVE FORMS LDL(TRAN) FACTORISATION INTO AKTS AND D
      ENDIF
C
      CALL SOLVCR(AKTS,D,GM,NV,IWRIT,IWR)
C     ABOVE KETS ITER. DISP. CHANGE IN GM
C
      DO 30 I=1,NV
      IF (IBC(I).EQ.0) THEN
      PT(I) = PT(I) + GM(I)
      ELSE
      PT(I)=QEX(I)
      ENDIF
   30 CONTINUE
C     ABOVE UPDATES DISPS.
```

```
      IF (IWRIT.EQ.1) WRITE (IWR,1004) (PT(I),I=1,NV)
 1004 FORMAT(/,1X,'TOTAL DISPS ARE',/,1X,6G13.5)
C
  100 CONTINUE
C
      WRITE (IWR,1002)
 1002 FORMAT(/,1X,'FAILED TO CONVERGE****')
      STOP 'ITER 100'
C
  200 CONTINUE
      RETURN
      END
```

## 2.5 A FLOWCHART AND COMPUTER PROGRAM FOR AN INCREMENTAL/ITERATIVE SOLUTION PROCEDURE USING FULL OR MODIFIED NEWTON–RAPHSON ITERATIONS

A computer program for a combined incremental/iterative solution can be generated by extending the concepts of Sections 1.2.3 and 1.3. The master program NONLTC (see below) is very similar to the incremental program NONLTA (see Section 2.3 and Figure 2.3) because it involves the generation of an incremental predictor prior to the application, via subroutine ITER (see Section 2.4 and Figure 2.3), of equilibrium iterations. The latter may be full or modified Newton–Raphson depending on a parameter, ITERTY, that is input in program NONLTC.

Figure 2.5 gives a flowchart for this program. Immediately after the beginning of the main incremental loop, the flowchart contains

$$\Delta \mathbf{q} = \Delta\lambda \mathbf{q}_{ef}\,(-\mathbf{g}). \tag{2.31}$$

The term in brackets is omitted (and a comment statement included in program NONLTC) if a '*pure* incremental predictor' is adopted. Generally, it is better to include the bracketed $-\mathbf{g}$ term because this ensures that the out-of-balance forces from the previous increment are included at the start of the current increment. Hence

$$\Delta \mathbf{q} = \lambda_{new}\mathbf{q}_{ef} - \mathbf{q}_i(\lambda = \lambda_{old}). \tag{2.32}$$

Clearly, if a very tight convergence tolerance is adopted, there will be no difference between the two formulations. If a very coarse convergence tolerance is adopted, no *real* iterations will be performed in subroutine ITER and a 'self-correcting incremental formulation' will be produced [H1.1]. However, the term 'self-correcting' is too strong for, although the procedure does make some allowance for the out-of-balance forces from the previous increment, it does not ensure a genuine 'equilibrium solution'. The reader may like to try this out by applying a coarse convergence tolerance (high $\beta$ = BETOK as input to program NONLTC) to the single-variable problem of Figure 1.1. The 'pure incremental' solution to this problem was obtained in Section 1.2.1 and is given in Figure 1.4. It will be found that a 'self-correcting incremental formulation' will only marginally improve the solution.

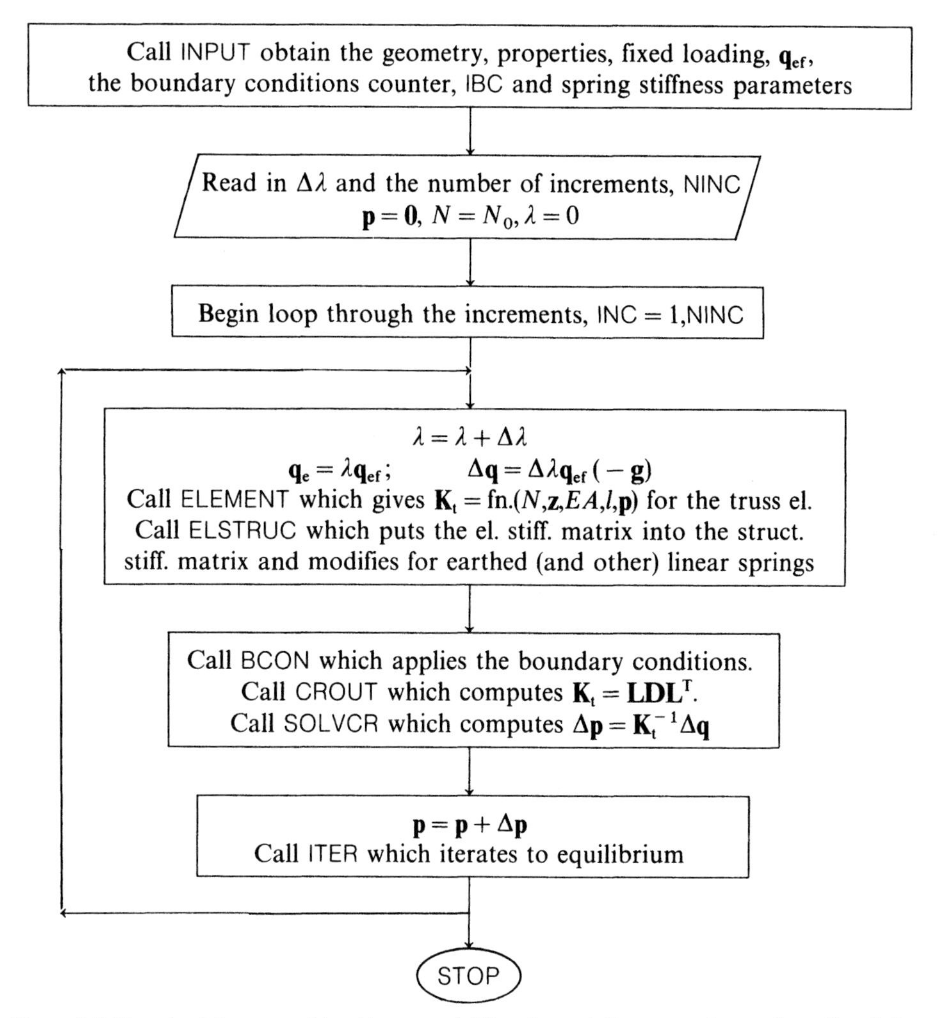


**Figure 2.5** Flowchart for a combined incremental/iterative solution procedure using either full or modified Newton–Raphson iterations.

### 2.5.1 **Program** NONLTC

```
      PROGRAM NONLTC
C
C     PERFORMS NON-LIN. INCREMENTAL/ITERATIVE SOLN. FOR TRUSS
C     NV=NO. OF VARIABLES(4 OR 5)
C     QFI=FIXED LOAD VECTOR
C     IBC=B. COND. COUNTER (0=FREE, 1=FIXED)
C     Z=Z COORDS OF NODES
C     QINC=INC. LOAD VECTOR
C     PT=TOTAL DISP. VECTOR
C     QEX=TOTAL (EXTERNAL) LOAD VECTOR
```

```
C         AKTS=STR. TAN. STIFF. MATRIX
C         AKTE=ELE. TAN. STIFF. MATRIX
C         FI=INTERNAL FORCES
C         D=DIAGONAL PIVOTS FROM LDL(TRAN) FACTORISATION
C         ID14S=VAR. NOS. (1-4) AT WHICH LINEAR EARTHED SPRINGS
C         AK14S=EQUIV. LINEAR SPRING STIFFNESS
C         AK15=LIN SPRING STIFF. BETWEEN VARBS. 1 AND 5 (IF NV=5)
C         GM=OUT-OF-BALANCE FORCES
C         REAC=REACTIONS
C         X=X COORDS
C         ARGUMENTS IN COMMON/DAT2/ AND ARRAY X NOT USED FOR SHALLOW TRUSS
C
          IMPLICIT DOUBLE PRECISION (A-H,O-Z)
          COMMON /DAT/ X(2),Z(2),E,ARA,AL,ID14S(4),AK14S(4),NDSP,ANIT,AK15
          COMMON /DAT2/ ARN,POISS,ALN,ITYEL
          DIMENSION QFI(5),IBC(5),QINC(5),PT(5),AKTE(4,4)
          DIMENSION FI(5),D(5),QEX(5),GM(5),AKTS(25),REAC(5)
C
          IRE=5
          IWR=6
          OPEN (UNIT=5,FILE=' ')
          OPEN (UNIT=6,FILE=' ')
C
          CALL INPUT(E,ARA,AL,QFI,X,Z,ANIT,IBC,IRE,IWR,AK14S,ID14S,NDSP,
        1            NV,AK15,
        2            POISS,ITYEL)
C         ARGUMENTS IN LINE ABOVE NOT USED FOR SHALLOW TRUSS
C         BELOW RELEVANT TO DEEP TRUSS BUT LEAVE FOR SHALLOW TRUSS
          ALN=AL
          ARN=ARA
C
          READ (IRE,*) FACI,NINC,IWRIT
          WRITE (IWR,1000) FACI,NINC,IWRIT
     1000 FORMAT(/,1X,'INCREMENTAL LOAD FACTOR=',G13.4,/,1X,
        1        'NO. OF INCS. (NINC)=  ',I5,/,1X,
        2        'WRITE CONTROL (IWRIT)=  ',15/13X,
        3        '0=LIMITED; 1=FULL')
          READ (IRE.*)BETOK,ITERTY
          WRITE (IWR,1003) BETOK,ITERTY
     1003 FORMAT(/,1X,'CONV. TOL FACTOR, BETOK=',G13.5,/,
        1        1X,'ITERATIVE SOLN. TYPE, ITERTY=',I5,/,
        2        5X,'=1, FULL N-R; =2, MOD. N-R')
          AN=ANIT
          FACT=0.D0
          DO 5 I=1,NV
          GM(I)=0.D0
        5 PT(I)=0.D0
C
C
          DO 100 INC=1,NINC
          FACT=FACT+FACI
```

```
C         FACI IS INC LOAD FACTOR, FACT IS TOTAL
          WRITE (IWR, 1001) INC,FACT
   1001   FORMAT(//,1X,'INC=  ',I5, 'LD. FACTOR=',G12.5)
          DO 10 I=1,NV
          QEX(I)=FACT*QFI(I)
          QINC(I)=FACI*QFI(I)
C         USE BELOW COMMENT LINE INSTEAD OF ABOVE TO INCLUDE
C         ALLOWANCE FOR PREVIOUS O.B. FORCES
C         QINC(I)=FACI*QFI(I)+GM(I)
     10   CONTINUE
C
C         BELOW FORMS EL. TAN. STIFF MATRIX AKTE
          CALL ELEMENT (FI,AKTE,AN,X,Z,PT,E,ARA,AL,IWRIT,IWR,2,
      1        ITYEL,ALN,ARN)
C         ARGUMENTS IN LINE ABOVE NOT USED FOR SHALLOW TRUSS
C         BELOW PUTS EL. STIFF. AKTE IN STRUCT. STIFF. AKTS
C         AND ADDS EFFECT OF VARIOUS LINEAR SPRINGS
          CALL ELSTRUC (AKTE,AKTS,NV,AK15,ID14S,AK14S,NDSP,FI,PT,
      1                2,IWRIT,IWR)
C
C
          CALL BCON(AKTS,IBC,NV,QINC,IWRIT,IWR)
C         ABOVE APPLIES B. CONDITIONS
          CALL CROUT(AKTS,D,NV,IWRIT,IWR)
C         ABOVE FORMS LDL(TRAN) FACTORISATION INTO AKTS AND D
C         BELOW CHECKS FOR NEGATIVE PIVOTS
          NEG=0
          DO 30 I=1,NV
          IF (D(I).LT.0.D0) NEG=NEG+1
     30   CONTINUE
          IF (NEG.GT.0) WRITE (IWR,1007) NEG
   1007   FORMAT(/,1X'*** WARNING NO. OF NEG. PIVOTS=',I5)
C
          CALL SOLVCR(AKTS,D,QINC,NV,IWRIT,IWR)
C         ABOVE SOLVES EQNS. AND GETS INC. DISPS IN QIN
C
          DO 20 I=1,NV
          IF (IBC(I).EQ.0) THEN
          PT(I)=PT(I) + QINC(I)
          ELSE
          PT(I)=QEX(I)
          ENDIF
     20   CONTINUE
C         ABOVE UPDATES TOTAL DISPS.
C
          WRITE (IWR,1002) (PT(I),I=1,NV)
   1002   FORMAT(/,1X,'TOTAL DISPS. AFTER TAN. SOLN ARE',/,1X,7G13.5)
C
C         BELOW ITERATES TO EQUILIBRIUM
          CALL ITER(PT,AN,BETOK,QEX,IBC,IWRIT,IWR,AKTS,D,ITERTY,NV,
      1              GM,FI,REAC)
```

```
C
      WRITE (IWR,1004) (PT(I),I=1,NV)
 1004 FORMAT(/.1X'FINAL TOTAL DISPLACEMENT ARE',/,1X,6G12.5)
      WRITE (IWR,1006) (REAC(I),I=1,NV)
 1006 FORMAT(/1X,'FINAL REACTIONS ARE',/,1X,5G12.5)
      WRITE (IWR,1005) AN
 1005 FORMAT(/,1X'AXIAL FORCE IN BAR IS', G12.5)
C
  100 CONTINUE
C
      STOP 'NONLTC'
      END
```

## 2.6 PROBLEMS FOR ANALYSIS

In this section, we will give the data for and describe the results from a number of bar–spring problems that can be solved with the aid of the previous computer programs. In a few instances, we will also give truncated versions of the output. In addition to the present problems, the reader can, of course, devise those of his own. The present problems are all based on the dimensions and properties of (1.12), so that

$$EA = 5 \times 10^7, l = 2500. \tag{2.33}$$

### 2.6.1 Single variable with spring

The following data relates to the single-variable problem of Section 1.2 (see 1.12)) and Figure 1.1 with $z = z_{21} = 25$ and $K_s = K_{s4} = 1.35$. The expected response is that shown in Figure 1.4. For this problem, the variables 1, 2 and 3 (Figures 2.1 and 2.2) are constrained to zero and a spring (of magnitude 1.35) is provided at variable 4. A negative loading is incremented at variable 4.

#### 2.6.1.1 *Incremental solution using program* NONLTA

The following data leads to 12 increments of a pure incremental solution which should illustrate the drift from equilibrium shown in Figure 1.4 and detailed for 2 increments in Section 1.2.1.

4 50000000. 2500. 0. ; data as in (2.33); solution as in Section 1.1.1
0. 25. ; $z = z_{21} = 25.$
0. 0. 0. −7. ; vert. load at variable 4 (node 2)
1 1 1 0 ; only free at variable 4
1
4
1.35 ; earthed spring of 1.35 at variable 4
1. 12 1 ; $\Delta\lambda = 1$, 12 incs., write control on

*2.6.1.2 Iterative solution using program* NONLTB

With a starting value of zero, program NONLTB takes five Newton–Raphson iterations to obtain an equilibrium solution of $w_2 = p_4 = -28.39$ for a load of $q_4 = -35$.

```
4  50000000. 2500. 0. ; data as in (2.33) and Figure 1.1
0. 25.  ; z = z21 = 25.
0. 0. 0. -35.  ; load of -35 at variable 4 (vertical at node 2)
1 1 1 0   ; only free at variable 4
1
4
1.35   ; earthed spring of 1.35 at variable 4
0.0 0.0 0.0 0.0 ; starting vector for N-R iteration
0.001 1  ; conv. tol., beta; write control on
```

*2.6.1.3 Incremental/iterative solution using program* NONLTC

With a fixed loading of $q_{ef}(4) = -7$, four increments with $\Delta\lambda = 2$ (leading to a final loading of $-28$) are applied. Using full Newton–Raphson (data as below), the number of iterations for the four increments are 2,2,4,2 while with modified Newton–Raphson iterations (data as below apart from changing the last variable from unity to zero), the first three increments require 5,9,4 iterations respectively and the fourth fails to converge.

```
4  50000000. 2500. 0. ; data as in (2.33) and Figure 1.1
0. 25. ; z = z21 = 25.
0. 0. 0. -7.0  ; load of -7 at variable 4 (vertical at node 2)
1 1 1 0 ; only variable 4 is free
1
4
1.35  ; earthed spring of 1.35 at variable 4
2. 4  0 ; 4 incs. of Δλ = 1.0 each
0.001 1 ; Conv. tol., beta, of 0.001; full N-R
```

### 2.6.2 Single variable; no spring

When the spring is removed, the 'exact' (the inverted commas are required because the solution is only exact within the context of shallow-truss theory) solution is governed by (1.11) with $K_s = 0$. In particular, the response is that shown in Figure 1.2(a) with $\bar{W} = EA(z/l)^3 = 50$ as the non-dimensionalising factor. The following data relates to the use of program NONLTC to obtain a load-controlled solution with increments of 0.1 $\bar{W}$.

```
4 50000000. 2500.0. ; data as in (2.33) and Figure 1.1(a), response 1.2(a)
0. 25. ; z = z21 = 25.
```

0. 0. 0. $-50.0$ ; load of $-50$ ($\bar{W}$) at variable 4 (vertical at node 2)
1 1 1 0 ; only variable 4 is free
0 ; no springs
0.1 3 0 ; 3 incs. of $\Delta\lambda = 0.1$ each, write control off
0.001 1 ; Conv. tol. beta, of 0.001; full N–R;

Using full N–R (as in the data above), the required number of iterations were 2,39,1. The 39 iterations† were required to jump over the limit points from a non-dimensional load (see Figure 1.2(a)) of $\bar{W} = 0.1$ to one of $\bar{W} = 0.2$. When the iterative procedure was changed to modified Newton–Raphson (by changing the 1 to a zero in the last data line above) 5, iterations were required for the first increment, and for the second, which encompasses the limit points, no convergence was obtained within 100 iterations.

In contrast to the previous difficulties, no problems are encountered when displacement control is used to analyse this problem. The following data relates to the application of seven increments of $\Delta w_4 = \Delta w = 0.3z$, which allows the complete response in Figure 1.2(a) to be traced. For this problem, the displacement controlled solution is, in fact, trivial with not only no iterations but also no real equation solving.

4 50000000. 2500. 0. ; data as in (2.33) and Figure 1.1(a), response 1.2(a)
0. 25. ; $z = z_{21} = 25.$
0. 0. 0. $-25.0$ ; disp. of $-25$ ($-z$) at variable 4 (vertical at node 2)
1 1 1 $-1$ ; variable 4 has presc. disp., restrained
0 ; no springs
0.3 7 0 ; 7 incs. of $\Delta\lambda = 0.3$ each, write control off
0.001 1 ; Conv. tol. beta, of 0.001; full N–R

### 2.6.3 Perfect buckling with two variables

The following data relates to the two-variable problem of Section 1.3 and Figure 1.1 with the data as in (2.33) with a spring stiffness, $K_s = K_{s4} = 4$ and $z = z_{21} = 0$ so that the truss element is flat. In these circumstances, the critical buckling load is given (see (1.68)) by

$$U_{cr} = lK_s = 2500 \times 4 = 10^4. \tag{2.34}$$

This load is applied to the incremental/iterative program NONLTC as $\mathbf{q}_f(4)$ with increments of $\Delta\lambda = 0.4$ using the following data:

4 50000000. 2500. 0. ; data as in (2.33) with $K_s = K_{s4} = 4$
0.0 0.0 $z = z_{21} = 0$ (perfect)
10000. 0.0 0.0 0.0 ; horizontal LHS buckling load (at $q_1$)
0 1 1 0 ; only LHS horiz. ($p_1$) and RHS vert ($p_4$) disps. free
1
4
4.00 ; earthed spring of 4.0 at RHS vertical variable (no. 4)

†To obtain these 39 iterations, NITMAX in subroutine ITER must be increased to, say, 101.

0.4 5 1 ; 5 incs. with $\Delta\lambda = 0.4$ ( × crit. buckl. ld.), write contrl. on
0.001 1 ; Conv. tol., $\beta$ of 0.001, full N–R

As anticipated, no iterations are required because the system remains flat and no out-of-balance forces are generated. Increment 3 takes the solution beyond the buckling load to $U = 1.2U_{cr}$ and this can be detected by the negative pivot in the $\mathbf{LDL}^T$ factorisation at the start of increment 4. The procedure then continues to climb the unstable equilibrium path (Figure 1.11). In fact, for this somewhat trivial problem, the negative pivots are obvious from the original tangent stiffness matrices which are of diagonal form, i.e. increment 2 ($\lambda = 0.8$) leads to

$$u_1 = u = 0.4, \qquad N = -EA\frac{u}{l} = -8000, \tag{2.35}$$

and, from (1.64), the tangent stiffness matrix at the start of increment 3 is

$$\mathbf{K}_t = \begin{bmatrix} 2000. & 0. \\ 0. & 0.8 \end{bmatrix} \tag{2.36}$$

while, at the end of increment 3 ($\lambda = 1.2$),

$$u_1 = u = 0.6, \qquad N = -EA\frac{u}{l} = -12\,000. \tag{2.37}$$

and the tangent stiffness matrix at the start of increment 4 is

$$\mathbf{K}_t = \begin{bmatrix} 2000. & 0. \\ 0. & -0.8 \end{bmatrix}. \tag{2.38}$$

It might be thought that the solutions would not continue to climb the unstable path if iterations are actually applied because these iterations will force the solution off the unstable path. The reader can easily amend the program to ensure one iteration at each increment even if the solution has converged. He or she will find that this has no effect and the unstable path is still followed.

### 2.6.4 Imperfect 'buckling' with two variables

As discussed in Section 1.3, the bifurcation in the previous problem can be removed by introducing an 'imperfection'. In particular, we will set $z = z_{21} = 25$ so that, in relation to the exact solutions of Section 1.2.1, $\beta$ of (1.74) = 0.5 and the possible equilibrium paths are shown in Figure 1.11.

#### 2.6.4.1 *Pure incremental solution using program* NONLTA

The following data leads to the application of five increments, each with $\Delta U/U_{cr} = 0.3$, using a pure incremental solution procedure.

4 50000000. 2500. 0. ; data as in (2.33) with $K_s = K_{s4} = 4$
0. 25.0 ; $z = z_{21} = 25.$

10000. 0.0 0.0 0.0 ; horizontal LHS critical buckling load (at $q_1$)
0 1 1 0 ; only LHS horiz. ($p_1$) and RHS vert ($p_4$) disp. free
1
4
4.00 ; earthed spring of 4.0 at RHS. vertical variable (no. 4)
0.3 5 0 ; 5 incs. with $\Delta\lambda = 0.3$ ( $\times$ crit. buckl. ld.), write contr, off

The resulting solutions are shown as squares in Figure 2.6. The drift from equilibrium can be clearly seen.

#### 2.6.4.2 *An incremental/iterative solution using program* NONLTC *with small increments*

The following data leads to the circles on Figure 2.6 which effectively lie on the exact curves (see Section 1.3.1).

4 50000000. 2500. 0. ; data as in (2.33) with $K_s = K_{s4} = 4$
0. 25.0 ; $z = z_{21} = 25.$
10000. 0.0 0.0 0.0 ; horizontal LHS critical buckling load (at $q_1$)
0 1 1 0 ; only LHS horiz. ($p_1$) and RHS vert ($p_4$) disps. free
1
4
4.00 ; earthed spring of 4.0 at RHS vertical (no. 4)
0.3 3 1 ; 3 incs. with $\Delta\lambda = 0.3$ ( $\times$ crit. buckl. ld.), write contr. on
0.001 1 ; Conv. tol., beta of 0.001, full N–R

The required number of iterations (using the full N–R method) was 2,3,5.

A truncated form of the output for the first increment of this problem is given below. The truncation has mainly involved only giving the response for the two active variables, $a_1 = p_1 = u_1$ and $a_2 = p_4 = w_2$ (Figures 2.1 and 2.2). Even without the computer program, the reader should be able to follow these results in conjunction with the theory of Sections 1.3 or 2.1 so as to understand the basis of the Newton–Raphson method for a two-variable geometrically non-linear problem. When studying these results, the spring stiffness (4.0) must be added for $K_{22}$ and the equivalent term of $4.0w_2$ must be added in to the out-of-balance force vector†:

```
INC=     1 LD. FACTOR=   .30000
TAN. STIFF. MATRIX IS*
  20000.        -200.00
  .00000        6.0000
DISP INCS. ARE
  .22500        7.5000
TOTAL DISPS. AFTER TAN. SOLN ARE
  .22500        7.5000
TOTAL DISPS. AFTER TAN. SOLN ARE
  .22500        7.5000
```

*Only the upper triangle of the printed tangent stiffness matrix is correct.

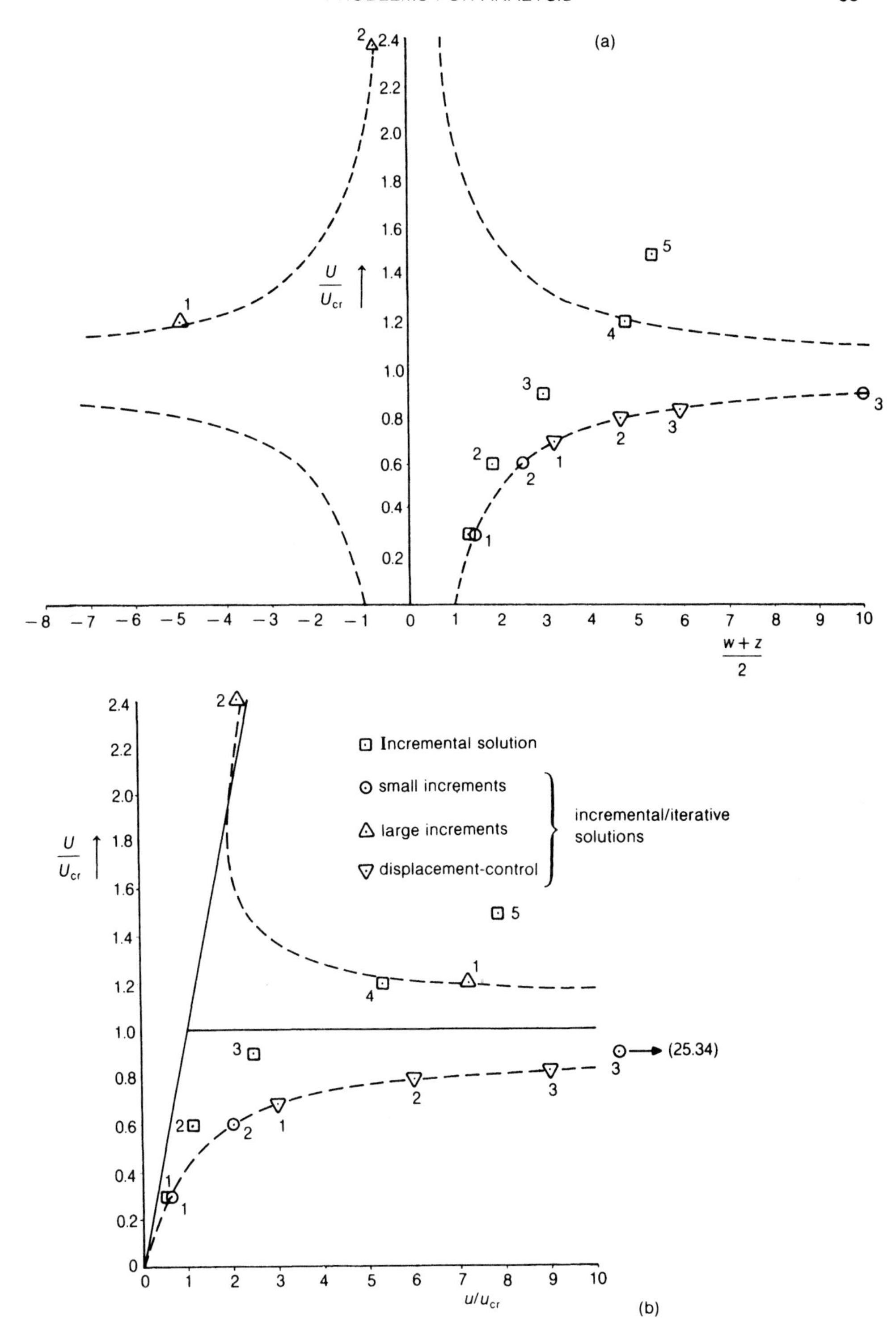


**Figure 2.6** Solutions for two-variable bar–spring problem: (a) transverse deflection; (b) shortening.

```
BEGIN ITERATIVE LOOP WITH ITE=   1
AXIAL FORCE AN=    -2775.0
OUT-OF-BAL. FORCE VECTOR=
  225.00        6.0750   ; CONV. FAC.=.75027E-01
TAN. STIFF. MATRIX IS
  20000.              -260.00
  .00000              6.2700
DISP INCS. ARE
  .51734E-01        3.1142
TOTAL DISPS ARE
  .27673         10.614
BEGIN ITERATIVE LOOP WITH ITE=   2
AXIAL FORCE AN=    -2961.2
OUT-OF-BAL. FORCE VECTOR=
  38.793         -.27235; CONV. FAC.=.12931E-01
TAN. STIFF. MAT. IS
  20000.        -284.91
  .00000        6.8743
DISP INCS. ARE
  .33577E-02        .99547E-01
TOTAL DISPS ARE
  .28009         10.714
BEGIN ITERATIVE LOOP WITH ITE=   3
AXIAL FORCE AN=    -3000.0
OUT-OF-BAL. FORCE VECTOR=
  .39639E-01        .97842E-03; CONV. FAC.=.13217E-04
FINAL TOTAL DISPLACEMENTS ARE
  .28009         10.714
FINAL REACTIONS ARE
  3000.0         -42.855
AXIAL FORCE IN BAR IS   -3000.0
```

Although the iterations are terminated within the iterative loop with ITE = 3, we have classified this increment as requiring two iterations. For, although the tangent stiffness matrix has been reformed for ITE = 3, it has not been used to solve for a new set of iterative displacements, because having computed the out-of-balance forces, the convergence factor has been found to be less than the required tolerance ($\beta = 0.001$—see (2.30)).

#### *2.6.4.3 An incremental/iterative solution using program* NONLTC *with large increments*

To some degree, we have been assisted in obtaining the previous solutions by knowing the answer! In particular, this knowledge helped guide our choice of initial increment.

As an early illustration of the possible pitfalls of non-linear analysis, we will now re-solve this problem using increment of $\Delta\lambda = U/U_{cr} = 1.2$. The results, which relate to the following data, are shown as the triangles in Figure 2.6.

4 50000000. 2500. 0. ; data as in (2.23) with $K_s = K_{s4} = 4$
0. 25.0 ; $z = z_{21} = 25.$
10000. 0.0 0.0 0.0 ; horizontal LHS critical buckling load (at $q_1$)
0 1 1 0 ; only LHS horiz. ($p_1$) and RHS vert ($p_4$) disps. free
1
4
4.00 ; earthed spring of 4.0 at RHS vertical variable (no. 4)
1.2 2 0 ; 3 incs. with $\Delta\lambda = 1.2$ ( × crit. buckl. ld.), write contr. on
0.0001 1 ; Conv. tol., beta of 0.001, full N–R

It will be seen that we have converged on to and stayed on the wrong equilibrium path! Luckily, there are warning signs. In particular, on factorising the stiffness matrix following convergence to point 1, the $\mathbf{LDL}^T$ factorisation indicates one negative pivot. The tangent stiffness is not positive definite because, as in the simpler example of Section 2.6.2, we have 'passed' a bifurcation point (more strictly, we have crossed a stable equilibrium line). The situation can be more confusing because negative pivots can also be caused by passing limit points (see Chapter 9). The number of full Newton–Raphson iterations to obtain the triangles in Figure 2.6 were 6 and 5.

#### *2.6.4.4 An incremental/iterative solution using program* NONLTC *with displacement control*

Instead of applying load control, we can apply displacement control at variable 1 (Figures 2.1 and 2.2). The critical buckling displacement, $u_{cr}$ is then

$$u_{cr} = \frac{l}{AE} U_{cr} = 0.5. \tag{2.39}$$

Applying increments of $\Delta\lambda = 3.0 = \Delta u/u_{cr}$ to this problem requires the following data which gives the solutions depicted by the inverted triangles in Figure 2.6. The results effectively lie on the exact equilibrium curves:

4 50000000. 2500. 0. ; data as in (2.33) with $K_s = K_{s4} = 4$
0. 25.0 ; $z = z_{21} = 25.$
0.5 0.0 0.0 0.0 ; horizontal LHS critical buckling displacement (at $p_1$)
−1 1 1 0 ; only RHS vert ($p_4$) disps. free, LHS horiz. disp. ($p_1$) prescribed
1
4
4.00 ; earthed spring of 4.0 at RHS vertical variable (no. 4)
3.0 3 1 ; 3 incs. with $\Delta\lambda = 3.0$ ( × crit. buckl. disp.), write contr. on
0.00001 1 ; Conv. tol., beta of 0.00001, full N–R

The required number of iterations (using the full Newton–Raphson method) for the three increments were 3,3,2.

## 2.7 SPECIAL NOTATION

$A$ = area of bar
$K_s$ = spring stiffness
$N$ = axial force in bar
$\mathbf{r}$ = reaction vector
$u$ = axial ($x$-direction) displacement in bar (function of $x$)
$u_1, u_2$ = nodal displacements for bar in $x$-direction
$u_{21} = u_2 - u_1$ (similar convention for $w_{21}$ and $z_{21}$)
$U_1, U_2$ = nodal forces corresponding to $u_1, u_2$
$w$ = vertical ($z$ direction) displacement in bar
$w_1, w_2$ = nodal displacements for bar in $z$-direction
$W_1, W_2$ = nodal forces corresponding to $w_1, w_2$
$z$ = initial vertical offset of bar
$z_1, z_2$ = nodal values of $z$
$l$ = initial length of bar
$\beta$ = geometric factor (equation (2.14)) or convergence tolerance factor (equation (2.30))
$\varepsilon$ = axial strain in bar

### Subscripts

f = free
p = prescribed

## 2.8 REFERENCES

[C1] Crisfield, M. A., Duxbury, P. G. & Hunt, G. W., Benchmark tests for geometric non-linearity, National Agency for Finite Element Methods and Standards Report SPNGL (1987)

[C2] Crisfield, M. A., *Finite Elements and Solution Procedures for Structural Analysis*, Vol. 1: *Linear analysis*, Pineridge Press, Swansea (1986).

[D1] Duxbury, P. G., Crisfield, M. A. & Hunt, G. W., Benchmark tests for geometric non-linearity, *Computers & Structures*, **33** (1), 21–29 (1981).

# 3 Truss elements and solutions for different strain measures

The developments of Chapters 1 and 2 have all been based on the 'shallow-truss strain relationships' of (1.51) and (2.3). In the present chapter, we will consider a number of alternative 'strain measures' that remain valid when the truss element is deep. In Sections 3.3–8, the new strain measures are used to derive finite element equations for a truss element. The detail is provided, not because of the intrinsic importance of truss elements, but rather because they provide a simple means of introducing some of the concepts that will later be used for continua or beams and shells. These concepts include 'total Lagrangian' and 'updated Lagrangian' techniques, 'corotational formulations', as well as 'equivalent constitutive laws'.

In Section 3.9, we provide three Fortran subroutines which will allow the computer program of Chapter 2 to be applied to deep truss elements using the various strain measures. Section 3.10 gives a range of problems for which data and, in some instances, results are given.

## 3.1 A SIMPLE EXAMPLE WITH ONE DEGREE OF FREEDOM

Before turning to finite elements, the alternative strain measures will be introduced in relation to the simple example of Section 1.2, which is reproduced (with slightly different notation) in Figure 3.1. For each strain measure, the starting point will be the virtual work relationship of Sections 1.3.2 and 2.1. For the example of Figure 3.1, this relationship is given by

$$V_{\mathrm{n}} = \int \sigma \, \delta\varepsilon_{\mathrm{v}} \, \mathrm{d}V_{\mathrm{n}} - q \, \delta w_{\mathrm{v}} \tag{3.1}$$

or

$$V_{\mathrm{o}} = \int \sigma \, \delta\varepsilon_{\mathrm{v}} \, \mathrm{d}V_{\mathrm{o}} - q \, \delta w_{\mathrm{v}} \tag{3.2}$$

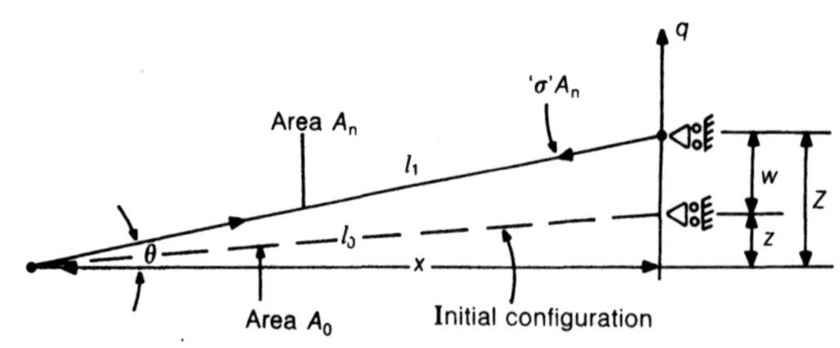


**Figure 3.1** Simple problem with one degree of freedom.

where (3.1) is related to the final configuration with volume, $V_n$ and (3.2) is related to the initial configuration with volume, $V_o$. (The clash of symbols with $V$ representing both the virtual work and the volume should cause no difficulties because the form of use should be obvious from the context.) The relationships in Sections 3.1.1 and 3.1.2 will be derived from the latter equation.

### 3.1.1 A rotated engineering strain

The strain measure

$$\varepsilon_E = \frac{l_n - l_o}{l_o} \tag{3.3}$$

measures the strain along the rotating bar so that the direction of $\varepsilon_E$ is continuously changing. From Figure 3.1,

$$l_n = ((z + w)^2 + x^2)^{1/2} = (Z^2 + x^2)^{1/2} = (Z^2 - z^2 + l_o^2)^{1/2}. \tag{3.4}$$

Hence, from (3.3) and (3.4),

$$\delta\varepsilon_E = \frac{\delta l_n}{l_o} = \frac{(z + w)}{l_n l_o}\delta w = \frac{Z\delta w}{l_n l_o}. \tag{3.5}$$

Substitution from (3.5) (with a virtual $\delta\varepsilon_v$) into (3.2) and integration over the original volume of the element leads to

$$q_E = \frac{\sigma_E A_o Z}{l_n} = \frac{\sigma_E A_o(z + w)}{l_n} \tag{3.6}$$

where the subscript E means 'Engineering'. *Assuming* a fixed '$E$-value', the relationship between the load, $q$, and the deflection, $w$, is given by

$$q_E = \frac{EA_o(z + w)(((z + w)^2 + x^2)^{1/2} - l_o)}{l_o((z + w)^2 + x^2)^{1/2}}. \tag{3.7}$$

Both here and frequently, during future work, we will talk of 'rotated' strain or stress. Such measures can be assumed to relate to a coordinate system that continuously rotates with the bar. Alternatively they can be thought of as the stresses or strains in the bar once the latter has been rigidly rotated back to its original position. Further work on these concepts will follow in Chapters 4 and 7.

### 3.1.2 Green's strain

Starting with (3.3), we can write

$$\varepsilon_{\mathrm{E}} = \frac{l_{\mathrm{n}} - l_{\mathrm{o}}}{l_{\mathrm{o}}} = \frac{(l_{\mathrm{n}} - l_{\mathrm{o}})(l_{\mathrm{n}} + l_{\mathrm{o}})}{l_{\mathrm{o}}(l_{\mathrm{n}} + l_{\mathrm{o}})} = \frac{l_{\mathrm{n}}^2 - l_{\mathrm{o}}^2}{l_{\mathrm{o}}^2(2 + \varepsilon_{\mathrm{E}})} \tag{3.8}$$

so that, if $\varepsilon_{\mathrm{E}}$ is small, we can write

$$\varepsilon_{\mathrm{G}} = \frac{l_{\mathrm{n}}^2 - l_{\mathrm{o}}^2}{2l_{\mathrm{o}}^2} \tag{3.9}$$

where $\varepsilon_{\mathrm{G}}$ is known as Green's strain. It is related to the rotated engineering strain of (3.3) via

$$\varepsilon_{\mathrm{G}} = \varepsilon_{\mathrm{E}}(1 + \tfrac{1}{2}\varepsilon_{\mathrm{E}}). \tag{3.10}$$

From (3.4) and (3.9),

$$\delta\varepsilon_{\mathrm{G}} = \frac{l_{\mathrm{n}}}{l_{\mathrm{o}}^2}\delta l_{\mathrm{n}} = \frac{(z + w)}{l_{\mathrm{o}}^2}\delta w = \frac{Z\delta w}{l_{\mathrm{o}}^2}. \tag{3.11}$$

Hence from (3.2) (using the virtual form of (3.11))

$$q_{\mathrm{G}} = \frac{\sigma_{\mathrm{G}} A_{\mathrm{o}} Z}{l_{\mathrm{o}}} = \frac{\sigma_{\mathrm{G}} A_{\mathrm{o}}(z + w)}{l_{\mathrm{o}}}. \tag{3.12}$$

When generalised to a continuum (Chapter 4), the stress $\sigma_{\mathrm{G}}$ is referred to as the second Piola–Kirchhoff stress. Using a fixed '$E$-value', from (3.12),

$$q_{\mathrm{G}} = \frac{EA_{\mathrm{o}}(z + w)(2zw + w^2)}{2l_{\mathrm{o}}^3} = \frac{EA_{\mathrm{o}}Z(2Zw - w^2)}{2l_{\mathrm{o}}^3}. \tag{3.13}$$

It will be noted that equation (3.13) is of a simpler form than (3.7). When the strains are small, $l_{\mathrm{n}} \simeq l_{\mathrm{o}}$ and from (3.8) and (3.9), the two strain measures coincide. In addition, the equilibrium relationships of (3.6) and (3.12) coincide. Hence, for small strains, the two load/deflection relationships of (3.7) and (3.13) coincide.

### 3.1.3 A rotated log-strain

For large strains, the adopted strain measure is often taken as the log-strain, which is basically of an incremental form, so that

$$\delta\varepsilon = \frac{\delta l}{l} \tag{3.14}$$

where $l$ is in the 'current configuration'. Hence

$$\varepsilon_L = \int_{l_o}^{l_n} \delta\varepsilon = \log_e\left(\frac{l_n}{l_o}\right) \tag{3.15}$$

which can be related to the previous strain measures ((3.3) and (3.9)) via

$$\varepsilon_L = \log_e(1 + \varepsilon_E) = \tfrac{1}{2}\log_e(1 + 2\varepsilon_G). \tag{3.16}$$

The kinematics of a small movement, $\delta w$, in Figure 3.1, ensure that

$$\delta l_n = \frac{Z}{l_n}\delta w. \tag{3.17}$$

Instead of using (3.2), it is now appropriate to adopt (3.1), which relates to the final configuration. However, for the present, we will assume no volume change, so that there is no difference between (3.1) and (3.2). In these circumstances, (3.14) and (3.17) can be substituted into (3.1) (or (3.2)) to give

$$q_L = \frac{\sigma_L A_o l_o Z}{l_n^2} = \frac{\sigma_L A_o l_o (z + w)}{l_n^2}. \tag{3.18}$$

If a fixed $E$-value is *assumed*, the relationship between the load, $q$, and the deflection, $w$, is given by

$$q_L = \frac{EA_o(z + w)l_o}{2((z + w)^2 + x^2)}\log_e\left(\frac{(z + w)^2 + x^2}{l_o^2}\right). \tag{3.19}$$

For small strains, the log-strain solutions coincide with the previous solutions involving rotated engineering strain (Section 3.1.1) and Green's strain (Section 3.1.2).

### 3.1.4 A rotated log-strain formulation allowing for volume change

It has been indicated that the log-strain formulation could be used for large strains. Hence for this formulation, we could include the effects of the changing volume. In these circumstances, we need to consider strain changes at right angles to the axis of the bar of magnitude $-\nu\,\delta\varepsilon$. Hence

$$A + \mathrm{d}A = A(1 - \nu\,\mathrm{d}\varepsilon)^2 \simeq A(1 - 2\nu\,\mathrm{d}\varepsilon) \tag{3.20}$$

so that

$$\int_{A_o}^{A_n} \frac{\mathrm{d}A}{A} = -2\nu\int_{l_o}^{l_n} \delta\varepsilon = -2\nu\int_{l_o}^{l_n} \frac{\mathrm{d}l}{l} \tag{3.21}$$

and

$$\log_e\left(\frac{A_n}{A_o}\right) = -2\nu\log_e\left(\frac{l_n}{l_o}\right) \tag{3.22}$$

or

$$\frac{A_n}{A_o} = \left(\frac{l_o}{l_n}\right)^{2\nu} \tag{3.23}$$

so that:

$$V_n = A_n l_n = A_o \frac{l_o^{2\nu}}{l_n^{2\nu - 1}} \tag{3.24}$$

and the formulation of Section 3.1.3, which assumed $V_n = V_o$ is only strictly valid for $\nu = 0.5$. Substitution from (3.14), (3.17) and (3.24) into (3.1) gives

$$q_L = \frac{\sigma_L A_n Z}{l_n} = \frac{\sigma_L A_o l_o^{2\nu} Z}{l_n^{1+2\nu}} \tag{3.25}$$

in place of (3.18). If a fixed $E$-value is *assumed*, (3.25) leads to the load/deflection relationship

$$q_L = \frac{A_o(z+w) l_o^{2\nu}}{2((z+w)^2 + x^2)^{(1+2\nu)/2}} \log_e\left(\frac{(z+w)^2 + x^2}{l_o^2}\right). \tag{3.26}$$

Solutions relating to a constant area can, from (3.23), be obtained by setting $\nu = 0$ or to the previous solution for a constant volume (Section 3.1.3), by setting $\nu = \frac{1}{2}$.

### 3.1.5 Comparing the solutions

Resolving vertically in Figure 3.1 gives:

$$q = \frac{A_n \text{‘}\sigma\text{’} Z}{l_n} \tag{3.27}$$

where ‘$\sigma$’ is the ‘true stress’ in the bar. (In a continuum context, the ‘true’ stress is often referred to as the Cauchy stress—see Chapter 4.) Equation (3.27) can be compared to equations (3.6), (3.12), (3.18) and (3.25) for the same load values, $q$ (i.e. removing the subscripts on $q$). Comparing the first form of (3.25) (with $A_n$ rather than $A_o$) with (3.27) shows that

$$\sigma_L = \text{‘}\sigma\text{’} \tag{3.28}$$

and hence the ‘log-stress’, $\sigma_L$, is the ‘true stress’.

If (3.6) is compared with (3.27),

$$\sigma_E = \text{‘}\sigma\text{’}\left(\frac{A_n}{A_o}\right) \tag{3.29}$$

while comparison of (3.12) with (3.27) gives

$$\sigma_G = \text{‘}\sigma\text{’} \frac{A_n l_o}{A_o l_n} = \frac{l_o}{l_n} \sigma_E. \tag{3.30}$$

Figure 3.2 plots the solutions to (3.7), (3.13) and (3.19) for the bar of Figure 3.1 with a *fixed* $E$ value and the properties

$$x = 2500, \qquad A_o = 100, \qquad E = 5 \times 10^5, \qquad z = 2500. \tag{3.31}$$

The solution with the rotated engineering strain is only a little more flexible than that obtained with the log-strain. However, Green's strain leads to a significantly

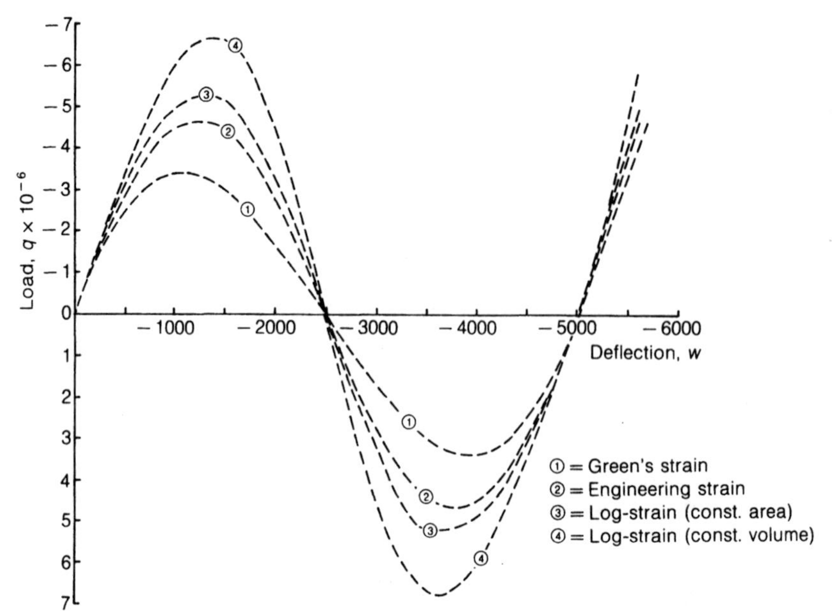


**Figure 3.2** Load/deflection relationships for deep truss.

more flexible response. All three solutions coincide in the early stages where the strains are small. If the rise, $z$, in Figure 3.1 were made small enough, the strains would remain small for the complete load/deflection response and all three solutions would coincide for the complete response. In the latter circumstances, while the angle $\theta$ in Figure 1.1 remained small, the solutions would coincide with those of Section 1.1 and Figure 1.2(a).

## 3.2 SOLUTIONS FOR A BAR UNDER UNIAXIAL TENSION OR COMPRESSION

The previous example (Figure 3.1) considered a rotating bar. Before proceeding to the full finite element formulations for a truss, it is instructive to consider the trivial example of a bar subject to a uniaxial load as in Figure 3.3. If we replace $\delta w_v$ by $\delta u_v$, the virtual work equations (3.1) and (3.2) still apply. This substitution will be assumed in the following. Using the engineering strain of (3.3), the virtual work relationship (3.2) can then be used to produce

$$q_E = A_o \sigma_E \tag{3.32}$$

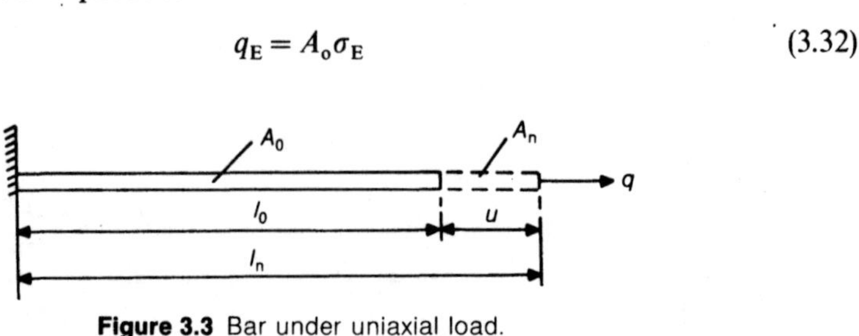


**Figure 3.3** Bar under uniaxial load.

and, *assuming* (throughout) a fixed $E$-value,

$$q_E = A_o E \frac{u}{l_o} = E A_o \bar{\varepsilon} \tag{3.33}$$

where $\bar{\varepsilon}$ is the engineering strain. Using (3.2), in conjunction with the Green's strain of (3.9), gives

$$q_G = \frac{A_o l_n}{l_o} \sigma_G \tag{3.34}$$

$$q_G = A_o E\left(1 + \frac{u}{l_o}\right)\left(\frac{u}{l_o} + \frac{1}{2}\left(\frac{u}{l_o}\right)^2\right) = A_o E(1 + \bar{\varepsilon})(\bar{\varepsilon} + \tfrac{1}{2}\bar{\varepsilon}^2) \tag{3.35}$$

while, for the log-strain, (3.1) and (3.24) can be used to produce

$$q_L = A\sigma_L = A_o\left(\frac{l_o}{l}\right)^{2\nu} \sigma_L \tag{3.36}$$

$$q_L = \frac{E A_o}{(1 + \bar{\varepsilon})^{2\nu}} \log_e(1 + \bar{\varepsilon}). \tag{3.37}$$

In terms of the 'true stress', '$\sigma$', the equilibrium relationship is

$$q = A_n \text{'}\sigma\text{'} \tag{3.38}$$

so that, from equilibrium, (3.27)–(3.29) again apply. The load/deflection relationships (3.33), (3.35) and (3.37) are plotted in Figure 3.4 for a bar with

$$l_o = 2500, \qquad A_o = 100, \qquad E = 5 \times 10^5. \tag{3.39}$$

Figure 3.4 demonstrates the potential unsuitability of Green's strain for work with large strains (*unless* appropriate modifications are made to the $\sigma$–$\varepsilon$ relationships). In particular, in compression, the formulation (see (3.35)) gives zero stress at $\bar{\varepsilon} = -1$ and $-2$ and artificial limit points at $\bar{\varepsilon} = -1 \pm 1/\sqrt{3}$. In contrast, the solution obtained with the engineering strain does not differ significantly (Figure 3.4) from that obtained with the log-strain provided the strains are only 'moderately large'.

It should be emphasised that the relationships in Figures 3.2 and 3.4 were obtained by *assuming* a constant $E$-value. The *same* load/deflection relationship could be obtained for each of the strain measures if the secant $E$-values were made functions of the strains, so that from (3.33), (3.35) and (3.37),

$$E_E = E_G \frac{(1 + \bar{\varepsilon})(\bar{\varepsilon} + \frac{1}{2}\bar{\varepsilon}^2)}{\bar{\varepsilon}} = E_L \frac{\log_e(1 + \bar{\varepsilon})}{(1 + \bar{\varepsilon})^{2\nu + 1}}. \tag{3.40}$$

We will return to this trivial example in Volume 2 when considering large strains in a more general environment.

### 3.2.1 Almansi's strain

Before leaving this example we will introduce Almansi's strain, which is often quoted

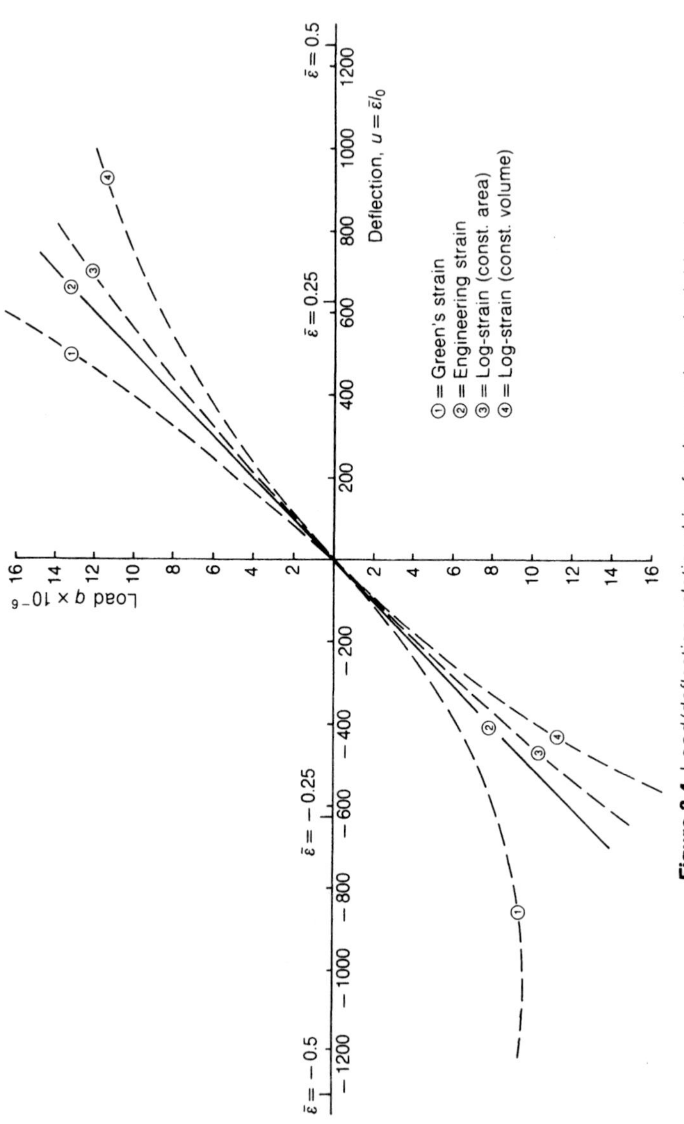


**Figure 3.4** Load/deflection relationships for bar under uniaxial load.

in the literature (if less often used). This strain is given by

$$\varepsilon_A = \frac{l_n^2 - l_o^2}{2l_n^2} \tag{3.41}$$

and is often quoted as being conjugate to the 'true' (or Cauchy—see Chapter 5) stress. The author finds this statement a little confusing for reasons that will be illustrated here.

From (3.41), the variation of $\varepsilon_A$ can be obtained as

$$\delta\varepsilon_A = \frac{l_o^2 \delta l_n}{l_n^3} = \frac{1}{(1+\bar{\varepsilon})^3}\delta\bar{\varepsilon} \tag{3.42}$$

where $\bar{\varepsilon}$ is the engineering strain (see (3.33)). Introducing equation (3.42) into the virtual work relationship of (3.2) gives

$$q_A = \sigma_A A_n \frac{1}{(1+\bar{\varepsilon})^2} \tag{3.43}$$

which, when compared with the equilibrium-based 'true stress' relationship of (3.38), shows that

$$\sigma_A = \text{`}\sigma\text{'}\frac{1}{(1+\bar{\varepsilon})^2} = \text{`}\sigma\text{'}\frac{l_o^2}{l_n^2}. \tag{3.44}$$

Hence the use of the Almansi strain does not lead to the relationship $\sigma_A = $ '$\sigma$'. Rather, as already demonstrated in Sections 3.1.3 and 3.1.5, the true stress is work conjugate to the log strain with variation $\delta\varepsilon = \delta l_n / l_n$ (see Section 4.6 for the continuum equivalent).

It has already been noted that there is no effective difference between any of the previous strain measures when the strains are small. This finding also relates to the Almansi strain. In these circumstances, it may be useful, for computational convenience, to use the Almansi strain (see Section 3.3.6).

## 3.3 A TRUSS ELEMENT BASED ON GREEN'S STRAIN

In devising the governing equations for the various truss elements, we will not necessarily adopt the most computationally efficient formulation. Instead, we intend to introduce the concepts in forms that can be readily extended to continua, beams and shells. Hence, we will adopt standard finite element procedures using shape functions etc., although such procedures are not strictly necessary for these simple elements. Detail will be given for two-dimensional 'planar truss elements', but it will be shown in Section 3.7 that the procedures and formulae are easily extendible to three-dimensional 'space truss elements'.

### 3.3.1 Geometry and the strain–displacement relationships

Figure 3.5 shows a truss element $P_oQ_o$ in its original configuration with a non-dimensional coordinate, $\xi$, being used to define the position of a point $A_o$ lying

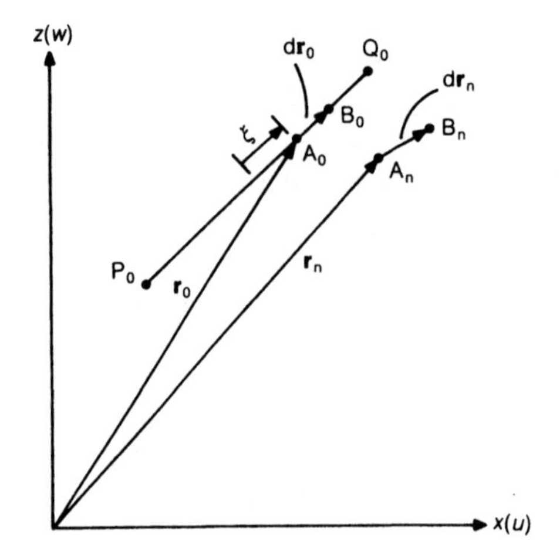


**Figure 3.5** Deformation of general truss element.

between $P_o$ and $Q_o$. As the truss experiences deformation, points $A_o$ and the adjacent $B_o$ move to $A_n$ and $B_n$ respectively. During this process, the position vector, $\mathbf{r}_o$, of point $A_o$ moves to the position vector, $\mathbf{r}_n$, of $A_n$, where:

$$\mathbf{r}_n = \mathbf{r}_o + \mathbf{u} \tag{3.45}$$

and, in two dimensions,

$$\mathbf{r}^T = \{x, z\}, \qquad \mathbf{u}^T = \{u, w\}. \tag{3.46}$$

Equivalent nodal coordinates will be written as

$$\mathbf{x}_n = \mathbf{x}' = \mathbf{x}_o + \mathbf{p} = \mathbf{x} + \mathbf{p} \tag{3.47}$$

where the initial coordinates $\mathbf{x}$ (or $\mathbf{x}_o$, but the subscript o will often be omitted) are

$$\mathbf{x}^T = (x_1, x_2, z_1, z_2) \tag{3.48}$$

and the nodal displacements are (see Figure 3.6)

$$\mathbf{p}^T = (u_1, u_2, w_1, w_2). \tag{3.49}$$

(Note the non-standard ordering of the components of $\mathbf{p}$ and see the footnote on page 25.)

In Figures 3.5 and 3.6, we have introduced the non-dimensional coordinate, $\xi$, for use with standard finite element shape functions. However, we will initially avoid the use of such shape functions which are not strictly necessary for these simple elements.

By Pythagoras' theorem, the initial length of the element is given by

$$l_o^2 = 4\alpha_o^2 = (x_{21}^2 + z_{21}^2) = \mathbf{x}_{21}^T \mathbf{x}_{21} \tag{3.50}$$

where

$$x_{21} = x_2 - x_1, \qquad z_{21} = z_2 - z_1 \tag{3.51}$$

and

$$\mathbf{x}_{21}^T = (x_{21}, z_{21}). \tag{3.52}$$

In (3.50), we have, for compatibility with later developments using shape functions (Section 3.3.4), introduced the original 'length parameter', $\alpha_o$, which is half the original

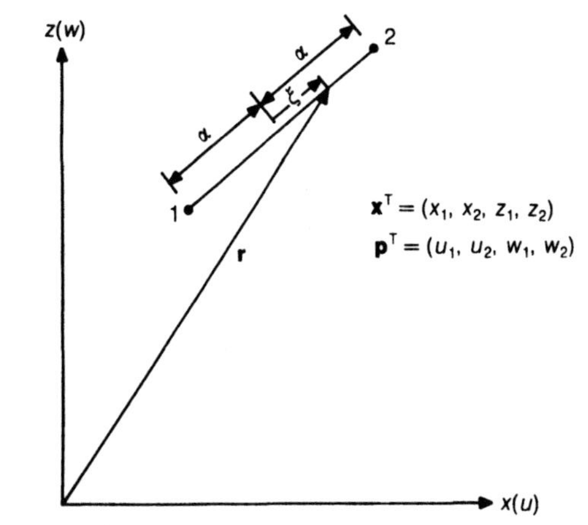


**Figure 3.6** Geometry and modes for general truss element.

length, $l_o$. For the current length, $l_n$, the equivalent of (3.50) is

$$l_n^2 = 4\alpha_n^2 = (x_{21} + u_{21})^2 + (z_{21} + w_{21})^2 = (\mathbf{x}_{21} + \mathbf{p}_{21})^T(\mathbf{x}_{21} + \mathbf{p}_{21}) \tag{3.53}$$

where in a similar manner to (3.52), $\mathbf{p}_{21}^T = (u_{21}, w_{21})$. Using (3.9), (3.50) and (3.53), Green's strain is given by

$$\varepsilon = \frac{l_n^2 - l_o^2}{2l_o^2} = \frac{(\mathbf{x}_{21} + \mathbf{p}_{21})^T(\mathbf{x}_{21} + \mathbf{p}_{21}) - \mathbf{x}_{21}^T\mathbf{x}_{21}}{2\mathbf{x}_{21}^T\mathbf{x}_{21}}$$

$$= \frac{1}{4\alpha_o^2}(\mathbf{x}_{21}^T\mathbf{p}_{21} + \tfrac{1}{2}\mathbf{p}_{21}^T\mathbf{p}_{21}). \tag{3.54}$$

Equation (3.54) can be re-expressed as

$$\varepsilon = \mathbf{b}_1^T\mathbf{p} + \frac{1}{2\alpha_o^2}\mathbf{p}^T\mathbf{A}\mathbf{p} \tag{3.55}$$

where

$$\mathbf{b}_1^T = \frac{1}{4\alpha_o^2}(-x_{21}, x_{21}, -z_{21}, z_{21}) = \frac{1}{4\alpha_o^2}\mathbf{c}(\mathbf{x})^T \tag{3.56}$$

and

$$A = \frac{1}{4}\begin{bmatrix} 1 & \text{symmetric} & & \\ -1 & 1 & & \\ 0 & 0 & 1 & \\ 0 & 0 & -1 & 1 \end{bmatrix}. \tag{3.57}$$

From (3.54)–(3.57), the incremental Green strain (caused by $\Delta\mathbf{p}$) is given by

$$\Delta\varepsilon = \frac{1}{4\alpha_o^2}((\mathbf{x}_{21} + \mathbf{p}_{21})^T\Delta\mathbf{p}_{21} + \tfrac{1}{2}\Delta\mathbf{p}_{21}^T\Delta\mathbf{p}_{21})$$

$$= (\mathbf{b}_1 + \mathbf{b}_2(\mathbf{p}))^T\Delta\mathbf{p} + \frac{1}{2\alpha_o^2}\Delta\mathbf{p}^T\mathbf{A}\,\Delta\mathbf{p} = \mathbf{b}(\mathbf{p})^T\Delta\mathbf{p} + \frac{1}{2\alpha_o^2}\Delta\mathbf{p}^T\mathbf{A}\,\Delta\mathbf{p} \tag{3.58}$$

where (compare (3.56))

$$\mathbf{b}_2(\mathbf{p})^{\mathrm{T}} = \frac{1}{4\alpha_o^2}(-u_{21}, u_{21}, -w_{21}, w_{21}) = \frac{1}{4\alpha_o^2}\mathbf{c}(\mathbf{p})^{\mathrm{T}} = \frac{1}{\alpha_o^2}(\mathbf{A}\mathbf{p})^{\mathrm{T}}. \tag{3.59}$$

Comparing (3.58) with a Taylor series expansion for $\Delta\varepsilon$,

$$\Delta\varepsilon = \frac{\partial\varepsilon}{\partial\mathbf{p}}\Delta\mathbf{p} + \frac{1}{2}\Delta\mathbf{p}^{\mathrm{T}}\frac{\partial\varepsilon}{\partial\mathbf{p}\,\Delta\mathbf{p}}\Delta\mathbf{p} = \mathbf{b}(\mathbf{p})^{\mathrm{T}}\Delta\mathbf{p} + \frac{1}{2}\Delta\mathbf{p}^{\mathrm{T}}\frac{\partial\mathbf{b}}{\partial\mathbf{p}}\Delta\mathbf{p} \tag{3.60}$$

we can see that $(1/\alpha_o^2)\mathbf{A}$ is the second partial derivative of $\boldsymbol{\varepsilon}$ with respect to the displacements, $\mathbf{p}$ or the first partial derivative of $\mathbf{b}$ with respect to $\mathbf{p}$.

For a small virtual displacement, with $\delta\mathbf{p}_v$ instead of $\Delta\mathbf{p}$, the last term in (3.58) becomes negligible and

$$\delta\boldsymbol{\varepsilon}_v = \frac{\partial\varepsilon}{\partial\mathbf{p}}\delta\mathbf{p}_v = (\mathbf{b}_1 + \mathbf{b}_2(\mathbf{p}))^{\mathrm{T}}\delta\mathbf{p}_v = \mathbf{b}(\mathbf{p})^{\mathrm{T}}\delta\mathbf{p}_v. \tag{3.61}$$

### 3.3.2 Equilibrium and the internal force vector

The principle of virtual work (Sections 1.3.2, 2.1 and 3.1) can now be used to provide internal nodal forces, $\mathbf{q}_i$, that are in a weighted average sense [C2.2], in equilibrium with a set of stresses, $\sigma$, that relate to total displacements, $\mathbf{p}$. To this end, using (3.61),

$$\sum_{e}\delta\mathbf{p}_v^{\mathrm{T}}\mathbf{q}_i = \sum_{e}\int\sigma_G\delta\varepsilon_v\,\mathrm{d}V_o = \sum_{e}\delta\mathbf{p}_v^{\mathrm{T}}\int\sigma_G\mathbf{b}\,\mathrm{d}V_o \tag{3.62}$$

where $\sum_e$ involves a 'summation' over the elements. For the following developments, we will drop this summation sign and hence will only directly deal with force vectors or stiffness matrices at the element level. The 'merging process' to the structural level is identical to that adopted for linear analysis [C2.2].

The strain–displacement vector $\mathbf{b}$ in (3.62) is given by (3.61) (with (3.56) and (3.59)) while the subscript G on $\sigma$ follows the work of Section 3.1.5, where it was shown that we must take note of the *type* of stress. The stress $\sigma_G$ is the stress that is work conjugate to the Green strain (later—Chapter 4—to be called the second Piola–Kirchhoff stress).

Equation (3.62) must stand for arbitrary $\delta\mathbf{p}_v$ and hence using (3.61), (3.56) and (3.59),

$$\mathbf{q}_i = \int\sigma_G\mathbf{b}\,\mathrm{d}V_o = 2\alpha_o A_o\sigma_G\mathbf{b} = 2\alpha_o A_o\sigma_G(\mathbf{b}_1 + \mathbf{b}_2(\mathbf{p})) = \frac{\sigma_G A_o}{2\alpha_o}(\mathbf{c}(\mathbf{x}) + \mathbf{c}(\mathbf{p})) = \mathbf{q}_{i1} + \mathbf{q}_{i2}. \tag{3.63}$$

Using (3.62), the procedure for computing the internal forces, $\mathbf{q}_i$, from a set of nodal displacements, $\mathbf{p}$, is as follows:

(1) compute the strain from (3.54) or (3.55);
(2) compute the stress, $\sigma_G$ (here, constant over the element), assuming a linear material response from $\sigma_G = E\varepsilon$;

(3) compute the internal forces, $\mathbf{q}_i$, from (3.63) with $\mathbf{b}_1$ and $\mathbf{b}_2$ being defined in (3.56) and (3.59).

### 3.3.3 The tangent stiffness matrix

From (2.20) and (3.63)

$$\mathbf{K}_t = \frac{\partial \mathbf{g}}{\partial \mathbf{p}} = \frac{\partial \mathbf{q}_i}{\partial \mathbf{p}} = 2\alpha_o A_o \mathbf{b} \frac{\partial \sigma_G}{\partial \mathbf{p}} + 2\alpha_o A_o \frac{\partial \mathbf{b}}{\partial \mathbf{p}} \sigma_G. \tag{3.64}$$

Using (3.64) and the non-virtual form of (3.61),

$$\frac{\partial \sigma_G}{\partial \mathbf{p}} = E \frac{\partial \varepsilon}{\partial \mathbf{p}} = E\{\mathbf{b}_1 + \mathbf{b}_2(\mathbf{p})\}^T = E\mathbf{b}(\mathbf{p})^T. \tag{3.65}$$

From (3.64) and (3.65), the first term of (3.64) can be written as

$$2\alpha_o A_o \mathbf{b} \frac{\partial \sigma_G}{\partial \mathbf{p}} = 2E\alpha_o A_o \mathbf{b}\mathbf{b}^T = \mathbf{K}_{t1} + \mathbf{K}_{t2} \tag{3.66}$$

where

$$K_{t1} = 2EA_o\alpha_o \mathbf{b}_1 \mathbf{b}_1^T = \frac{EA}{8\alpha_o^3} \mathbf{c}(\mathbf{x})\mathbf{c}(\mathbf{x})^T \tag{3.67}$$

$$\mathbf{K}_{t2} = 2EA\alpha_o[\mathbf{b}_1\mathbf{b}_2^T + \mathbf{b}_2\mathbf{b}_1^T + \mathbf{b}_2\mathbf{b}_2^T] = \mathbf{K}_{t2a} + \mathbf{K}_{t2a}^T + \mathbf{K}_{t2b}. \tag{3.68}$$

Equation (3.67) provides the standard linear stiffness matrix while (3.68) gives the 'initial displacement (or slope) matrix' (compare (1.10)). The 'geometric' or 'initial-stress matrix' (Section 1.2) comes from the second term in (3.64). Noting that, of the constituents of $\mathbf{b}$ (see (3.58)), only $\mathbf{b}_2$ is a function of $\mathbf{p}$, from (3.64) and (3.59),

$$\mathbf{K}_{t\sigma} = 2\alpha_o A_o \frac{\partial \mathbf{b}}{\partial \mathbf{p}} \sigma_G = 2\alpha_o A_o \frac{\partial \mathbf{b}_2}{\partial \mathbf{p}} \sigma_G = \frac{2A_o\sigma_G}{\alpha_o} \mathbf{A} = \frac{A_o\sigma_G}{2\alpha_o} \begin{bmatrix} 1 & \text{symmetric} & & \\ -1 & 1 & & \\ 0 & 0 & 1 & \\ 0 & 0 & -1 & 1 \end{bmatrix}. \tag{3.69}$$

Equations (3.67) and (3.68) can be expanded to give

$$\mathbf{K}_{t1} = \frac{EA}{8\alpha_o^3} \begin{bmatrix} x_{21}^2 & & \text{symmetric} & \\ -x_{21}^2 & x_{21}^2 & & \\ x_{21}z_{21} & -x_{21}z_{21} & z_{21}^2 & \\ -x_{21}z_{21} & x_{21}z_{21} & -z_{21}^2 & z_{21}^2 \end{bmatrix} = \frac{EA_o}{8\alpha_o^3} \mathbf{c}(\mathbf{x})\mathbf{c}(\mathbf{x})^T \tag{3.70}$$

$$\mathbf{K}_{t2a} = 2EA_o\alpha_o \mathbf{b}_1 \mathbf{b}_2^T = \frac{EA_o}{8\alpha_o^3} \begin{bmatrix} x_{21}u_{21} & -x_{21}u_{21} & x_{21}w_{21} & -x_{21}w_{21} \\ -x_{21}u_{21} & x_{21}u_{21} & -x_{21}w_{21} & x_{21}w_{21} \\ z_{21}u_{21} & -z_{21}u_{21} & z_{21}w_{21} & -z_{21}w_{21} \\ -z_{21}u_{21} & z_{21}u_{21} & -z_{21}w_{21} & z_{21}w_{21} \end{bmatrix} \tag{3.71}$$

$$\mathbf{K}_{t2b} = \frac{EA}{8\alpha_o^3} \begin{bmatrix} u_{21}^2 & & & \text{symmetric} \\ -u_{21}^2 & u_{21}^2 & & \\ u_{21}w_{21} & -u_{21}w_{21} & w_{21}^2 & \\ -u_{21}w_{21} & u_{21}w_{21} & -w_{21}^2 & w_{21}^2 \end{bmatrix} \tag{3.72}$$

with the final tangent stiffness matrix being given by

$$\mathbf{K}_t = \mathbf{K}_{t1} + \mathbf{K}_{t2} + \mathbf{K}_{t\sigma} = \mathbf{K}_{t1} + \mathbf{K}_{t2a} + \mathbf{K}_{t2a}^T + \mathbf{K}_{t2b} + \mathbf{K}_{t\sigma}. \tag{3.73}$$

The internal force vector, $\mathbf{q}_i$, tangent stiffness matrix, $\mathbf{K}_t$, and strain/displacement relationships that have just been derived can be incorporated into a computer program using a very similar procedure to that adopted for a shallow-truss theory in Chapter 2. (This is discussed further in Section 3.9.) The technique is known as 'total Lagrangian' because all measures are related back to the initial configuration. While the detail has been given in relation to a two-dimensional analysis, the concepts are equally valid in three dimensions—see Section 3.7.

### 3.3.4 Using shape functions

While shape functions are unnecessary for the current elements, with a view to more complex elements, it is useful to apply them. To this end, in relation to Figure 3.5, we define the incremental vector, $\mathbf{dr}_o$ along $A_oB_o$ as

$$\mathbf{dr}_o = \frac{\mathbf{dr}_o}{d\xi} d\xi \tag{3.74}$$

where $\mathbf{r}_o$ was given in (3.46). In a similar fashion, the new incremental vector is, with the aid of (3.45),

$$\mathbf{dr}_n = \frac{d(\mathbf{r}_o + \mathbf{u})}{d\xi} d\xi \tag{3.75}$$

where $\mathbf{u}$ has been defined in (3.46). Hence the length of $\mathbf{dr}_n$ is

$$dr_n = \|\mathbf{dr}_n\| = \left(\frac{\mathbf{dr}_o^T}{d\xi}\frac{\mathbf{dr}_o}{d\xi} + 2\frac{\mathbf{dr}_o^T}{d\xi}\frac{\mathbf{du}}{d\xi} + \frac{\mathbf{du}^T}{d\xi}\frac{\mathbf{du}}{d\xi}\right)^{1/2} d\xi = \alpha_n\, d\xi \tag{3.76}$$

where for the current simple elements, $2\alpha_n$ is the final length of the bar. In a similar fashion, the length of $\mathbf{dr}_o$ is

$$dr_o = \|\mathbf{dr}_o\| = \left(\frac{\mathbf{dr}_o^T}{d\xi}\frac{\mathbf{dr}_o}{d\xi}\right)^{1/2} d\xi = \alpha_o\, d\xi \tag{3.77}$$

where, again for the current element, $2\alpha_o$ is the original length of the bar.

From (3.9), the Green's strain can be expressed as

$$\varepsilon = \frac{dr_n^2 - dr_o^2}{2dr_o^2}. \tag{3.78}$$

Substitution from (3.76) and (3.77) into (3.78) gives

$$\varepsilon = \frac{1}{\alpha_o^2}\frac{d\mathbf{r}_o^T}{d\xi}\frac{d\mathbf{u}}{d\xi} + \frac{1}{2\alpha_o^2}\frac{d\mathbf{u}^T}{d\xi}\frac{d\mathbf{u}}{d\xi} \tag{3.79}$$

and an increment, $\Delta\varepsilon$ relating to a displacement increment $\Delta\mathbf{u}$, can be expressed as

$$\Delta\varepsilon = \frac{1}{\alpha_o^2}\frac{d\mathbf{r}_o^T}{d\xi}\frac{d\Delta\mathbf{u}}{d\xi} + \frac{1}{\alpha_o^2}\frac{d\mathbf{u}^T}{d\xi}\frac{d\Delta\mathbf{u}}{d\xi} + \frac{1}{2\alpha_o^2}\frac{d\Delta\mathbf{u}^T}{d\xi}\frac{d\Delta\mathbf{u}}{d\xi}. \tag{3.80}$$

For the displacement-based finite element method, shape functions are used to relate both the geometry and the displacements to nodal values so that, in relation to Figure 3.6, simple linear expansions give

$$\mathbf{r}_o = \begin{pmatrix} x \\ z \end{pmatrix} = \begin{bmatrix} \frac{1}{2}(1-\xi) & \frac{1}{2}(1+\xi) & 0 & 0 \\ 0 & 0 & \frac{1}{2}(1-\xi) & \frac{1}{2}(1+\xi) \end{bmatrix} \begin{pmatrix} x_1 \\ x_2 \\ z_1 \\ z_2 \end{pmatrix} = \begin{bmatrix} \mathbf{h}^T & \mathbf{0}^T \\ \mathbf{0}^T & \mathbf{h}^T \end{bmatrix} \mathbf{x} = \mathbf{H}\mathbf{x}. \tag{3.81}$$

In a similar fashion, the displacements, $\mathbf{u}$, can be expressed as

$$\mathbf{u} = \begin{pmatrix} u \\ w \end{pmatrix} = [\mathbf{H}] \begin{pmatrix} u_1 \\ u_2 \\ w_1 \\ w_2 \end{pmatrix} = [\mathbf{H}]\mathbf{p} \tag{3.82}$$

where the vector, $\mathbf{p}$, contains the nodal displacements as given in (3.49). Differentiation of (3.81) gives

$$\frac{d\mathbf{r}_o}{d\xi} = \mathbf{r}_{o\xi} = \begin{bmatrix} \frac{1}{2} & \frac{1}{2} & 0 & 0 \\ 0 & 0 & -\frac{1}{2} & \frac{1}{2} \end{bmatrix} \mathbf{x} = \begin{bmatrix} \mathbf{h}_\xi^T & \mathbf{0}^T \\ \mathbf{0}^T & \mathbf{h}_\xi^T \end{bmatrix} \mathbf{x} = \mathbf{H}_\xi \mathbf{x} \tag{3.83}$$

while differentiation of (3.82) gives

$$\frac{d\mathbf{u}}{d\xi} = \mathbf{u}_\xi = \mathbf{H}_\xi \mathbf{p} \tag{3.84}$$

so that, from (3.79),

$$\varepsilon = \left(\frac{1}{\alpha_o^2}\mathbf{r}_{o\xi}^T\mathbf{H}_\xi\right)\mathbf{p} + \left(\frac{1}{2\alpha_o^2}\mathbf{p}^T\mathbf{H}_\xi^T\mathbf{H}_\xi\mathbf{p}\right) = \mathbf{b}_1^T\mathbf{p} + \frac{1}{2\alpha_o^2}\mathbf{p}^T\mathbf{A}\mathbf{p} \tag{3.85}$$

where the explicit form of $\mathbf{b}_1$ has already been given in (3.56) and of the matrix $\mathbf{A}$ in (3.57). From (3.80) (or (3.85))

$$\begin{aligned}\Delta\varepsilon &= \left(\frac{1}{\alpha_o^2}\mathbf{r}_{o\xi}^T\mathbf{H}_\xi\right)\Delta\mathbf{p} + \left(\frac{1}{\alpha_o^2}\mathbf{u}_\xi^T\mathbf{H}_\xi\right)\Delta\mathbf{p} + \left(\frac{1}{2\alpha_o^2}\Delta\mathbf{p}^T\mathbf{H}_\xi^T\mathbf{H}_\xi\Delta\mathbf{p}\right) \\ &= \mathbf{b}_1^T\Delta\mathbf{p} + \mathbf{b}_2^T\Delta\mathbf{p} + \frac{1}{2\alpha_o^2}\Delta\mathbf{p}^T\mathbf{A}\,\Delta\mathbf{p}\end{aligned} \tag{3.86}$$

where the explicit form of $\mathbf{b}_2(\mathbf{p})$ has been given in (3.59). The virtual strain then follows as in (3.61) and one can proceed, as in Section 3.3.1, to apply virtual work as in (3.62) to obtain the internal force vector $\mathbf{q}_i$ as in (3.63). Using the shape function approach, in performing the integrals in (3.63), one would apply

$$\int dV_o = \int A_o\, dr_o = \int A_o \alpha_o\, d\xi = 2A_o\alpha_o. \tag{3.87}$$

The derivation of the tangent stiffness matrix follows as in (3.64) but to maintain the shape function approach, one would write

$$\mathbf{K}_t = \frac{\partial \mathbf{g}}{\partial \mathbf{p}} = \frac{\partial \mathbf{q}_i}{\partial \mathbf{p}} = \int \mathbf{b}\frac{\partial \sigma_G}{\partial \mathbf{p}}\, dV_o + \int \frac{\partial \mathbf{b}}{\partial \mathbf{p}}\sigma_G\, dV_o \tag{3.88}$$

which coincides with (3.64) and leads, as before, to (3.73).

### 3.3.5 Alternative expressions involving updated coordinates

The updated coordinates, $\mathbf{x}'$, can be expressed as

$$\mathbf{x}' = \mathbf{x} + \mathbf{p} \qquad \text{or} \qquad \mathbf{r}_n = \mathbf{r}_o + \mathbf{u} \tag{3.89}$$

where the first expression relates to nodal variables (Figure 3.6) while the second relates to a general point (see Figure 3.5). Using these updated coordinates, the strain increment of (3.58) can be re-expressed as

$$\Delta\varepsilon = \mathbf{b}_1(\mathbf{x}')^T\Delta\mathbf{p} + \frac{1}{2\alpha_o^2}\Delta\mathbf{p}^T\mathbf{A}\,\Delta\mathbf{p} = \frac{1}{4\alpha_o^2}\mathbf{c}(\mathbf{x}')^T\Delta\mathbf{p} + \frac{1}{2\alpha_o^2}\Delta\mathbf{p}^T\mathbf{A}\,\Delta\mathbf{p} \tag{3.90}$$

where $\mathbf{c}(\mathbf{x}')$ follows the same form as the $\mathbf{c}(\mathbf{x})$ and $\mathbf{c}(\mathbf{p})$ vectors defined in (3.56) and (3.59). In place of (3.61), the virtual strain can be expressed as

$$\delta\varepsilon_v = \mathbf{b}_1(\mathbf{x}')^T\delta\mathbf{p}_v. \tag{3.91}$$

Using (3.91), an alternative expression to (3.63) can easily be derived for the internal forces, $\mathbf{q}_i$, whereby

$$\mathbf{q}_i = \int \sigma_G \mathbf{b}_1(\mathbf{x}')\, dV_o = 2A_o\sigma_G\alpha_o\mathbf{b}_1(\mathbf{x}') = \frac{A_o\sigma_G}{2\alpha_o}\mathbf{c}(\mathbf{x}') = \frac{A_o\sigma_G}{2\alpha_o}\begin{pmatrix} -x'_{21} \\ x'_{21} \\ -z'_{21} \\ z'_{21} \end{pmatrix} = \mathbf{q}'_{i1}. \tag{3.92}$$

In comparison with (3.63), there is now no $\mathbf{q}'_{i2}$ term.

Again using updated coordinates, $\mathbf{K}_{t1}$ ((3.67) and (3.70)) as well as $\mathbf{K}_{t2}$ ((3.68), (3.71) and (3.72)) can be combined to give

$$\mathbf{K}'_{t1} = \int EA_o\mathbf{b}_1(\mathbf{x}')\mathbf{b}_1(\mathbf{x}')^T dV_o = \frac{EA_o}{8\alpha_o^3}\mathbf{c}(\mathbf{x}')\mathbf{c}(\mathbf{x}')\mathbf{T} \tag{3.93}$$

or

$$\mathbf{K}_{t1}+\mathbf{K}_{t2a}+\mathbf{K}_{t2a}^{T}+\mathbf{K}_{t2b}=\frac{EA_o}{8\alpha_o^3}\begin{bmatrix} x_{21}'^2 & & & \text{symmetric} \\ -x_{21}'^2 & x_{21}'^2 & & \\ x_{21}'z_{21}' & -x_{21}'z_{21}' & z_{21}'^2 & \\ -x_{21}'z_{21}' & x_{21}'z_{21}' & -z_{21}'^2 & z_{21}'^2 \end{bmatrix}=\mathbf{K}_{t1}' \tag{3.94}$$

while the geometric stiffness matrix of (3.69) is unaltered. Consequently, if an updated coordinate system is adopted, both the internal force vector (3.92) and the tangent stiffness matrix (3.94) involve the standard linear terms although these are related to the new coordinates. However, the 'linear' tangent stiffness matrix must always be supplemented by the 'geometric' or 'initial stress' matrix.

The introduction of updated coordinates can simply be considered as an alternative way of expressing the 'Green-strain system' which avoids the $\mathbf{b}_2(\mathbf{p})$ terms and hence ommits $\mathbf{q}_{i2}$ (3.63) and $\mathbf{K}_{t2a}$ (3.71) and $\mathbf{K}_{t2b}$ (3.72).

### 3.3.6 An updated Lagrangian formulation

The procedure in Section 3.3.5 is simply an alternative way of writing the previous total Lagrangian formulation, but it is very closely related to a so-called updated Lagrangian formulation. Using such a technique, after the coordinates had been updated using (3.89), the datum would be re-set so that the new configuration would become the old configuration (o). Before proceeding to the next increment or iteration, the second Piola–Kirchhoff stresses ($\sigma_G$ in the notation of Section 3.1), which related to the old configuration, must be converted to 'true stresses' relating to the new configuration so that, from (3.30),

$$\text{`}\sigma\text{'}=\frac{A_o l_n}{A_n l_o}\sigma_G=\text{`}\sigma\text{'}=\frac{A_o\alpha_n}{A_n\alpha_o}\sigma_G \tag{3.95}$$

at which stage, with respect to the new configuration, the displacements $\mathbf{p}$ are zero and we must use $\alpha_n$ and $A_n$. With these differences, we may use the standard total Lagrangian formulae of Sections 3.3.2 and 3.3.3. Hence, from (3.63) (with $\mathbf{p}=\mathbf{0}$), the internal force vector is given by

$$\mathbf{q}_i=\int \text{`}\sigma\text{'}\,\mathbf{b}_1(\mathbf{x}')\,dV_n=\frac{A_n\text{`}\sigma\text{'}}{2\alpha_n}\mathbf{c}(\mathbf{x}'). \tag{3.96}$$

Substituting from (3.95) into (3.92) leads to (3.96) so that the updated Lagrangian procedure leads to an identical internal force vector to that obtained with the standard total Lagrangian formulation. In a similar fashion, the tangent stiffness matrix would, from (3.73) with $\mathbf{p}=\mathbf{0}$, be given by

$$\mathbf{K}_{t1}=\frac{E'A_n}{8\alpha_n^2}\mathbf{c}(\mathbf{x}')\mathbf{c}(\mathbf{x}')^T+\frac{2A_n\text{`}\sigma\text{'}}{\alpha_n}\mathbf{A}. \tag{3.97}$$

The second term in (3.97) gives the geometric stiffness matrix which, with '$\sigma$' from

(3.95) coincides with (3.69). However, the first term only corresponds with (3.94) if

$$\frac{E'A_n}{\alpha_n^3} = \frac{EA_o}{\alpha_o^3} \tag{3.98}$$

where both of the $E$-values in (3.98) are tangential, but $E'$ is appropriate to the updated Lagrangian formulation while $E$ relates to the total Lagrangian procedure. From the considerations of Section 3.2, we should not be surprised that different $E$-values are required if different formulations are to give the same answers. However, as before, *for small strains*, there is no need either to introduce the tangent $E$ transformations of (3.98) or the stress measure transformations of (3.95). Both the previous (Section 3.3.5) use of updated coordinates in a total Lagrangian framework and the present updated Lagrangian formulation avoid the $\mathbf{b}_2(\mathbf{p})$ terms and hence omit $\mathbf{q}_{i2}$ (3.73) and $\mathbf{K}_{t2a}$ (3.71) and $\mathbf{K}_{t2b}$ (3.72).

We have so far introduced an updated Lagrangian formulation whereby all measures are related to the current updated configuration. However, the Green-strain measure of (3.55) is related to the initial configuration. Using (3.59) and (3.89), this expression can be re-stated in terms of the current coordinates so that

$$\begin{aligned}\varepsilon &= \mathbf{b}_1(\mathbf{x})^T\mathbf{p} + \mathbf{b}_2(\mathbf{p})^T\mathbf{p} + \frac{1}{2\alpha_o^2}\mathbf{p}^T\mathbf{A}\mathbf{p} \\ &= \mathbf{b}_1(\mathbf{x}')^T\mathbf{p} - \frac{1}{2\alpha_o^2}\mathbf{p}^T\mathbf{A}\mathbf{p} = \frac{1}{4\alpha_o^2}\mathbf{c}(\mathbf{x}')^T\mathbf{p} - \frac{1}{2\alpha_o^2}\mathbf{p}^T\mathbf{A}\mathbf{p}.\end{aligned} \tag{3.99}$$

Equation (3.99) corresponds to a re-expression of (3.79) as

$$\varepsilon = \frac{1}{\alpha_o^2}\frac{d(\mathbf{r}_o + \mathbf{u})^T}{d\xi}\frac{d\mathbf{u}}{d\xi} - \frac{1}{2\alpha_o^2}\frac{d\mathbf{u}^T}{d\xi}\frac{d\mathbf{u}}{d\xi} = \frac{1}{\alpha_o^2}\frac{d\mathbf{r}_n^T}{d\xi}\frac{d\mathbf{u}}{d\xi} - \frac{1}{2\alpha_o^2}\frac{d\mathbf{u}^T}{d\xi}\frac{d\mathbf{u}}{d\xi}. \tag{3.100}$$

Neither (3.99) nor (3.100) is fully related to the current configuration because of the terms $\alpha_o$. However, we can generalise the expression in (3.41) for the Almansi strain to obtain a strain measure involving $\alpha_n$. In comparison to (3.78), the Almansi strain involves

$$\varepsilon_A = \frac{dr_n^2 - dr_o^2}{2dr_n^2} = \frac{\alpha_n^2 - \alpha_o^2}{2\alpha_n^2}. \tag{3.101}$$

Substituting from (3.76) and (3.77) into (3.101) gives

$$\varepsilon_A = \frac{1}{\alpha_n^2}\frac{d(\mathbf{r}_o + \mathbf{u})^T}{d\xi}\frac{d\mathbf{u}}{d\xi} - \frac{1}{2\alpha_n^2}\frac{d\mathbf{u}^T}{d\xi}\frac{d\mathbf{u}}{d\xi} = \frac{1}{\alpha_n^2}\frac{d\mathbf{r}_n^T}{d\xi}\frac{d\mathbf{u}}{d\xi} - \frac{1}{2\alpha_n^2}\frac{d\mathbf{u}^T}{d\xi}\frac{d\mathbf{u}}{d\xi} \tag{3.102}$$

which is identical to (3.100) apart from having $\alpha_n$s rather than $\alpha_o$s in the denominator. A similar relationship would follow (via (3.53) and (3.50)) for the Almansi form of (3.99) for which $\alpha_n$s would appear in the denominator. As shown in Section 3.2, we can expect the same answers from the two measures provided the strains (but not necessarily the rotations) are small.

Further discussion on the total and updated Lagrangian formulations will be given in a continuum context in Chapters 5. Before moving to alternative strain measures, we should emphasise that the updating system that has been discussed here and in

Section 3.3.5 involves *updated* but *unrotated* coordinates. We will later discuss rotated coordinates.

## 3.4 AN ALTERNATIVE FORMULATION USING A ROTATED ENGINEERING STRAIN

The previous derivation (Section 3.3) was based on the use of Green's strain. Sections 3.1 and 3.2 considered a number of alternative strain measures and we will now consider the rotated engineering strain. A natural derivation would involve a rotated coordinate system. This will be described in Section 3.6 but we will firstly maintain a fixed cartesian coordinate system. Throughout this section, the subscript E for engineering is implied but, for brevity, omitted both on the stresses and the strains.

Extending the definition of (3.3), the rotated engineering strain, which relates to the direction of the rotating bar, is given by

$$\varepsilon = \frac{\mathrm{d}r_n - \mathrm{d}r_o}{\mathrm{d}r_o} = \frac{\alpha_n - \alpha_o}{\alpha_o} \tag{3.103}$$

where $\alpha_o$ is given by (3.50) or (3.77) and $\alpha_n$ by (3.53) or (3.76). With the aid of (3.57), $\alpha_n$ from (3.53) can be re-expressed by

$$\alpha_n^2 = (\mathbf{x} + \mathbf{p})^T \mathbf{A}(\mathbf{x} + \mathbf{p}) = \mathbf{x}'^T \mathbf{A}\mathbf{x}'. \tag{3.104}$$

From (3.103) and (3.104),

$$\mathbf{b} = \frac{\partial \varepsilon^T}{\partial \mathbf{p}} = \frac{1}{\alpha_o}\frac{\partial \alpha_n^T}{\partial \mathbf{p}} = \frac{1}{\alpha_o \alpha_n}\mathbf{A}\mathbf{x}' = \frac{1}{4\alpha_o\alpha_n}\mathbf{c}(\mathbf{x}') \tag{3.105}$$

where we have used the relationship,

$$\mathbf{c}(\mathbf{x}') = 4\mathbf{A}\mathbf{x}' \tag{3.106}$$

(see (3.59) for an identical relationship between $\mathbf{c}(\mathbf{p})$ and $4\mathbf{A}\mathbf{p}$). Hence, from the principle of virtual work,

$$\mathbf{q}_i = \int \sigma \frac{\partial \varepsilon^T}{\partial \mathbf{p}} \mathrm{d}V_o = 2A_o\alpha_o\sigma\mathbf{b} = \frac{\sigma A_o}{2\alpha_n}\mathbf{c}(\mathbf{x}') = \lambda\frac{A_o\sigma}{2\alpha_o}\mathbf{c}(\mathbf{x}') \tag{3.107}$$

where

$$\lambda = \left(\frac{\alpha_o}{\alpha_n}\right). \tag{3.108}$$

Equation (3.107) can be compared with the Green's strain solution of (3.92).

In order to obtain the tangent stiffness matrix, (3.107) is differentiated so that

$$\mathbf{K}_t = \frac{\partial \mathbf{q}_i}{\partial \mathbf{p}} = \frac{A_o}{2\alpha_n}\mathbf{c}(\mathbf{x}')\frac{\partial \sigma^T}{\partial \mathbf{p}} + \frac{\sigma A_o}{2\alpha_n}\frac{\partial \mathbf{c}(\mathbf{x}')}{\partial \mathbf{p}} - \frac{\sigma A_o}{2\alpha_n^2}\mathbf{c}(\mathbf{x}')\frac{\partial \alpha_n}{\partial \mathbf{p}} \tag{3.109}$$

where the first two terms have parallels in (3.64). From (3.105),

$$\frac{\partial \sigma}{\partial \mathbf{p}} = E\frac{\partial \varepsilon}{\partial \mathbf{p}} = \frac{E}{4\alpha_o\alpha_n}\mathbf{c}(\mathbf{x}')^T = E\mathbf{b}^T \tag{3.110}$$

so that the first term from (3.109) is given by

$$K_{t1} = \frac{EA_o}{8\alpha_n^2\alpha_o}\mathbf{c}(\mathbf{x}')\mathbf{c}(\mathbf{x}')^T = \frac{EA_o\lambda^2}{8\alpha_o^3}\mathbf{c}(\mathbf{x}')\mathbf{c}(\mathbf{x}')^T \tag{3.111}$$

which differs by the $\lambda^2$ term from the matrix $\mathbf{K}'_{t1}$ of (3.93).

From (3.106),

$$\frac{\partial \mathbf{c}(\mathbf{x}')}{\partial \mathbf{p}} = 4\mathbf{A} \tag{3.112}$$

so that the second term from (3.109) gives

$$\mathbf{K}_{t\sigma 1} = \frac{2\sigma A_o}{\alpha_n}\mathbf{A} = \frac{2\sigma A_o\lambda}{\alpha_o}\mathbf{A} \tag{3.113}$$

which differs by a factor $\lambda$ from the 'initial-stress matrix', $\mathbf{K}_{t\sigma}$ of (3.69). Finally, with the aid of (3.105), the last term in (3.109) can be expressed as

$$\mathbf{K}_{t\sigma 2} = -\frac{\sigma A_o}{8\alpha_n^3}\mathbf{c}(\mathbf{x}')\mathbf{c}(\mathbf{x}')^T = -\frac{\sigma A_o\lambda^3}{8\alpha_o^3}\mathbf{c}(\mathbf{x}')\mathbf{c}(\mathbf{x}')^T. \tag{3.114}$$

If $\lambda$ (3.108) is assumed to be unity, the present formulation gives identical equations to the Green's strain formulation of Section 3.3, apart from the $\mathbf{K}_{t\sigma 2}$ matrix of (3.114) which has no counterpart in the Green's strain formulation.

In applying a formulation based on a rotated engineering strain, equation (3.103) is an inaccurate way of computing the strain because it involves the small difference between two relatively large numbers. It is computationally better to relate the engineering strain of (3.103) to the Green strain of (3.54), so that

$$\varepsilon_E = \frac{\alpha_n - \alpha_o}{\alpha_o} = \frac{2\alpha_o}{\alpha_n + \alpha_o}\varepsilon_G(3.54). \tag{3.115}$$

## 3.5 AN ALTERNATIVE FORMULATION USING A ROTATED LOG-STRAIN

In Section 3.1.3, we introduced a rotated log-strain and showed in Section 3.1.5 that the corresponding stress is the 'true stress'. In relation to the current truss elements, the log-strain of (3.15) is

$$\varepsilon = \log_e\left(\frac{\alpha_n}{\alpha_o}\right). \tag{3.116}$$

In (3.116) and throughout this section, a subscript, L, for log is implied on all the stress and strain terms. With the aid of (3.104), (3.116) can be differentiated to give

$$\frac{\partial \varepsilon}{\partial \mathbf{p}} = \frac{1}{\alpha_n}\frac{\partial \alpha_n}{\partial \mathbf{p}} = \lambda\frac{\partial \varepsilon}{\partial \mathbf{p}}(3.105) \tag{3.117}$$

where $\lambda$ is given by (3.108) and $(\partial\varepsilon/\partial\mathbf{p})(3.105)$ is the relationship in (3.105) for the engineering strain.

Hence, applying the principle of virtual work in relation to the *current* configuration,

$$\mathbf{q}_{\mathrm{i}} = \int \sigma \frac{\partial \varepsilon^{\mathrm{T}}}{\partial \mathbf{p}} \mathrm{d}V_{\mathrm{n}} = \frac{\lambda A_{\mathrm{n}} \sigma}{2\alpha_{\mathrm{o}}} \mathbf{c}(\mathbf{x}') = \lambda^{1+2v} \frac{A_{\mathrm{o}} \sigma}{2\alpha_{\mathrm{o}}} \mathbf{c}(\mathbf{x}'). \tag{3.118}$$

In deriving the last expression in (3.118), it has been assumed, from (3.24), that

$$\frac{A_{\mathrm{n}}}{A_{\mathrm{o}}} = \left(\frac{\alpha_{\mathrm{o}}}{\alpha_{\mathrm{n}}}\right)^{2v} = \lambda^{2v}. \tag{3.119}$$

By including this relationship, we include both the solution for no volume change (with $v = 0.5$) and volume change ($v \neq 0.5$).

Differentiation of (3.118) follows the lines previously adopted in Section 3.4 and leads to

$$\mathbf{K}_{t1} = \frac{EA_{\mathrm{o}}\lambda^{3+2v}}{8\alpha_{\mathrm{o}}^4} \mathbf{c}(\mathbf{x}')\mathbf{c}(\mathbf{x}')^{\mathrm{T}} = \frac{EA_{\mathrm{n}}\lambda^3}{8\alpha_{\mathrm{o}}^3} \mathbf{c}(\mathbf{x}')\mathbf{c}(\mathbf{x}')^{\mathrm{T}} \tag{3.120}$$

$$\mathbf{K}_{t\sigma 1} = \frac{2\sigma A_{\mathrm{o}}\lambda^{1+2v}}{\alpha_{\mathrm{o}}} \mathbf{A} = \frac{2\sigma A_{\mathrm{n}}\lambda}{\alpha_{\mathrm{o}}} \times \mathbf{A} \tag{3.121}$$

$$\mathbf{K}_{t\sigma 2} = -\frac{(1+2v)\sigma A_{\mathrm{o}}^{3+2v}}{8\alpha_{\mathrm{o}}^3} \mathbf{c}(\mathbf{x}')\mathbf{c}(\mathbf{x}')^{\mathrm{T}} = -\frac{(1+2v)\sigma A_{\mathrm{n}}\lambda^3}{8\alpha_{\mathrm{o}}^3} \mathbf{c}(\mathbf{x}')\mathbf{c}(\mathbf{x}')^{\mathrm{T}}. \tag{3.122}$$

## 3.6 AN ALTERNATIVE COROTATIONAL FORMULATION USING ENGINEERING STRAIN

In all of the previous developments, the coordinate axes $x, z$ (and $y$) have remained fixed in direction even if, as in Sections 3.3.5 and 3.3.6, we have updated the coordinates. We will now apply a 'corotational' formulation and will show that it gives the same results as those previously obtained in Section 3.4. The procedure adopts a set of corotational axes ($x_l$, $z_l$—Figure 3.7) which rotate with the element. In these circumstances, the engineering strain is given by

$$\varepsilon = \frac{1}{2\alpha_{\mathrm{o}}} \begin{bmatrix} -1 \\ 1 \\ 0 \\ 0 \end{bmatrix}^{\mathrm{T}} \begin{bmatrix} u_1 \\ u_2 \\ w_1 \\ w_2 \end{bmatrix} = \mathbf{b}_l^{\mathrm{T}} \mathbf{p}_l = \frac{1}{2\alpha_{\mathrm{o}}} \mathbf{c}_l^{\mathrm{T}} \mathbf{p}_l. \tag{3.123}$$

In the above equation and throughout this section a subscript E for engineering will be implied but omitted on all strain and stress measures. Equation (3.123) is obvious but it could be derived by relating the shape-function approaches of the previous sections to the local coordinate system. Following from (3.123), the principle of virtual work gives

$$\mathbf{q}_{\mathrm{il}} = \int \frac{\sigma}{2\alpha_{\mathrm{o}}} \mathbf{c}_l \,\mathrm{d}V_{\mathrm{o}} = A_{\mathrm{o}} \sigma \mathbf{c}_l. \tag{3.124}$$

We can now apply standard transformation procedures [C2.2], to give

$$\mathbf{q}_i = \mathbf{T}^T\mathbf{q}_{il} = A_o\sigma\mathbf{T}^T\mathbf{c}_l \tag{3.125}$$

where the transformation matrix, $\mathbf{T}$, relates the local displacements, $\mathbf{p}_l$ to the 'global' cartesian displacements, $\mathbf{p}$, so that

$$\mathbf{p}_l = \mathbf{T}\mathbf{p} = \begin{bmatrix} c & 0 & s & 0 \\ 0 & c & 0 & s \\ -s & 0 & c & 0 \\ 0 & -s & 0 & c \end{bmatrix} \mathbf{p} = \frac{1}{2\alpha_n} \begin{bmatrix} x'_{21} & 0 & z'_{21} & 0 \\ 0 & x'_{21} & 0 & z'_{21} \\ -z'_{21} & 0 & x'_{21} & 0 \\ 0 & -z'_{21} & 0 & x'_{21} \end{bmatrix} \mathbf{p}. \tag{3.126}$$

The terms $c$ and $s$ in (3.126) are $\cos\theta$ and $\sin\theta$ respectively, where $\theta$ is illustrated in Figure 3.7. If $\mathbf{T}^T$ is multiplied by $\mathbf{c}_l$ from (3.123), it can be shown that

$$\mathbf{T}^T\mathbf{c}_l = \frac{1}{2\alpha_n}\mathbf{c}(\mathbf{x}') \tag{3.127}$$

where $\mathbf{c}(\mathbf{x}')$ is given in (3.92). Hence substitution into (3.125) gives

$$\mathbf{q}_i = \frac{A_o\sigma}{2\alpha_n}\mathbf{c}(\mathbf{x}') \tag{3.128}$$

which coincides with (3.107), which was obtained with the aid of 'fixed coordinates'.

We could now proceed to differentiate (3.128) to obtain the tangent stiffness matrix given by the components (3.111), (3.113) and (3.114). However, we will instead adopt the spirit of the corotational approach and firstly differentiate (3.124) to obtain a 'local tangent stiffness matrix'. From (3.123) and (3.124) this gives

$$\mathbf{K}_{tl} = \frac{\partial \mathbf{q}_{il}}{\partial \mathbf{p}_l} = EA_o\mathbf{c}_l\frac{\partial \varepsilon_l}{\partial \mathbf{p}_l} = \frac{EA_o}{2\alpha_o}\mathbf{c}_l\mathbf{c}_l^T. \tag{3.129}$$

In order to relate this local stiffness matrix to the fixed cartesian coordinate system,

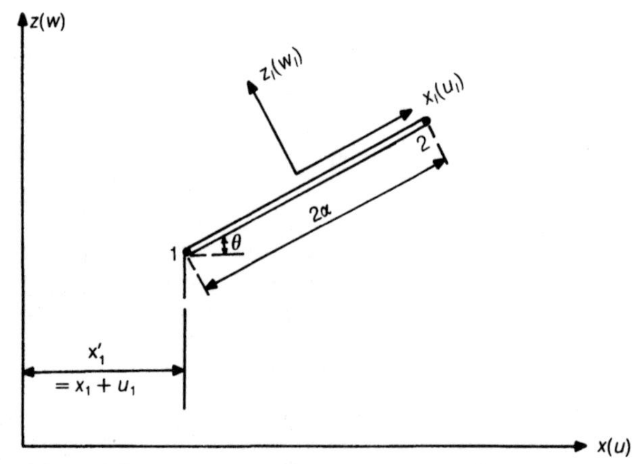


**Figure 3.7** Local (rotating) and global coordinate systems.

(3.125) can be differentiated to give

$$\delta \mathbf{q}_i = \mathbf{T}^T \frac{\partial \mathbf{q}_{il}}{\partial \mathbf{p}_l} \delta \mathbf{p}_l + \delta \mathbf{T}^T \mathbf{q}_{il} = \mathbf{T}^T \mathbf{K}_{tl} \mathbf{T} \delta \mathbf{p} + \delta \mathbf{T}^T \mathbf{q}_{il} = \mathbf{K}_t \delta \mathbf{p} \tag{3.130}$$

where use has been made of (3.126). Substitution from (3.127) into the first of the two stiffness terms in (3.130) gives

$$\mathbf{K}_{tl} = \frac{EA}{8\alpha_n^2 \alpha_o} \mathbf{c}(\mathbf{x}')\mathbf{c}(\mathbf{x}')^T \tag{3.131}$$

which coincides with (3.111).

In order to deal with the second stiffness term in (3.130), the $\mathbf{T}$ matrix in (3.126) can be differentiated so that

$$\delta \mathbf{T}^T = \begin{bmatrix} -s & 0 & -c & 0 \\ 0 & -s & 0 & -c \\ c & 0 & -s & 0 \\ 0 & c & 0 & -s \end{bmatrix} \delta\theta. \tag{3.132}$$

From Figure 3.8, a unit vector normal to the rotating element is given by

$$\mathbf{n} = \frac{1}{2\alpha_n} \begin{pmatrix} -z'_{21} \\ x'_{21} \end{pmatrix} \tag{3.133}$$

which is orthogonal to the truss vector, $\mathbf{x}'_{21}$. The infinitesimal relative displacement vector (Figure 3.8) can be expressed as

$$\delta \mathbf{p}_{21} = \begin{pmatrix} \delta u_{21} \\ \delta w_{21} \end{pmatrix}. \tag{3.134}$$

Resolving this vector in the direction $\mathbf{n}$ gives a scalar length:

$$\delta a = \mathbf{n}^T \delta \mathbf{p}_{21} = \mathbf{n}^T \begin{pmatrix} \delta u_{21} \\ \delta w_{21} \end{pmatrix} = \frac{1}{2\alpha_n} \begin{pmatrix} -z'_{21} \\ x'_{21} \end{pmatrix}^T \begin{pmatrix} \delta u_{21} \\ \delta w_{21} \end{pmatrix}. \tag{3.135}$$

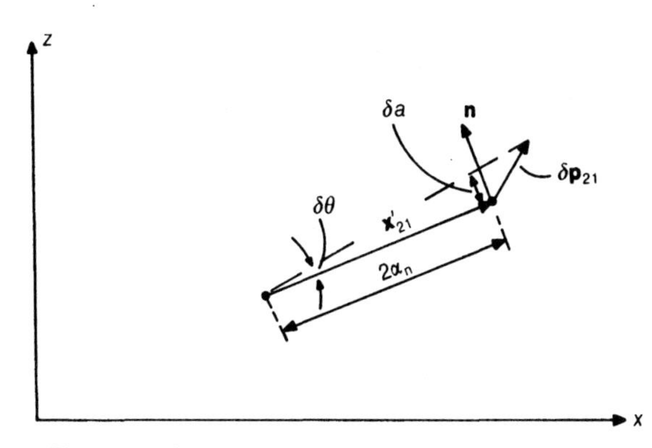


**Figure 3.8** Small movement from new configuration.

Consequently, the angle $\delta\theta$ (Figure 3.8) is given by

$$\delta\theta = \frac{\delta a}{2\alpha_n} = \frac{1}{4\alpha_n^2}\begin{pmatrix} -z'_{21} \\ x'_{21} \end{pmatrix}^T \begin{pmatrix} \partial u_{21} \\ \partial w_{21} \end{pmatrix} = \frac{1}{4\alpha_n^2}\begin{pmatrix} z'_{21} \\ -z'_{21} \\ x'_{21} \\ -x'_{21} \end{pmatrix}^T \Delta\mathbf{p} = \frac{1}{4\alpha_n^2}\mathbf{z}^T\delta\mathbf{p}. \tag{3.136}$$

Hence, using (3.124) and (3.132), the second stiffness term in (3.130) is given by

$$\delta\mathbf{q}_i = \frac{A_o\sigma}{8\alpha_n^3}\begin{bmatrix} -z'_{21} & 0 & -x'_{21} & 0 \\ 0 & -z'_{21} & 0 & -x'_{21} \\ x'_{21} & 0 & -z'_{21} & 0 \\ 0 & x'_{21} & 0 & -z'_{21} \end{bmatrix}\mathbf{c}_l\mathbf{z}^T\delta\mathbf{p} \tag{3.137}$$

with $\mathbf{c}_l$ from (3.123). Alternatively

$$\delta\mathbf{q}_i = \frac{A_o\sigma}{8\alpha_n^3}\mathbf{z}\mathbf{z}^T\delta\mathbf{p} = \frac{A\sigma}{8\alpha_n^3}\begin{bmatrix} z'^2_{21} & & \text{symmetric} & \\ -z'^2_{21} & z'^2_{21} & & \\ -x'_{21}z'_{21} & z'_{21}x'_{21} & x'^2_{21} & \\ z'_{21}x'_{21} & -z'_{21}x'_{21} & -x'^2_{21} & x'^2_{21} \end{bmatrix}\delta\mathbf{p} = \mathbf{K}_{t\sigma}\delta\mathbf{p}. \tag{3.138}$$

It can easily be shown that the matrix $\mathbf{K}_{t\sigma}$ in (3.138) coincides with the sum of $\mathbf{K}_{t\sigma1}$ and $\mathbf{K}_{t\sigma2}$ from (3.113) and (3.114). Hence, identical solutions are produced by the two formulations using (a) a fixed cartesian system and (b) a rotating (corotational) coordinate system. A similar correspondence can be shown for the log-strain formulation.

## 3.7 SPACE TRUSS ELEMENTS

The detailed workings of the previous sections have related to the 'planar truss element' of Figure 3.6 and 3.7. However, the theory is readily extendible to the 'space truss element' of Figure 3.9. In these circumstances, the vectors $\mathbf{r}$ and $\mathbf{u}$ of (3.46) become

$$\mathbf{r}^T = (x, y, z), \qquad \mathbf{u}^T = (u, v, w) \tag{3.139}$$

while the nodal vectors $\mathbf{x}$ and $\mathbf{p}$ of (3.48) and (3.49) become

$$\mathbf{x}^T = (x_1, x_2, y_1, y_2, z_1, z_2) \qquad \mathbf{p}^T = (u_1, u_2, v_1, v_2, w_1, w_2). \tag{3.140}$$

Allowing for these new definitions, most of the formulae in Section 3.3.1 remain valid, although the matrix $\mathbf{A}$ of (3.57) becomes

$$\mathbf{A} = \frac{1}{4}\begin{bmatrix} 1 & & & \text{symmetric} & & \\ -1 & 1 & & & & \\ 0 & 0 & 1 & & & \\ 0 & 0 & -1 & 1 & & \\ 0 & 0 & 0 & 0 & 1 & \\ 0 & 0 & 0 & 0 & -1 & 1 \end{bmatrix}. \tag{3.141}$$

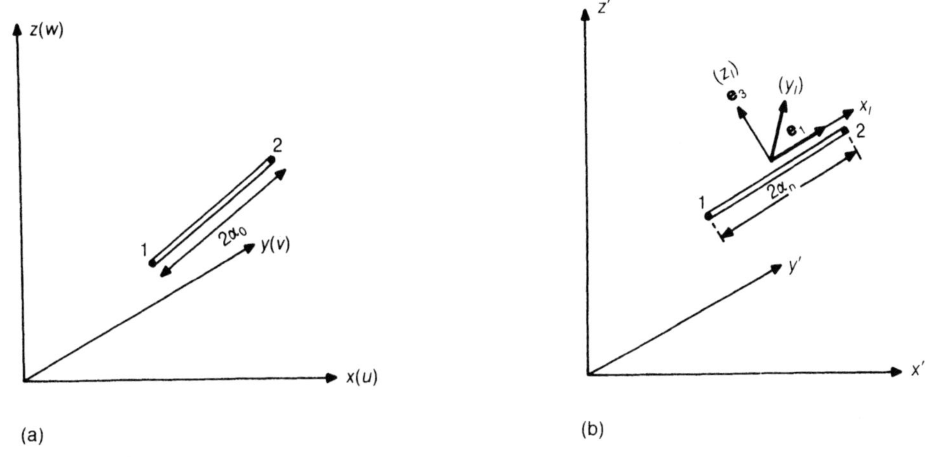


**Figure 3.9** Space-truss element: (a) initial coordinates: (b) updated $(x', y', z')$ and corotated $(x_l, y_l, z_l)$ coordinates.

For the Green's strain formulation, (3.63) remains valid for the internal force vector while (3.94) and (3.69) (but with $\mathbf{A}$ from (3.141)) still apply for the tangent stiffness matrix. However, in (3.94), the vector $\mathbf{c}(\mathbf{x}')$ of (3.92) must be extended to

$$\mathbf{c}(\mathbf{x}')^{\mathrm{T}} = (-x'_{21}, x'_{21}, -y'_{21}, y'_{21}, -z'_{21}, z'_{21}) \tag{3.142}$$

or

$$\mathbf{c}(\mathbf{x}')^{\mathrm{T}} = 2\alpha_{\mathrm{n}}(-\mathbf{e}_1(1), \mathbf{e}_1(1), -\mathbf{e}_1(2), \mathbf{e}_1(2), -\mathbf{e}_1(3), \mathbf{e}_1(3)). \tag{3.143}$$

The alternative form of (3.142) given in (3.143) employs the local unit base vector $\mathbf{e}_1$ (Figure 3.9) and will be used in later developments. In the latter equation, $\mathbf{e}_1(1)$, for example, is the component of the unit vector $\mathbf{e}_1$ in the $x$ (or 1) direction. With the extended definitions of (3.141) and (3.142), the formulae in Sections 3.4 and 3.5 remain valid for space truss elements.

For the corotated formulation of Section 3.6, the three-dimensional equivalent of the transformation matrix $\mathbf{T}^{\mathrm{T}}$ given in (3.126) is best expressed in terms of the three 'local unit base vectors' $\mathbf{e}_1, \mathbf{e}_2, \mathbf{e}_3$ shown in Figure 3.9 (although it will be shown later that, for truss elements, there is no need to explicitly compute $\mathbf{e}_2$ and $\mathbf{e}_3$). Hence, in place of (3.125) and (3.126),

$$\mathbf{q}_{\mathrm{i}} = \mathbf{T}^{\mathrm{T}}\mathbf{q}_{\mathrm{i}l} = A_{\mathrm{o}}\sigma \begin{bmatrix} \mathbf{e}_1(1) & 0 & \mathbf{e}_2(1) & 0 & \mathbf{e}_3(1) & 0 \\ 0 & \mathbf{e}_1(1) & 0 & \mathbf{e}_2(1) & 0 & \mathbf{e}_3(1) \\ \mathbf{e}_1(2) & 0 & \mathbf{e}_2(2) & 0 & \mathbf{e}_3(2) & 0 \\ 0 & \mathbf{e}_1(2) & 0 & \mathbf{e}_2(2) & 0 & \mathbf{e}_3(2) \\ \mathbf{e}_1(3) & 0 & \mathbf{e}_2(3) & 0 & \mathbf{e}_3(3) & 0 \\ 0 & \mathbf{e}_1(3) & 0 & \mathbf{e}_2(3) & 0 & \mathbf{e}_3(3) \end{bmatrix} \begin{bmatrix} -1 \\ 1 \\ 0 \\ 0 \\ 0 \\ 0 \end{bmatrix}. \tag{3.144}$$

Clearly, only the first two columns (involving $\mathbf{e}_1$—Figure 3.9) in the $\mathbf{T}^{\mathrm{T}}$ matrix of (3.144) are required both in the above and in the three-dimensional equivalent of the first stiffness term in (3.130). The three-dimensional equivalent of the second stiffness term in (3.130) involves

$$\delta \mathbf{q}_i = \delta \mathbf{T}^T \mathbf{q}_{il} = A_o \sigma \begin{bmatrix} -\delta \mathbf{e}_1(1) \\ \delta \mathbf{e}_1(1) \\ -\delta \mathbf{e}_1(2) \\ \delta \mathbf{e}_1(2) \\ -\delta \mathbf{e}_1(3) \\ \delta \mathbf{e}_1(3) \end{bmatrix} = A_o \sigma \begin{bmatrix} -1 & 0 & 0 \\ 1 & 0 & 0 \\ 0 & -1 & 0 \\ 0 & 1 & 0 \\ 0 & 0 & -1 \\ 0 & 0 & 1 \end{bmatrix} \delta \mathbf{e}_1$$

$$= A_o \sigma \mathbf{F} \begin{bmatrix} \delta \mathbf{e}_1(1) \\ \delta \mathbf{e}_1(2) \\ \delta \mathbf{e}_1(3) \end{bmatrix} = A_o \sigma \mathbf{F} \delta \mathbf{e}_1 \tag{3.145}$$

where the Boolean matrix $\mathbf{F}$ is such that:

$$\delta \mathbf{p}_{21} = \mathbf{F}^T \delta \mathbf{p}. \tag{3.145a}$$

The $\mathbf{e}_1$ vector (Figure 3.9(b)) can be written as

$$\mathbf{e}_1 = \frac{\mathbf{x}_{21} + \mathbf{p}_{21}}{2\alpha_n} = \frac{\mathbf{x}'_{21}}{2\alpha_n} \tag{3.146}$$

so that differentiation leads to

$$\delta \mathbf{e}_1 = \frac{\delta \mathbf{p}_{21}}{2\alpha_n} - \mathbf{e}_1 \frac{\delta \alpha_n}{\alpha_n}. \tag{3.147}$$

From (3.104) and (3.106),

$$\alpha_n \delta \alpha_n = \mathbf{x}'^T \mathbf{A} \delta \mathbf{p} = \tfrac{1}{4} \mathbf{c}(\mathbf{x}')^T \delta \mathbf{p} \tag{3.148}$$

Hence, using (3.143),

$$\delta \mathbf{e}_1 = \frac{\delta \mathbf{p}_{21}}{2\alpha_n} - \frac{\mathbf{e}_1 \mathbf{c}(\mathbf{x}')^T \delta \mathbf{p}}{4\alpha_n^2} = \frac{1}{2\alpha_n} [\mathbf{I} - \mathbf{e}_1 \mathbf{e}_1^T] \delta \mathbf{p}_{21}. \tag{3.149}$$

Substitution into (3.145) and using (3.145a) gives

$$\delta \mathbf{q}_i = \frac{A_o \sigma}{2\alpha_n} [\mathbf{F}\mathbf{F}^T - (\mathbf{F}\mathbf{e}_1)(\mathbf{F}\mathbf{e}_1)^T] \delta \mathbf{p} = \frac{A_o \sigma}{2\alpha_n} \left[ \mathbf{A} - \frac{1}{4\alpha_n^2} \mathbf{c}(\mathbf{x}')\mathbf{c}(\mathbf{x}')^T \right] \delta \mathbf{p} \tag{3.150}$$

which corresponds with the combination of (3.113) and (3.114).

## 3.8 MID-POINT INCREMENTAL STRAIN UPDATES

The Green-strain increment of (3.58) and (3.80) can be rewritten as

$$\Delta \varepsilon = \frac{1}{4\alpha_o^2} (\mathbf{x}'_{21} + \tfrac{1}{2} \Delta \mathbf{p}_{21})^T \Delta \mathbf{p}_{21} = \frac{1}{\alpha_o^2} \frac{\mathrm{d}(\mathbf{r}_n + \frac{1}{2} \Delta \mathbf{u})^T}{\mathrm{d}\xi} \frac{\mathrm{d}\Delta \mathbf{u}}{\mathrm{d}\xi} \tag{3.151}$$

where $\mathbf{x}'_{21}$ and $\mathbf{r}_n = \mathbf{r}_o + \mathbf{u}$ relate to the configuration at the beginning of the increment prior to the imposition of an incremental displacement, $\Delta \mathbf{u}$. In direct terms of the nodal variables, instead of (3.90)

$$\Delta\varepsilon = \frac{1}{4\alpha_o^2}\mathbf{c}(\mathbf{x}_m)^T\Delta\mathbf{p} = \frac{1}{4\alpha_o^2}\mathbf{c}(\mathbf{x}' + \tfrac{1}{2}\Delta\mathbf{p})^T\Delta\mathbf{p} = \mathbf{b}_1(\mathbf{x}_m)^T\Delta\mathbf{p} \tag{3.152}$$

where $\mathbf{c}(\mathbf{x}_m)$ is a 'mid-point geometric vector', given by

$$\mathbf{c}(\mathbf{x}_m)^T = (-x'_{21} - \tfrac{1}{2}\Delta u_{21}, x'_{21} + \tfrac{1}{2}\Delta u_{21}, -z'_{21} - \tfrac{1}{2}\Delta w_{21}, z'_{21} + \tfrac{1}{2}\Delta w_{21}) \tag{3.153}$$

and the $x'$- and $z'$-coordinates in (3.153) relate to the updated coordinates at the *beginning* of the increment. Equation (3.152) is of the form of a 'linear strain increment' (i.e. similar to $\mathbf{b}_1^T\Delta\mathbf{p}$ in (3.58)), yet it is exact. A similar approach can be applied to the other strain measures so that, in relation to (3.103),

$$\Delta\varepsilon_E \simeq \frac{1}{\alpha_o\alpha_m}\frac{d(\mathbf{r}_n + \tfrac{1}{2}\Delta\mathbf{u})^T}{d\xi}\frac{d\Delta\mathbf{u}}{d\xi} = \frac{1}{4\alpha_o\alpha_m}\mathbf{c}(\mathbf{x}_m)^T\Delta\mathbf{p} = \frac{1}{2\alpha_o}\mathbf{e}_m^T\Delta\mathbf{p} \tag{3.154}$$

where (Figure 3.10), $\mathbf{e}_m$ is a unit vector relating to the mid-point configuration. In relation to the log-strain relationship of (3.116),

$$\Delta\varepsilon_L \simeq \frac{1}{\alpha_m^2}\frac{d(\mathbf{r}_n + \tfrac{1}{2}\Delta\mathbf{u})^T}{d\xi}\frac{d\Delta\mathbf{u}}{d\xi} = \frac{1}{4\alpha_m^2}\mathbf{c}(\mathbf{x}_m)^T\Delta\mathbf{p} = \frac{1}{2\alpha_m}\mathbf{e}_m^T\Delta\mathbf{p}. \tag{3.155}$$

In contrast to (3.152), which is exact, (3.154) and (3.155), which involve the mid-point

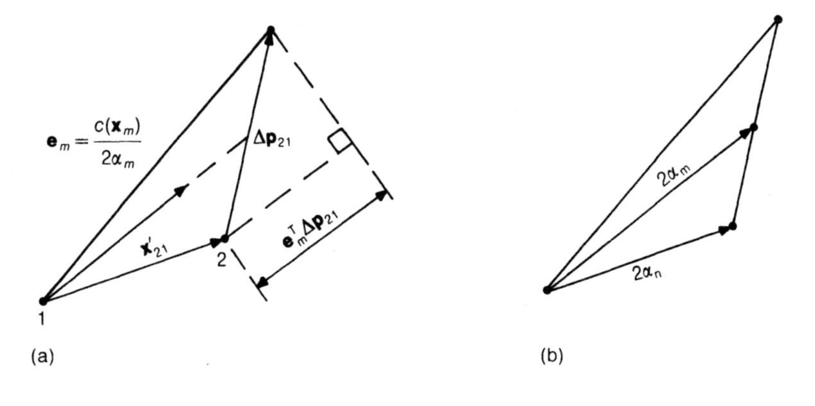


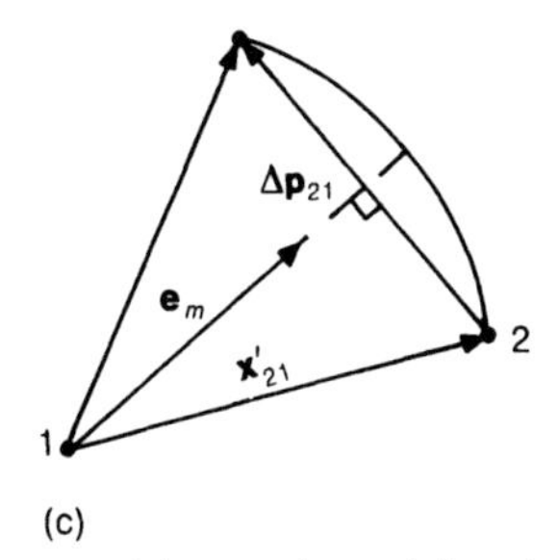


**Figure 3.10** Mid-point incremental procedures: (a) vectors and displacements; (b) lengths; (c) under a rigid-body rotation.

'length parameter' $\alpha_m$, are approximate. For example, Figure 3.11 illustrates the approximations inherent in using (3.155) to integrate the simple stretching of a bar. For this illustration, the increment has been assumed to start from the initial configuration with length $l_o = 2\alpha_o$, although in general a succession of mid-point incremental approximations would be used. Nonetheless, (3.154) and (3.155) are easy to compute since they are in the form of simple 'linear strain terms'. In addition, they

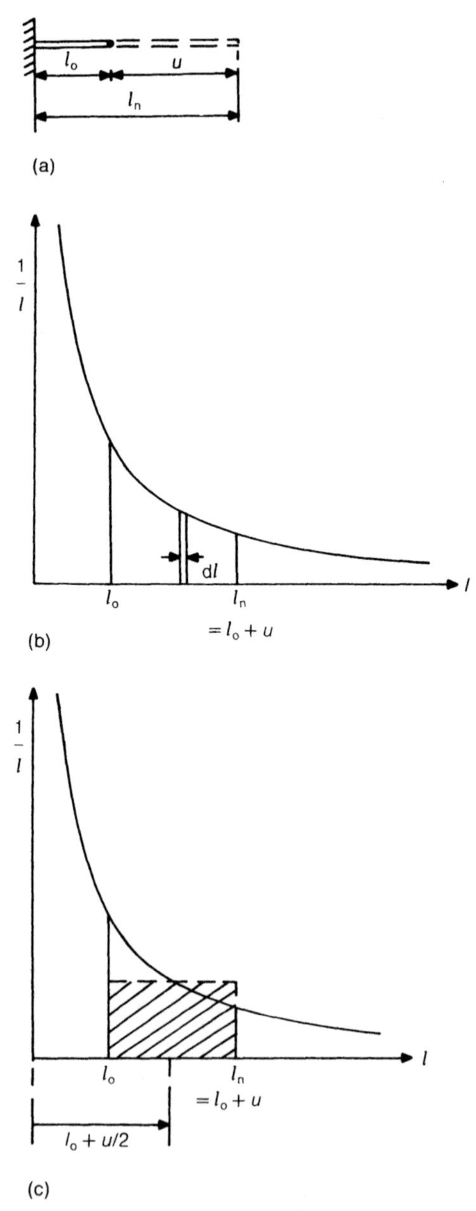


**Figure 3.11** Mid-point procedure for bar under uniaxial load: (a) Bar; (b) Exact integration of $\int_{l_o}^{l_n}(dl/l = \log_e(l_n/l_o)$ (c) Approx. integration of $\int_{l_o}^{l_n}(dl/l) = u/(l_o + u/2)$

give zero strain when the incremental displacements $\Delta\mathbf{p}$ relate to a rigid-body rotation (Figure 3.10(c)).

For the present truss elements, there is little to be gained by using an 'incremental mid-point algorithm' for updating the strains, since an exact solution using total strains can easily be obtained. However, the mid-point incremental algorithms can be very useful for more complex structures such as shells or continua subject to large strains.

## 3.9 FORTRAN SUBROUTINES FOR GENERAL TRUSS ELEMENTS

In Chapter 2, we gave Fortran computer programs for the analysis of shallow trusses. These programs were designed so that they could also be used for the deep truss elements of the present chapter, provided a set of new subroutines are used. These subroutines are given below.

### 3.9.1 Subroutine ELEMENT

The following subroutine should be used in place of the subroutine ELEMENT of Section 2.2.1

```
      SUBROUTINE ELEMENT (FI,AKT,AN,X,Z,P,E,ARA,ALO,IWRIT,IWR,IMOD,
     1   ITY,ALN,ARN)
C
C       FOR GENERAL TRUSS ELEMENT
C       IMOD=1 COMPUTES INT. LD. VECT. FI
C       IMOD=2 COMPUTES TAN. STIFF. AKT
C       IMOD=3 COMPUTES BOTH
C
C       AN=INPUT TOTAL FORCE IN BAR
C       Z=INPUT=Z COORD VECTOR; X=INPUT=X COORDS
C       P=INPUT=TOTAL DISP. VECTOR
C       ALO=INPUT=ORIGINAL LENGTH OF ELEMENT; ALN(IN)=NEW LENGTH
C       E=INPUT=YOUNG'S MOD: ARN=INPUT=CURRENT AREA
C
C       ITY=1, GREEN:  =2, ENG.,  =3, LOG:  =4 LOG WITH VOL CHANGE
C
C       IF IWRIT.NE.0 WRITES OUT FI AND/OR AKT ON CHANNEL IWR
C
        IMPLICIT DOUBLE PRECISION(A-H,O-Z)
        DIMENSION AKT(4,4),FI(4),Z(2),P(4),X(2),C(4)
C
        IF (ITY.EQ.3) POISS=0.5DO
        ALAM=ALO/ALN
        X21D=X(2)-X(1)+P(2)-P(1)
        Z21D=Z(2)-Z(1)+P(4)-P(3)
C
        IF (IMOD.NE.2) THEN
```

```
C       COMPUTES INT. FORCE VECTOR
C       SEE (3.92) FOR GREEN; (3.107) FOR ROTATED ENGNG.,
C       (3.118) FOR LOG-STRAIN
        CON=AN/ALO
        IF (ITY.GT.1) CON=ALAM*CON
        FI(1)=-CON*X21D
        FI(2)=-FI(1)
        FI(3)=-CON*Z21D
        FI(4)=-FI(3)
        IF (IWRIT.NE.0) THEN
        WRITE (IWR, 1000) (FI(I),I=1,4)
 1000   FORMAT(/,1X,'INT. FORCE VECT. FOR TRUSS EL IS ',/,1X,4G13.5)
        ENDIF
C
        ENDIF
C
        IF (IMOD.NE.1) THEN
C       COMPUTES TAN STIFF. MATRIX (UPPER TRIANGLE)
C       SEE (3.92) FOR C
        C(1)=-X21D
        C(2)=X21D
        C(3)=-Z21D
        C(4)=Z21D
        EA=E*ARN
        CON1=1./(ALO**3)
        IF (ITY.EQ.1) CON1=EA*CON1
        IF (ITY.EQ.2) CON1=ALAM*ALAM*CON1*(EA-AN*ALAM)
        IF (ITY.GE.3) CON1=CON1*ALAM**4*(EA-(1.0D0+0.5D0*POISS)*AN)
C       SEE (3.93) FOR GREEN, (3.111) AND (3.114) FOR ROTATED ENGNG.,
C       SEE (3.120) AND (3.122) FOR LOG-STRAIN
        DO 3 I=1,4
        DO 3 J=I,4
    3   AKT(I,J)=CON1*C(I)*C(J)
C
        CON2=AN/ALO
        IF (ITY.GE.2) CON2=CON2*ALAM
C       SEE (3.69) FOR GREEN, (3.113) FOR ROTATED ENGNG.,
C       (3.121) FOR LOG-STRAIN
        AKT(1,1)=AKT(1,1)+CON2
        AKT(1,2)=AKT(1,2)-CON2
        AKT(2,2)=AKT(2,2)+CON2
        AKT(3,3)=AKT(3,3)+CON2
        AKT(3,4)=AKT(3,4)-CON2
        AKT(4,4)=AKT(4,4)+CON2
C
        IF (IWRIT.NE.0) THEN
        WRITE (IWR,1001)
 1001   FORMAT(/,1X,'TAN. STIFF. MATRIX FOR TRUSS EL. IS',/)
        DO 14 I=1,4
   14   WRITE (IWR,67) (AKT(I,J),J=1,4)
   67   FORMAT(1X,7G13.5)
        ENDIF
```

```
C
      ENDIF

      RETURN
      END
```

### 3.9.2 Subroutine INPUT

This subroutine should be used in place of subroutine INPUT of Section 2.2.2. In addition to the previous data which relates to the bar–spring system of Figure 2.2, this routine inputs the type of strain measure (via the parameter ITYE). Also, in contrast to the work of the previous section, the initial area of the element, $A_o$, is required as well as the $E$-value, $E$. Also, Poisson's ratio is required although, in practice, it is not used unless the specified type of non-linearity is 'log-strain with volume changes' (ITYE = 4). (Solutions obtained using the log strain without volume changes (ITYE = 3) should be the same as those obtained with $\nu = 0.5$ and ITYE = 4.)

```
      SUBROUTINE INPUT(E,ARA,AL,QFI,X,Z,ANIT,IBC,IRE,IWR,AK14S,ID14S,
     1                 NDSP,NV,AK15,
     2                 POISS,ITYE)
C
C     READS INPUT FOR DEEP TRUSS ELEMENT
C
      IMPLICIT DOUBLE PRECISION (A-H,O-Z)
      DIMENSION X(2),Z(2),QFI(NV),IBC(NV),AK14S(4),ID14S(4)
C
      READ (IRE,*) NV,ITYE,E,ARA,POISS,ANIT
      WRITE (IWR,1000) NV,E,ARA,POISS,ANIT
 1000 FORMAT(/,1X,'NV=NO. OF VARBLS.=  ',I5,/,1X,
     1  'E=  ',G13.5,/,1X,
     2  'ARA=EL. INIT. AREA=  ',G13.5,/,1X,
     3  'POISS=  ',G13.5,/,1X,
     3 'ANIT=INIT. FORCE=  ',G13.5)
      WRITE (6,1101) ITYE
 1101 FORMAT(/,1X,'ELEMENT TYPE=  ',I5,/,
     1 3X,'=1, GREENS STRAIN',/,
     2 3X,'=2, ENGNG. STRAIN',/,
     3 3X,'=3, LOG STRAIN',/,
     4 3X,'=4, LOG STRAIN WITH VOLUME CHANGES')
      IF (NV.NE.4.AND.NV.NE.5) STOP 'INPUT 1000'
      READ (IRE,*) X(1),X(2)
      READ (IRE,*) Z(1),Z(2)
      WRITE (IWR,1001) X(1),X(2)
 1001 FORMAT(/,1X,'X CO-ORD OF NODE 1=  ',G13.5,1X,
     1  'X CO-ORD OF NODE 2=  ',G13.5)
      WRITE (IWR,1006) Z(1),Z(2)
 1006 FORMAT(/,1X,'Z CO-ORD OF NODE 1=  ',G13.5,1X,
     1  'Z CO-ORD OF NODE 2=  ',G13.5)
      AL=(X(2)-X(1))**2+(Z(2)-Z(1))**2
      AL=DSQRT(AL)
```

```
      READ (IRE,*) (QFI(I),I=1,NV)
      WRITE (IWR,1002) (QFI(I),I=1,NV)
 1002 FORMAT(/,1X,'FIXED LOAD OR DISP. VECTOR,QFI=  ',/,1X,5G13.5)
      WRITE (IWR,1008)
 1008 FORMAT(/,1X,'IF IBC(I)—SEE BELOW—=0, VARIABLE=A LOAD',/,1X,
     2    'IF IBC(I)—SEE BELOW—=-1, VARIABLE=A DISP.')
      READ (IRE,*) (IBC(I),I=1,NV)
      WRITE (IWR,1003) (IBC(I),I=1,NV)
 1003 FORMAT(/,1X,'BOUND. COND. COUNTER, IBC',/,1X,
     1    '=0, FREE:  =1, REST. TO ZERO:  =-1 REST. TO NON-ZERO',/,1X,
     2    5I5)
      READ (IRE,*) NDSP
      IF (NDSP.NE.0) THEN
      READ (IRE,*) (ID14S(I),I=1,NDSP)
      READ (IRE,*) (AK14S(I),I=1,NDSP)
      DO 40 I=1,NDSP
      WRITE (IWR,1004) AK14S(I),ID14S(I)
 1004 FORMAT(/1X,'LINEAR SPRING OF STIFFNESS ',G13.5,/,1X,
     1   'ADDED AT VAR. NO. ',I5)
   40 CONTINUE
      ENDIF
C
      IF (NV.EQ.5) THEN
      READ (IRE,*) AK15
      WRITE (IWR,1005) AK15
 1005 FORMAT(/,1X,'LINEAR SPRING BETWEEN VARBLS. 1 AND 5 OF STIFF  ',
     1    G13.5)
      ENDIF
C
      RETURN
      END
```

### 3.9.3 **Subroutine** FORCE

This subroutine should be used in place of subroutine FORCE of Section 2.2.3. It not only computes the force in the bar but also the new area ($A_n$ = ARN) although the latter should only differ from the original area, $A_o$, for the log-strain measures (ITYE = 3 or 4).

```
      SUBROUTINE FORCE(AN,ANIT,E,ARO,ALO,X,Z,P,IWRIT,IWR,
     1  ITY,ARN,ALN,POISS)
C
C       FOR GENERAL TRUSS ELEMENT, COMPUTES:
C                A) INTERNAL FORCE,AN
C                B) NEW LENGTH OF ELEMENT, ALN
C                C) NEW AREA OF ELEMENT, ARN
C       INPUTS:
C                E=YOUNG'S MOD, ARO=ORIGINAL AREA, ALO=ORIG. LENGTH
C                POISS=POISSON'S RATIO, Z=Z-COORDS, X=X-COORDS,
C                P=TOTAL DISPS., IWRIT=WRITE CONTROL, IWR=WRITE CHANNEL
```

```
C             ITY=1, GREEN'S STRAIN,  =2 ENGNG. STRAIN,  =3 LOG-STRAIN
C                =4 LOG-STRAIN WITH VOLUME CHANGE
C
      IMPLICIT DOUBLE PRECISION(A-H,O-Z)
      DIMENSION Z(2),P(4),X(2),B1(4)
C
C     COMPUTES NEW LENGTH
      X21D=X(2)-X(1)+P(2)-P(1)
      Z21D=Z(2)-Z(1)+P(4)-P(3)
      ALN=X21D*X21D+Z21D*Z21D
      ALN=SQRT(ALN)
C
      IF (ITY.LE.2) THEN
C     GREEN OR ENG. STRAIN
      X21=X(2)-X(1)
      Z21=Z(2)-Z(1)
      U21=P(2)-P(1)
      W21=P(4)-P(3)
      ALO2=ALO*ALO
C     SEE (3.56) FOR B1
      B1(1)=-X21/ALO2
      B1(2)=-B1(1)
      B1(3)=-Z21/ALO2
      B1(4)=-B1(3)
      EGR=0.D0
C     LINEAR PART OF GREEN STRAIN (SEE (3.55))
      DO 1 I=1,4
    1 EGR=EGR+B1(I)*P(I)
C     ADD-IN NON-LIN PART (SEE (3.55) OR (3.54))
      EGR=EGR+0.5*(U21*U21+W21*W21)/ALO2
      EST=EGR
C     SEE (3.115) FOR ROTATED ENGNG. STRAIN
      IF (ITY.EQ.2) EST=2.*ALO*EGR/(ALN+ALO)
C
      ELSE
C     LOG-STRAIN (SEE (3.116))
      EST=ALOG(ALN/ALO)
      ENDIF
C
      ARN=ARO
      IF (ITY.EQ.4) THEN
C     ALLOWS FOR VOLUME CHANGE (SEE (3.23))
      POW=2.DO*POISS
      RAT=ALO/ALN
      ARN=ARO*RAT**POW
      ENDIF
C
      AN=ANIT+E*ARN*EST
C
      IF (IWRIT.NE.0) WRITE (IWR,1000) AN,ALN,ARN
 1000 FORMAT(/,1X,'AXIAL FORCE AN=  ',G13.5,/,
```

```
     1          1X,'NEW LENGTH, ALN=  ',G13.4,/,
     2          1X,'NEW AREA, ARN=  ',G13.4)
      RETURN
      END
C
C
```

## 3.10 PROBLEMS FOR ANALYSIS

The following problems mainly relate to the NAFEMS (National Agency of Finite Elements) tests [C1.2,D1.2]. The problem numbers will be related to those used in the latter document. i.e. 3.10.4 will refer to NAFEMS Example 4. Whenever exact solutions are given, they will relate to the 'rotated engineering strain'. Apart from the problems headed 'large strain', there should be little difference between the solutions obtained with the different strain measures. For a number of problems, although the exact governing equations are given, the detail of their solution, which involved the use of Laguerre's method, is not included here. Full details are given in [C1.2] as are tabulated solutions including both the primary and secondary equilibrium paths. Similar responses related to a simple structure can be found in [P1].

### 3.10.1 Bar under uniaxial load (large strain)

This is the problem previously discussed in Section 3.2 and defined in (3.39) and Figure 3.3. The responses are as shown in Figure 3.4.

The following data relates to a solution using Green's strain for the compressive regime. It is obtained using displacement control so that no structural equations are solved.

4 1 500000. 100. 0. 0. ; NV,ITYE(Green),E,ARA,POIS,ANIT,
0. 2500. ; $x$-coords.
0. 0.00 : $z$-coords.
0. −1000. 0. 0. ; fixed displ. vector
1 −1 1 1 ; Bdry condn. code.
0 ; no earthed springs
0.2 6 0 ; load inc. factor, no. of incs., write control
0.001 1 ; convergence tol., iteration. type (N–R)

### 3.10.2 Rotating bar

This problem has previously been used in Chapters 1 and 2 and involves the configuration of Figure 3.12(a) with $K_{s4} = 0$ and a negative loading $q_4$.

#### *3.10.2.1 Deep truss (large strains) (Example 2.1)*

This is the problem previously discussed in Section 3.1.5 and defined in (3.31) and Figure 3.12(a). The responses are as shown in Figure 3.2.

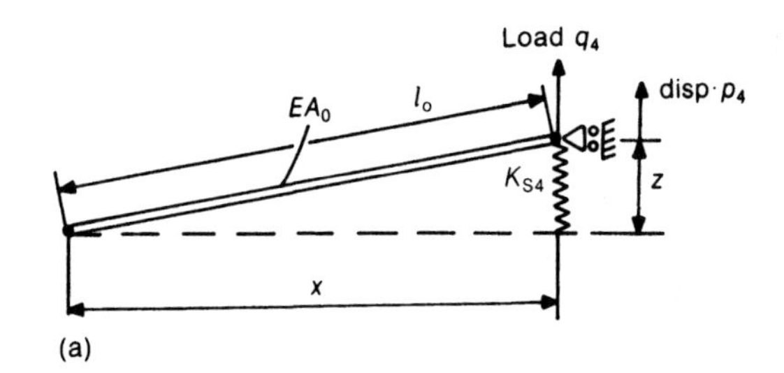


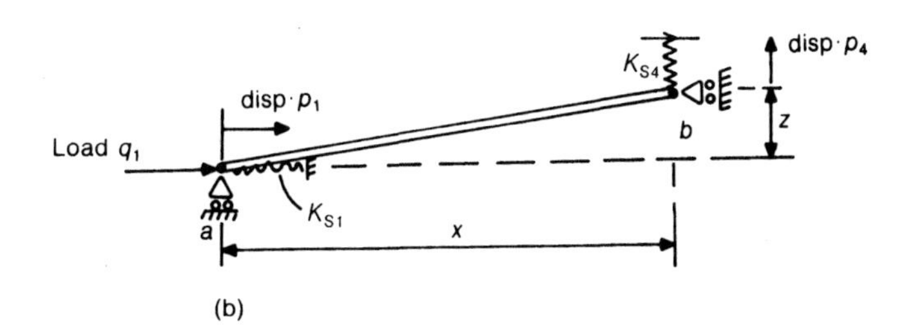


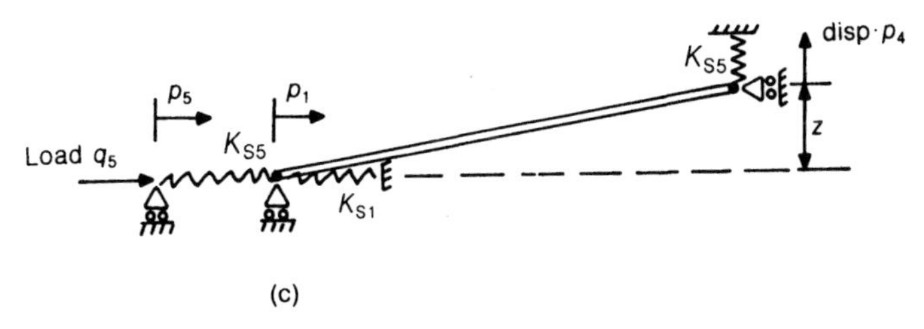


**Figure 3.12** Configurations for bar–spring problems: (a) single degree of freedom; (b) two degrees of freedom; (c) three degrees of freedom.

Data for a displacement-controlled solution using a rotated log-strain with Poisson's ratio of 0.5 (constant volume) is given below:

4 4 500000. 100. 0.5 0.0 ; NV,ITYE=Log,E,ARA,POIS,ANIT,
0. 2500. ; *x*-coords.
0. 2500. ; *z*-coords.
0. 0. 0. −1000. ; fixed displ. vector
1 1 1 −1 ; Bdry. condn. code
0 ; no earthed springs
0.5 12 0 ; Load inc. factor, no. of incs., write control
0.001 1 ; convergence tol., iterative type (N–R)

*3.10.2.2 Shallow truss (small strains) (Example 2.2)*

This problem is identical to that of 3.10.2.1 apart from the provision of a lower 'eccentricity' so that $z$ of Figure 3.12(a) is 25. With such an eccentricity, the response will be close to that of the shallow truss of Sections 1.1 (Figure 1.2) and 2.6.2. In contrast to the shallow solution of ((1.11), the exact load/deflection relationship, assuming a rotated engineering strain, is ([C1.2])

$$q_4 = \frac{EA_o x}{l_o}\left(\frac{(1+\alpha^2)^{1/2}}{(1+(\bar{p}_4+\alpha)^2)^{1/2}} - 1\right) - K_{s4} x \bar{p}_4 \tag{3.156}$$

where

$$\alpha = z/x \tag{3.157}$$

and the symbols are from Figure 3.12(a)) while the bar implies that the variable has been non-dimensionalised with respect to $x$ (Figure 3.12(a)) i.e. $\bar{p}_4 = p_4/x$. Also, for the current problem, $K_{s4} = 0$. This spring term has been included in order to provide the solution for the following problem.

The following data relates to a load-controlled solution using the modified Newton–Raphson method, for which the solutions are the points shown in Figure 3.13.

4 2 50000000. 1. 0. 0. ; NV, ITYE (rot. eng.), E, ARA, POIS, ANIT
0. 2500. ; $x$-coords.
0. 25. ; $z$-coords.
0. 0. 0. −1.0 ; load of −1.0 at variable 4 (vertical at node 2)
1 1 1 0 ; only variable 4 is free
0 ; no earthed springs
1.9 6 0 ; Load inc. factor, no. of incs., write control
0.001 2 ; Convergence tol., Iterative type (mN–R)

Table 3.1 compares the iterative performance of the mN–R solution with that obtained (by changing the 2 to a 1 in the last line of the previous data) for the N–R method.

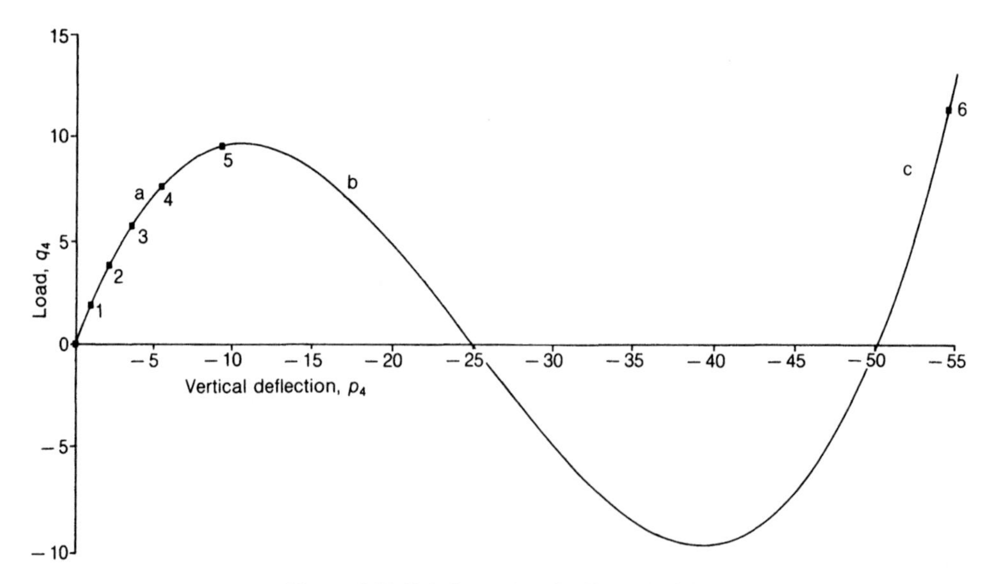


**Figure 3.13** Solution points for Example 2.2.

**Table 3.1** Iterative performance for Problem 3.10.2.2 (see Figures 3.12(a) and 3.13) ($\Delta q_4 = -1.9$).

| Method | Iterations at load step 1 | 2 | 3 | 4 | 5 | 6 |
|---|---|---|---|---|---|---|
| mN–R | 2 | 2 | 3 | 3 | 12 | fail |
| N–R | 1 | 1 | 1 | 2 | 3 | fail |

Although the full N–R method gave a better performance than the mN–R method, it was unable to take the jump from point 5 to point 6. For this trivial, one-dimensional problem, the solution could be obtained by displacement-control. Other methods will be discussed in Chapter 9.

### 3.10.3 Hardening problem with one variable (Example 3)

For this problem (Figure 3.12(a)), a linear spring, $K_{s4} = 1.125$ has been added so that the response is continuously hardening although with a softening and then a stiffening region. The response will be a little stiffer than that shown in Figure 1.2 for $K_s = EAz^2/2l^3$. (Here, $K_s \simeq 1.125EAz^2/2l^3$.) The load/deflection response is governed by Equation (3.156).

The following data relates to a load-controlled solution using the full N–R method and produced the points on Figure 3.14.

```
4  2  50000000. 1. 0. 0. ; NV, ITYE (rot. eng.), E, ARA, POIS, ANIT
0. 2500. ; x-coords.
0. 25.   ; z-coords.
```

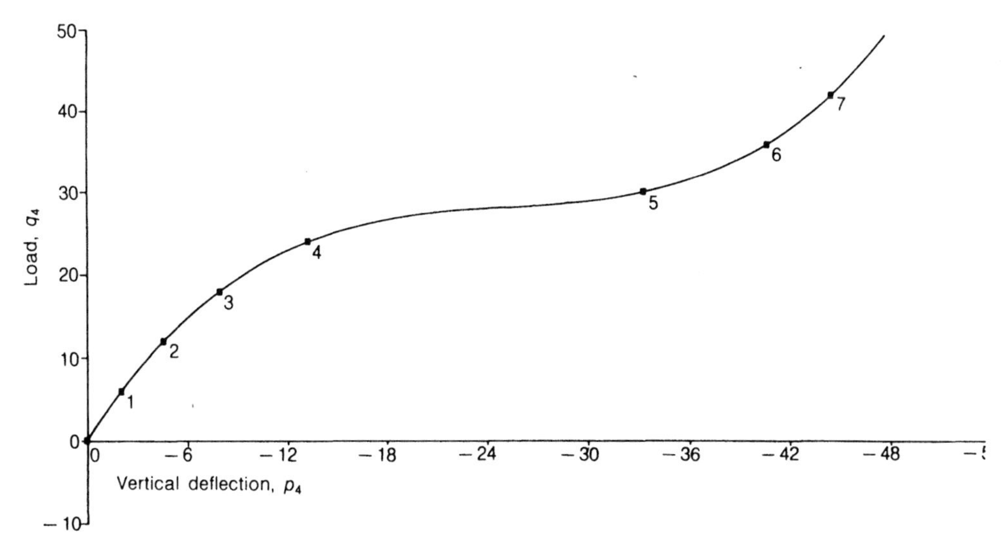


**Figure 3.14** Solution points for Example 3.

**Table 3.2** Iterative performance for Problem 3.10.3 (see Figures 3.12(a) and 3.14) ($\Delta q_4 = -6$).

| Method | Iterations at load step 1 | 2 | 3 | 4 | 5 | 6 | 7 |
|---|---|---|---|---|---|---|---|
| mN–R | 3 | 3 | 3 | 5 | 10 | fail | |
| N–R | 1 | 1 | 2 | 2 | 2 | 3 | 2 |

0. 0. 0. −1.0 ; load of −1 at variable 4 (vertical at node 2)
1 1 1 0 ; only variable 4 is free
1 one earthed spring
4 at variable 4
1.125 ; of magnitude 1.125
6. 7 0 ; Load inc. factor, no. of incs., write control
0.001 1 ; Convergence tol., Iterative type (N–R)

Table 3.2 compares the iterative performance of the full N–R solution with that obtained (by changing the 1 to a 2 in the last line of the previous data) for the modified N–R method.

### 3.10.4 Bifurcation problem (Example 4)

This problem is very similar to that discussed in Section 1.3 and further in Section 2.6.3. The configuration is that shown in Figure 3.12(b) with $z = 0$ and

$$EA_o = 5 \times 10^7, \qquad x = 2500. \tag{3.158}$$

In addition, $K_{s4} = 1.5$ so that, from (1.68), the buckling load is 3750.

In reality, the length term $l$ in (1.67) and (1.68) should be the current, rather than the original length. Hence

$$U_{cr} = q_{1cr} = l_o K_{s4}\left(1 - \frac{q_{1cr}}{EA_o}\right) \tag{3.159}$$

from which

$$q_{1cr} = l_o K_{s4}\left(1 + \frac{K_{s4} l_o}{EA_o}\right)^{-1} = l_o K_{s4}\left(1 + \frac{K_{s4}}{K_{sb}}\right)^{-1} \simeq l_o K_{s4} \tag{3.160}$$

where we have introduced the notation

$$K_{sb} = \frac{EA_o}{l_o} \tag{3.161}$$

to represent the stiffness of the rotating bar elements in Figures 3.12(a)–(c). (The approximation sign in (3.160) relates to the configurations used for these examples.)

Figure 1.11 plotted the fundamental and post-buckling paths for a perfect shallow truss. Figure 3.15 plots the equivalent solutions for a rotated engineering strain. In contrast to the shallow formulation, the current formulation leads to a 'falling' or

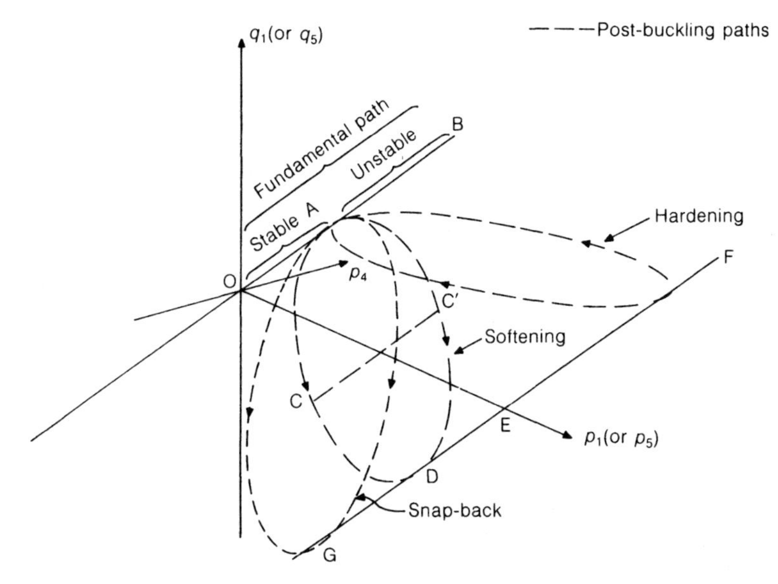


**Figure 3.15** Load/deflection relationships for 'perfect' bar–spring system.

'softening' post-buckling path. However, this path can be made to harden by the addition of the spring $K_{s1}$ in Figure 3.12(b). In addition the response (now at variable 5—see Figure 3.12(c)) can be made to 'snap back' by adding a spring $K_{s5}$ so that the problem has three variables. It can be shown that the critical buckling load relating to Figure 3.12(b) (with $z = 0$ and $x = l_o$) is

$$q_{1\text{cr}} = q_{5\text{cr}} = l_o K_{s4} \frac{(K_{sb} + K_{s1})}{(K_{sb} + K_{s4})} \simeq l_o K_{s4} \tag{3.162}$$

where the approximation again relates to the configurations adopted for the current examples. Equation (3.162) also applies to $q_{5\text{cr}}$, the critical buckling load for the perfect form of the configuration in Figure 3.12(c) (with $z = 0$ so that $x = l_o$).

Without the provision for post-buckling analysis (see Volume 2), the computer program will only be able to follow the basic fundamental path OAB of Figure 3.15. However, as discussed in Sections 1.3 and 2.6.3, the instability of the solution beyond the bifurcation point (point A, Figure 3.15) should be apparent from the presence of a negative pivot following the $\mathbf{LDL}^T$ factorisation of the tangent stiffness matrix for equilibrium points on the portion AB of Figure 3.15.

While the current computer program may be unable to trace the post-buckling paths such as ACD or AC'D (this would only be indirectly possible by adopting a very small 'imperfection', $z$ or a small destabilising lateral force at $q_4$), it is nonetheless worth briefly discussing the alternative paths in Figure 3.15. In particular, we will concentrate on the system of Figure 3.12(a) (with $z = 0$) with $K_{s1} = 0$, for which the solution under displacement control (with monotonically increasing $p_1$ in Figures 3.12(b) and 3.13) may be assumed to follow the path OA until bifurcation, at which point it may follow either AC or AC'. The bar then follows the configurations illustrated in Figure 3.16. Assuming that CC' is horizontal and has $P = 0$, point C in

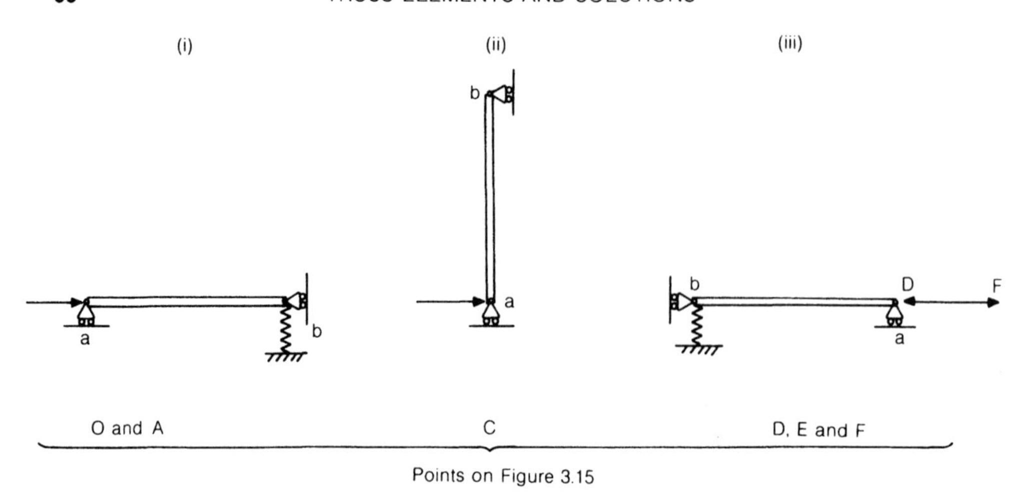


**Figure 3.16** Bar–spring configurations in relation to the 'softening' solution in Figure 3.15.

Figure 3.15 relates to the configuration (ii) in Figure 3.16 with 'b' lying vertically above 'a' and so that there is no force in the bar and no loading, $q_5$. However, at this stage there is a considerable force in the compressed spring $K_{s4}$, so that when the deflection, $p_1$, is taken beyond this configuration (with point 'a' of Figure 3.12(b) now to the right of point 'b'), a compressive stress is required in the bar element and hence a negative force, $q_5$. The equilibrium path therefore follows CD in Figure 3.15. At point D, points 'b' and 'a' (Figure 3.12(b)) now lie on a horizontal line but with point 'a' to the right of point 'b'. This configuration is illustrated in Figure 3.16(iii). From this stage onwards, $q_5$ is stretching the bar. Up to point E in Figure 3.15, this stretching is merely reducing the compression already in the bar while from point E to point F the bar is pulled into tension.

### 3.10.5 Limit point with two variables (Example 5)

An 'imperfection' is added to the previous example by setting $z = 25$ so that the bifurcation is eliminated. Assuming shallow truss theory, the response for this structure is that previously described in Section 1.3.1 and illustrated in Figure 1.11. When the theory is extended beyond the shallow truss assumptions, the response becomes that illustrated by the dotted lines in Figure 3.17, where the primary imperfect path is the 'true path' that should be followed by an increasing shortening ($p_1$) and the secondary (or complementary) imperfect path is the equivalent alternative equilibrium state to that discussed in Section 1.3.1 and reached when large increments were adopted in Section 2.6.4.3. For the current deep truss theory, the primary imperfect path has a limit point beyond which the load $q_1$ reduces.

Using Pythagoras's theorem, from Figure 3.12(b) in conjunction with (3.157) and (3.161), the force in the bar is given by

$$N = K_{sb}(((\bar{p}_4 + \alpha)^2 + (1 - \bar{p}_1)^2)^{1/2} - (1 + \alpha^2)^{1/2})x = K_{sb}(\varphi - (1 + \alpha^2)^{1/2})x \quad (3.163)$$

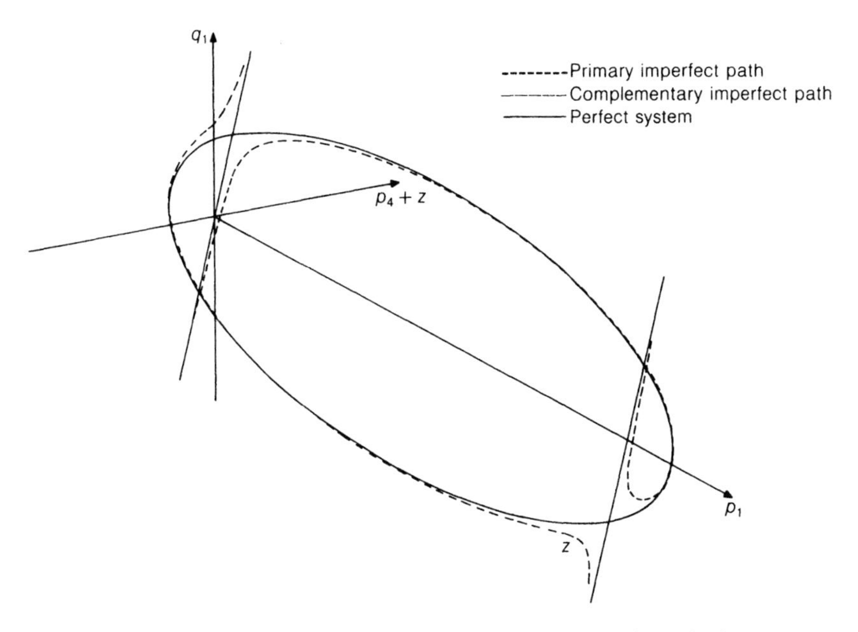


**Figure 3.17** Decomposition of perfect system due to an imperfection.

where the bar again indicates that the variable has been non-dimensionalised with respect to $x$ (Figure 3.12(b)). The exact equilibrium equations can be obtained from equilibrium by resolving vertically and horizontally. This will be shown in the next section, which involves a generalisation of the present problem.

For the NAFEMS problems, load control was firstly adopted with constant increments of $\Delta q_1 = 760$ and should have produced the points marked 1–5 on

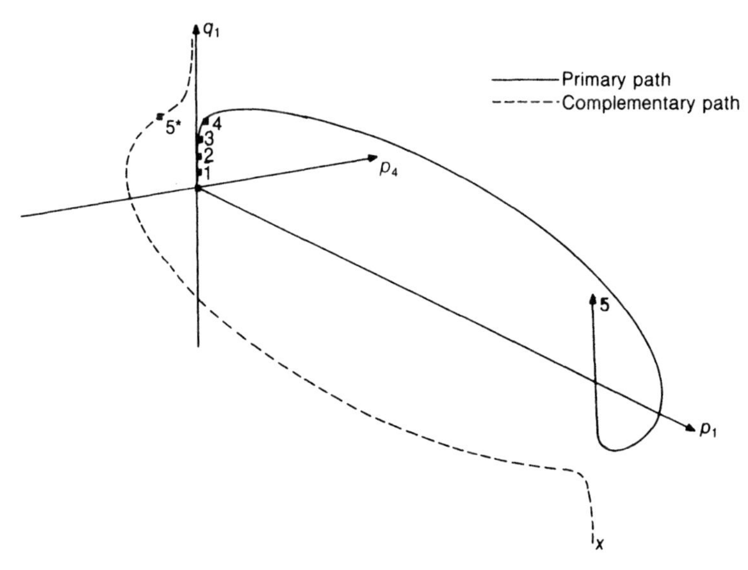


**Figure 3.18** Solution points for Example 5.

**Table 3.3** Iterative performance for Problem 3.10.5 (see Figures 3.12(b) and 3.18) ($\Delta q_1 = 760$).

| | Iterations at load step | | | | | |
|---|---|---|---|---|---|---|
| **Method** | **1** | **2** | **3** | **4** | **5** | **6** |
| mN–R | 5 | 8 | fail | | | |
| N–R | 2 | 2 | 3 | 3 | 7* | 5* |

*Solution converged onto secondary imperfect path.

Figure 3.18. In practice, the modified Newton–Raphson method was successful for only the first two increments, while the full Newton–Raphson method easily achieved solutions for steps 1–4 (see Table 3.3). The data for the latter solution is given below.

```
4 2 50000000. 1. 0.0 0.0 ; NV, ITYE (rot. eng.), E, ARA, POIS, ANIT
0. 2500.  ; x-coords.
0. 25.    ; z-coords.
760. 0. 0. 0. ; fixed load vector
0  1  1  0 ; Bdry. condn. code
1  ; one earthed spring
4  ; at Variable 4
1.5 ; of magnitude 1.5
1.0 6 0 ; Load inc. factor, no. of incs., write control
0.001 1 ; Convergence tol., iterative type (N-R)
```

Knowing that Step 5 would take the solution to $q_1 = 3800$, while the critical buckling load for the perfect solution (see Section 3.10.4) was 3750, trouble could be anticipated on Step 5. Indeed, even with the full N–R method, convergence was not achieved using Green's strain while convergence to the 'wrong path' (point 5* in Figure 3.18) was obtained with the rotating engineering strain. Displacement control with large steps of $\Delta p_1 = 250$ gave satisfactory solutions with full N–R but failure with mN–R (from the very first step). Reducing the step size to $\Delta p_1 = 0.3$ gave converged solutions with either solution procedure. Further work on this example will be given in Section 9.9.5.

### 3.10.6 Hardening solution with two variables (Example 6)

For this problem, a spring $K_{s1} = 2.0$ was added so that the limit points were removed and the response was continuously hardening (Figure 3.15 and 3.19). Using a similar procedure to that of Section 1.3 (see equation (1.54)), we can resolve firstly horizontally and then vertically to obtain the equilibrium equations:

$$g_1 = -N\cos\theta + K_{s1}p_1 - q_1 = K_{sb}\frac{(1-\bar{p}_1)(\varphi - (1+\alpha^2)^{1/2})x}{\varphi} + K_{s1}\bar{p}_1 x - q_1 = 0 \quad (3.164)$$

$$g_4 = N\sin\theta + K_{s4}p_4 = K_{sb}\frac{(\bar{p}_4+\alpha)(\varphi - (1+\alpha^2)^{1/2})x}{\varphi} + K_{s4}\bar{p}_4 x = 0 \quad (3.165)$$

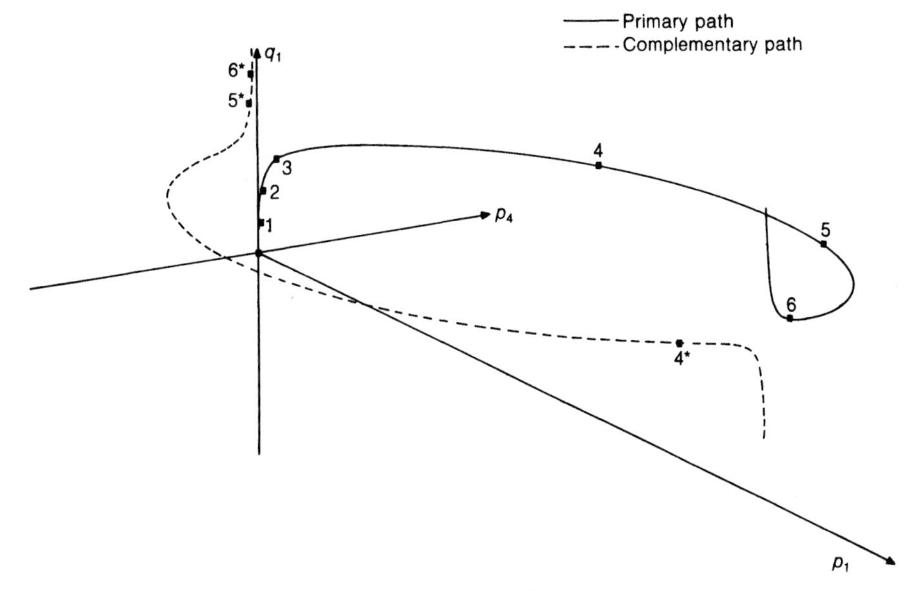


**Figure 3.19** Solution points for Example 6.

where $\theta$ relates to the displaced configuration (Figures 1.1 and 3.1) and use has been made of (3.162), (3.167) and (3.161).

With load control, and steps of $\Delta q_1 = 1100$, the mN–R method was only successful only for Step 1, while the full N–R method had difficulties from Step 4 onwards. With Green's strain, convergence was not achieved within a specified maximum of 12 iterations while with the rotated engineering strain, Steps 4–6 ended on the wrong equilibrium path (see Figure 3.19). The data for the latter solution is given below and the convergence characteristics in Table 3.4:

```
4 2 50000000. 1. 0.0 0.0 ; NV,ITYE=Engng.,E,ARA,POIS,ANIT
0. 2500.  ; x coords.
0. 25.    ; z coords.
1100. 0. 0. 0. ; Fixed load vector, loading at variable 1
0  1  1  0 ; Bdry condn. code; rest. at varbls. 2 and 3
2  ; Two earthed springs
1 4 ; At varbls. 1 and 4
2.0 1.5 ; of mag 2.0 and 1.5 respectively
1.0 6 0  ; Load inc. factor, no. of incs., write control
0.001 1 21 0 ; Convergence tol., iterative type (N–R)
```

If the 'imperfection', $z$, is reduced from 25 to 2.5, both strain measures will lead to solutions that swap branches onto the wrong equilibrium path from Step 4 onwards. With all these solutions in which such branch-swapping has occurred there is a potential warning in that a negative pivot is found on the $\mathbf{LDL}^T$ factorisation for points on the 'wrong' path. This warning will be output by the computer program (see Section 2.5). When applying displacement control with steps of $\Delta p_1 = 500$, the

**Table 3.4** Iterative performance for Problem 3.10.6 (see Figures 3.12(b) and 3.19) ($\Delta q_1 = 1100$).

| | Iterations at load step | | | | | |
|---|---|---|---|---|---|---|
| **Method** | **1** | **2** | **3** | **4** | **5** | **6** |
| mN–R | 8 | fail | | | | |
| N–R (Eng.) | 2 | 3 | 5 | 5* | 3* | 5* |
| N–R (Green) | 2 | 3 | 5 | fail | | |

*Solution converged onto secondary imperfect path.

mN–R method failed from the first step while the full N–R method gave satisfactory solutions.

### 3.10.7 Snap-back (Example 7)

By adding a third spring (Figure 3.12(c)) it is possible to create a 'snap-back'. The governing equilibrium equations are then given by (3.165) in conjunction with

$$g_1 = -N\cos\theta + K_{s1}p_1 + K_{s5}(p_1 - p_5)$$
$$= K_{sb}\frac{(1-\bar{p}_1)(\varphi - (1+\alpha^2)^{1/2})x}{\varphi} + K_{s1}\bar{p}_1 x - K_{s5}(\bar{p}_1 - \bar{p}_5)x = 0 \qquad (3.166)$$

$$g_5 = -K_{s5}(p_1 - p_5) - q_5 = -K_{s5}(\bar{p}_1 - \bar{p}_5)x - q_5 x = 0. \qquad (3.167)$$

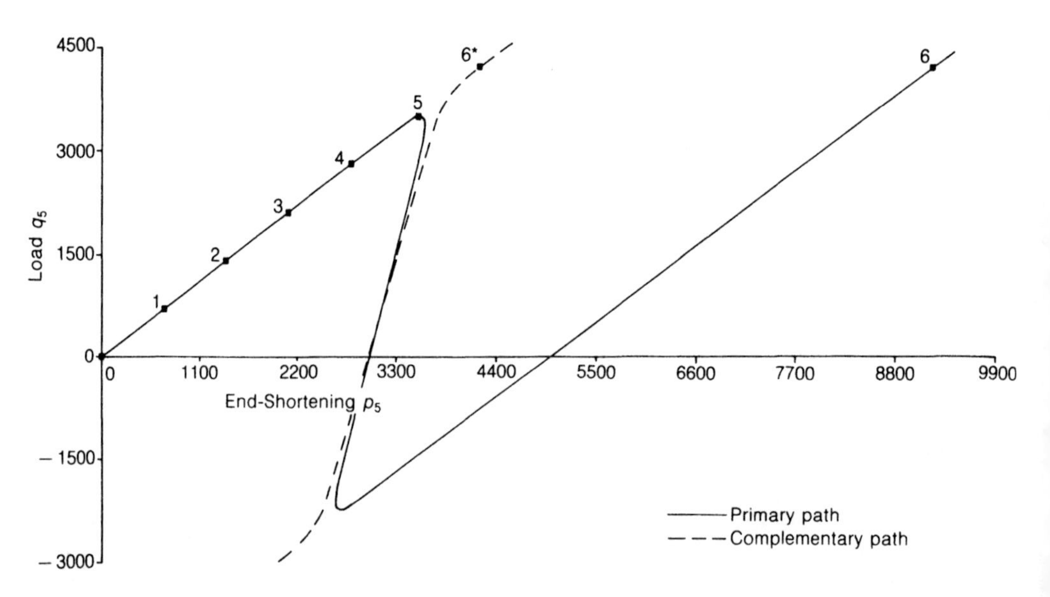


**Figure 3.20** Solution points for Example 7.

For the numerical example, the geometry was maintained as before with $z = 25$, but the adopted spring stiffnesses were

$$K_{s1} = 0.25, \qquad K_{s4} = 1.5, \qquad K_{s5} = 1.0. \tag{3.168}$$

From (3.162) the critical buckling load is again approximately 3750 (see Section 3.10.4). In this situation, the relationship between the load, $q_5$, and the shortening displacement, $p_5$, is that shown in Figures 3.15 and 3.20. Adopting load control, with increment of $\Delta q_5 = 700$, the solution should involve Steps 1–6 shown on that figure. However, the mN–R method achieved convergence only up to point 2 (see Table 3.4), while the full N–R method was successful for the first five steps but then either failed to converge within the specified maximum number of iterations (with Green's strain) or converged onto the wrong equilibrium path (point 6*—with rotated engineering strain). The data for the latter solution is given below.

```
5 2 50000000. 1. 0.0 0.0 ; NV,ITYE=Engng.,E,ARA,POIS,ANIT
0. 2500.  ; x coords.
0. 25.    ; z coords.
0. 0. 0. 0. 700. Fixed load vector, loading at variable 5
0  1  1  0  0 ; Bdry code; rest. at varbls. 2 and 3
2  ; Two earthed springs
1 4 ; At varbls. 1 and 4
0.25 1.5 ; of mag 0.25 and 1.5 respectively
1.0   ; linear spring of stiff 1.0 between variables 1 and 5
1.0  7  0 ; Load inc. factor, no. of incs., write control
0.001 1 ; Convergence tol., iterative type (N–R)
```

With hindsight, this is not surprising because between Steps 5 ($q_5 = 3500$) and Steps 6 ($q_5 = 4200$), the solution passes the critical buckling load of 3750. Displacement controlled solutions with $\Delta p_5 = 1000$ failed on Step 2 with mN–R. With full N–R, from Step 4 onwards, the Green's strain solutions again failed to converge within the specified maximum number of steps while the rotated engineering strain converged onto the wrong path. Cutting the step size with displacement control did not overcome the difficulties because of the snap-back. However, it will be shown in Chapter 9 that these problems can easily be overcome using a modified form of displacement control called the 'arc-length method'.

**Table 3.5** Iterative performance for Problem 3.10.7 (see Figures 3.12(c) and 3.20) ($\Delta q_5 = 700$).

| | Iterations at load step | | | | | | |
|---|---|---|---|---|---|---|---|
| **Method** | **1** | **2** | **3** | **4** | **5** | **6** | **7** |
| mN–R | 5 | 7 | fail | | | | |
| N–R (Eng.) | 2 | 2 | 3 | 4 | 5 | 9* | 3* |
| N–R (Green) | 2 | 2 | 3 | 4 | 9 | fail | |

*Solution converged onto secondary imperfect path.

## 3.11 SPECIAL NOTATION

$A$ = area of bar
$\mathbf{A}$ = matrix (3.57) or (3.141)
$\mathbf{b}_1$ = vector connecting linear part of $\boldsymbol{\varepsilon}$ to $\mathbf{p}$
$\mathbf{b}$ = vector connecting $\delta\boldsymbol{\varepsilon}$ to $\delta\mathbf{p}$; $\mathbf{b} = \mathbf{b}_1 + \mathbf{b}_2$
$c = \cos\theta$
$\mathbf{c}$ = vector (3.56) such that $\mathbf{c}(\mathbf{x}) = 4\mathbf{A}\mathbf{x}$ or $\mathbf{c}(\mathbf{p}) = 4\mathbf{A}\mathbf{p}$
$\mathbf{c}_l$ = vector (3.123)
$\mathbf{e}_1$ = unit vector lying along the truss element
$\mathbf{e}_2, \mathbf{e}_3$ = unit vectors orthogonal to $\mathbf{e}_1$
$\mathbf{e}_\mathrm{m}$ = unit vector relating to configuration at mid-increment
$\mathbf{F}$ = Boolean matrix (see (3.145))
$\mathbf{h}$ = shape function array
$\mathbf{h}_\xi$ = differential of $\mathbf{h}$ w.r.t. $\xi$
$\mathbf{H}$ = shape function matrix
$\mathbf{H}_\xi$ = differential of $\mathbf{H}$ w.r.t. $\xi$
$K_\mathrm{s}$ = spring stiffness
$K_\mathrm{sb}$ = stiffness of bar element (see (3.161)) (Section 3.10)
$N$ = axial force in bar
$\mathbf{n}$ = unit vector orthogonal to truss element
$\mathbf{p}$ = nodal displacement vector
ordering for 2-D truss is $\mathbf{p}^\mathrm{T} = (u_1, u_2, w_1, w_2)$
ordering for 3-D truss is $\mathbf{p}^\mathrm{T} = (u_1, u_2, v_1, v_2, w_1, w_2)$
$\mathbf{p}_{21}$ = vector such that $\mathbf{p}_{21}^\mathrm{T} = (u_{21}, w_{21})$ (or equivalent in 3-D)
$q$ = load at end of bar (Sections 3.1 and 3.2)
$\mathbf{r}$ = position vector: initial position vector = $\mathbf{r}_\mathrm{o}$
$\mathbf{r}_{\mathrm{o}\xi} = \dfrac{\mathrm{d}\mathbf{r}_\mathrm{o}}{\mathrm{d}\xi}$
$s = \sin\theta$
$\mathbf{T}$ = transformation matrix (see (3.126) or (3.144))
$u$ = axial ($x$-direction) displacement at end of bar (Section 3.2)
$\mathbf{u}$ = displacement vector such that $\mathbf{u}^\mathrm{T} = (u, w)$ (or, in 3-D—see (3.139))
$\mathbf{u}_\xi = \dfrac{\mathrm{d}\mathbf{u}}{\mathrm{d}\xi}$
$u_1, u_2$ = nodal displacements for bar in $x$-direction
$u_{21} = u_2 - u_1$
$w$ = vertical ($z$ direction) displacement at end of bar (Section 3.1)
$w_1, w_2$ = nodal displacements for bar in $z$-direction
$w_{21} = w_2 - w_1$
$x$ = horizontal length (Sections 3.1 and 3.10)
$x_1, x_2$ = nodal values of $x$
$x_{21} = x_2 - x_1$
$\mathbf{x} = \mathbf{x}_\mathrm{o}$ = vector of initial nodal coordinates (3.48)
$\mathbf{x}' = \mathbf{x}_\mathrm{n}$ = vector of current nodal coordinates (3.47)
$\mathbf{x}_{21}$ = vector (3.52) (or equivalent in 3-D)
$\mathbf{x}_\mathrm{m}$ = vector of coordinates at mid-increment (see (3.153))

$z$ = initial vertical offset at end of bar (Sections 3.1 and 3.10)
$Z = z + w$ (Section 3.1)
$z_1, z_2$ = nodal values of $z$
$x_{21} = z_2 - z_1$
$l_n$ = current length of bar
$l_o$ = initial length of bar
$\alpha$ = 'length parameter' (half-length of bar)
$\alpha = z/x$ (Section 3.10)
'$\sigma$' = 'true stress'
$\varepsilon$ = axial strain in bar
$\bar{\varepsilon}$ = engineering strain
$\lambda$ = length ratio (see (3.108))
$\theta$ = angular orientation of bar
$\varphi$ = non-dimensional constant (see (3.163))

### Subscripts

A = Almansi
E = engineering
G = Green
$l$ = local
L = log
o = old or original
n = new or current

### Superscripts

$-$ = quantity divided by $x$ (Section 3.10)

## 3.12 REFERENCES

[P1] Pecknold, D. A., Ghaboussi, J. & Healey, T. J., Snap-through and bifurcation in a simple structure, *J. Engng. Mech.*, **111** (7), 909–922 (1985).

# 4 Basic continuum mechanics

The previous chapters have all been related to one-dimensional problems. Before moving on to two- and three-dimensional problems, we will include a chapter on basic continuum mechanics. Many specialist texts are available either directly on continuum mechanics [H1, M1, M3, M4, S1, W2] or containing basic background information in relation to elasticity [T1, L1, G1] or mechanics of solids [B3, W1]. The present chapter will be far less rigorous and is intended to introduce sufficient continuum mechanics for the following chapters which concentrate on the finite element discretisation or solution procedures. It should also pave the way for some of the more advanced work in Volume 2.

This chapter starts by introducing both vector and tensor [Y1] notations for stress and strain. In the first instance, the 'tensor form' is introduced simply as a matrix. In the early sections, the distinction between the vector and tensor forms will be indicated by adding a subscript 2 (second order tensor) on the latter. This procedure will later be dropped as the distinction becomes obvious. Indicial notation is also introduced although for many developments such notation is completely avoided.

Sections 4.1, 4.2 and 4.4 provide the basis for the finite element work of Chapter 5 which uses the total Lagrangian formulations. Section 4.3 involves transformations between one coordinate system and another (not required for the main theme of Chapter 5) while Sections 4.5 and 4.6 introduces the Cauchy stress and hence are strictly required for updated Lagrangian formulations which are also considered in Chapter 5. Section 4.7 briefly discusses the relationship between the various stress and strain measures while Section 4.8 introduces the polar decomposition which is used in Section 4.9 to relate the Green and Almansi strains to the principal stretches and in Section 4.10 to give a simple explanation of the second Piola–Kirchhoff stresses. Section 4.11 gives a very brief overview of constitutive models with the aim that the finite element work of Chapter 5 should make sense in relation to concepts such as plasticity, although this subject will not be treated in detail until later (Chapter 6).

The reader wishing to get straight into finite element work with only the minimum continuum mechanics could try reading only Sections 4.1, 4.2 and 4.4.

## 4.1 STRESS AND STRAIN

We can either represent the stress at a point (Figure 4.1) by a vector.

$$\boldsymbol{\sigma}^{\mathrm{T}} = (\sigma_{xx}, \sigma_{yy}, \sigma_{zz}, \tau_{xy}, \tau_{xz}, \tau_{yz}) \tag{4.1}$$

or by a matrix

$$\boldsymbol{\sigma}_2 = \begin{bmatrix} \sigma_{xx} & \tau_{xy} & \tau_{xz} \\ \tau_{yx} & \sigma_{yy} & \tau_{yz} \\ \tau_{zx} & \tau_{zy} & \sigma_{zz} \end{bmatrix} = \begin{bmatrix} \sigma_{11} & \sigma_{12} & \sigma_{13} \\ \sigma_{21} & \sigma_{22} & \sigma_{23} \\ \sigma_{31} & \sigma_{32} & \sigma_{33} \end{bmatrix} \tag{4.2}$$

where the subscript 2 on $\boldsymbol{\sigma}$ is used to distinguish this two-dimensional representation from the one-dimensional vector representation of (4.1). The matrix, $\boldsymbol{\sigma}_2$, in (4.2) can be represented using either letters or numbers. In either case, there are only six independent quantities since, from rotational equilibrium, for example, $\sigma_{12} = \tau_{xy} = \sigma_{21} = \tau_{yx}$. The quantity $\boldsymbol{\sigma}_2$ in (4.2) is more than a matrix, it is a tensor which transforms to new axes according to certain laws [Y1, M1] which will be discussed in Section 4.3.

With the stresses being expressed by (4.1), the strains at the same point can be expressed as

$$\boldsymbol{\varepsilon}^{\mathrm{T}} = (\varepsilon_{xx}, \varepsilon_{yy}, \varepsilon_{zz}, \gamma_{xy}, \gamma_{xz}, \gamma_{yz}). \tag{4.3}$$

Alternatively, tensor notation can be adopted so that

$$\boldsymbol{\varepsilon}_2 = \begin{bmatrix} \varepsilon_{xx} & 0.5\gamma_{xy} & 0.5\gamma_{xz} \\ 0.5\gamma_{yx} & \varepsilon_{yy} & 0.5\gamma_{yz} \\ 0.5\gamma_{zx} & 0.5\gamma_{zy} & \varepsilon_{zz} \end{bmatrix} = \begin{bmatrix} \varepsilon_{xx} & \varepsilon_{xy} & \varepsilon_{xz} \\ \varepsilon_{yx} & \varepsilon_{yy} & \varepsilon_{yz} \\ \varepsilon_{zx} & \varepsilon_{zy} & \varepsilon_{zz} \end{bmatrix} = \begin{bmatrix} \varepsilon_{11} & \varepsilon_{12} & \varepsilon_{13} \\ \varepsilon_{21} & \varepsilon_{22} & \varepsilon_{23} \\ \varepsilon_{31} & \varepsilon_{32} & \varepsilon_{33} \end{bmatrix} \tag{4.4}$$

where again there are only six independent quantities and the tensor is symmetric.

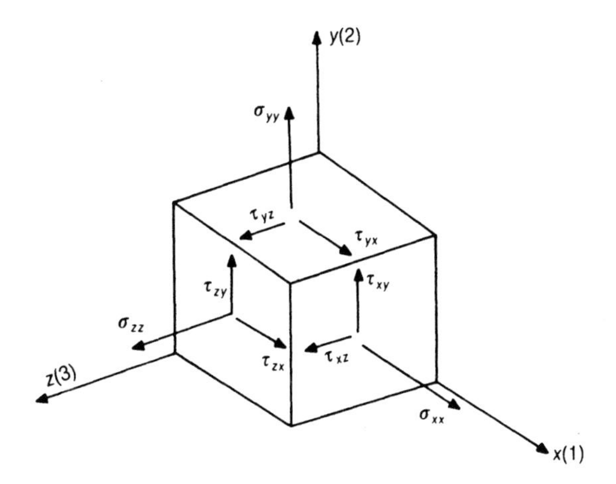


**Figure 4.1** The stresses.

For small strains,

$$\varepsilon_{xx} = \frac{\partial u}{\partial x}, \qquad \varepsilon_{yy} = \frac{\partial y}{\partial y}; \qquad \varepsilon_{zz} = \frac{\partial w}{\partial z}, \qquad \gamma_{xy} = \left(\frac{\partial u}{\partial y} + \frac{\partial v}{\partial x}\right),$$

$$\gamma_{xz} = \left(\frac{\partial u}{\partial z} + \frac{\partial w}{\partial x}\right), \qquad \gamma_{yz} = \left(\frac{\partial v}{\partial z} + \frac{\partial w}{\partial y}\right). \tag{4.5}$$

It should be noted that the tensor shear-strains in (4.4) are half the 'engineering' shear strains in (4.3). (The symbol $\gamma$ is used for the latter.) This ensures that both notations can be used to express the (linear) strain energy, $\varphi$, per unit volume via 'scalar products' so that

$$\varphi = \tfrac{1}{2}\boldsymbol{\sigma}^{\mathrm{T}}\boldsymbol{\varepsilon} = \tfrac{1}{2}\boldsymbol{\sigma}_2 : \boldsymbol{\varepsilon}_2 \tag{4.6}$$

where the $\boldsymbol{\sigma}^{\mathrm{T}}\boldsymbol{\varepsilon}$ involves the familiar dot (or inner product) and the contraction symbol: implies a similar scalar product with every term in $\boldsymbol{\sigma}_2$ being multiplied by its equivalent term in $\boldsymbol{\varepsilon}_2$, i.e. from (4.2) and (4.4):

$$\varphi = \tfrac{1}{2}(\sigma_{xx}\varepsilon_{xx} + 0.5\tau_{xy}\gamma_{xy} + 0.5\tau_{xz}\gamma_{xz} + 0.5\tau_{yx}\gamma_{yx} + \sigma_{yy}\varepsilon_{yy} + \cdots + \sigma_{zz}\varepsilon_{zz}). \tag{4.7}$$

The stress tensor notation allows the stresses to be very simply related to equilibrating external forces (Figure 4.2) via

$$\boldsymbol{\sigma}_2\mathbf{A} = \begin{bmatrix} \sigma_{xx} & \tau_{xy} & \tau_{xz} \\ \tau_{yx} & \sigma_{yy} & \tau_{yz} \\ \tau_{zx} & \tau_{zy} & \sigma_{zz} \end{bmatrix} \begin{bmatrix} A_x \\ A_y \\ A_z \end{bmatrix} = \mathbf{F} = \begin{bmatrix} F_x \\ F_y \\ F_z \end{bmatrix} = A \begin{bmatrix} t_x \\ t_y \\ t_z \end{bmatrix} \tag{4.8}$$

where $A_x$, $A_y$ and $A_z$ are the components of the area $\mathbf{A}$ (Figure 4.2) in the directions $x, y, z$. If the area $A$ is unity, (4.8) becomes

$$\boldsymbol{\sigma}_2\mathbf{n} = \mathbf{t} \tag{4.9}$$

where $\mathbf{n}$ is the unit normal vector (Figure 4.2) and $\mathbf{t}$ is the vector of equilibrating external tractions per unit area.

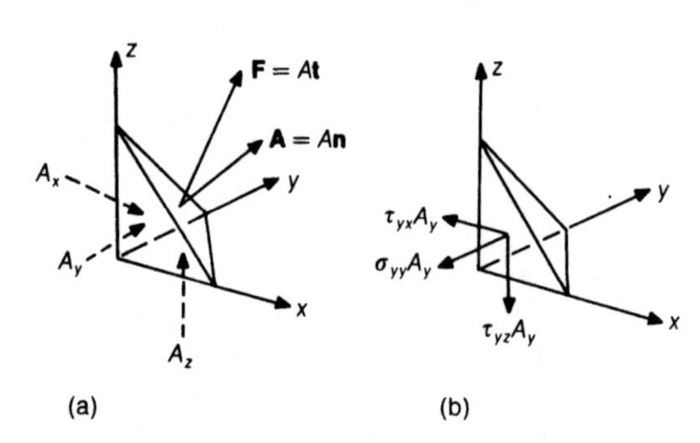


**Figure 4.2** Relationship between the external forces, $\mathbf{F} = \mathbf{At}$ and the stresses: (a) forces and areas; (b) stresses on face $A_y$.

## 4.2 STRESS–STRAIN RELATIONSHIPS

For a linear elastic, isotropic, material, the stresses and strains are related by

$$\begin{bmatrix} \sigma_x \\ \sigma_y \\ \sigma_z \\ \tau_{xy} \\ \tau_{xz} \\ \tau_{yz} \end{bmatrix} = \frac{E}{(1+\nu)(1-2\nu)} \begin{bmatrix} (1-\nu) & \nu & \nu & & & \\ \nu & (1-\nu) & \nu & & & \\ \nu & \nu & (1-\nu) & & & \\ & & & \frac{1}{2}(1-2\nu) & & \\ & & & & \frac{1}{2}(1-2\nu) & \\ & & & & & \frac{1}{2}(1-2\nu) \end{bmatrix} \times \begin{bmatrix} \varepsilon_x \\ \varepsilon_y \\ \varepsilon_z \\ \gamma_{xy} \\ \gamma_{xz} \\ \gamma_{yz} \end{bmatrix} \tag{4.10}$$

where $E$ is Young's modulus and $\nu$ is Poisson's ratio. Equation (4.10) can be re-expressed as

$$\boldsymbol{\sigma} = \mathbf{C}_2 \boldsymbol{\varepsilon} \tag{4.11}$$

where $\mathbf{C}_2$ is the constitutive matrix. Alternatively, we can write

$$\boldsymbol{\sigma}_2 = \mathbf{C}_4 \mathbin{:} \boldsymbol{\varepsilon}_2 \tag{4.12}$$

where $\mathbf{C}_4$ is the equivalent four-dimensional or fourth-order tensor [Y1] to the two-dimensional matrix $\mathbf{C}_2$ in (4.11).

Rather than attempt to visualise a four-dimensional tensor, many tensor equations can be viewed in terms of vectors with nine terms (three repeating), so that

$$\boldsymbol{\sigma}^{\mathrm{T}} = (\sigma_{xx}, \sigma_{yy}, \sigma_{zz}, \tau_{xy}, \tau_{xz}, \tau_{yz}, \tau_{xy}, \tau_{xz}, \tau_{yz}) \tag{4.13}$$

and

$$\boldsymbol{\varepsilon}^{\mathrm{T}} = (\varepsilon_{xx}, \varepsilon_{yy}, \varepsilon_{zz}, 0.5\gamma_{xy}, 0.5\gamma_{xz}, 0.5\gamma_{yz}, 0.5\gamma_{xy}, 0.5_{xz}, 0.5\gamma_{yz}). \tag{4.14}$$

Using such vectors, the connecting two-dimensional $\mathbf{C}_2$ would contain the same upper three-by-three submatrix as in (4.10) but the remaining diagonal terms would be doubled.

### 4.2.1 Plane strain, axial symmetry and plane stress

The three-dimensional formulation (Figure 4.1) can be simply reduced to two-dimensions (Figure 4.3) by setting

$$\tau_{xz} = \tau_{yz} = \gamma_{xz} = \gamma_{yz} = 0 \tag{4.15}$$

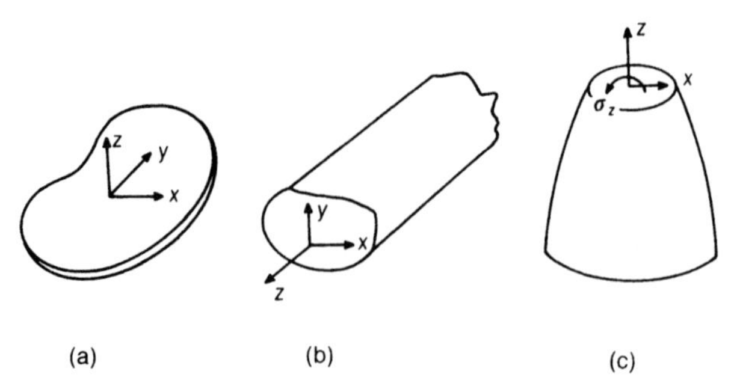


**Figure 4.3** Various two-dimensional stress states: (a) plane stress ($\sigma_z = 0$); (b) plane strain ($\varepsilon_z = 0$); (c) axial symmetry.

so that equations (4.12) reduce to

$$\boldsymbol{\sigma} = \begin{bmatrix} \sigma_x \\ \sigma_y \\ \sigma_z \\ \tau_{xy} \end{bmatrix} = \frac{E}{(1+\nu)(1-1\nu)} \begin{bmatrix} (1-\nu) & \nu & \nu & \\ \nu & (1-\nu) & \nu & \\ \nu & \nu & (1-\nu) & \\ & & & \frac{1}{2}(1-2\nu) \end{bmatrix} \begin{bmatrix} \varepsilon_x \\ \varepsilon_y \\ \varepsilon_z \\ \gamma_{xy} \end{bmatrix} = \mathbf{C}_2\boldsymbol{\varepsilon}. \tag{4.16}$$

For axial symmetry (Figures (4.3(c)), $\sigma_z$ can be taken as the hoop stress with $z = \theta$, while for plane strain, $\varepsilon_z$ is set to zero.

For plane stress, equations (4.16) are supplemented by, $\sigma_z = 0$ and (4.16) reduces to

$$\begin{bmatrix} \sigma_x \\ \sigma_y \\ \tau_{xy} \end{bmatrix} = \frac{E}{1-\nu^2} \begin{bmatrix} 1 & \nu & 0 \\ \nu & 1 & 0 \\ 0 & 0 & (1-\nu)/2 \end{bmatrix} \begin{bmatrix} \varepsilon_x \\ \varepsilon_y \\ \gamma_{xy} \end{bmatrix}. \tag{4.17}$$

### 4.2.2 Decomposition into volumetric and deviatoric components

An alternative form of the elastic stress–strain laws involves a decomposition of both the stresses and strains into their volumetric (or mean—subscript m) and deviatoric (subscript d) components. This decomposition is best applied with the tensor stress and strain measures, so that for the stresses

$$\boldsymbol{\sigma}_2 = \sigma_{\mathrm{m}}\mathbf{I} + \boldsymbol{\sigma}_{2\mathrm{d}} = \begin{bmatrix} \sigma_{\mathrm{m}} & 0 & 0 \\ 0 & \sigma_{\mathrm{m}} & 0 \\ 0 & 0 & \sigma_{\mathrm{m}} \end{bmatrix} + \begin{bmatrix} \sigma_{xx} - \sigma_{\mathrm{m}} & \tau_{xy} & \tau_{xz} \\ \tau_{yx} & \sigma_{yy} - \sigma_{\mathrm{m}} & \tau_{yz} \\ \tau_{zx} & \tau_{zy} & \sigma_{zz} - \sigma_{\mathrm{m}} \end{bmatrix} \tag{4.18}$$

where the deviatoric stresses $\boldsymbol{\sigma}_{2\mathrm{d}}$ are often written as $\mathbf{s}$. For the strains,

$$\boldsymbol{\varepsilon}_2 = \varepsilon_{\mathrm{m}}\mathbf{I} + \boldsymbol{\varepsilon}_{2\mathrm{d}} = \begin{bmatrix} \varepsilon_{\mathrm{m}} & 0 & 0 \\ 0 & \varepsilon_{\mathrm{m}} & 0 \\ 0 & 0 & \varepsilon_{\mathrm{m}} \end{bmatrix} + \begin{bmatrix} \varepsilon_{xx} - \varepsilon_{\mathrm{m}} & 0.5\gamma_{xy} & 0.5\gamma_{xz} \\ 0.5\gamma_{yx} & \varepsilon_{yy} - \varepsilon_{\mathrm{m}} & 0.5\gamma_{yz} \\ 0.5\gamma_{zx} & 0.5\gamma_{zy} & \varepsilon_{zz} - \varepsilon_{\mathrm{m}} \end{bmatrix} \tag{4.19}$$

where the deviatoric strains $\boldsymbol{\varepsilon}_{2\mathrm{d}}$ are often written as $\mathbf{e}$. In (4.18) and (4.19),

$$\sigma_{\mathrm{m}} = \tfrac{1}{3}(\sigma_{xx} + \sigma_{yy} + \sigma_{zz}), \qquad \varepsilon_{\mathrm{m}} = \tfrac{1}{3}(\varepsilon_{xx} + \varepsilon_{yy} + \varepsilon_{zz}). \tag{4.20}$$

The stress–strain laws are then given by

$$\boldsymbol{\sigma}_{2\mathrm{d}} = \mathbf{s}_2 = 2\mu\,\boldsymbol{\varepsilon}_{2\mathrm{d}} = 2\mu\mathbf{e}_2 \tag{4.21}$$

and

$$\sigma_{\mathrm{m}} = 3k\,\varepsilon_{\mathrm{m}} \tag{4.22}$$

where the shear modulus $\mu$ (often written as G) and bulk modulus, $k$, are related to $E$ and $\nu$ via [T1]:

$$\mu = \frac{E}{2(1+\nu)}, \qquad k = \frac{E}{3(1-2\nu)}. \tag{4.23}$$

### 4.2.3 An alternative expression using the Lamé constants

From (4.18), (4.19) and (4.21),

$$\boldsymbol{\sigma}_2 = (\sigma_{\mathrm{m}} - 2\mu\varepsilon_{\mathrm{m}})\mathbf{I} + 2\mu\boldsymbol{\varepsilon}_2 \tag{4.24}$$

and from (4.22),

$$\boldsymbol{\sigma}_2 = (3k - 2\mu)\varepsilon_{\mathrm{m}}\mathbf{I} + 2\mu\boldsymbol{\varepsilon}_2. \tag{4.25}$$

From (4.20), we can write

$$\varepsilon_{\mathrm{m}} = \tfrac{1}{3}\mathrm{tr}(\boldsymbol{\varepsilon}_2) \tag{4.26}$$

where the operation 'trace' (or tr) involves summing the diagonal elements. Hence in (4.25),

$$\boldsymbol{\sigma}_2 = \left(\frac{3k - 2\mu}{3}\right)\mathrm{tr}(\boldsymbol{\varepsilon}_2)\mathbf{I} + 2\mu\boldsymbol{\varepsilon}_2 = \lambda\,\mathrm{tr}(\boldsymbol{\varepsilon}_2)\mathbf{I} + 2\mu\boldsymbol{\varepsilon}_2. \tag{4.27}$$

Here, $\lambda$ and $\mu$ are the 'Lamé constants'. With suffix notation, (4.27) becomes

$$\sigma_{ij} = \lambda\delta_{ij}\varepsilon_{kk} + 2\mu\varepsilon_{ij} \tag{4.28}$$

where $\delta_{ij}$ is the Kronecker delta ($=1,\ i=j;\ =0,\ i \neq j$). (All of the work in this chapter will be related to a rectangular cartesian reference frame so that all indices can be written as subscripts, there being no distinction, for such a system, between co- and contravariant components. The latter will be considered in Volume 2.) Alternatively, we can write

$$\sigma_{ij} = C_{ijkl}\varepsilon_{kl} \tag{4.29}$$

where $C_{ijkl}$ are the components of a fourth-order tensor ($\mathbf{C}_4$) which, assuming linear-elastic isotropic conditions (as in (4.28)) is given by

$$C_{ijkl} = \mu(\delta_{ik}\delta_{jl} + \delta_{il}\delta_{jk}) + \lambda\delta_{ij}\delta_{kl}. \tag{4.30}$$

This equation is sometimes written as

$$\mathbf{C} = \mathbf{C}_4 = 2\mu\mathbf{I} + \lambda(\mathbf{1}\otimes\mathbf{1}) = 2\mu\mathbf{I}_4 + \lambda(\mathbf{1}_2\otimes\mathbf{1}_2) \tag{4.31}$$

where, as indicated, the subscripts which relate to the order of the tensor are usually omitted. In equation (4.31), the symbol $\otimes$ means tensor product. The symbol $\mathbf{I}$ is in some work reserved exclusively for the fourth-order unit tensor (as in (4.31)) with **1** being used for the second-order unit tensor. We will sometimes follow this procedure but may also use **I** more conventionally as the second-order unit tensor or unit matrix.

## 4.3 TRANSFORMATIONS AND ROTATIONS

We have already mentioned in Section 4.1 that the stress tensor, $\boldsymbol{\sigma}_2$, transforms in certain special ways to changes of axes and rotations. For most of this section, the stresses will be written in matrix or tensor notation and the subscript 2 will be dropped.

### 4.3.1 Transformation to a new set of axes

In this section, a line element or stress tensor will remain *fixed* but be represented in relation to a new set of axes. Starting in two dimensions (Figure 4.4), the line element **r** can either be represented in 'global' coordinates as $\mathbf{r}_g$ or in 'local' coordinates

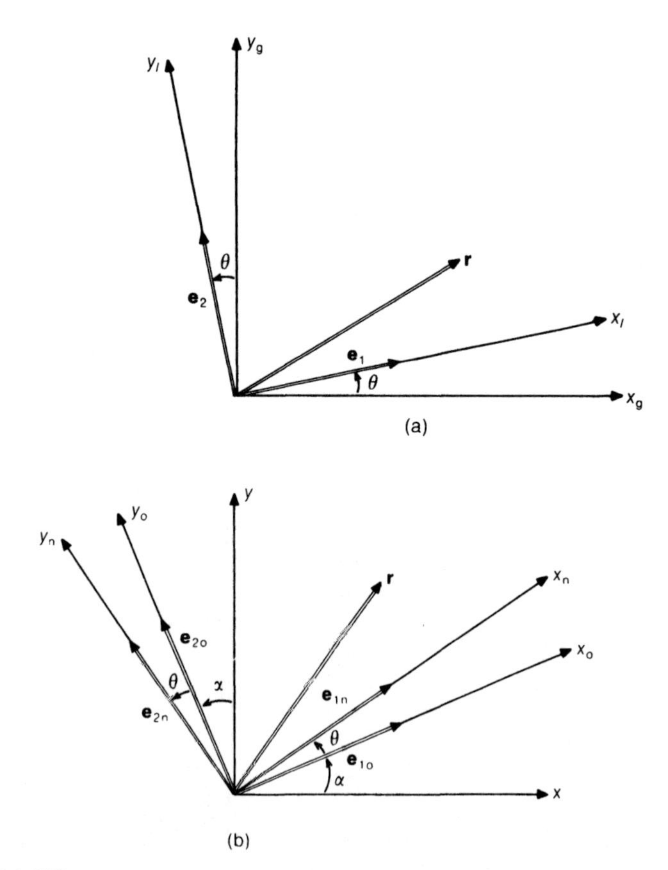


**Figure 4.4** Different axes for transformations: (a) 'local' and 'global'; (b) 'old' and 'new'.

as $\mathbf{r}_l$. Then, from Figure 4.4(a),

$$\mathbf{r}_l = \mathbf{T}\mathbf{r}_g \tag{4.32}$$

where, for two dimensions,

$$\mathbf{T} = \begin{bmatrix} \cos\theta & \sin\theta \\ -\sin\theta & \cos\theta \end{bmatrix} = \begin{bmatrix} \mathbf{e}_1^T \\ \mathbf{e}_2^T \end{bmatrix} \tag{4.33}$$

and $\mathbf{e}_1$ and $\mathbf{e}_2$ are the unit vectors of the $x_l$-, $y_l$-axes (Figure 4.4(a)) expressed in relation to the global coordinates and, hence, for example,

$$x_l = \mathbf{r}_l(1) = \cos\theta x_g + \sin\theta\, y_g \tag{4.34}$$

where $\gamma_l(1)$ is the first component of $\mathbf{r}_l$.

For three dimensions,

$$\mathbf{T} = \begin{bmatrix} \mathbf{e}_1^T \\ \mathbf{e}_2^T \\ \mathbf{e}_3^T \end{bmatrix}. \tag{4.35}$$

The components of the stress tensor, $\boldsymbol{\sigma}_g$, transform to local coordinates via

$$\boldsymbol{\sigma}_l = \mathbf{T}\boldsymbol{\sigma}_g\mathbf{T}^T. \tag{4.36}$$

Using the vector representation of stress in two dimensions, the equivalent expression to (4.36) is

$$\begin{bmatrix} \sigma_x \\ \sigma_y \\ \tau_{xy} \end{bmatrix}_l = \begin{bmatrix} c^2 & s^2 & 2sc \\ s^2 & c^2 & -2sc \\ -sc & sc & (c^2 - s^2) \end{bmatrix} \begin{bmatrix} \sigma_x \\ \sigma_y \\ \tau_{xy} \end{bmatrix}_g \tag{4.37}$$

where $c = \cos\theta$ and $s = \sin\theta$. This relationship is easily proved using the simple stress block of Figure 4.5. (The reader may also like to verify (most easily in two dimensions) that (4.37) corresponds to (4.36).) Physically, the stress remain the same but their

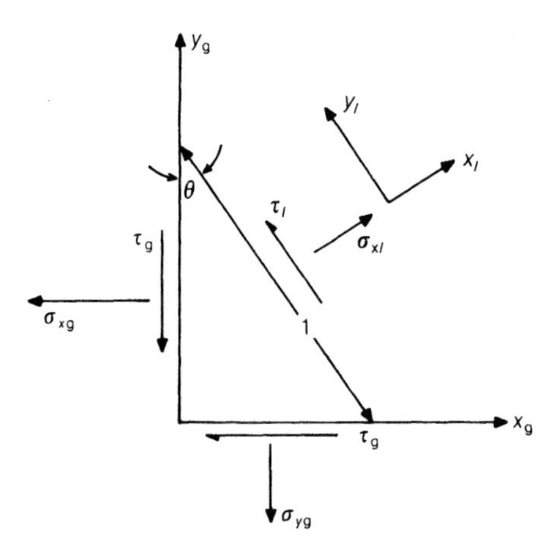


**Figure 4.5** Stresses in local and global coordinate systems.

components change because they are now written with regard to a new set of axes. It is easy to show that $\mathbf{T}$ is an orthogonal matrix such that

$$\mathbf{T}^{-1} = \mathbf{T}^{\mathrm{T}} \tag{4.38}$$

Instead of considering 'local' and 'global' axes, we could consider 'new' and 'old' (local) axes (Figure 4.4(b)). In this case, the same vector $\mathbf{r}$ (Figure 4.4(b)) can be expressed either in relation to the initial (global) coordinates or with respect to either of the two local coordinate systems so that

$$\mathbf{r}_{\mathrm{n}} = \mathbf{T}_{\mathrm{n}}\mathbf{r}_{\mathrm{g}} \tag{4.39}$$

while

$$\mathbf{r}_{\mathrm{o}} = \mathbf{T}_{\mathrm{o}}\mathbf{r}_{\mathrm{g}} \tag{4.40}$$

and hence

$$\mathbf{r}_{\mathrm{n}} = \mathbf{T}_{\mathrm{n}}\mathbf{T}_{\mathrm{o}}^{\mathrm{T}}\mathbf{r}_{\mathrm{o}} = \mathbf{T}\mathbf{r}_{\mathrm{o}} \tag{4.41}$$

and

$$\boldsymbol{\sigma}_{\mathrm{n}} = \mathbf{T}\boldsymbol{\sigma}_{\mathrm{o}}\mathbf{T}^{\mathrm{T}} \tag{4.42}$$

where $\boldsymbol{\sigma}_{\mathrm{n}}$ contains the components of stress with respect to the new local system and $\boldsymbol{\sigma}_{\mathrm{o}}$ with respect to the old local system. In two dimensions (Figure 4.4(b)),

$$\mathbf{T} = \begin{bmatrix} \cos\theta & \sin\theta \\ -\sin\theta & \cos\theta \end{bmatrix} = \begin{bmatrix} \mathbf{e}_{1\mathrm{n}}^{\mathrm{T}}\mathbf{e}_{1\mathrm{o}} & \mathbf{e}_{1\mathrm{n}}^{\mathrm{T}}\mathbf{e}_{2\mathrm{o}} \\ \mathbf{e}_{2\mathrm{n}}^{\mathrm{T}}\mathbf{e}_{1\mathrm{o}} & \mathbf{e}_{2\mathrm{n}}^{\mathrm{T}}\mathbf{e}_{2\mathrm{o}} \end{bmatrix} = \mathbf{T}_{\mathrm{n}}\mathbf{T}_{\mathrm{o}}^{\mathrm{T}} \tag{4.43}$$

where $\mathbf{e}_{1\mathrm{o}}$ and $\mathbf{e}_{2\mathrm{o}}$ are the unit base vectors of the old local system (written with respect to the global system) and $\mathbf{e}_{1\mathrm{n}}$ and $\mathbf{e}_{2\mathrm{n}}$ are the same for the new local system. (In confirming (4.43), the reader should note the transposes on the $\mathbf{e}$s in the definition of $\mathbf{T}$ in (4.33).) From Figure 4.4(b),

$$\mathbf{e}_{1\mathrm{o}} = \begin{bmatrix} \cos\alpha \\ \sin\alpha \end{bmatrix}, \qquad \mathbf{e}_{2\mathrm{o}} = \begin{bmatrix} -\sin\alpha \\ \cos\alpha \end{bmatrix} \tag{4.44}$$

$$\mathbf{e}_{1\mathrm{n}} = \begin{bmatrix} \cos(\alpha+\theta) \\ \sin(\alpha+\theta) \end{bmatrix}, \qquad \mathbf{e}_{2\mathrm{n}} = \begin{bmatrix} -\sin(\alpha+\theta) \\ \cos(\alpha+\theta) \end{bmatrix} \tag{4.45}$$

so that

$$\mathbf{e}_{1\mathrm{o}}^{\mathrm{T}}\mathbf{e}_{1\mathrm{n}} = \mathbf{e}_{2\mathrm{o}}^{\mathrm{T}}\mathbf{e}_{2\mathrm{n}} = \cos(\alpha+\theta)\cos\alpha + \sin(\alpha+\theta)\sin\alpha = \cos\theta \tag{4.46}$$

$$\mathbf{e}_{1\mathrm{o}}^{\mathrm{T}}\mathbf{e}_{2\mathrm{n}} = -\cos\alpha\sin(\alpha+\theta) + \sin\alpha\cos(\alpha+\theta) = -\sin\theta \tag{4.47}$$

$$\mathbf{e}_{2\mathrm{o}}^{\mathrm{T}}\mathbf{e}_{1\mathrm{n}} = -\sin\alpha\cos(\alpha+\theta) + \cos\alpha\sin(\alpha+\theta) = \sin\theta \tag{4.48}$$

and (4.43) is confirmed. For three dimensions, the transformation matrix, $\mathbf{T}$, is easily extended from (4.43) to give

$$\mathbf{T} = \begin{bmatrix} \mathbf{e}_{1\mathrm{n}}^{\mathrm{T}}\mathbf{e}_{1\mathrm{o}} & \mathbf{e}_{1\mathrm{n}}^{\mathrm{T}}\mathbf{e}_{2\mathrm{o}} & \mathbf{e}_{1\mathrm{n}}^{\mathrm{T}}\mathbf{e}_{3\mathrm{o}} \\ \mathbf{e}_{2\mathrm{n}}^{\mathrm{T}}\mathbf{e}_{1\mathrm{o}} & \mathbf{e}_{2\mathrm{n}}^{\mathrm{T}}\mathbf{e}_{2\mathrm{o}} & \mathbf{e}_{2\mathrm{n}}^{\mathrm{T}}\mathbf{e}_{3\mathrm{o}} \\ \mathbf{e}_{3\mathrm{n}}^{\mathrm{T}}\mathbf{e}_{1\mathrm{o}} & \mathbf{e}_{3\mathrm{n}}^{\mathrm{T}}\mathbf{e}_{2\mathrm{o}} & \mathbf{e}_{3\mathrm{n}}^{\mathrm{T}}\mathbf{e}_{3\mathrm{o}} \end{bmatrix} = \mathbf{T}_{\mathrm{n}}\mathbf{T}_{\mathrm{o}}^{\mathrm{T}}. \tag{4.49}$$

Using suffix notation, equation (4.49) can be expressed as

$$T_{ij} = \mathbf{e}_{i\mathrm{n}}^{\mathrm{T}}\mathbf{e}_{j\mathrm{o}} \tag{4.50}$$

while (4.42) would be

$$\sigma^{\mathrm{n}}_{ij} = T_{ia}\sigma^{\mathrm{o}}_{ab}T^{\mathrm{T}}_{bj} = T_{ia}\sigma^{\mathrm{o}}_{ab}T_{jb} = T_{ia}T_{jb}\sigma^{\mathrm{o}}_{ab} \tag{4.51}$$

where the 'n' and 'o' (for new and old) have been moved to be superscripts so as to ease the congestion. In a similar fashion, the strains can be expressed as

$$\varepsilon^{\mathrm{n}}_{ij} = T_{ia}T_{jb}\varepsilon^{\mathrm{o}}_{ab}, \qquad \varepsilon^{\mathrm{o}}_{ij} = T_{ai}T_{bj}\varepsilon^{\mathrm{n}}_{ab}. \tag{4.52}$$

Consequently, a constitutive relationship,

$$\sigma^{\mathrm{o}}_{ab} = C^{\mathrm{o}}_{abcd}\varepsilon^{\mathrm{o}}_{cd} \tag{4.53}$$

would (using (4.51) and the second part of (4.52) with *cdkl* instead of *ijab*) transform to

$$\sigma^{\mathrm{n}}_{ij} = C^{\mathrm{n}}_{ijkl}\varepsilon^{\mathrm{n}}_{kl} \tag{4.54}$$

where

$$C^{\mathrm{n}}_{ijkl} = T_{ia}T_{jb}T_{kc}T_{ld}C^{\mathrm{o}}_{abcd} \tag{4.55}$$

is the new constitutive tensor resulting from the transformation of coordinates.

### 4.3.2 A rigid-body rotation

In the previous section, the line element remained fixed but was related to different sets of axes: in the present, the line element, $\mathbf{r}$, will be physically *rotated* from $\mathbf{r}_{\mathrm{o}}$ to $\mathbf{r}_{\mathrm{n}}$ (Figure 4.6). We can then write:

$$\mathbf{r}_{\mathrm{n}} = \mathbf{R}\mathbf{r}_{\mathrm{o}} \tag{4.56}$$

In relation to Figure 4.6, when $\mathbf{r}_{\mathrm{o}}$ is rotated through $\theta$ degrees to $\mathbf{r}_{\mathrm{n}}$,

$$\mathbf{R} = \begin{bmatrix} \cos\theta & -\sin\theta \\ \sin\theta & \cos\theta \end{bmatrix} = [\mathbf{e}_1, \mathbf{e}_2] \tag{4.57}$$

where $\mathbf{e}_1$ and $\mathbf{e}_2$ are the unit vectors caused by the rotation of the original $\mathbf{x}$ and $\mathbf{y}$ vectors, $\mathbf{i}_1 = (1, 0)$ and $\mathbf{i}_2 = (0, 1)$. In three dimensions,

$$\mathbf{e}_{\mathrm{i}} = \mathbf{R}\mathbf{i}_{\mathrm{i}}, \qquad \mathrm{i} = 1, 3 \tag{4.58}$$

with the obvious extension of $\mathbf{i}_{1-3}$ and

$$\mathbf{R} = [\mathbf{e}_1, \mathbf{e}_2, \mathbf{e}_3]. \tag{4.59}$$

A comparison of (4.41) and (4.33) with (4.56) and (4.57) shows that

$$\mathbf{R} = \mathbf{T}^{\mathrm{T}} \tag{4.60}$$

where $\mathbf{T}$ is the transformation matrix of the previous section and $\mathbf{R}$ is the current rotation matrix. Equation (4.60) applies whether the rotation involves two or three dimensions.

In general, if a vector $\mathbf{r}_{\mathrm{o}}$ (Figure 4.7) is rotated through $\theta$ degrees to $\mathbf{r}_{\mathrm{n}}$, the 'global' coordinate axes can be thought of as also rotating through $\theta$ to 'local' axes (Figure 4.7). Clearly, $\mathbf{r}_{\mathrm{o}}$ with respect to the 'global' system (i.e. $\mathbf{r}_{\mathrm{o,g}}$) equals $\mathbf{r}_{\mathrm{n}}$ with respect to the

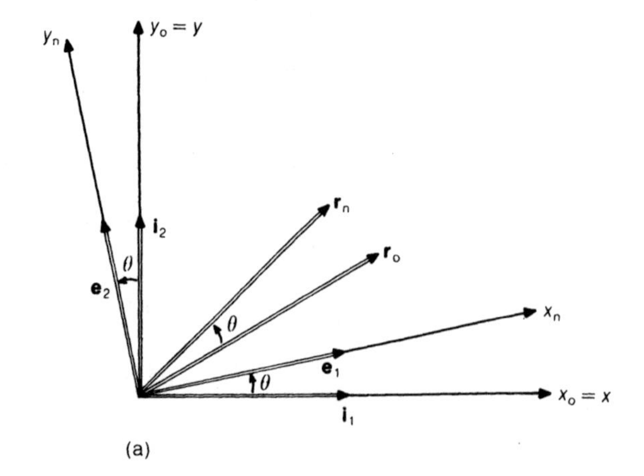


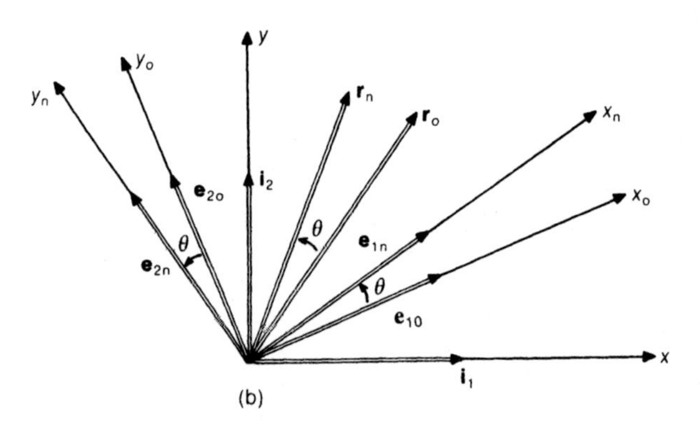


**Figure 4.6** Applying a rotation: (a) with $x = x_o$, $y = y_o$; (b) with $x \neq x_o$, $y \neq y_o$.

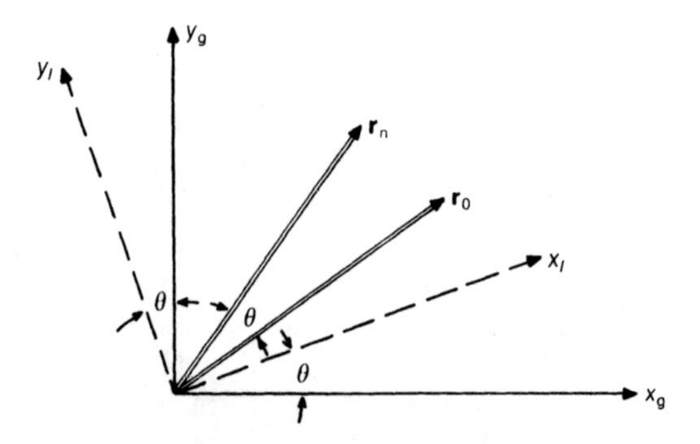


**Figure 4.7** Illustrating rotations and transformations.

'local' system ($\mathbf{r}_{n,l}$). Using (4.56), it follows that

$$\mathbf{r}_{n,g} = \mathbf{R}\mathbf{r}_{o,g} = \mathbf{R}\mathbf{r}_{n,l}. \tag{4.61}$$

But we also know from (4.32) that

$$\mathbf{r}_{n,g} = \mathbf{T}^{T}\mathbf{r}_{n,l}. \tag{4.62}$$

Hence, equation (4.60) is confirmed.

In a similar fashion, if a set of stress in the 'global' coordinate system is rotated through $\theta$ degrees (Figure 4.8), which corresponds to the use of the rotation matrix of (4.56) and (4.57), the global system can be assumed to have rotated through $\theta$ (or, more generally, via $\mathbf{R}$). Clearly, the new stress with respect to the local system, $\boldsymbol{\sigma}_{n,l}$, are equal to the old stress with respect to the global system, $\boldsymbol{\sigma}_{o,g}$. The components of the new stresses, $\boldsymbol{\sigma}_{n,l}$, in the old, global system (i.e. $\boldsymbol{\sigma}_{n,g}$) are (from (4.36) with $\mathbf{T}$ changed to $\mathbf{T}^{T}$ because we are moving from local to global rather than global to local):

$$\boldsymbol{\sigma}_{n,g} = \mathbf{T}^{T}\boldsymbol{\sigma}_{n,l}\mathbf{T} = \mathbf{T}^{T}\boldsymbol{\sigma}_{o,g}\mathbf{T} = \mathbf{R}\boldsymbol{\sigma}_{o,g}\mathbf{R}^{T}. \tag{4.63}$$

Equation (4.63) defines a new stress-state, $\boldsymbol{\sigma}_{n,g}$, caused by a rotation, $\mathbf{R}$, of the stressed body while the coordinate system (g) remains fixed. In contrast, the earlier equation (4.36) involves a change of the axis system within which the stresses are represented, while the stressed body remained fixed.

In the following, all terms will be related to global coordinates. If, in these circumstances, a set of old and new axes, $\mathbf{e}_n$ and $\mathbf{e}_o$, are given by

$$\mathbf{e}_{in} = \mathbf{R}_n\mathbf{i}_i, \quad \mathbf{e}_{io} = \mathbf{R}_o\mathbf{i}_i, \qquad i = 1,3 \tag{4.64}$$

with also

$$\mathbf{r}_n = \mathbf{R}_n\mathbf{r}, \qquad \mathbf{r}_o = \mathbf{R}_o\mathbf{r} \tag{4.65}$$

representing rotations of any vector $\mathbf{r}$ into the two frames, it follows that

$$\mathbf{r}_n = \mathbf{R}_n\mathbf{R}_o^{T}\mathbf{r}_o = \mathbf{R}\mathbf{r}_o \tag{4.66}$$

where $\mathbf{R}$ represents the rotation from one frame to the other. From (4.59) and (4.66),

$$\mathbf{R} = \mathbf{e}_{1n}\mathbf{e}_{1o}^{T} + \mathbf{e}_{2n}\mathbf{e}_{2o}^{T} + \mathbf{e}_{3n}\mathbf{e}_{3o}^{T}. \tag{4.67}$$

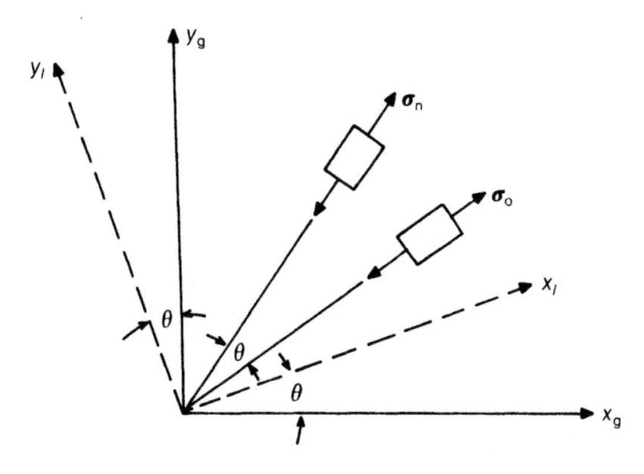


**Figure 4.8** Rotating a stress state.

## 4.4 GREEN'S STRAIN

The strains in (4.5) are related to small deflections. In Sections 3.1.2 and 3.3.1 we derived a convenient strain measure for describing strains in a bar when the deflections were large. These techniques can be generalised to a continuum so that (Figure 4.9) in place of (3.9),

$$\mathbf{dr}_n^2 - \mathbf{dr}_o^2 = \mathbf{dx}^2 - \mathbf{dX}^2 = 2\mathbf{dX}^T\mathbf{E}_2\mathbf{dX} \tag{4.68}$$

where $\mathbf{dr}_n^2 = \mathbf{dx}^2$ is the squared final length of a line element originally of squared length $\mathbf{dr}_o^2 = \mathbf{dX}^2$. The matrix $\mathbf{E}_2$ then defines the 'Green–Lagrange' (usually just referred to as Green) strain tensor which is valid for large deflections and replaces the small-deflection tensor, $\boldsymbol{\varepsilon}_2$ of (4.4) with components obtained from (4.5). The matrix $\mathbf{E}_2$ measures the strain in an element $\mathbf{dX}$ as the original coordinates of a point $\mathbf{X}$ are moved to new coordinates $\mathbf{x}$ by the addition of displacements $\mathbf{u}$, i.e.

$$\mathbf{x} = \mathbf{X} + \mathbf{u} \tag{4.69}$$

(see Figure 4.9). The latter equation can be differentiated to give

$$\mathbf{dx} = \frac{\partial \mathbf{x}}{\partial \mathbf{X}}\mathbf{dX} = \mathbf{F}\,\mathbf{dX} = \frac{\partial(\mathbf{X}+\mathbf{u})}{\partial \mathbf{X}}\mathbf{dX} \tag{4.70}$$

or:

$$\mathbf{dx} = \mathbf{F}\,\mathbf{dX} = \begin{bmatrix} \frac{\partial x}{\partial X} & \frac{\partial x}{\partial Y} & \frac{\partial x}{\partial Z} \\ \frac{\partial y}{\partial X} & \frac{\partial y}{\partial Y} & \frac{\partial y}{\partial Z} \\ \frac{\partial z}{\partial X} & \frac{\partial z}{\partial Y} & \frac{\partial z}{\partial Z} \end{bmatrix} \begin{bmatrix} \mathrm{d}X \\ \mathrm{d}Y \\ \mathrm{d}Z \end{bmatrix}$$

$$= \begin{bmatrix} 1+\frac{\partial u}{\partial X} & \frac{\partial u}{\partial Y} & \frac{\partial u}{\partial Z} \\ \frac{\partial v}{\partial X} & 1+\frac{\partial y}{\partial Y} & \frac{\partial v}{\partial Z} \\ \frac{\partial w}{\partial X} & \frac{\partial w}{\partial Y} & 1+\frac{\partial w}{\partial Z} \end{bmatrix} \begin{bmatrix} \mathrm{d}X \\ \mathrm{d}Y \\ \mathrm{d}Z \end{bmatrix} = [\mathbf{I}+\mathbf{D}]\,\mathbf{dX} \tag{4.71}$$

where $\mathbf{F}$ is the deformation gradient and the matrix, $\mathbf{D}$, in (4.71) is the displacement–derivative matrix:

$$\mathbf{D} = \begin{bmatrix} \frac{\partial u}{\partial X} & \frac{\partial u}{\partial Y} & \frac{\partial u}{\partial Z} \\ \frac{\partial v}{\partial X} & \frac{\partial v}{\partial Y} & \frac{\partial v}{\partial Z} \\ \frac{\partial w}{\partial X} & \frac{\partial w}{\partial Y} & \frac{\partial w}{\partial Z} \end{bmatrix} = \begin{bmatrix} u_{1,1} & u_{1,2} & u_{1,3} \\ u_{2,1} & u_{2,2} & u_{2,3} \\ u_{3,1} & u_{3,2} & u_{3,3} \end{bmatrix} = [u_{i,j}] = \frac{\partial \mathbf{u}}{\partial \mathbf{X}}. \tag{4.72}$$

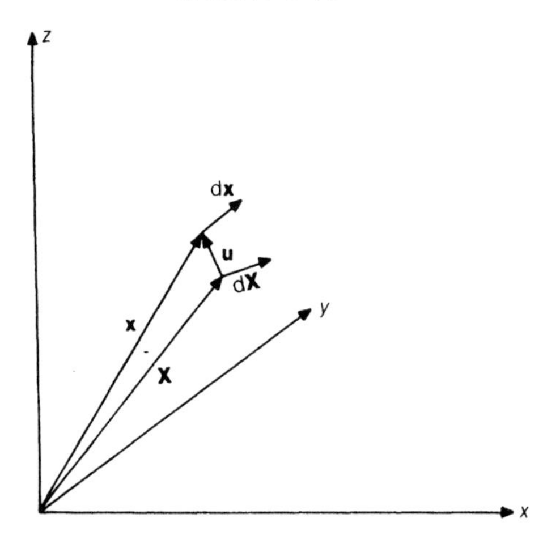


**Figure 4.9** Position vectors and displacements.

Substitution from (4.71) into (4.68) gives

$$2\mathbf{E}_2 = \mathbf{F}^{\mathrm{T}}\mathbf{F} - \mathbf{I} = [\mathbf{I} + \mathbf{D}]^{\mathrm{T}}[\mathbf{I} + \mathbf{D}] - \mathbf{I} \tag{4.73}$$

so that

$$\mathbf{E}_2 = \tfrac{1}{2}[\mathbf{F}^{\mathrm{T}}\mathbf{F} - \mathbf{I}] = \tfrac{1}{2}[\mathbf{D} + \mathbf{D}^{\mathrm{T}}] + \tfrac{1}{2}\mathbf{D}^{\mathrm{T}}\mathbf{D}. \tag{4.74}$$

The first term in (4.74) is a linear function of the displacement derivatives and corresponds exactly to the small-displacement matrix $\boldsymbol{\varepsilon}_2$ of (4.4), with (4.5) for the individual terms. From (4.72) and (4.73), the vector strain **E** can be expressed as

$$\mathbf{E} = \begin{bmatrix} \varepsilon_{xx} \\ \varepsilon_{yy} \\ \varepsilon_{zz} \\ \gamma_{xy} \\ \gamma_{xz} \\ \gamma_{yz} \end{bmatrix} = \begin{bmatrix} \dfrac{\partial u}{\partial X} \\ \dfrac{\partial v}{\partial Y} \\ \dfrac{\partial w}{\partial Z} \\ \dfrac{\partial u}{\partial Y} + \dfrac{\partial v}{\partial X} \\ \dfrac{\partial u}{\partial Z} + \dfrac{\partial w}{\partial X} \\ \dfrac{\partial v}{\partial Z} + \dfrac{\partial w}{\partial Y} \end{bmatrix} + \begin{bmatrix} \dfrac{1}{2}\left(\left(\dfrac{\partial u}{\partial X}\right)^2 + \left(\dfrac{\partial v}{\partial X}\right)^2 + \left(\dfrac{\partial w}{\partial X}\right)^2\right) \\ \dfrac{1}{2}\left(\left(\dfrac{\partial u}{\partial Y}\right)^2 + \left(\dfrac{\partial v}{\partial Y}\right)^2 + \left(\dfrac{\partial w}{\partial Y}\right)^2\right) \\ \dfrac{1}{2}\left(\left(\dfrac{\partial u}{\partial Z}\right)^2 + \left(\dfrac{\partial v}{\partial Z}\right)^2 + \left(\dfrac{\partial w}{\partial Z}\right)^2\right) \\ \left(\dfrac{\partial u}{\partial X}\right)\left(\dfrac{\partial u}{\partial Y}\right) + \left(\dfrac{\partial v}{\partial X}\right)\left(\dfrac{\partial v}{\partial Y}\right) + \left(\dfrac{\partial w}{\partial X}\right)\left(\dfrac{\partial w}{\partial Y}\right) \\ \left(\dfrac{\partial u}{\partial X}\right)\left(\dfrac{\partial u}{\partial Z}\right) + \left(\dfrac{\partial v}{\partial X}\right)\left(\dfrac{\partial v}{\partial Z}\right) + \left(\dfrac{\partial w}{\partial X}\right)\left(\dfrac{\partial w}{\partial Z}\right) \\ \left(\dfrac{\partial u}{\partial Y}\right)\left(\dfrac{\partial u}{\partial Z}\right) + \left(\dfrac{\partial v}{\partial Y}\right)\left(\dfrac{\partial v}{\partial Z}\right) + \left(\dfrac{\partial w}{\partial Y}\right)\left(\dfrac{\partial w}{\partial Z}\right) \end{bmatrix}. \tag{4.75}$$

Clearly (4.74) is a neater representation.

If we apply a rigid-body rotation so that, from (4.56), $\mathrm{d}\mathbf{x} = \mathbf{R}\mathrm{d}\mathbf{X}$ and we compare the solution with (4.71), we can observe that, for such a rigid rotation $\mathbf{D} = \mathbf{R} - \mathbf{I}$.

Substitution into (4.74), then leads (after noting that $\mathbf{R}^T\mathbf{R} = \mathbf{I}$) to $\mathbf{E}_2 = 0$ and, consequently, the Green strains are zero as a result of rigid-body rotation. However, if we neglect the $\frac{1}{2}\mathbf{D}^T\mathbf{D}$ term in (4.74) and effectively use a linear strain measure which is equivalent to the linear strain $\varepsilon_2$ of (4.4) and (4.5), we would obtain:

$$\boldsymbol{\varepsilon}_2 = \tfrac{1}{2}[\mathbf{R} + \mathbf{R}^T] - \mathbf{I}$$

which is not zero.

Consider the two-dimensional case and let $\mathbf{R}$ be given by (4.57). It is then easy to show that:

$$\boldsymbol{\varepsilon}_2 = (\cos\theta - 1)\mathbf{I} \simeq \frac{\theta^2}{2}\mathbf{I}$$

which is very nearly zero for small rotations, $\theta$.

### 4.4.1 Virtual work expressions using Green's strain

The stress measure that is conjugate the Green's strain tensor is the second Piola–Kirchhoff stress tensor, $\mathbf{S}_2$, or its vector equivalent $\mathbf{S}$. Following the developments of Chapters 2 and 3, the finite element formulation usually spring from a virtual work expression. Using the Green's strain, such an expression is given by

$$V = V_i - V_e = \int \mathbf{S}^T \delta\mathbf{E}_v \, dV_o - V_e = \int \mathbf{S}_2 : \delta\mathbf{E}_{v2} \, dV_o - V_e \tag{4.76}$$

where $V_e$ is the external virtual work (see Sections 3.12 and 3.3 for similar developments with the truss element).

With the aid of differentiation, from (4.73), the change in Green's strain can be obtained as

$$\delta\mathbf{E}_2 = \tfrac{1}{2}[\delta\mathbf{D} + \delta\mathbf{D}^T] + \tfrac{1}{2}\mathbf{D}^T\delta\mathbf{D} + \tfrac{1}{2}\delta\mathbf{D}^T\mathbf{D} + (\tfrac{1}{2}\delta\mathbf{D}^T\delta\mathbf{D})_h \tag{4.77}$$

or

$$\delta\mathbf{E}_2 = \tfrac{1}{2}\mathbf{F}^T\delta\mathbf{D} + \tfrac{1}{2}\delta\mathbf{D}^T\mathbf{F} + (\tfrac{1}{2}\delta\mathbf{D}^T\delta\mathbf{D})_h. \tag{4.78}$$

The $\delta\mathbf{D}$ terms are simply obtained from (4.72) by replacing $\partial u/\partial X$ by $\partial\delta u/\partial X$, etc. The final bracketed terms (marked h) become negligible when the infinitesimal, virtual changes are applied and

$$\delta\mathbf{E}_{v2} = \tfrac{1}{2}\mathbf{F}^T\delta\mathbf{D}_v + \tfrac{1}{2}\delta\mathbf{D}_v^T\mathbf{F}. \tag{4.79}$$

As shown in Chapter 2 and 3, the tangent stiffness matrix can be obtained from the variation of the virtual work expression $V$. From (4.76), this leads to

$$\delta V = \int (\delta\mathbf{S}^T\delta\mathbf{E}_v + \mathbf{S}^T\delta(\delta\mathbf{E}_v)) \, dV_o = \int (\delta\mathbf{S}_2 : \delta\mathbf{E}_{v2} + \mathbf{S} : \delta(\delta\mathbf{E}_{v2})) \, dV_o. \tag{4.80}$$

From (4.71) and (4.79),

$$\delta(\delta\mathbf{E}_{v2}) = \tfrac{1}{2}[\delta\mathbf{D}_v^T\delta\mathbf{D} + \delta\mathbf{D}^T\delta\mathbf{D}_v] \tag{4.81}$$

If we assume that the changes in second Piola–Kirchhoff stress, $\delta \mathbf{S}$ can be related to the changes in Green's strain $\delta \mathbf{E}$ via:

$$\delta \mathbf{S} = \mathbf{C}_{t2} \delta \mathbf{E}, \qquad \delta \mathbf{S}_2 = \mathbf{C}_{t4} : \delta \mathbf{E}_2 \tag{4.82}$$

where the first form involves a matrix or second order constitutive tensor, $\mathbf{C}_2$, while the second form involves a fourth order constitutive tensor, $\mathbf{C}_4$ (see Section 4.2). From (4.80)–(4.82),

$$\delta V = \int (\delta \mathbf{E}_{v2} : \mathbf{C}_{t4} : \delta \mathbf{E}_2 + \mathbf{S} : \delta \mathbf{D}_v^{\mathrm{T}} \delta \mathbf{D}) \, \mathrm{d} V_o = \int (\delta \mathbf{E}^{\mathrm{T}} \mathbf{C}_{t2} \delta \mathbf{E} + \mathbf{S}^{\mathrm{T}} \delta(\delta \mathbf{E}_{v2})) \, \mathrm{d} V_o. \tag{4.83}$$

The equations of this section are sufficient for the derivation of a total Lagrangian finite element formulation for a continuum (see Sections 5.1 and 5.2).

### 4.4.2 Work expressions using von Kármán's non-linear strain–displacement relationships for a plate

The membrane part of von Kármán's strain–displacement relationships for moderately large-deflection analysis can be considered as a special case of the Green strain. If, for a plate in the $x$–$y$ plane, the 'in-plane strain terms', $(\partial u/\partial X)^2$ and $\partial v/\partial X)^2$ etc., are considered negligible, the Green strain vector of (4.75) degenerates to

$$\mathbf{E} = \begin{bmatrix} \dfrac{\partial u}{\partial X} \\[2mm] \dfrac{\partial v}{\partial Y} \\[2mm] \dfrac{\partial u}{\partial Y} + \dfrac{\partial v}{\partial X} \end{bmatrix} + \begin{bmatrix} \dfrac{1}{2}\left(\dfrac{\partial w}{\partial X}\right)^2 \\[2mm] \dfrac{1}{2}\left(\dfrac{\partial w}{\partial Y}\right)^2 \\[2mm] \left(\dfrac{\partial w}{\partial X}\right)\left(\dfrac{\partial w}{\partial Y}\right) \end{bmatrix} \tag{4.84}$$

which are von Kármán's equations [V1, T2] for the membrane strains. The latter can be modified to Marguerre's equations [M2, T2] for use with shallow shells (Chapter 8). Differentiation of (4.84) leads to a degenerated form of (4.79), whereby

$$\delta \mathbf{E} = \begin{bmatrix} \dfrac{\partial \delta u}{\partial X} \\[2mm] \dfrac{\partial \delta v}{\partial Y} \\[2mm] \dfrac{\partial \delta u}{\partial Y} + \dfrac{\partial \delta v}{\partial X} \end{bmatrix} + \begin{bmatrix} \dfrac{\partial w}{\partial X} & 0 \\[2mm] 0 & \dfrac{\partial w}{\partial Y} \\[2mm] \dfrac{\partial w}{\partial Y} & \dfrac{\partial w}{\partial X} \end{bmatrix} \begin{bmatrix} \dfrac{\partial \delta w}{\partial X} \\[2mm] \dfrac{\partial \delta w}{\partial Y} \end{bmatrix} + \begin{bmatrix} \dfrac{1}{2}\left(\dfrac{\partial \delta w}{\partial X}\right)^2 \\[2mm] \dfrac{1}{2}\left(\dfrac{\partial \delta w}{\partial Y}\right)^2 \\[2mm] \left(\dfrac{\partial \delta w}{\partial X}\right)\left(\dfrac{\partial \delta w}{\partial Y}\right) \end{bmatrix}_{\mathrm{h}} . \tag{4.85}$$

As with (4.77), the higher order (marked h) terms in (4.85) becomes negligible when the infinitesimal virtual displacements are involved. One final expression, $\delta(\delta \mathbf{E}_v)$, is required (see 4.83) before we have the basis for a finite element formulation (Chapter 8).

From (4.85),

$$\delta(\delta \mathbf{E}_v) = \begin{bmatrix} \left(\dfrac{\partial \delta w}{\partial X}\right)\left(\dfrac{\partial \delta w}{\partial X}\right)_v \\ \left(\dfrac{\partial \delta w}{\partial Y}\right)\left(\dfrac{\partial \delta w}{\partial Y}\right)_v \\ \left(\dfrac{\partial \delta w}{\partial X}\right)\left(\dfrac{\partial \delta w}{\partial Y}\right)_v + \left(\dfrac{\partial \delta w}{\partial X}\right)_v \left(\dfrac{\partial \delta w}{\partial Y}\right) \end{bmatrix}. \tag{4.86}$$

Hence, from (4.80),

$$\mathrm{d}V = \int \left[ \delta \mathbf{S}^{\mathrm{T}} \delta \mathbf{E}_v + \begin{bmatrix} \dfrac{\partial \delta w}{\partial X} \\ \dfrac{\partial \delta w}{\partial Y} \end{bmatrix}_v^{\mathrm{T}} \mathbf{S}_2 \begin{bmatrix} \dfrac{\partial \delta w}{\partial X} \\ \dfrac{\partial \delta w}{\partial Y} \end{bmatrix} \right] \mathrm{d}V_o \tag{4.87}$$

where

$$\mathbf{S}^{\mathrm{T}} = \{\sigma_x, \sigma_y, \tau_{xy}\}, \qquad \mathbf{S}_2 = \begin{bmatrix} \sigma_x & \tau_{xy} \\ \tau_{xy} & \sigma_y \end{bmatrix} \tag{4.88}$$

and in (4.87), for convenience, we have mixed vector and tensor notation.

The equations in this section provide the basis for the derivation of a shallow-shell element (see Section 8.1).

## 4.5 ALMANSI'S STRAIN

In Section 3.2.1, we introduced Almansi's strain (also sometimes called the Eulerian strain) for a one-dimensional truss element. The continuum extension of (3.41) can be written as

$$\mathrm{d}\mathbf{r}_n^2 - \mathrm{d}\mathbf{r}_o^2 = \mathrm{d}\mathbf{x}^2 - \mathrm{d}\mathbf{X}^2 = 2\mathrm{d}\mathbf{x}^{\mathrm{T}} \mathbf{A}\, \mathrm{d}\mathbf{x} \tag{4.89}$$

which can be contrasted with (4.68) for Green's strain. In order to compute the Almansi strain, $\mathbf{A}$, the original length-increment vector, $\mathrm{d}\mathbf{X}$, must now be related to the final length-increment vector, $\mathrm{d}\mathbf{x}$, using the inverse of (4.70) or (4.71), so that

$$\mathrm{d}\mathbf{X} = \frac{\partial \mathbf{X}}{\partial \mathbf{x}} \mathrm{d}\mathbf{x} = \mathbf{F}^{-1} \mathrm{d}\mathbf{x} = \left[ \mathbf{I} - \frac{\partial \mathbf{u}}{\partial \mathbf{x}} \right] \mathrm{d}\mathbf{x}. \tag{4.90}$$

The last relationship in (4.90) follows from (4.69) with $\partial \mathbf{u}/\partial \mathbf{x}$ being of the same form as $\mathbf{D} = \partial \mathbf{u}/\partial \mathbf{X}$ in (4.72). Substitution from (4.90) into (4.89) gives

$$\mathbf{A} = \tfrac{1}{2}[\mathbf{I} - \mathbf{F}^{-\mathrm{T}} \mathbf{F}^{-1}] = \frac{1}{2}\left[ \frac{\partial \mathbf{u}}{\partial \mathbf{x}} + \frac{\partial \mathbf{u}^{\mathrm{T}}}{\partial \mathbf{x}} \right] - \frac{1}{2} \frac{\partial \mathbf{u}^{\mathrm{T}}}{\partial \mathbf{x}} \frac{\partial \mathbf{u}}{\partial \mathbf{x}} \tag{4.91}$$

in place of (4.74). The strain measure can be related to the Green strain, $\mathbf{E}$, of (4.74) via

$$\mathbf{F}^{\mathrm{T}} \mathbf{A} \mathbf{F} = \tfrac{1}{2}[\mathbf{F}^{\mathrm{T}} \mathbf{F} - \mathbf{I}] = \mathbf{E}. \tag{4.92}$$

## 4.6 THE TRUE OR CAUCHY STRESS

If, in a finite element context, we adopt an updated coordinate system (as in Sections 3.3.6 and 5.3), but maintain the directions of the original rectangular cartesian system, we will need to use a stress measure that relates to this new (or current) system. Even if we adopt a Green strain second Piola–Kirchhoff system (as in Section 4.4) we may wish to interpret our final stresses in relation to the final geometry because (Figure 4.10) without additional knowledge concerning the deformations, the second Piola–Kirchhoff stresses are difficult to interpret. In either case, the solution involves the Cauchy or true stress, $\boldsymbol{\sigma}$, which is the tensor equivalent of the 'true' stress '$\sigma$', introduced in Section 3.1.5.

In very simple terms, the Cauchy stress is force/final area rather than force/original area and is related to the current configuration, while the second Piola–Kirchhoff stress (work-conjugate to the Green's strain) relates to the original configuration. For the rest of the chapter, we will omit the subscript 2 implying tensor or matrix since we will usually use this notation rather than the vector forms for stress and strain.

In Section 3.1.5, it was shown that the true stress was work conjugate to the virtual strain measure $\delta l_v/l$ (see (3.14)). The equivalent continuum form is

$$\delta\boldsymbol{\varepsilon}_v = \left[\frac{\partial\delta\mathbf{u}_v}{\partial\mathbf{x}} + \frac{\partial\delta\mathbf{u}_v^{\mathrm{T}}}{\partial\mathbf{x}}\right] \tag{4.93}$$

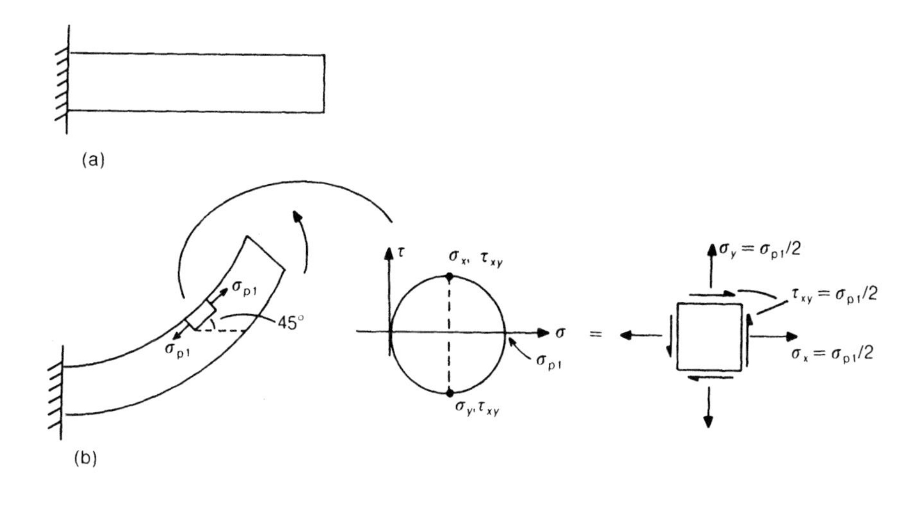


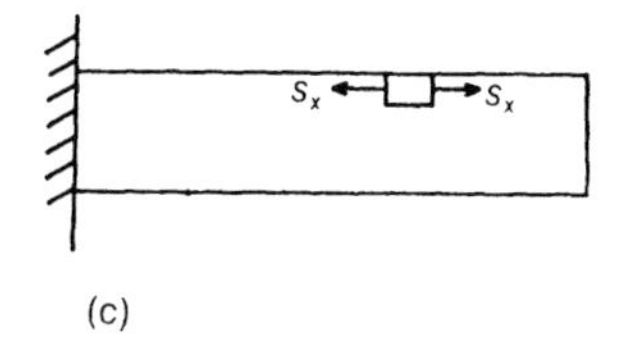


**Figure 4.10** Cauchy and second Piola–Kirchhoff stresses: (a) initial unloaded state; (b) loaded state with Cauchy stresses; (c) unloaded state with second Piola–Kirchhoff stresses.

where $\mathbf{x}$ contains the current coordinates. This strain measure is effectively in the form of a small 'engineering strain', but related to the current configuration. It is sometimes known as the linear Almansi strain *increment* or linear Euler strain increment but, as will be discussed later, it is *not* the virtual variation of the Almansi strain (4.91) but is more closely related (as for the truss—see Section 3.1.5) to a log-strain measure (Volume 2). The strain measure in (4.93) is also known as the Rivlin–Erikson rate of Almansi strain [B1].

In order to relate the Cauchy stress to the second Piola–Kirchhoff stress, we now adopt equivalent work concepts, so that

$$V_i = \int \mathbf{S}:\delta\mathbf{E}_v \, dV_o = \int \boldsymbol{\sigma}:\delta\boldsymbol{\varepsilon}_v \, dV \tag{4.94}$$

where $\boldsymbol{\sigma}$ is the Cauchy stress.

We can relate the current incremental volume $dV$ to the old volume $dV_o$ via

$$dV = dx\,dy\,dz = J\,dX\,dY\,dZ = J\,dV_o \tag{4.95}$$

where:

$$J = \det(\mathbf{F}) = \det\left(\frac{\partial \mathbf{x}}{\partial \mathbf{X}}\right). \tag{4.96}$$

(The two-dimensional form of this relationship is illustrated in Figure 4.11 with an initial element of side $dX$, $dY$ being moved to an element with sides $d\mathbf{x}_a$, $d\mathbf{x}_b$, so that

$$dA\mathbf{i}_3 = d\mathbf{x}_a \times d\mathbf{x}_b = \begin{bmatrix} \dfrac{\partial x}{\partial X} \\[2mm] \dfrac{\partial y}{\partial X} \end{bmatrix} dX \times \begin{bmatrix} \dfrac{\partial x}{\partial Y} \\[2mm] \dfrac{\partial y}{\partial Y} \end{bmatrix} dY = \det(\mathbf{F})\,dX\,dY\mathbf{i}_3 = \det(\mathbf{F})\,dA_o\mathbf{i}_3 \tag{4.97}$$

where $\mathbf{i}_3$ is the unit vector orthogonal to the $x$–$y$ plane.)

Substitution from (4.95) into (4.94) gives

$$V_i = \int \boldsymbol{\sigma}:\delta\boldsymbol{\varepsilon}_v \, dV = \int J\boldsymbol{\sigma}:\delta\boldsymbol{\varepsilon}_v \, dV_o = \int \boldsymbol{\tau}:\delta\boldsymbol{\varepsilon}_v \, dV_o = \int \mathbf{S}:\delta\mathbf{E}_v \, dV_o \tag{4.98}$$

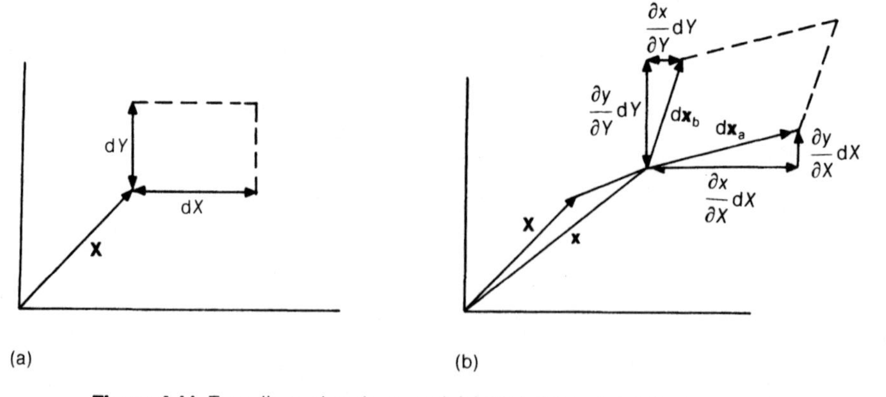


**Figure 4.11** Two-dimensional areas: (a) initial element; (b) final element.

where

$$\boldsymbol{\tau} = \mathbf{J}\boldsymbol{\sigma} \tag{4.99}$$

is known as the Kirchhoff or nominal stress.

In order to complete the relationship between the stress measures, $\boldsymbol{\sigma}$ and $\mathbf{S}$, we must establish a relationship between the virtual strain measures, $\delta\boldsymbol{\varepsilon}_v$ (see (4.93) and $\delta\mathbf{E}_v$ (see (4.79)) as used in (4.98). To this end, with the aid of (4.70) and (4.72),

$$\delta\mathbf{D}_v = \frac{\partial\delta\mathbf{u}_v}{\partial\mathbf{X}} = \frac{\partial\delta\mathbf{u}_v}{\partial\mathbf{x}}\frac{\partial\mathbf{x}}{\partial\mathbf{X}} = \frac{\partial\delta\mathbf{u}_v}{\partial\mathbf{x}}\mathbf{F}. \tag{4.100}$$

Substituting from (4.100) into (4.79) and comparing the result with (4.93) leads to

$$\delta\mathbf{E}_v = \mathbf{F}^{\mathrm{T}}\delta\boldsymbol{\varepsilon}_v\mathbf{F}. \tag{4.101}$$

Substitution from (4.101) into (4.98) gives

$$V_i = \int \mathbf{J}\boldsymbol{\sigma}:\delta\boldsymbol{\varepsilon}_v\,\mathrm{d}V_o = \int \mathbf{S}:[\mathbf{F}^{\mathrm{T}}\delta\boldsymbol{\varepsilon}_v\mathbf{F}]\,\mathrm{d}V_o = \int [\mathbf{F}\mathbf{S}\mathbf{F}^{\mathrm{T}}]:\delta\boldsymbol{\varepsilon}_v\,\mathrm{d}V_o. \tag{4.102}$$

The last relationship in (4.102) can be derived by noting the symmetry of $\mathbf{S}$ and $\delta\boldsymbol{\varepsilon}_v$ and twice making use of the matrix relationship

$$\mathbf{AB}:\mathbf{C}^{\mathrm{T}} = \mathbf{CA}:\mathbf{B}^{\mathrm{T}} = \mathbf{BC}:\mathbf{A}^{\mathrm{T}}. \tag{4.103}$$

Hence, the final relationships between the true, Cauchy stresses and the second Piola–Kirchhoff stresses are given by

$$\boldsymbol{\sigma} = \frac{1}{J}\mathbf{F}\mathbf{S}\mathbf{F}^{\mathrm{T}} \tag{4.104}$$

or

$$\mathbf{S} = J\mathbf{F}^{-1}\sigma\mathbf{F}^{-\mathrm{T}}. \tag{4.105}$$

At this point, we should re-emphasise an issue first raised in Section 3.2.1. The Cauchy stress, $\boldsymbol{\sigma}$, is work conjugate to the virtual strain measure of (4.93). This term is not the virtual variation of the Almansi strain. In this sense, one should not really consider the Cauchy stress as being conjugate to the Almansi strain. Indeed from (4.91), one can show that

$$\delta\mathbf{A}_v = \delta\boldsymbol{\varepsilon}_v - \left[\mathbf{A}^{\mathrm{T}}\frac{\partial\delta\mathbf{u}_v}{\partial\mathbf{x}} + \frac{\partial\delta\mathbf{u}_v^{\mathrm{T}}}{\partial\mathbf{x}}\mathbf{A}\right]. \tag{4.106}$$

Although we have specifically omitted dynamics from the scope of this book, a stricter derivation of the previous equations requires the introduction of a time measure with a superimposed dot representing differentiation with respect to time. Because a lot of published work introduces this time element, we will give a brief summary at this stage. The strain-rate tensor, $\dot{\boldsymbol{\varepsilon}}$, can then be considered as:

$$\dot{\boldsymbol{\varepsilon}} = \frac{1}{\mathrm{d}t}\delta\boldsymbol{\varepsilon} \tag{4.107}$$

(strictly in the limit). As already indicated, the Euler strain increment, $\delta\boldsymbol{\varepsilon}$ (see (4.93)), is related to the current geometry $\mathbf{x}$ rather than the original geometry, $\mathbf{X}$. Hence there are no 'initial displacement (**D**) terms' as in (4.77) and $\delta\boldsymbol{\varepsilon}$ takes a similar form to the engineering strain increment, so that

$$\dot{\boldsymbol{\varepsilon}} = \frac{1}{\mathrm{d}t}\delta\boldsymbol{\varepsilon} = \frac{1}{\mathrm{d}t}\left[\frac{\partial \delta \mathbf{u}}{\partial \mathbf{x}} + \frac{\partial \delta \mathbf{u}^{\mathrm{T}}}{\partial \mathbf{x}}\right] = \tfrac{1}{2}[\mathbf{L} + \mathbf{L}^{\mathrm{T}}] \tag{4.108}$$

where $\dot{\boldsymbol{\varepsilon}}$ is often referred to as the rate of deformation tensor or velocity strain tensor. It is a function of **L**, where

$$\mathbf{L} = \frac{1}{\mathrm{d}t}\frac{\partial \delta \mathbf{u}}{\partial \mathbf{x}} = \frac{\partial \dot{\mathbf{x}}}{\partial \mathbf{x}} = \frac{\partial \mathbf{v}}{\partial \mathbf{x}} \tag{4.109}$$

and $\mathbf{v}$ is the velocity. **L** is known as the velocity gradient. Equation (4.109) is of a similar form to (4.72) and in component form involves

$$\mathbf{L} = \begin{bmatrix} \dfrac{\partial \dot{x}_1}{\partial x_1} & \dfrac{\partial \dot{x}_1}{\partial x_2} & \dfrac{\partial \dot{x}_1}{\partial x_3} \\ \dfrac{\partial \dot{x}_2}{\partial x_1} & \dfrac{\partial \dot{x}_2}{\partial x_2} & \dfrac{\partial \dot{x}_2}{\partial x_3} \\ \dfrac{\partial \dot{x}_3}{\partial x_1} & \dfrac{\partial \dot{x}_3}{\partial x_2} & \dfrac{\partial \dot{x}_3}{\partial x_3} \end{bmatrix}. \tag{4.110}$$

From (4.73),

$$\dot{\mathbf{E}}_{\mathrm{v}} = \tfrac{1}{2}[\dot{\mathbf{F}}_{\mathrm{v}}^{\mathrm{T}}\mathbf{F} + \mathbf{F}^{\mathrm{T}}\dot{\mathbf{F}}_{\mathrm{v}}] \tag{4.111}$$

where

$$\dot{\mathbf{F}}_{\mathrm{v}} = \frac{\partial \dot{\mathbf{x}}_{\mathrm{v}}}{\partial \mathbf{X}} = \frac{\partial \dot{\mathbf{x}}_{\mathrm{v}}}{\partial \mathbf{x}}\frac{\partial \mathbf{x}}{\partial \mathbf{X}} = \mathbf{L}_{\mathrm{v}}\mathbf{F}. \tag{4.112}$$

Hence

$$\dot{\mathbf{E}}_{\mathrm{v}} = \tfrac{1}{2}[\mathbf{F}^{\mathrm{T}}\mathbf{L}_{\mathrm{v}}\mathbf{F} + \mathbf{F}^{\mathrm{T}}\mathbf{L}_{\mathrm{v}}^{\mathrm{T}}\mathbf{F}] = \tfrac{1}{2}\mathbf{F}^{\mathrm{T}}[\mathbf{L}_{\mathrm{v}} + \mathbf{L}_{\mathrm{v}}^{\mathrm{T}}]\mathbf{F} = \mathbf{F}^{\mathrm{T}}\dot{\boldsymbol{\varepsilon}}_{\mathrm{v}}\mathbf{F} \tag{4.113}$$

which, with (4.107), coincides with (4.101). If the $\delta\boldsymbol{\varepsilon}_{\mathrm{v}}$ and $\delta\mathbf{E}_{\mathrm{v}}$ terms in (4.101) are now replaced by $\dot{\boldsymbol{\varepsilon}}_{\mathrm{v}}$ and $\dot{\mathbf{E}}_{\mathrm{v}}$ respectively, (4.104) and (4.105) can be derived as before.

## 4.7 SUMMARISING THE DIFFERENT STRESS AND STRAIN MEASURES

We have not yet introduced the first Piola–Kirchoff stress tensor, **P**. Referring back to (4.9) and Figure 4.2, the external tractions can be expressed either in terms of the original configuration (via **P**) or the final configuration (via, $\boldsymbol{\sigma}$, the Cauchy stress). Hence

$$\mathbf{t}_{\mathrm{o}}\,\mathrm{d}S_{\mathrm{o}} = \mathbf{P}\mathbf{N}\,\mathrm{d}S_{\mathrm{o}} = \mathbf{t}\,\mathrm{d}S = \boldsymbol{\sigma}\mathbf{n}\,\mathrm{d}S \tag{4.114}$$

where **n** is the final and **N** the initial unit normal vectors. Referring to Figure 4.12, the unit area $\mathrm{d}S_{\mathrm{o}}$ can be expressed as (see [M1] for details):

$$\mathbf{N}\,\mathrm{d}S_{\mathrm{o}} = \mathrm{d}\mathbf{X}_a \times \mathrm{d}\mathbf{X}_b = \mathbf{F}^{-1}\,\mathrm{d}\mathbf{x}_a \times \mathbf{F}^{-1}\,\mathrm{d}\mathbf{x}_b = \frac{1}{J}\mathbf{F}^{\mathrm{T}}(\mathrm{d}\mathbf{x}_a \times \mathrm{d}\mathbf{x}_b) = \frac{1}{J}\mathbf{F}^{\mathrm{T}}\mathbf{n}\,\mathrm{d}S \tag{4.115}$$

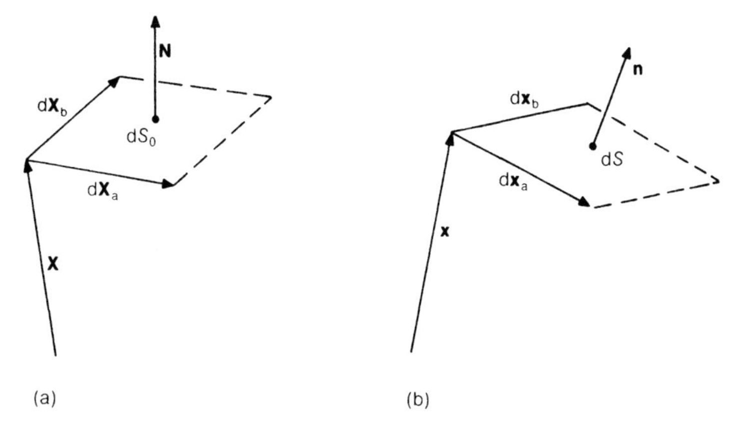


**Figure 4.12** Areas in the initial and final three-dimensional configurations: (a) initial; (b) final.

with $J$ being given by (4.96). Equation (4.115) is known as Nanson's formula. Substitution into (4.114) gives

$$\mathbf{P} = J\boldsymbol{\sigma}\mathbf{F}^{-\mathrm{T}} = \mathbf{FS} \tag{4.116}$$

where the last expression in (4.116) is obtained with the aid of (4.104).

The first Piola–Kirchhoff stress, $\mathbf{P}$, which is non-symmetric, is work-conjugate to the infinitesimal virtual displacement gradient, $\delta\mathbf{D}_v$ (related to (4.72)). This can be demonstrated with the aid of (4.76) and (4.79), i.e.

$$\begin{aligned} V_i &= \int \mathbf{S}:\delta\mathbf{E}_v \, dV_o = \tfrac{1}{2}\int \mathbf{S}:[\mathbf{F}^{\mathrm{T}}\delta\mathbf{D}_v + \delta\mathbf{D}_v^{\mathrm{T}}\mathbf{F}]\, dV_o \\ &= \int \mathbf{S}:[\mathbf{F}^{\mathrm{T}}\delta\mathbf{D}_v]\, dV_n = \int \mathbf{FS}:\delta\mathbf{D}_v \, dV_o = \int \mathbf{P}:\delta\mathbf{D}_v \, dV_o. \end{aligned} \tag{4.117}$$

We can now summarise the relationship between the various stress and strain measures using the principle of virtual power which is effectively equivalent to the principle of virtual work, but with virtual velocities instead of virtual displacements. Using the principle,

$$\begin{aligned} \dot{V} &= \frac{V_i}{dt} = \frac{1}{dt}\int \mathbf{S}:\delta\mathbf{E}_v \, dV_o = \int \mathbf{S}:\dot{\mathbf{E}}_v \, dV_o = \frac{1}{dt}\int \mathbf{P}:\delta\mathbf{D}_v \, dV_o \\ &= \int \mathbf{P}:\dot{\mathbf{F}}_v \, dV_o = \frac{1}{dt}\int \boldsymbol{\tau}:\delta\boldsymbol{\varepsilon}_v \, dV_o = \int \boldsymbol{\tau}:\dot{\boldsymbol{\varepsilon}}_v \, dV_o = \frac{1}{dt}\int \boldsymbol{\sigma}:\delta\boldsymbol{\varepsilon}_v \, dV = \int \boldsymbol{\sigma}:\dot{\boldsymbol{\varepsilon}}_v \, dV \end{aligned} \tag{4.118}$$

where

the first Piola–Kirchhoff stress: $\mathbf{P} = \mathbf{FS} = [\mathbf{I} + \mathbf{D}]\mathbf{S} = \det(\mathbf{F})\boldsymbol{\sigma}\mathbf{F}^{-\mathrm{T}}$ (4.119)

the second Piola–Kirchhoff stress: $\mathbf{S} = \det(\mathbf{F})\mathbf{F}^{-1}\boldsymbol{\sigma}\mathbf{F}^{-\mathrm{T}}$ (4.120)

the true or Cauchy stress: $\boldsymbol{\sigma} = \dfrac{1}{\det(\mathbf{F})}\mathbf{FSF}^{\mathrm{T}}$ (4.121)

the Kirchhoff or nominal stress: $\boldsymbol{\tau} = \det(\mathbf{F})\boldsymbol{\sigma} = \mathbf{FSF}^{\mathrm{T}}$. (4.122)

The stress tensors **S**, **P** and $\boldsymbol{\tau}$ are work-conjugate to their complementary virtual 'strain measures' $\delta\mathbf{E}_v$, $\delta\mathbf{D}_v$ (or $\dot{\mathbf{F}}_v$) and $\dot{\boldsymbol{\varepsilon}}_v$ (or $\delta\boldsymbol{\varepsilon}_v$) in relation to the original volume $V_0$ while the Cauchy stress, $\boldsymbol{\sigma}$, is conjugate to $\dot{\boldsymbol{\varepsilon}}_v$ (or $\delta\boldsymbol{\varepsilon}_v$) in relation to the final volume $V$.

## 4.8 THE POLAR-DECOMPOSITION THEOREM

The polar decomposition theorem is useful for (1) large-strain and large-rotation applications and (2) applications with corotational or convective coordinates [B2]. It can also be used to provide a simple physical explanation of the second Piola–Kirchhoff stresses. The theorem states that the deformation gradient of (4.71) can be decomposed (Figures 4.13 and 4.14) into a set of stretches followed by a rigid rotation.

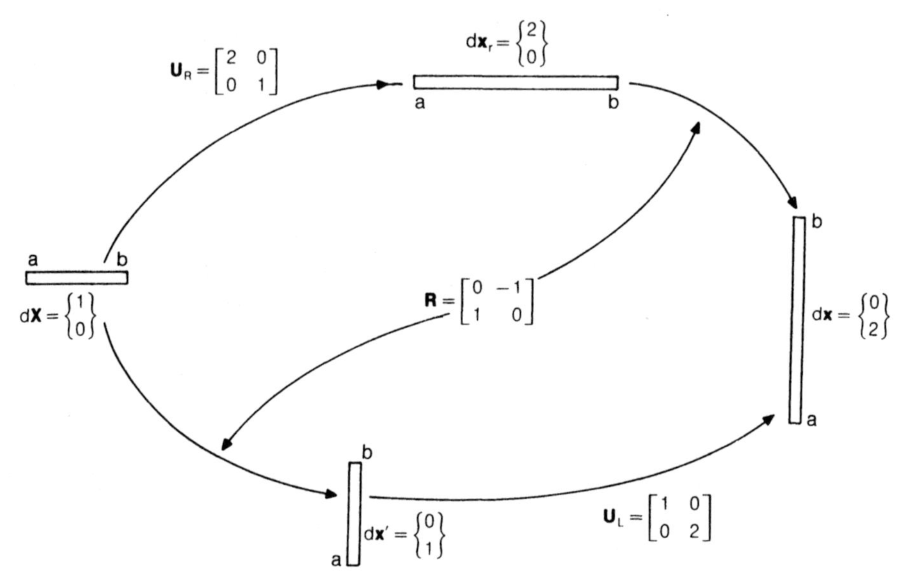


**Figure 4.13** A simple example of polar decomposition.

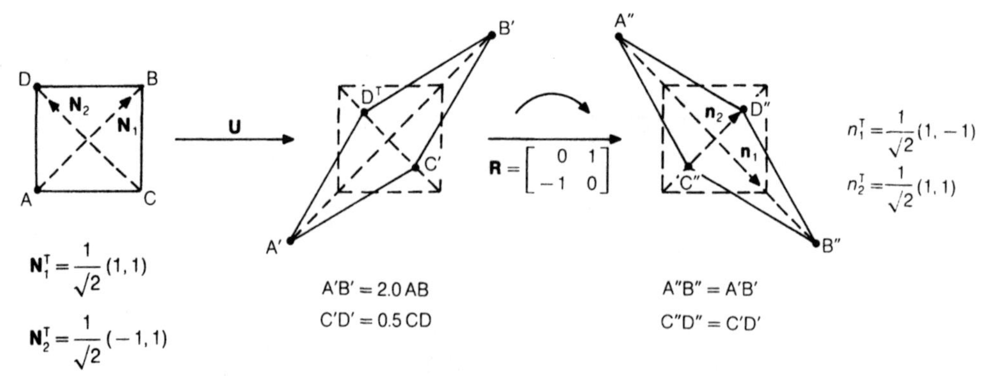


**Figure 4.14** A more complex example of polar decomposition.

The stretches involve (Figure 4.13)

$$\mathbf{dx}_r = \frac{\partial \mathbf{x}_r}{\mathbf{dX}}\mathbf{dX} = \mathbf{U}_R\,\mathbf{dX} \tag{5.123}$$

and are followed (Figure 4.13) by the rotation

$$\mathbf{dx} = \frac{\partial \mathbf{x}}{\partial \mathbf{x}_r}\mathbf{dx}_r = \mathbf{R}\,\mathbf{dx}_r \tag{4.124}$$

and hence

$$\mathbf{dx} = \frac{\partial \mathbf{x}}{\partial \mathbf{X}}\mathbf{dX} = \mathbf{F}\,\mathbf{dX} = \frac{\partial \mathbf{x}}{\partial \mathbf{x}_r}\frac{\partial \mathbf{x}_r}{\partial \mathbf{X}}\mathbf{dX} = \mathbf{R}\mathbf{U}_R\,\mathbf{dX}. \tag{4.125}$$

Equally (Figure 4.13), the rotation can be followed by the stretch and hence

$$\mathbf{F} = \mathbf{R}\mathbf{U}_R = \mathbf{U}_L\mathbf{R} = \mathbf{V}\mathbf{R}. \tag{4.126}$$

We will usually use the first part of (4.126) and will often omit the subscript R (for right). For the simple two-dimensional example of Figure 4.13,

$$\mathbf{F} = \mathbf{R}\mathbf{U}_R = \begin{bmatrix} 0 & -1 \\ 2 & 0 \end{bmatrix} = \begin{bmatrix} 0 & -1 \\ 1 & 0 \end{bmatrix}\begin{bmatrix} 2 & 0 \\ 0 & 1 \end{bmatrix}. \tag{4.127}$$

In general, the decomposition is more complicated (Figure 4.14), because the principal stretch directions must be computed. In Figure 4.14, the material is sheared and then rotated. In relation to the original body, the principal directions involve AB (or A′B′) and CD (or C′D′) and have unit vectors given by

$$\mathbf{N}_1^T = \frac{1}{\sqrt{2}}\{1, 1\} \qquad \mathbf{N}_2^T = \frac{1}{\sqrt{2}}\{-1, 1\}. \tag{4.128}$$

The $\mathbf{N}_1$ direction is stretched by a factor of 2.0 and the $\mathbf{N}_2$ direction by a factor of 0.5 (i.e. compressed). hence, the stretch matrix, $\mathbf{U}$, can be written as

$$\mathbf{U} = 2.0\,\mathbf{N}_1\mathbf{N}_1^T + 0.5\,\mathbf{N}_2\mathbf{N}_2^T = \begin{bmatrix} 1.25 & 0.75 \\ 0.75 & 1.25 \end{bmatrix} \tag{4.129}$$

and the final deformation-gradient matrix $\mathbf{F}$ is given by

$$\mathbf{F} = \mathbf{R}\mathbf{U} = \begin{bmatrix} 0 & -1 \\ 1 & 0 \end{bmatrix}\begin{bmatrix} 1.25 & 0.75 \\ 0.75 & 1.25 \end{bmatrix}. \tag{4.130}$$

The concepts should become clearer once the theory has been formalised.

In (4.74) and (4.91), we introduced the strain measures $\mathbf{E}$ (Green) and $\mathbf{A}$ (Almansi). A more direct 'stretching measure' is given by

$$\lambda = \frac{dr_n}{dr_o} = \left(\frac{\mathbf{dx}^T\mathbf{dx}}{\mathbf{dX}^T\mathbf{dX}}\right)^{1/2} = \left\|\frac{\mathbf{dx}}{\mathbf{dX}}\right\| \tag{4.131}$$

so that

$$\lambda^2 = \frac{\mathbf{dx}^T\mathbf{dx}}{\mathbf{dX}^T\mathbf{dX}} = \mathbf{N}^T\mathbf{F}^T\mathbf{F}\mathbf{N} \tag{4.132}$$

where

$$\mathbf{N} = \frac{d\mathbf{X}}{(d\mathbf{X}^T d\mathbf{X})^{1/2}} = \frac{d\mathbf{X}}{\|d\mathbf{X}\|} \tag{4.133}$$

is a unit vector in the direction of $d\mathbf{X}$. The stretch measure, $\lambda$, will be unity for a rigid-body rotation.

It is useful to vary the directions of $\mathbf{N}$ and find the principal stretch values and their corresponding directions. To this end, consider the functional

$$\phi = \mathbf{N}^T\mathbf{F}^T\mathbf{F}\mathbf{N} - \alpha(\mathbf{N}^T\mathbf{N} - 1) \tag{4.134}$$

where $\alpha$ is a Lagrangian multiplier provided to ensure that $\mathbf{N}$ remains a unit vector. For $\delta\phi = 0$, variations with respect to $\mathbf{N}$ give

$$[\mathbf{F}^T\mathbf{F} - \alpha\mathbf{I}]\mathbf{N} = \mathbf{0}. \tag{4.135}$$

Pre-multiplying (4.135) by $\mathbf{N}^T$ and comparing with (4.132) shows that $\alpha = \lambda^2$ and hence (4.135) gives

$$[\mathbf{F}^T\mathbf{F} - \lambda^2\mathbf{I}]\mathbf{N} = [\mathbf{U}^T\mathbf{U} - \lambda^2\mathbf{I}]\mathbf{N} \tag{4.136}$$

where the last relationship follows from (4.126) (dropping the subscript R) having noted that $\mathbf{R}$ is a rotation matrix so that $\mathbf{R}^T = \mathbf{R}^{-1}$. Equation (4.136) is an eigenvalue problem from which $\lambda_1^2 - \lambda_3^2$ and hence $\lambda_1 - \lambda_3$ can be obtained along with the principal directions $\mathbf{N}_1 - \mathbf{N}_3$. Hence, we can write

$$\mathbf{F}^T\mathbf{F} = \mathbf{U}^T\mathbf{U} = \lambda_1^2\mathbf{N}_1\mathbf{N}_1^T + \lambda_2^2\mathbf{N}_2\mathbf{N}_2^T + \lambda_3^2\mathbf{N}_3\mathbf{N}_3^T = \mathbf{Q}(\mathbf{N})\,\mathrm{Diag}\,(\lambda^2)\mathbf{Q}(\mathbf{N})^T \tag{4.137}$$

where

$$\mathbf{Q}(\mathbf{N}) = [\mathbf{N}_1, \mathbf{N}_2, \mathbf{N}_3] \tag{4.138}$$

contains the eigenvectors $\mathbf{N}_1$–$\mathbf{N}_3$. These eigenvectors can be used to express the stretch matrix, $\mathbf{U}$ in terms of $\lambda_1$–$\lambda_3$ and $\mathbf{N}_1$–$\mathbf{N}_3$ via

$$\mathbf{U} = \lambda_1\mathbf{N}_1\mathbf{N}_1^T + \lambda_2\mathbf{N}_2\mathbf{N}_2^T + \lambda_3\mathbf{N}_3\mathbf{N}_3^T = \mathbf{Q}(\mathbf{N})\,\mathrm{Diag}\,(\lambda)\mathbf{Q}(\mathbf{N})^T = \mathbf{Q}(\mathbf{N})\begin{bmatrix} \lambda_1 & & \\ & \lambda_2 & \\ & & \lambda \end{bmatrix}\mathbf{Q}(\mathbf{N})^T \tag{4.139}$$

which is the solution to the eigenvalue problem

$$[\mathbf{U} - \lambda\mathbf{I}]\mathbf{N} = \mathbf{0}. \tag{4.140}$$

Clearly (4.139) is compatible with (4.137). The eigenvectors or principal stretch directions $\mathbf{N}_1$–$\mathbf{N}_3$ satisfy $\mathbf{N}_1^T\mathbf{N}_2 = \mathbf{N}_1^T\mathbf{N}_3 = \mathbf{N}_2^T\mathbf{N}_3 = 0$ and define a rectangular orthogonal system of unit vectors referred to as the 'Lagrangian triad' or 'material axes'.

The equivalent of (4.133) in the current (spatial) configuration is

$$\mathbf{n} = \frac{d\mathbf{x}}{(d\mathbf{x}^T d\mathbf{x})^{1/2}} = \frac{d\mathbf{x}}{\|d\mathbf{x}\|}. \tag{4.141}$$

From (4.141), (4.70) and (4.133),

$$\mathbf{n} = \frac{\mathbf{dx}}{\|\mathbf{dx}\|} = \frac{\mathbf{F}\,\mathbf{dX}}{\|\mathbf{dx}\|} = \left\|\frac{\mathbf{dX}}{\mathbf{dx}}\right\| \mathbf{FN} = \frac{1}{\lambda}\mathbf{FN}. \tag{4.142}$$

Substitution from (4.142) into (4.136) gives

$$\lambda \mathbf{F}^{\mathrm{T}}\mathbf{n} - \lambda^3 \mathbf{F}^{-1}\mathbf{n} = \mathbf{0}. \tag{4.143}$$

Multiplying (4.143) by $(1/\lambda)\mathbf{F}$ (assuming $\lambda \neq 0$) gives

$$[\mathbf{FF}^{\mathrm{T}} - \lambda^2 \mathbf{I}]\mathbf{n} = [\mathbf{VV}^{\mathrm{T}} - \lambda^2 \mathbf{I}]\mathbf{n} \tag{4.144}$$

where for the second relationship in (4.144), we have used the **VR** decomposition in (4.126). Equation (4.144) is the spatial equivalent of (4.136) while the equivalents of (4.139) and (4.140) are

$$\mathbf{V} = \lambda_1 \mathbf{n}_1 \mathbf{n}_1^{\mathrm{T}} + \lambda_2 \mathbf{n}_2 \mathbf{n}_2^{\mathrm{T}} + \lambda_3 \mathbf{n}_3 \mathbf{n}_3^{\mathrm{T}} = \mathbf{Q}(\mathbf{n})\,\mathrm{Diag}\,(\lambda)\mathbf{Q}(\mathbf{n})^{\mathrm{T}} \tag{4.145}$$

and

$$[\mathbf{V} - \lambda \mathbf{I}]\mathbf{n} = \mathbf{0}. \tag{4.146}$$

The unit vectors $\mathbf{n}_1$–$\mathbf{n}_3$ define the 'Eulerian triad'.

Because the rotation matrix, $\mathbf{R}$, defines the movement from $\mathbf{N}_i$ to $\mathbf{n}_i$, from (4.67), it is given by

$$\mathbf{R} = \mathbf{Q}(\mathbf{n})\mathbf{Q}(\mathbf{N})^{\mathrm{T}}. \tag{4.147}$$

Hence, from (4.126), (4.139) and (4.147):

$$\mathbf{F} = \mathbf{RU} = \mathbf{Q}(\mathbf{n})\,\mathrm{Diag}\,(\lambda)\mathbf{Q}(\mathbf{N})^{\mathrm{T}} = \lambda_1 \mathbf{n}_1 \mathbf{N}_1^{\mathrm{T}} + \lambda_2 \mathbf{n}_2 \mathbf{N}_2^{\mathrm{T}} + \lambda_3 \mathbf{n}_3 \mathbf{N}_3^{\mathrm{T}}. \tag{4.148}$$

### 4.8.1 Example

To help understand these concepts, it may help the reader to return to the example of Figure 4.14. A question could be formulated as follows.

Given

$$\mathbf{D} = \begin{bmatrix} \dfrac{\partial u}{\partial X} & \dfrac{\partial u}{\partial Y} \\ \dfrac{\partial v}{\partial X} & \dfrac{\partial v}{\partial Y} \end{bmatrix} = \begin{bmatrix} 0.25 & 0.75 \\ 0.75 & 0.26 \end{bmatrix} \tag{4.149}$$

calculate, in order,

(a) $\mathbf{F}$ and $\mathbf{F}^{\mathrm{T}}\mathbf{F}$
(b) $\lambda_1^2, \lambda_2^2, \mathbf{N}_1, \mathbf{N}_2$ from the eigenvalue problem of (4.136)
(c) $\mathbf{U}$ from (4.139)
(d) $\mathbf{U}^{-1}$ indirectly from (4.139) using

$$\mathbf{U}^{-1} = \frac{1}{\lambda_1}\mathbf{N}_1\mathbf{N}_1^{\mathrm{T}} + \frac{1}{\lambda_2}\mathbf{N}_2\mathbf{N}_2^{\mathrm{T}} = \mathbf{Q}(\mathbf{N})\,\mathrm{Diag}\left(\frac{1}{\lambda}\right)\mathbf{Q}(\mathbf{N})^{\mathrm{T}} \tag{4.150}$$

(e) $\mathbf{R}$ from (4.126) via $\mathbf{R} = \mathbf{F}\mathbf{U}^{-1}$.
(f) $\mathbf{Q(n)}$ and hence $\mathbf{n}_1$ and $\mathbf{n}_2$ using (4.147) so that $\mathbf{Q(n)} = \mathbf{RQ(N)}$
(g) the stretch $\mathbf{V}$ from (4.145).
The solutions have already been given in Figure 4.14 and (4.127)–(4.130).
Although the information has already been obtained could also compute
(h) $\mathbf{FF}^T$ and hence $\lambda_1^2, \lambda_2^2$, $\mathbf{N}_1, \mathbf{N}_2$ from the eigenvalue problem of (4.144).

## 4.9 GREEN AND ALMANSI STRAINS IN TERMS OF THE PRINCIPAL STRETCHES

In principal strain space, the Green strain components (see (4.68) and (3.9)) can be written as

$$E_i = \tfrac{1}{2}(\lambda_i^2 - 1), \qquad i = 1, 3 \tag{4.151}$$

while the Almansi strains (see (4.89) and (3.41)) can be written as

$$A_i = \tfrac{1}{2}\left(1 - \frac{1}{\lambda_i^2}\right), \qquad i = 1, 3 \tag{4.152}$$

and using (4.74) for $\mathbf{E}$, and (4.137) for $\mathbf{F}^T\mathbf{F}$,

$$\mathbf{E} = \tfrac{1}{2}(\mathbf{F}^T\mathbf{F} - \mathbf{I}) = \tfrac{1}{2}(\mathbf{U}^T\mathbf{U} - \mathbf{Q(N)Q(N)}^T) = \left[\mathbf{Q(N)}\,\mathrm{Diag}\left(\frac{\lambda^2 - 1}{2}\right)\mathbf{Q(N)}^T\right] \tag{4.153}$$

where we have used the orthogonality of $\mathbf{Q(N)}$ so that $\mathbf{Q(N)Q(N)}^T = \mathbf{I}$. Equation (4.153) confirms (4.151) and shows that the directions of the principal Green strain coincide with the Lagrangian (or material) triad, $\mathbf{N}_1$–$\mathbf{N}_3$.

In a similar fashion, using (4.91) for the Almansi strain, $\mathbf{A}$ and (4.145) for $\mathbf{V}$,

$$\mathbf{A} = \tfrac{1}{2}(\mathbf{I} - \mathbf{F}^T\mathbf{F}^{-1}) = \tfrac{1}{2}(\mathbf{Q(n)Q(n)}^T - \mathbf{V}^{-T}\mathbf{V}^{-1}) = \left[\mathbf{Q(n)}\,\mathrm{Diag}\left(\frac{\lambda^2 - 1}{2\lambda^2}\right)\mathbf{Q(n)}^T\right] \tag{4.154}$$

which not only confirms (4.152) but also shows that the directions of the principal Almansi strain coincide with the Eulerian (or spatial) triad, $\mathbf{n}_1$–$\mathbf{n}_3$.

Suppose a Green strain, $\mathbf{E}_1$, is created by a stretch $\mathbf{U}$, followed by a rotation $\mathbf{R}_1$ (see Figures 4.13 and 4.14) while $\mathbf{E}_2$ is created by $\mathbf{U}$ being followed by $\mathbf{R}_2$. It follows from (4.74) and (4.126) that

$$\mathbf{E}_1 = \mathbf{E}_2 = \tfrac{1}{2}(\mathbf{F}^T\mathbf{F} - \mathbf{I}) = \tfrac{1}{2}(\mathbf{U}^T\mathbf{U} - \mathbf{I}) \tag{4.155}$$

and is therefore unaltered by a rotation change from $\mathbf{R}_1$ to $\mathbf{R}_2$. The same conclusion can be drawn by observing that the material triad, $\mathbf{N}_1$–$\mathbf{N}_3$ is unaltered by $\mathbf{R}$ and hence from the right-hand side of (4.153), $\mathbf{E}$ is unaltered.

In contrast, from (4.147), the Eulerian triad, $\mathbf{Q(n)}$, does change with $\mathbf{R}$ and hence the Almansi strain does change as a result of a rigid rotation.

## 4.10 A SIMPLE DESCRIPTION OF THE SECOND PIOLA–KIRCHHOFF STRESS

If the strains are small, $\lambda_i \simeq 1$ and, from (4.139) we can write

$$\mathbf{U} \simeq \mathbf{I} \tag{4.156}$$

so that, from the decomposition theorem of (4.126),

$$\mathbf{F} \simeq \mathbf{R}. \tag{4.157}$$

Because of (4.157),

$$J = \det(\mathbf{F}) \simeq 1. \tag{4.158}$$

Hence, given a second Piola–Kirchhoff stress, $\mathbf{S}$, obtained from the Green strain in fixed (global) axes, the latter can be converted to a Cauchy stress (still related to fixed, global, axes) so that with the aid of (4.104), (4.157) and (4.158),

$$\boldsymbol{\sigma}_g = \mathbf{R}\mathbf{S}\mathbf{R}^T. \tag{4.159}$$

This stress can now be related to the new (local) rotated axes via (4.36) so that with $\mathbf{T}^T = \mathbf{R}$ (see (4.60)),

$$\boldsymbol{\sigma}_l = \mathbf{R}^T\boldsymbol{\sigma}_g\mathbf{R} = \mathbf{R}^T\mathbf{R}\mathbf{S}\mathbf{R}^T\mathbf{R} = \mathbf{S}. \tag{4.160}$$

Hence, for small strains, the second Piola–Kirchhoff stress can be interpreted as the Cauchy stress related to axes that rotate with the material.

## 4.11 COROTATIONAL STRESSES AND STRAINS

Although the coincidence with the second Piola–Kirchhoff stress is valid only for small strains, the concept of a 'rotated Cauchy stress' is also useful for shell analysis [B2]. Figure 4.15 illustrates the ideas.

Suppose the stress state in Figure 4.15(a) is rotated to the state in Figure 4.15(c). Clearly, in local coordinates, the state is unchanged. However, from Mohr's circle (Figure 4.15(d)), the stress state in global coordinates is as shown in Figure 4.15(e).

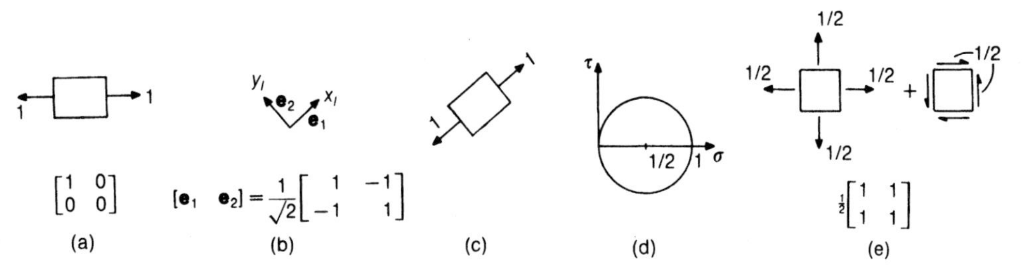


**Figure 4.15** Some concepts with rotating coordinates: (a) $\boldsymbol{\sigma}_a$; (b) $\mathbf{R}$; (c) $\boldsymbol{\sigma}_l$; (d) Mohr's circle; (e) $\boldsymbol{\sigma}_g$.

This relationship can, alternatively, be found from the use of (4.160), so that

$$\boldsymbol{\sigma}_{\rm g} = \mathbf{R}\boldsymbol{\sigma}_{\rm a}\mathbf{R}^{\rm T} = \frac{1}{2}\begin{bmatrix} 1 & -1 \\ 1 & 1 \end{bmatrix}\begin{bmatrix} 1 & 0 \\ 0 & 0 \end{bmatrix}\begin{bmatrix} 1 & 1 \\ -1 & 1 \end{bmatrix} = \frac{1}{2}\begin{bmatrix} 1 & 1 \\ 1 & 1 \end{bmatrix}. \tag{4.161}$$

If we assume that $\boldsymbol{\sigma}_{\rm g}$ (Figure 4.15(e)) is now a given Cauchy stress, a 'corotated stress', $\boldsymbol{\sigma}_{\rm l}$, related to the rotated system (Figure 4.15(c)) can be computed via

$$\boldsymbol{\sigma}_{\rm l} = \mathbf{T}\boldsymbol{\sigma}_{\rm g}\mathbf{T}^{\rm T} = \mathbf{R}^{\rm T}\boldsymbol{\sigma}_{\rm g}\mathbf{R}. \tag{4.162}$$

Hence, this 'corotated stress' can either be thought of as the original Cauchy stress, $\boldsymbol{\sigma}_{\rm g}$, expressed in 'local coordinates' that rotate with the body or, alternatively, as the original stresses, $\boldsymbol{\sigma}_{\rm g}$, still related to 'global coordinates' but rotated by $\mathbf{R}^{\rm T}$ (see 4.63)) i.e. rotated back to the original configuration (as in Figure 4.15(a)) via $\mathbf{R}^{\rm T}$.

## 4.12 MORE ON CONSTITUTIVE LAWS

In the present books, we will be mainly concerned with three types of stress–stain law:

(a) linear elastic
(b) hyperelastic
(c) hypoelastic
(d) elasto-plastic.

We have already discussed (a) in Section 4.2. A chapter of Volume 2 will be devoted to (b) while Chapter 6 of the current volume and further chapters of Volume 2 will be devoted to plasticity. Here we will give a very brief summary of the different relationships providing only sufficient detail to enable an understanding of the following Chapter 5 on finite elements and continua.

Hyperelastic models are essentially higher-order forms of linear elastic models in which the stresses are some functions of the total strains or stretches [D1]. The obvious example of a hyperelastic material is rubber. If we consider a 'Green elastic materials' this total relationship is derivable from an elastic potential. If the strain energy/unit volume can be expressed as $\varphi$, then the change of strain energy is

$$\delta\varphi = \boldsymbol{\sigma}{:}\delta\boldsymbol{\varepsilon} = \frac{\partial\varphi}{\partial\boldsymbol{\varepsilon}}{:}\delta\boldsymbol{\varepsilon} \tag{4.163}$$

while the stresses can be obtained from

$$\varphi = \int\delta\varphi = \int\boldsymbol{\sigma}{:}\delta\boldsymbol{\varepsilon} \tag{4.164}$$

via

$$\boldsymbol{\sigma} = \frac{\partial\varphi}{\partial\boldsymbol{\varepsilon}}, \qquad \sigma_{ij} = \frac{\partial\varphi}{\partial\varepsilon_{ij}} \tag{4.165}$$

and the tangent relationship follows from a further differentiation, so that

$$\dot{\boldsymbol{\sigma}} = \mathbf{C}_{\rm t}{:}\dot{\boldsymbol{\varepsilon}} = \frac{\partial^2\varphi}{\partial\boldsymbol{\varepsilon}\partial\boldsymbol{\varepsilon}}{:}\dot{\boldsymbol{\varepsilon}}, \qquad \dot{\sigma}_{ij} = \frac{\partial^2\varphi}{\partial_{ij}\partial\varepsilon_{kl}}\dot{\varepsilon}_{kl}. \tag{4.166}$$

(We might have written $\delta\boldsymbol{\sigma}$ and $\delta\boldsymbol{\varepsilon}$ rather than $\dot{\boldsymbol{\sigma}}$ and $\dot{\boldsymbol{\varepsilon}}$.) As a consequence of (4.166), $\mathbf{C}_t$ will be symmetric $(\partial\varphi/\partial x\partial y = \partial\varphi/\partial y)$.

A one-dimensional example could involve a second-order parabolic relationship for $\sigma$ so that

$$\sigma = f(\varepsilon) = E\left(\varepsilon - \frac{\varepsilon^2}{2\varepsilon_o}\right) \tag{4.167}$$

where $\varepsilon_o$ is the strain at the peak response. From (4.164)

$$\varphi = \frac{E\varepsilon^2}{2} - \frac{E\varepsilon^3}{6\varepsilon_o} \tag{4.168}$$

while from (4.166)

$$\dot{\sigma} = E_t\dot{\varepsilon} = \frac{\partial^2\varphi}{\partial\varepsilon^2}\dot{\varepsilon} = E\left(\varepsilon - \frac{\varepsilon^2}{2\varepsilon_o}\right)\dot{\varepsilon} \tag{4.169}$$

This illustrates that the $\mathbf{C}_t$ matrix is generally not constant (see also Sections 3.2). Within a finite element context, we will need the $\mathbf{C}_t$ matrix (as in (4.82) or (4.166)) for the structural tangent stiffness matrix, but the total stresses and hence the internal force vector will, for a hyperelastic material, come directly from (4.165) and not from the integration of the 'rate equation' of the form of (4.166).

In contrast, for hypoelastic materials [D1], we have no such total relationships and are forced to start with a 'rate (or incremental) relationship' of the form:

$$\dot{\boldsymbol{\sigma}} = \mathbf{C}_t\dot{\boldsymbol{\sigma}}, \qquad \delta\boldsymbol{\sigma} = \mathbf{C}_t\delta\varepsilon \tag{4.170}$$

in which $\mathbf{C}_t$ (which may or may not be constant) will be given. Such relationships are often used for geomechanical materials. For a hypoelastic material, the $\mathbf{C}_t$ matrix is not only required for the structural tangent stiffness matrix but also it must be 'integrated' (at the Gauss-point level) to obtain the total stresses and hence the internal force vector. On applying a closed cycle of strain (which ends up with a total strain of zero), a hyperelastic model will give zero stress. This does not necessarily follow for a hypoelastic model (as will be discussed further in Volume 2).

The reader will be aware of the main aspects of plasticity which is treated in Chapter 6. For the present we need merely state that plasticity leads to a 'rate-form' of constitutive law along the lines of (4.170). Generally, the tangent $\mathbf{C}_t$ matrix will not only be a function of some material parameters but also of the current stresses and possibly 'internal variables' (see Section 6.4.1). The previous comments on hypoelastic materials apply also to plasticity.

We have so far not mentioned the issue of large or small strains. Hyperelastic materials such as rubber inevitably involve large strains. Some metal plasticity also involves large strains. Many previous finite element formulations effectively treated such matters via hypoelastic relationships coupled with plasticity, so that the integration of the rate equations was relevant both in relationship to plasticity and to the 'large-strain rate measures' (as will be discussed in Volume 2).

Before leaving this section we should re-emphasise the observation drawn in Section 3.2 that to obtain the same solution with different stress and strain measures, the stress–strain laws will need to be changed. This issue is discussed further in Sections 5.3 and 5.4.

## 4.13 SPECIAL NOTATION

$\mathbf{1}$ (or $\mathbf{1}_2$) = unit second-order tensor (see (4.30) and (4.31))
$\mathbf{e}_1, \mathbf{e}_2, \mathbf{e}_3$ = orthogonal unit vectors
$\mathbf{A}$ = vector area with components $\mathbf{A}_x, \mathbf{A}_y, \mathbf{A}_z$ (Section 4.1)
$\mathbf{A}$ = Almani's strain (see (4.91))
$\mathbf{C}_2, \mathbf{C}_4$ = second- and fourth-order constitutive tensor (or matrices)
$\mathbf{D}$ = displacement-derivative matrix (see (4.72))
$\mathbf{E}$ = Green's strain (as vector or tensor; the latter is sometimes $\mathbf{E}_2$)
$\mathbf{F}$ = external force vector (Section 4.1)
$\mathbf{F}$ = deformation gradient (see (4.71))
$\mathbf{I}$ = unit fourth-order tensor (see (4.30) and (4.31)) or (sometimes) unit matrix (or second-order tensor)
$J = \det(\mathbf{F})$
$k$ = bulk modulus
$\mathbf{L}$ = velocity gradient (see (4.110))
$\mathbf{n}$ = unit normal vector (Section 4.1)
$\mathbf{n}_1, \mathbf{n}_2, \mathbf{n}_3$ = unit vectors defining directions of principal stretch in final configuration (defines the Eulerian triad)
$\mathbf{N} = \mathbf{n}$ in initial configuration (Section 4.7)
$\mathbf{N}_1, \mathbf{N}_2, \mathbf{N}_3$ = unit vectors defining directions of principal stretch in initial configuration (defines the Lagrangian triad)
$\mathbf{P}$ = first Piola–Kirchhoff stresses
$\mathbf{Q}$ = orthogonal matrix containing principal directions, **N**s or **n**s
$\mathbf{r}$ = line element
$\mathbf{R}$ = rotation matrix
$\mathbf{S}$ = second Piola–Kirchhoff stresses (as vector or tensor; the latter is sometimes $\mathbf{S}_2$)
$\mathbf{t}$ = vector of equilibrating external tractions/unit area (Section 4.1)
$\mathbf{T}$ = transformation matrix
$\mathbf{u}$ = displacements
$\mathbf{U}$ (sometimes $\mathbf{U}_R$) = right-stretch matrix (from polar decomposition)
$\mathbf{U}_L$ = left-stretch matrix (sometimes $\mathbf{V}$)
$\mathbf{V}$ = left-stretch matrix (from polar decomposition)
$\mathbf{x}$ = final coordinates ($\mathbf{x} = \mathbf{X} + \mathbf{u}$)
$\mathbf{X}$ = initial coordinates
$\gamma_{xy}$, etc. = shear strains $= \varepsilon_{xy}/2 = \varepsilon_{12}/2$
$\varepsilon_m$ = mean strain
$\boldsymbol{\varepsilon}$ = vector or tensor of strains (latter sometimes $\boldsymbol{\varepsilon}_2$)
$\boldsymbol{\varepsilon}_d$ (or $\boldsymbol{\varepsilon}_{2d}$ or $\mathbf{e}$) = deviatoric strains
$\lambda$ = Lamé's constant (Section 4.2.3)
$\lambda$ = stretch scalar (see 4.131))
$\mu$ = shear modulus
$\sigma_m$ = mean stress
$\boldsymbol{\sigma}$ = stress (as vector or tensor; the latter is sometimes $\boldsymbol{\sigma}_2$); note: $\boldsymbol{\sigma}$ is sometimes used specifically for the Cauchy stress
$\boldsymbol{\sigma}_{2d}$ (or $\mathbf{s}$) = deviatoric stresses

$\tau_{xy}$, etc. = shear stress
$\tau$ = Kirchhoff or nominal stresses
$\varphi$ = strain energy

### Subscripts

g = global
$l$ = local
n = new
o = old

## 4.14 REFERENCES

[B1] Belytschko, T., An overview of semidiscretization and time integration procedures, *Computational Methods for Transient Problems*, ed. T. Belytschko *et al.*, North-Holland, Amsterdam, p. 1–65 (1983).
[B2] Belytschko, T. & Hseih, J., Non-linear transient finite element analysis with convected co-ordinates, *Int. J. Num. Meth. Engng.*, **7**, 255–271 (1973).
[B3] Bisplinghoff, R. L. Mar, J. M. & Pian, T. H. H., *Statics of Deformable Solids*, Addison-Wesley, Reading, Mass. (1965).
[D1] Desai, C. S. & Siriwardane, H. J., *Constitutive Laws for Engineering Materials*, Prentice Hall (1984).
[G1] Green, A. E. & Zerna, W., *Theoretical Elasticity*, Clarendon Press, Oxford (1954).
[H1] Hunter, S. C., *Mechanics of Continuous Media*, 2nd edition, Ellis Horwood, Chichester (1983).
[L1] Love, A. E. H., *A Treatise on the Mathematical Theory of Elasticity*, 4th edition, Dover, New York (1944).
[M1] Malvern, L. E., *Introduction to the Mechanics of a Continuous Medium*, Prentice Hall, Englewood Cliffs (1969).
[M2] Marguerre, K., Zur theorie der gekrummten platte grosser formanderung, *Proc. Fifth Int. Congress of Appl. Mech.*, Wiley, New York (1938).
[M3] Mase, G. E., *Theory and Problems of Continuum mechanics*, Schaum's Outline Series, McGraw-Hill, New York (1970).
[M4] Mattiasson, K., Continuum mechanics principles for large deformation problems in solid and structural mechanics, Pub. 81.6, Dept. of Struct. Mech., Chalmers University, Sweden (1981).
[S1] Spencer, A. J. M., *Continuum Mechanics*, Longmans, Hong-Kong (1980).
[T1] Timoshenko, S. & Goodier, J. N., *Theory of Elasticity*, 2nd edition, McGraw-Hill, New York (1951).
[T2] Timoshenko, S. P. & Woinowsky-Krieger, S., *Theory of Plates and Shells*, McGraw-Hill (1959).
[V1] Von Karman, T., Festigkeitsprobleme im maschinenbau, *Encyklopadie der Mathematishen Wissenschaften*, IV/4, C, pp. 311–385 (1910).
[W1] Wempner, G., *Mechanics of Solids*, McGraw-Hill, New York (1973).
[W2] Wood, R., Lecture notes on nonlinear continuum mechanics and associated finite element formulations, Dept. of Civil Engng., University of Swansea (1986).
[Y1] Young, E. C., *Vector and Tensor Analysis*, Marcel Decker, New York (1978).

# 5 Basic finite element analysis of continua

## 5.1 INTRODUCTION AND THE TOTAL LAGRANGIAN FORMULATION

In the present chapter, we will apply some of the continuum mechanics of Chapter 4 to the development of finite element formulations for two- and three-dimensional continua. We will mainly concentrate on the total Lagrangian formulation [Z1, B1, B2] but will also consider the updated Lagrangian technique [B1, B2]. Although the work is closely related to that described by Bathe [B2], there are some significant differences in approach.

The total Lagrangian method is appropriate for large rotations and small strains but may also be applied to large elastic strains (such as occur in rubber) if an appropriate hyperelastic material model (see Section 4.11 and Volume 2) is used. The method can be used for elasto-plastic problems (Chapter 6) with small strains but large rotations. Readers wishing for a brief introduction could omit the sections on the updated Lagrangian technique.

The total and updated Lagrangian procedures, which uses the Green strains and second Piola–Kirchhoff stresses (Section 4.4), have already been applied in Section 3.3 to truss elements. The present work is closely related to these earlier developments.

Generally, in this chapter, we will adopt the procedure of Chapter 4, whereby upper-case $X, Y$ and $Z$ (collectively $\mathbf{X}$) will relate to the initial coordinates, while lower-case $x, y, z$ (collectively $\mathbf{x}$) relate to the current configuration. However, throughout the present section on the total Lagrangian procedure, such a distinction is unnecessary since we will always be referring to the initial configuration. Hence, purely to aid a neater presentation, we will here violate this convention and use lower case.

We start by summarising the main results from Section 4.4 on the Green strain which (see (4.74)) is given by

$$\mathbf{E}_2 = \tfrac{1}{2}[\mathbf{F}^{\mathrm{T}}\mathbf{F} - \mathbf{I}] = \tfrac{1}{2}[\mathbf{D} + \mathbf{D}^{\mathrm{T}}] + \tfrac{1}{2}\mathbf{D}^{\mathrm{T}}\mathbf{D} \tag{5.1}$$

where $\mathbf{F} = \mathbf{I} + \mathbf{D}$ (see (4.71)). Instead, we could use the vector form of (4.75). From (4.78) the change, $\delta\mathbf{E}$, is

$$\delta\mathbf{E}_2 = \tfrac{1}{2}\mathbf{F}^{\mathrm{T}}\delta\mathbf{D} + \tfrac{1}{2}\mathbf{D}^{\mathrm{T}}\mathbf{F} + [\tfrac{1}{2}\delta\mathbf{D}^{\mathrm{T}}\delta\mathbf{D}]_{\mathrm{h}} \tag{5.2}$$

where the square-bracketed 'higher-order' terms vanish in the case of infinitesimal, virtual changes. From (4.76), the virtual work can be expressed as

$$V = \int \mathbf{S}_2 : \delta \mathbf{E}_{\mathrm{v}2} \, \mathrm{d}V_{\mathrm{o}} - V_{\mathrm{e}} = \int \mathbf{S}^{\mathrm{T}} \delta \mathbf{E}_{\mathrm{v}} \, \mathrm{d}V_{\mathrm{o}} - V_{\mathrm{e}}, \tag{5.3}$$

where $V_{\mathrm{e}}$ contains the virtual work performed by the external loads. Both the tensor and vector forms are included in (5.3) (see (Section 4.1)) because they will both be required in the following.

In equations (5.1)–(5.3), we have adopted the approach, introduced in the early part of Chapter 4, whereby a subscript 2 (second order) is introduced for the tensor (or matrix) stress and strain forms in order to distinguish them from the vector forms (which have no suffix). We will often, later, drop the subscript, 2, because it should be obvious from the context which form is being used. In particular, the tensor form will be required when the contraction symbol, :, is used (see discussion below equation (4.6)) while the vector form will be associated with the 'transpose', T (see equation (5.3)).

Equation (5.3) will allow the formation of the out-of-balance force vector, $\mathbf{g}$, while to form the tangent stiffness matrix, we can (but do not have to—see Section 5.1.2) use the change in (5.3). From (4.83), this is given by

$$\delta V = \int \{ \delta \mathbf{E}_{\mathrm{v}}^{\mathrm{T}} \mathbf{C}_{\mathrm{t}_2} \delta \mathbf{E} + \mathbf{S}_2 : \delta \mathbf{D}_{\mathrm{v}}^{\mathrm{T}} \delta \mathbf{D} \} \, \mathrm{d}V_{\mathrm{o}} \tag{5.4}$$

where, for future convenience, we have used the vector form for the first term and the tensor form for the second.

### 5.1.1 Element formulation

We will firstly consider a two-dimensional formulation in which the displacements $u$ and $v$ are related to nodal values $\mathbf{u}$ and $\mathbf{v}$ via shape functions $\mathbf{h}$ which involve the non-dimensional coordinates $\xi$ and $\eta$, so that

$$u = \mathbf{h}(\xi, \eta)^{\mathrm{T}} \mathbf{u}, \qquad v = \mathbf{h}(\xi, \eta)^{\mathrm{T}} \mathbf{v}. \tag{5.5}$$

In the standard isoparametric manner, the shape functions are also used to relate the coordinates $x$ and $y$ to nodal coordinates $\mathbf{x}$ and $\mathbf{y}$. Then at any point (in particular, a Gauss point), the Jacobian is obtained as

$$\begin{pmatrix} \dfrac{\partial}{\partial \xi} \\ \dfrac{\partial}{\partial \eta} \end{pmatrix} = \begin{bmatrix} \dfrac{\partial x}{\partial \xi} & \dfrac{\partial y}{\partial \xi} \\ \dfrac{\partial x}{\partial \eta} & \dfrac{\partial y}{\partial \eta} \end{bmatrix} \begin{pmatrix} \dfrac{\partial}{\partial x} \\ \dfrac{\partial}{\partial y} \end{pmatrix} = \mathbf{J} \begin{pmatrix} \dfrac{\partial}{\partial x} \\ \dfrac{\partial}{\partial y} \end{pmatrix}. \tag{5.6}$$

From (5.5) and (5.6), one can obtain

$$\frac{\partial u}{\partial x} = \mathbf{J}^{-1}(1,1) \frac{\partial u}{\partial \xi} + \mathbf{J}^{-1}(1,2) \frac{\partial u}{\partial \eta} \tag{5.7}$$

and equivalent terms so that a vectorised form of the displacement derivative tensor,

$\mathbf{D}$, can be obtained as

$$\boldsymbol{\theta} = \begin{bmatrix} \dfrac{\partial u}{\partial x} \\ \dfrac{\partial u}{\partial y} \\ \dfrac{\partial v}{\partial x} \\ \dfrac{\partial v}{\partial y} \end{bmatrix} = \begin{bmatrix} \mathbf{J}^{-1}(1,1)\mathbf{h}_\xi^{\mathrm{T}} + \mathbf{J}^{-1}(1,2)\mathbf{h}_\eta^{\mathrm{T}} & \mathbf{0}^{\mathrm{T}} \\ \mathbf{J}^{-1}(2,1)\mathbf{h}_\xi^{\mathrm{T}} + \mathbf{J}^{-1}(2,2)\mathbf{h}_\eta^{\mathrm{T}} & \mathbf{0}^{\mathrm{T}} \\ \mathbf{0}^{\mathrm{T}} & \mathbf{J}^{-1}(1,1)\mathbf{h}_\xi^{\mathrm{T}} + \mathbf{J}^{-1}(1,2)\mathbf{h}_\eta^{\mathrm{T}} \\ \mathbf{0}^{\mathrm{T}} & \mathbf{J}^{-1}(2,1)\mathbf{h}_\xi^{\mathrm{T}} + \mathbf{J}^{-1}(2,2)\mathbf{h}_\eta^{\mathrm{T}} \end{bmatrix} \begin{bmatrix} \mathbf{u} \\ \mathbf{v} \end{bmatrix} = \mathbf{G}\mathbf{p} \quad (5.8)$$

where $\mathbf{p}$ contains the full vector of nodal displacement. (See the footnote on page 25 regarding the ordering of the variables). A similar equation relates $\delta\boldsymbol{\theta}$ (equivalent to $\delta\mathbf{D}$) to $\delta\mathbf{p}$ so that

$$\delta\boldsymbol{\theta} = \mathbf{G}\,\delta\mathbf{p} \quad (5.9)$$

and we can add the subscript 'v' for virtual to both $\delta\boldsymbol{\theta}$ and $\delta\mathbf{p}$ in (5.9). Using (5.8), the Green's strain of (5.1) can be written in vector form (see (4.75)) as

$$\mathbf{E} = \mathbf{E}_l + \mathbf{E}_{nl} = \begin{bmatrix} \dfrac{\partial u}{\partial x} \\ \dfrac{\partial v}{\partial y} \\ \dfrac{\partial u}{\partial y} + \dfrac{\partial v}{\partial x} \end{bmatrix} + \frac{1}{2} \begin{bmatrix} \dfrac{\partial u}{\partial x} & 0 & \dfrac{\partial v}{\partial x} & 0 \\ 0 & \dfrac{\partial u}{\partial y} & 0 & \dfrac{\partial v}{\partial y} \\ \dfrac{\partial u}{\partial y} & \dfrac{\partial u}{\partial x} & \dfrac{\partial v}{\partial y} & \dfrac{\partial v}{\partial x} \end{bmatrix} \begin{bmatrix} \dfrac{\partial u}{\partial x} \\ \dfrac{\partial u}{\partial y} \\ \dfrac{\partial v}{\partial x} \\ \dfrac{\partial v}{\partial y} \end{bmatrix} = \mathbf{E}_l + \tfrac{1}{2}\mathbf{A}(\boldsymbol{\theta})\boldsymbol{\theta} \quad (5.10)$$

or

$$\mathbf{E} = \mathbf{E}_l + \mathbf{E}_{nl} = [\mathbf{H} + \tfrac{1}{2}\mathbf{A}(\boldsymbol{\theta})]\boldsymbol{\theta} \quad (5.11)$$

where

$$\mathbf{H} = \begin{bmatrix} 1 & 0 & 0 & 0 \\ 0 & 0 & 0 & 1 \\ 0 & 1 & 1 & 0 \end{bmatrix}. \quad (5.12)$$

The change in Green's strain (5.2) can be expressed in vector form as

$$\delta\mathbf{E} = \delta\mathbf{E}_l + \tfrac{1}{2}\mathbf{A}(\boldsymbol{\theta})\delta\boldsymbol{\theta} + \tfrac{1}{2}\delta\mathbf{A}(\boldsymbol{\theta})\boldsymbol{\theta} + O(\delta\boldsymbol{\theta}^2) \quad (5.13)$$

where $\delta\mathbf{E}_l$ is as $\mathbf{E}_l$ in (5.10) but with terms such as $\partial\delta u/\partial x$ in place of $\partial u/\partial x$. From the expression for $\mathbf{A}(\boldsymbol{\theta})$ in (5.10), it is clear the $\delta\mathbf{A}$ will be $\mathbf{A}(\delta\boldsymbol{\theta})$, which is of the same form as $\mathbf{A}(\boldsymbol{\theta})$ in (5.10) but again with $\partial u/\partial x$ replaced by $\partial\delta u/\partial x$, etc. Because it can be shown that

$$\mathbf{A}(\delta\boldsymbol{\theta})\boldsymbol{\theta} = \mathbf{A}(\boldsymbol{\theta})\delta\boldsymbol{\theta} \tag{5.14}$$

equation (5.13) can be re-expressed as

$$\begin{aligned}\delta\mathbf{E} &= \delta\mathbf{E}_l + \mathbf{A}(\boldsymbol{\theta})\delta\boldsymbol{\theta} + O(\delta\boldsymbol{\theta}^2)\\ &= [\mathbf{H} + \mathbf{A}(\theta)]\delta\boldsymbol{\theta} + O(\delta\boldsymbol{\theta}^2) = [\mathbf{H} + \mathbf{A}(\boldsymbol{\theta})]\mathbf{G}\delta\mathbf{p} + O(\delta\boldsymbol{\theta}^2).\end{aligned} \tag{5.15}$$

Equation (5.15) can be rewritten as

$$\delta\mathbf{E} = (\mathbf{B}_l + \mathbf{A}(\boldsymbol{\theta})\mathbf{G})\delta\mathbf{p} + O(\delta\mathbf{p}^2) = \mathbf{B}_{nl}\delta\mathbf{p} + O(\delta\mathbf{p}^2) = (\mathbf{H} + \mathbf{A}(\theta))\mathbf{G}\delta\mathbf{p} + O(\delta\mathbf{p}^2). \tag{5.16}$$

For small virtual displacements, (5.16) becomes

$$\delta\mathbf{E}_{\mathrm{v}} = \mathbf{B}_{nl}(\mathbf{p})\delta\mathbf{p}_{\mathrm{v}}. \tag{5.17}$$

Substituting from (5.17) into (5.3) gives

$$V = \delta\mathbf{p}_{\mathrm{v}}^{\mathrm{T}}\int \mathbf{B}_{nl}^{\mathrm{T}}(\mathbf{p})\mathbf{S}\,\mathrm{d}V_{\mathrm{o}} - \delta^{\mathrm{PT}}_{\mathrm{v}}\mathbf{q}_{\mathrm{e}} = \delta\mathbf{p}_{\mathrm{v}}^{\mathrm{T}}\mathbf{g} \tag{5.18}$$

from which the out-of-balance force vector, $\mathbf{g}$, is given by

$$\mathbf{g} = \mathbf{q}_{\mathrm{i}} - \mathbf{q}_{\mathrm{e}} = \int \mathbf{B}_{nl}^{\mathrm{T}}(\mathbf{p})\mathbf{S}\,\mathrm{d}V_{\mathrm{o}} - \mathbf{q}_{\mathrm{e}} = \int \mathbf{G}^{\mathrm{T}}[\mathbf{H} + \mathbf{A}(\boldsymbol{\theta})]^{\mathrm{T}}\mathbf{S}\,\mathrm{d}V_{\mathrm{o}} - \mathbf{q}_{\mathrm{e}} \tag{5.19}$$

where, for the two-dimensional case,

$$\mathbf{S}^{\mathrm{T}} = (S_{xx}, S_{yy}, S_{xy}) = (S_{11}, S_{22}, S_{12}). \tag{5.20}$$

### 5.1.2 The tangent stiffness matrix

From (1.80)

$$\delta V = \delta\mathbf{p}_{\mathrm{v}}^{\mathrm{T}}\frac{\partial\mathbf{g}}{\partial\mathbf{p}}\delta\mathbf{p} = \delta\mathbf{p}_{\mathrm{v}}^{\mathrm{T}}\mathbf{K}_{\mathrm{t}}\delta\mathbf{p} \tag{5.21}$$

or from (5.4) and (5.16) and (5.17):

$$\delta V = \delta\mathbf{p}_{\mathrm{v}}^{\mathrm{T}}\mathbf{K}_{\mathrm{t}}\delta\mathbf{p} = \delta\mathbf{p}_{\mathrm{v}}^{\mathrm{T}}(\mathbf{K}_{\mathrm{t}1} + \mathbf{K}_{\mathrm{t}\sigma})\delta\mathbf{p} = \delta\mathbf{p}_{\mathrm{v}}^{\mathrm{T}}\int \mathbf{B}_{nl}^{\mathrm{T}}(\mathbf{p})\mathbf{C}_{\mathrm{t}}(\mathbf{S})\mathbf{B}_{nl}\mathrm{d}V_{\mathrm{o}}\delta\mathbf{p} + \int \mathbf{S}{:}\delta\mathbf{D}_{\mathrm{v}}^{\mathrm{T}}\delta\mathbf{D}\,\mathrm{d}V_{\mathrm{o}}. \tag{5.22}$$

In the two-dimensional case, the second term in (5.22) is given by

$$\delta\mathbf{p}_{\mathrm{v}}^{\mathrm{T}}\mathbf{K}_{\mathrm{t}\sigma}\delta\mathbf{p} = \int \mathbf{S}{:}\delta\mathbf{D}_{\mathrm{v}}^{\mathrm{T}}\delta\mathbf{D}\,\mathrm{d}V_{\mathrm{o}} = \int \begin{bmatrix} S_{11} & S_{12} \\ S_{12} & S_{22}\end{bmatrix} : \begin{bmatrix} \dfrac{\partial\delta u}{\partial x} & \dfrac{\partial\delta v}{\partial x} \\ \dfrac{\partial\delta u}{\partial y} & \dfrac{\partial\delta v}{\partial y}\end{bmatrix}_{\mathrm{v}} \begin{bmatrix} \dfrac{\partial\delta u}{\partial x} & \dfrac{\partial\delta u}{\partial y} \\ \dfrac{\partial\delta v}{\partial x} & \dfrac{\partial\delta v}{\partial y}\end{bmatrix} \mathrm{d}V_{\mathrm{o}} \tag{5.23}$$

where $\mathbf{K}_{t\sigma}$ is the 'geometric' or 'initial stress' contribution to the tangent stiffness matrix (see also Sections 1.3 and 2.1). Equation (5.23) can be re-expressed as

$$\delta \mathbf{p}_v^T \mathbf{K}_{t\sigma} \delta \mathbf{p} = \int \begin{bmatrix} \frac{\partial \delta u}{\partial x} \\ \frac{\partial \delta u}{\partial y} \\ \frac{\partial \delta v}{\partial x} \\ \frac{\partial \delta v}{\partial y} \end{bmatrix}^T \begin{bmatrix} \begin{bmatrix} S_{11} & S_{12} \\ S_{12} & S_{22} \end{bmatrix} & \begin{bmatrix} 0 & 0 \\ 0 & 0 \end{bmatrix} \\ \begin{bmatrix} 0 & 0 \\ 0 & 0 \end{bmatrix} & \begin{bmatrix} S_{11} & S_{12} \\ S_{12} & S_{22} \end{bmatrix} \end{bmatrix} \begin{bmatrix} \frac{\partial \delta u}{\partial x} \\ \frac{\partial \delta u}{\partial y} \\ \frac{\partial \delta v}{\partial x} \\ \frac{\partial \delta v}{\partial y} \end{bmatrix} \mathrm{d}V_o = \int \delta \boldsymbol{\theta}_v^T \hat{\mathbf{S}} \delta \boldsymbol{\theta} \, \mathrm{d}V_o. \tag{5.24}$$

From (5.9) (in real and virtual form), it follows that

$$\delta \mathbf{p}_v^T \mathbf{K}_{t\sigma} \delta \mathbf{p} = \delta \mathbf{p}_v^T \int \mathbf{G}^T \hat{\mathbf{S}} \mathbf{G} \, \mathrm{d}V_o \, \delta \mathbf{p}. \tag{5.25}$$

Hence, from (5.22) and (5.25), the full tangent stiffness matrix is

$$\mathbf{K}_t = \mathbf{K}_{t1} + \mathbf{K}_{t\sigma} = \int (\mathbf{B}_{nl}^T(\mathbf{p}) \mathbf{C}_t \mathbf{B}_{nl}(\mathbf{p}) + \mathbf{G}^T \hat{\mathbf{S}} \mathbf{G}) \, \mathrm{d}V_o. \tag{5.26}$$

We have just derived the tangent stiffness matrix by starting (see (5.4)) with the continuum form of the change in virtual work (derived in Section 4.4.1). We can instead work directly from the discretised out-of-balance or internal force vectors in (5.19). Using this approach, the change of the internal force vector $\mathbf{q}_i$, in (5.19) is given by

$$\delta \mathbf{q}_i = \int (\mathbf{B}_{nl}^T \delta \mathbf{S} + \delta \mathbf{B}_{nl}^T \mathbf{S}) \, \mathrm{d}V_o. \tag{5.27}$$

The first term of (5.27) leads directly to the standard tangent stiffness matrix, $\mathbf{K}_{t1}$ (see (5.26)), while the second leads to the geometric stiffness matrix. Considering the expression for $\mathbf{B}_{nl}$ in (5.16), it is clear that the changes in both $\mathbf{H}$ (see (5.12)) and $\mathbf{G}$ (see (5.8)) are zero. (For the latter observation, one should note that $\mathbf{J}$ in (5.6) involves the initial, fixed, coordinates.) We are left with a change to $\mathbf{A}(\boldsymbol{\theta})$ which, as previously discussed, is $\mathbf{A}(\delta\boldsymbol{\theta})$. It is then easy to show that:

$$\delta \mathbf{B}_{nl}^T \mathbf{S} = \mathbf{A}(\delta \boldsymbol{\theta})^T \mathbf{S} = \hat{\mathbf{S}} \delta \boldsymbol{\theta} = \hat{\mathbf{S}} \mathbf{G} \delta \mathbf{p} \tag{5.28}$$

where $\hat{\mathbf{S}}$ is given in (5.24). In conjunction with (5.27), (5.28) leads to the geometric tangent stiffness matrix, $\mathbf{K}_{t\sigma}$ previously derived in (5.24).

### 5.1.3 Extension to three dimensions

The extension to three dimensions is straightforward. In particular, we have:

$$\boldsymbol{\theta}^T = \left( \frac{\partial u}{\partial x}, \frac{\partial u}{\partial v}, \frac{\partial u}{\partial z}, \frac{\partial v}{\partial x}, \frac{\partial v}{\partial y}, \frac{\partial v}{\partial z}, \frac{\partial w}{\partial x}, \frac{\partial w}{\partial y}, \frac{\partial w}{\partial z} \right) \tag{5.29}$$

$$\mathbf{p}^{\mathrm{T}} = (\mathbf{u}, \mathbf{v}, \mathbf{w})^{\mathrm{T}} \tag{5.30}$$

so that the $\mathbf{G}$ matrix takes a very similar form to that of (5.8) but instead, for example, the second row would read

$$\mathbf{G}(2, j) = ((\mathbf{J}^{-1}(2,1)\mathbf{h}_{\xi}^{\mathrm{T}} + \mathbf{J}^{-1}(2,2)\mathbf{h}_{\eta}^{\mathrm{T}} + \mathbf{J}^{-1}(2,3)\mathbf{h}_{\zeta}^{\mathrm{T}}) \quad \mathbf{0}^{\mathrm{T}} \quad \mathbf{0}^{\mathrm{T}}) \tag{5.31}$$

where $\xi, \eta, \zeta$ are the three non-dimensional coordinates. In place of the terms in (5.10),

$$\mathbf{E}_{\mathrm{l}}^{\mathrm{T}} = \left(\frac{\partial u}{\partial x}, \frac{\partial v}{\partial y}, \frac{\partial w}{\partial z}, \frac{\partial u}{\partial y} + \frac{\partial v}{\partial x}, \frac{\partial u}{\partial z} + \frac{\partial w}{\partial x}, \frac{\partial v}{\partial z} + \frac{\partial w}{\partial y}\right) \tag{5.32}$$

and the matrix, $\mathbf{H}$, of (5.12) is replaced by

$$\mathbf{H} = \begin{bmatrix} 1 & 0 & 0 & 0 & 0 & 0 & 0 & 0 & 0 \\ 0 & 0 & 0 & 0 & 1 & 0 & 0 & 0 & 0 \\ 0 & 0 & 0 & 0 & 0 & 0 & 0 & 0 & 1 \\ 0 & 1 & 0 & 1 & 0 & 0 & 0 & 0 & 0 \\ 0 & 0 & 1 & 0 & 0 & 0 & 1 & 0 & 0 \\ 0 & 0 & 0 & 0 & 0 & 1 & 0 & 1 & 0 \end{bmatrix} \tag{5.33}$$

with the $\mathbf{A}(\boldsymbol{\theta})$ matrix of (5.10) being replaced by

$$\mathbf{A}(\boldsymbol{\theta}) = \begin{bmatrix} \frac{\partial u}{\partial x} & 0 & 0 & \frac{\partial v}{\partial x} & 0 & 0 & \frac{\partial w}{\partial x} & 0 & 0 \\ 0 & \frac{\partial u}{\partial y} & 0 & 0 & \frac{\partial v}{\partial y} & 0 & 0 & \frac{\partial w}{\partial y} & 0 \\ 0 & 0 & \frac{\partial u}{\partial z} & 0 & 0 & \frac{\partial v}{\partial z} & 0 & 0 & \frac{\partial w}{\partial z} \\ \frac{\partial u}{\partial y} & \frac{\partial u}{\partial x} & 0 & \frac{\partial v}{\partial y} & \frac{\partial v}{\partial x} & 0 & \frac{\partial w}{\partial y} & \frac{\partial w}{\partial x} & 0 \\ \frac{\partial u}{\partial z} & 0 & \frac{\partial u}{\partial x} & \frac{\partial v}{\partial z} & 0 & \frac{\partial v}{\partial x} & \frac{\partial w}{\partial z} & 0 & \frac{\partial w}{\partial x} \\ 0 & \frac{\partial u}{\partial z} & \frac{\partial u}{\partial y} & 0 & \frac{\partial v}{\partial z} & \frac{\partial v}{\partial y} & 0 & \frac{\partial w}{\partial z} & \frac{\partial w}{\partial y} \end{bmatrix}. \tag{5.34}$$

In addition, the vector $\mathbf{S}$ of (5.20) becomes

$$\mathbf{S}^{\mathrm{T}} = (S_{xx}, S_{yy}, S_{zz}, S_{xy}, S_{yz}, S_{xz}) = (S_{11}, S_{22}, S_{33}, S_{12}, S_{23}, S_{13}) \tag{5.35}$$

while $\hat{\mathbf{S}}$ of (5.24) becomes

$$\hat{\mathbf{S}} = \begin{bmatrix} \bar{\mathbf{S}} & \mathbf{0} & \mathbf{0} \\ \mathbf{0} & \bar{\mathbf{S}} & \mathbf{0} \\ \mathbf{0} & \mathbf{0} & \bar{\mathbf{S}} \end{bmatrix}, \qquad \bar{\mathbf{S}} = \begin{bmatrix} S_{11} & S_{12} & S_{13} \\ S_{12} & S_{22} & S_{23} \\ S_{13} & S_{23} & S_{33} \end{bmatrix}. \tag{5.36}$$

### 5.1.4 An axisymmetric membrane

We will now consider a special form of continuum—an axisymmetric membrane—and will derive the finite element equations for a simple two-noded element (Figure 5.1) using a total Lagrangian formulation. We could derive such an element by specialising the general formulation of Sections 5.1.1–3 but instead will restart from first principles.

The element has much in common with the total Lagrangian truss element of Section 3.3. In particular, the displacements and radius, $R = x$, are interpolated via

$$u = \mathbf{h}^{\mathrm{T}}\mathbf{u} = \frac{1}{2}\begin{pmatrix} 1-\xi \\ 1+\xi \end{pmatrix}^{\mathrm{T}} \begin{pmatrix} u_1 \\ u_2 \end{pmatrix}, \qquad w = \mathbf{h}^{\mathrm{T}}\mathbf{w}, \qquad R = \mathbf{h}^{\mathrm{T}}\mathbf{R}. \tag{5.37}$$

As for the two-dimensional formulation of Sections 5.1.1–2, we will assume that the vector of total displacements can be written as $\mathbf{p}^{\mathrm{T}} = (\mathbf{u}^{\mathrm{T}}, \mathbf{w}^{\mathrm{T}})$.

The strain along the membrane, $E_1$, is precisely the same as the strain, $\varepsilon$, of Section 3.3. Hence, from that chapter (in particular, equation (3.55)), or directly from (5.37), we can write

$$E_1 = \frac{l_{\mathrm{n}}^2 - l_{\mathrm{o}}^2}{2l_{\mathrm{o}}^2} = \mathbf{b}_1^{\mathrm{T}}\mathbf{p} + \frac{1}{2\alpha_{\mathrm{o}}^2}\mathbf{p}^{\mathrm{T}}\bar{\mathbf{A}}\mathbf{p} \tag{5.38}$$

where $l_{\mathrm{n}}$ is the final length of the element, $l_{\mathrm{o}} = 2\alpha_{\mathrm{o}}$ is the original length and (see (3.56)):

$$\mathbf{b}_1^{\mathrm{T}} = \frac{1}{4\alpha_{\mathrm{o}}^2}(-x_{21}, x_{21}, -z_{21}, z_{21}) \tag{5.39}$$

where, as in Chapters 1–3, terms such as $x_{21}$ means $x_2 - x_1$. The matrix $\bar{\mathbf{A}}$ in (5.38)

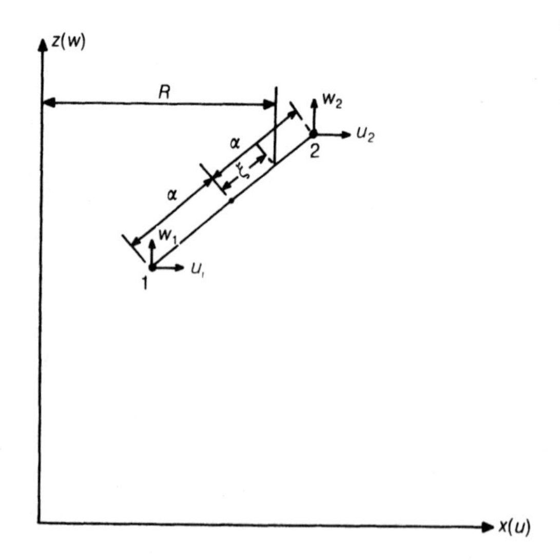


**Figure 5.1** A simple axisymmetric membrane element.

is given by

$$\bar{\mathbf{A}} = \frac{1}{4}\begin{bmatrix} 1 & -1 & 0 & 0 \\ -1 & 1 & 0 & 0 \\ 0 & 0 & 1 & -1 \\ 0 & 0 & -1 & 1 \end{bmatrix}. \tag{5.40}$$

(We have used the symbol $\bar{\mathbf{A}}$ here in contrast to the symbol $\mathbf{A}$ in (3.57) because of the alternative use in this chapter of $\mathbf{A}$—see (5.10).) Also (see (3.61)):

$$\delta E_1 = \mathbf{b}_1^{\mathrm{T}}\delta\mathbf{p} + \mathbf{b}_2^{\mathrm{T}}\delta\mathbf{p} = \mathbf{b}_{nl}(\mathbf{p})^{\mathrm{T}}\delta\mathbf{p} \tag{5.41}$$

where (again using the convention $u_{21} = u_2 - u_1$, etc.):

$$\mathbf{b}_2^{\mathrm{T}} = \frac{1}{4\alpha_{\mathrm{o}}^2}(-u_{21}, u_{21}, -w_{21}, w_{21}). \tag{5.42}$$

The second strain, $E_2$, is the hoop strain which is given by

$$E_2 = \frac{c_{\mathrm{n}}^2 - c_{\mathrm{o}}^2}{2c_{\mathrm{o}}^2} = \frac{u}{R} + \frac{1}{2}\left(\frac{u}{R}\right)^2 = E_{2l} + \tfrac{1}{2}E_{2l}^2. \tag{5.43}$$

The terms $c_{\mathrm{n}}$ and $c_{\mathrm{o}}$ are the final and original circumferential lengths. From the adopted shape function, we can write

$$E_{2l} = u/R = \mathbf{d}^{\mathrm{T}}\mathbf{p} = \frac{1}{2R}(1-\xi, 1+\xi, 0, 0)^{\mathrm{T}}\mathbf{p}; \quad \delta E_{2l} = \mathbf{d}^{\mathrm{T}}\delta\mathbf{p}. \tag{5.44}$$

From (5.43) and (5.44),

$$\delta E_2 = (1 + E_{2l})\delta E_{2l} = (1 + E_{2l})\mathbf{d}^{\mathrm{T}}\delta\mathbf{p}. \tag{5.45}$$

Combining (5.41) and (5.45),

$$\delta\mathbf{E} = \begin{pmatrix} \delta E_1 \\ \delta E_2 \end{pmatrix} = \begin{bmatrix} \mathbf{b}_{nl}(\mathbf{p})^{\mathrm{T}} \\ (1 + E_{2l})\mathbf{d}^{\mathrm{T}} \end{bmatrix}\delta\mathbf{p} = \mathbf{B}_{nl}\delta\mathbf{p}. \tag{5.46}$$

Hence, as in (5.19), virtual work gives

$$\mathbf{q}_i = \int \mathbf{B}_{nl}^{\mathrm{T}}\mathbf{S}\,\mathrm{d}V_{\mathrm{o}} \tag{5.47}$$

where the vector $\mathbf{S}$ contains the two second Piola–Kirchhoff stresses corresponding to the Green strains $E_1$ and $E_2$ respectively. Also.

$$\mathrm{d}V_{\mathrm{o}} = 2\pi t_{\mathrm{o}} R(\xi)\alpha_{\mathrm{o}}\,\mathrm{d}\xi \tag{5.48}$$

where $t_{\mathrm{o}}$ is the (initial) thickness.

The tangent stiffness matrix follows from differentiation of (5.47), which leads to an equation of the form of (5.27) with $\mathbf{K}_{t1}$ as in (5.26) and $\mathbf{C}_t$ as the $2 \times 2$ tangential modular matrix relating the small changes in $\mathbf{S}$ to the small changes in $\mathbf{E}$ (see Volume 2 for a $\mathbf{C}_t$ matrix appropriate to a rubber membrane). The geometric stiffness matrix

$\mathbf{K}_{t\sigma}$, follows from the $\delta\mathbf{B}_{nl}^{T}\mathbf{S}$ term in (5.27) so that, using (5.46),

$$\mathbf{K}_{t\sigma}\delta\mathbf{p} = \int S_1 \frac{\partial \mathbf{b}_{nl}}{\partial \mathbf{p}} \mathrm{d}V_o \delta\mathbf{p} + \int S_2 \mathbf{d}\delta E_{2l} \mathrm{d}V_o \tag{5.49}$$

from which, with the aid of (5.41) and (5.42) for $\partial\mathbf{b}_{nl}/\partial\mathbf{p}$) (noting that $\mathbf{b}_1$ in (5.41) is constant) and (5.44) for $\delta E_{2l}$

$$\mathbf{K}_{t\sigma} = \int \frac{1}{\alpha_o^2} S_1 \bar{\mathbf{A}} \mathrm{d}V_o + \int S_2 \mathbf{d}\mathbf{d}^T \mathrm{d}V_o \tag{5.50}$$

where $\bar{\mathbf{A}}$ has been defined in (5.40).

## 5.2 IMPLEMENTATION OF THE TOTAL LAGRANGIAN METHOD

For a linear elastic material, the implementation would follow very similar lines to those already discussed in Chapters 2–3 for truss elements (the shallow truss formulation of Chapter 2 can be considered as a special form of the total Lagrangian technique using a special form of the Green strain.) In particular, if a linear elastic analysis were performed with a fixed modular matrix so that $\mathbf{C}_t = \mathbf{C}$, the stresses, $\mathbf{S}$, in (5.20) or (5.35) could be computed directly from the total strains using $\mathbf{S} = \mathbf{C}\mathbf{E}$ with the Green strains, $\mathbf{E}$, being computed in tensor form directly from (5.1), where $\mathbf{D}$ (see (4.72)), would be obtained from the components of $\boldsymbol{\theta}$ (see (5.8) or (5.29)). Alternatively, in vector form, we would use (5.10). Extensions of the total Lagrangian procedure to cover hyperelastic materials (see Section 4.12) will be considered in Volume 2.

### 5.2.1 With an elasto-plastic or hypoelastic material

As already discussed in Section 4.10 with an elasto-plastic or hypoelastic material, the stress–strain relationships take a 'rate' or 'incremental form' (see (4.170)) which, for the present purposes, can be considered to involve

$$\Delta\mathbf{S} = \mathbf{C}_t(\mathbf{S}_o)\Delta\mathbf{E} \qquad \text{or} \qquad \Delta\mathbf{S} = \mathrm{fn}\,(\mathbf{S}_o, \Delta\mathbf{E}) \tag{5.51}$$

where $\mathbf{C}_t$ would not only be a function of the material properties but also of the current stresses, $\mathbf{S}$. Strictly, the material relationship will involve 'rates' or very small changes (as in (4.170)). To overcome this problem, some form of integration procedure (Chapter 6) would be used at the Gauss-point level in order that $\Delta\mathbf{S}$ could be computed from $\Delta\mathbf{E}$. Hence, the right-hand form in (5.51) would apply. In these circumstances, the 'new' stresses, $\mathbf{S}_n$, would be obtained from

$$\mathbf{S}_n = \mathbf{S}_o + \Delta\mathbf{S}(\Delta\mathbf{E}(\Delta\mathbf{p})) \tag{5.52}$$

where $\mathbf{S}_o$ are the 'old' stresses stored at the end of the last converged increment and $\Delta\mathbf{S}$ would be computed from $\Delta\mathbf{E}$ where $\Delta\mathbf{E}$ is the *incremental* strain. (The reason for working with incremental rather than iterative strains is discussed at the beginning of the next chapter.) The incremental strain, $\Delta\mathbf{E}$, would be computed from the total incremental displacements obtained from the 'predictor', $\Delta\mathbf{p}_o$ in conjuction with a number of iterative changes, $\delta\mathbf{p}_1$, $\delta\mathbf{p}_2$, etc., so that

$$\Delta\mathbf{p} = \Delta\mathbf{p}_o + \delta\mathbf{p}_1 + \delta\mathbf{p}_2 + \cdots \tag{5.53}$$

(or from the difference $\mathbf{p}_n - \mathbf{p}_o$ with $\mathbf{p}_n$ as the current total displacement and $\mathbf{p}_o$ as the displacements at the end of the previous increment).

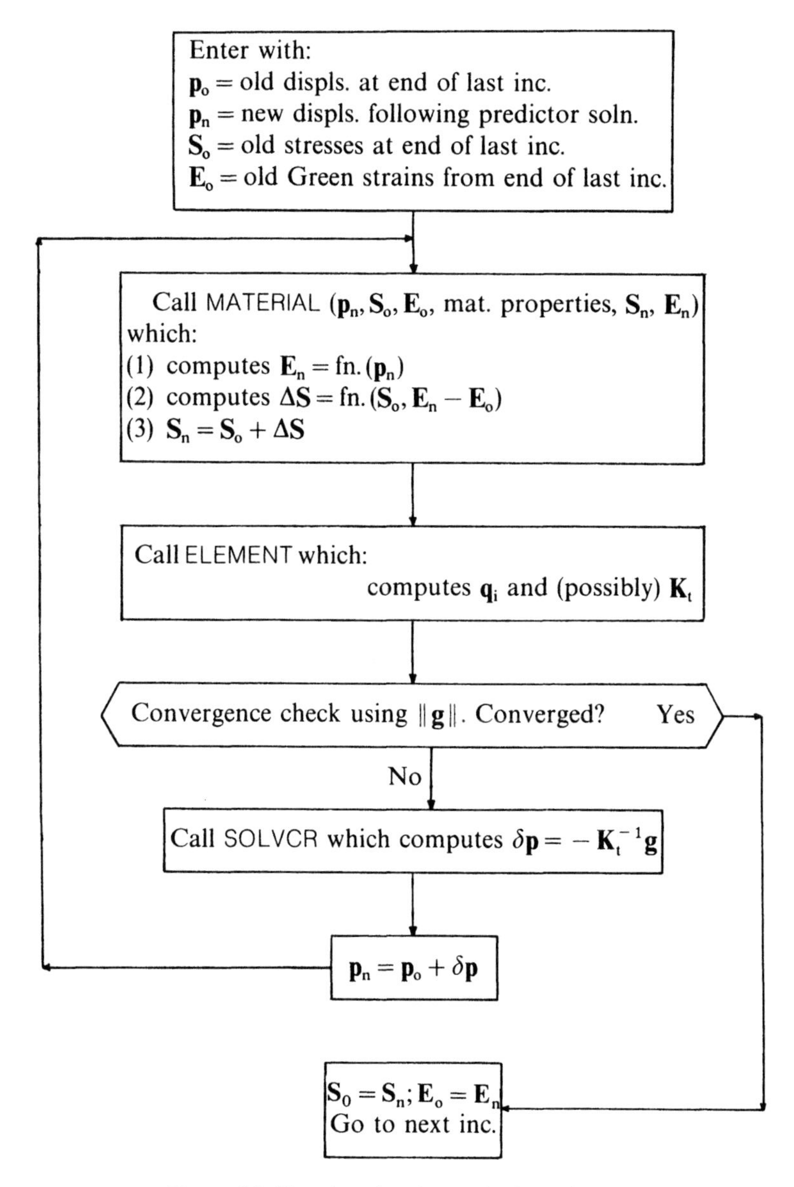


**Figure 5.2** Flowchart for elasto-plastic update.

Given $\Delta\mathbf{p}$, $\Delta\mathbf{E}$, for use in (5.52) can be computed from

$$\Delta\mathbf{E} = \mathbf{E}_n(\mathbf{p}_o + \Delta\mathbf{p}) - \mathbf{E}_o(\mathbf{p}_o) \tag{5.54}$$

where the total strains $\mathbf{E}_o$ and $\mathbf{E}_n$ would be computed using the procedure described in the previous section. The old Green strain, $\mathbf{E}_o$, might well be stored along with the old second Piola–Kirchhoff stress, $\mathbf{S}_o$. Alternatively, $\Delta\mathbf{E}$ could be computed directly from (5.2) (with $\Delta$s instead of $\delta$s). But it would be essential to include the higher-order bracketed term, so that

$$\Delta\mathbf{E}_2 = \tfrac{1}{2}\mathbf{F}^T\Delta\mathbf{D} + \tfrac{1}{2}\Delta\mathbf{D}^T\mathbf{F} + \tfrac{1}{2}\Delta\mathbf{D}^T\Delta\mathbf{D} \tag{5.55}$$

with the components of $\mathbf{D}$ (for $\mathbf{F}$) and $\Delta\mathbf{D}$ being computed from (5.8). To minimise the accumulation of round-off error, (5.54) might be preferred to (5.55).

Figure 5.2 gives a flowchart containing one possible scheme. This chart can be considered to relate to the iterative subroutine ITER of Section 2.4.2 and would be entered following an incremental (predictor solution) which would have obtained $\Delta\mathbf{p}_o$ and hence $\mathbf{p}_n = \mathbf{p}_o + \Delta\mathbf{p}_o$.

## 5.3 THE UPDATED LAGRANGIAN FORMULATION

It has been shown in Section 3.3.6, for truss elements, that it is relatively simple to change a total Lagrangian formulation to an updated Lagrangian formulation. A very similar procedure can be used for continuum elements. As pointed out in Section 3.3.6, if used in its *pure* form, the procedure should lead to the same solution as that of the total Lagrangian technique (see also [B2]). Gains in computational efficiency appear possible but it will be argued that, to take full advantage, approximations are required. Nonetheless, the updated Lagrangian procedure is often quoted and hence it is worth studying. Also there may be advantages for non-continuum applications such as shells when the two formulations could differ as a result of different shape function approximations.

The essence of the updated Lagrangian procedure is that the reference system would be periodically updated so that

$$\mathbf{x} = \mathbf{X} + \mathbf{p} \tag{5.56}$$

where $\mathbf{x}$ contains the 'new' (current coordinates) that are to be used as the new reference configuration. (Recall that in Section 5.1, we have been using this symbol $\mathbf{x}$ for the initial coordinates where strictly we should have used $\mathbf{X}$.) Having updated the coordinates, we require the stresses with respect to the current configuration. As discussed in Sections 3.3.6 and 4.6, the relevant stresses with respect to the current configuration are 'true' or Cauchy stresses. Hence we must modify the previous second Piola–Kirchhoff stresses from the previous configuration to Cauchy stresses using (4.104) (details will be given in the next section). With respect to the current configuration, the displacement, $\mathbf{p}$, are zero and hence, in place of (5.19), we have the simpler

$$\mathbf{g} = \int \mathbf{B}_l(\mathbf{x})^T \boldsymbol{\sigma}\, \mathrm{d}V_n - \mathbf{q}_e. \tag{5.57}$$

Depending on the precise manner in which we apply the updates, the stresses $\boldsymbol{\sigma}$ in (5.57) may, as indicated, be Cauchy stresses or they may be second Piola–Kirchhoff stresses with respect to the new configuration (details in the next section).

In a similar manner, the tangent stiffness matrix of (5.26) would become

$$\mathbf{K}_{\mathrm{t}} = \int (\mathbf{B}_l^{\mathrm{T}} \mathbf{C}_{\mathrm{t}}' \mathbf{B}_l + \mathbf{G}^{\mathrm{T}} \hat{\boldsymbol{\sigma}} \mathbf{G})\, \mathrm{d}V_{\mathrm{n}} \tag{5.58}$$

with $\mathbf{B}_l(\mathbf{x})$ replacing $\mathbf{B}_{\mathrm{n}l}(\mathbf{X}, \mathbf{p})$. In computing $\mathbf{G}$ from (5.8) (and, hence, $\mathbf{B}_l$ from (5.15) and (5.16)), the Jacobian matrix $\mathbf{J}$ (5.6) would, for the total Lagrangian formulation, always involve the initial geometry, i.e. $\mathbf{J}(\mathbf{X})$, while in the updated procedure, it would involve $\mathbf{J}(\mathbf{x})$. In a similar fashion, the $\mathrm{d}V_{\mathrm{o}}$ terms in (5.19) and (5.26) would involve $\det(\mathbf{J}(\mathbf{X}))$ while in (5.57) and (5.58), the $\mathrm{d}V_{\mathrm{n}}$ term would involve $\det(\mathbf{J}(\mathbf{x}))\,\mathrm{d}\xi\,\mathrm{d}\zeta\,\mathrm{d}\eta$.

From the work on truss elements in Section 3.3.6, it can be inferred that the tangent modular matrix for the updated Lagrangian formulation procedure should strictly differ from that for total Lagrangian formulation. For this reason, the prime has been added to the $\mathbf{C}_{\mathrm{t}}$ matrix in (5.58). Again, as discussed in Section 3.3.6, such a modification is unnecessary if the strains are small.

## 5.4 IMPLEMENTATION OF THE UPDATED LAGRANGIAN FORMULATION

The updated Lagrangian procedure can be implemented in a number of different ways. In each case, at a certain stage, the reference configuration would be updated and 'frozen'. This contrasts with the so-called 'Eulerian' or 'spatial' formulations discussed in Volume 2. The latter procedures have some links with the 'rotated engineering strain' and 'rotated log strain' procedures discussed, for truss elements, in Chapter 3. In these techniques, the reference coordinates continuously change and are never 'frozen'. Because of this freezing process, the essence of the updated Lagrangian technique is the same as that of the total Lagrangian procedure (see Section 3.3).

### 5.4.1 Incremental formulation involving updating after convergence

Having converged with a set of second Piola–Kirchhoff stresses, $\mathbf{S}$, these stresses would be transformed to Cauchy stresses relating to the new configuration using (4.104) so that

$$\boldsymbol{\sigma} = \frac{1}{\det(\mathbf{F})} \mathbf{F} \mathbf{S} \mathbf{F}^{\mathrm{T}} \tag{5.59}$$

where

$$\mathbf{F} = \mathbf{I} + \frac{\partial \Delta \mathbf{u}}{\partial \mathbf{x}_{\mathrm{o}}} = \mathbf{I} + \Delta \mathbf{D}(\Delta \mathbf{p}) \tag{5.60}$$

with $\Delta\mathbf{u}$ (and equivalent nodal values, $\Delta\mathbf{p}$) are incremental displacements from the previous (converged) configuration,

$$\mathbf{x}_o = \mathbf{X} + \mathbf{p}_o. \tag{5.61}$$

At this stage, the nodal coordinates would be updated via

$$\mathbf{x}_n = \mathbf{x}_o + \Delta\mathbf{p} = \mathbf{X} + (\mathbf{p}_o + \Delta\mathbf{p}) = \mathbf{X} + \mathbf{p}_n. \tag{5.62}$$

This would now become the new reference configuration 'o' (i.e. via $\mathbf{x}_o = \mathbf{x}_n$). Equation (5.58) would now be used for the tangent stiffness matrix, $\mathbf{K}_t$. (With regard to the new fixed reference configuration, the Cauchy stresses and second Piola–Kirchhoff stresses at the start of the increment would coincide because with $\Delta\mathbf{p} = \mathbf{0}$, $\mathbf{F}$ of (5.60) would equal $\mathbf{I}$.) This computation would involve $\mathbf{J}(\mathbf{x}_o)$. With the reference configuration kept fixed, the incremental green strain would be computed from, say, (5.55) and the stresses at the beginning of the increment would be updated to those at the end of the increment via

$$\mathbf{S}_n = \boldsymbol{\sigma}_o + \Delta\mathbf{S}(\Delta\mathbf{E}(\Delta\mathbf{p})) \tag{5.63}$$

which replaces (5.52) with the minor change that $\boldsymbol{\sigma}_o$ are the 'old' Cauchy stresses previously obtained via (5.59). The out-of-balance forces would then be computed from (5.19) (although now related to a reference configuration at the beginning of the increment). Using this procedure, the advantage whereby (5.57) uses $\mathbf{B}_l$ instead of $\mathbf{B}_{nl}$ (as in (5.19)) would not be gained. Also, if full Newton–Raphson iterations were used, $\mathbf{K}_t$ would, in a similar fashion, need to be computed from a form similar to (5.26). Hence for the remainder of the increment, there would be little difference between the total and updated Lagrangian formulation except that the latter would use $\mathbf{x}_o$ (referring to the configuration at the beginning of the increment) instead of $\mathbf{X}$ as the reference configuration.

### 5.4.2 A total formulation for an elastic response

In a *total* formulation, the reference configuration could always be the updated configuration so that advantages could be taken of (5.57) and (5.58). Strictly, a total formulation should follow similar lines to those described in the previous section so that $\mathbf{S} = \mathbf{C}\mathbf{E}$ should be followed by (5.59) with $\mathbf{F} = \mathbf{I} + \mathbf{D}(\mathbf{p})$ involving the total nodal displacements, $\mathbf{p}$. Equations (5.57) and (5.58) would then be used for the out-of-balance forces and tangent stiffness matrix, in each case using the Jacobian, $\mathbf{J}(\mathbf{x}_n) = \mathbf{J}(\mathbf{X} + \mathbf{p})$.

If the modular matrix, $\mathbf{C}$, is isotropic and small strains (but, possibly large rotations) are considered, the transformation of (5.59) can be avoided [B2] and, instead, the Almansi strain (see Section 4.5) can be adopted so that

$$\boldsymbol{\sigma} \simeq \mathbf{C}\boldsymbol{\varepsilon}_a \tag{5.64}$$

where $\boldsymbol{\varepsilon}_a$ are the Almansi strains of (4.91) (the symbol $\boldsymbol{\varepsilon}_a$ is now being used in place of the symbol $\mathbf{A}$ in Chapter 4 because $\mathbf{A}$ now has an alternative use (see (5.10)). The Almansi strain, $\boldsymbol{\varepsilon}_a$, relates to the current configuration and, in place of (5.10), is given by

$$\boldsymbol{\varepsilon}_a = \frac{1}{2}\left[\frac{\partial\mathbf{u}}{\partial\mathbf{x}} + \frac{\partial\mathbf{u}^T}{\partial\mathbf{x}}\right] - \frac{1}{2}\frac{\partial\mathbf{u}^T}{\partial\mathbf{x}}\frac{\partial\mathbf{u}}{\partial\mathbf{x}} = \frac{1}{2}\left[\frac{\partial\mathbf{u}}{\partial\mathbf{x}} + \frac{\partial\mathbf{u}^T}{\partial\mathbf{x}}\right] - \tfrac{1}{2}\mathbf{A}(\boldsymbol{\theta})\boldsymbol{\theta} = \mathbf{E}_l - \tfrac{1}{2}\mathbf{A}(\boldsymbol{\theta})\boldsymbol{\theta}. \tag{5.65}$$

This formulation would not give identical solutions to those obtained from a total Lagrangian formulation with $\mathbf{S} = \mathbf{CE}$ but they would be very similar if the strains were small and $\mathbf{C}$ were isotropic.

In these circumstances, the directions of principal stress and strain coincide and hence starting with a formulation based on Green's strain, from Section 4.9, we can write

$$\mathbf{S} = \mathbf{Q}(\mathbf{N}) \begin{bmatrix} f_1(E_{1-3}) & & \\ & f_2(E_{1-3}) & \\ & & f_3(E_{1-3}) \end{bmatrix} \mathbf{Q}(\mathbf{N})^{\mathrm{T}} \tag{5.66}$$

where $E_{1-3}$ are the three principle Green strains and, for example,

$$f_2(E_{1-3}) = S_2 = \frac{E}{1 - \nu^2}(E_2 + \nu E_1 + \nu E_3). \tag{5.67}$$

If the strains are small, $\det(\mathbf{F}) \simeq 1$, and the stretch, $\mathbf{U}$, is approximately the identity matrix (see (4.156)) so that $\mathbf{F} \simeq \mathbf{R}$ (see (4.157) and with the aid of (5.59) and (4.147)

$$\boldsymbol{\sigma} \simeq \mathbf{Q}(\mathbf{n}) \begin{bmatrix} f_1(E_{1-3}) & & \\ & f_2(E_{1-3}) & \\ & & f_3(E_{1-3}) \end{bmatrix} \mathbf{Q}(\mathbf{n})^{\mathrm{T}}. \tag{5.68}$$

But, when the strains are small (and the stretches nearly unity), the principal Green strains $E_i$ are approximately equal to the principal Almansi strains $\boldsymbol{\varepsilon}_{ai}$ because via (4.151) and (4.152):

$$E_i = \tfrac{1}{2}(\lambda_i^2 - 1) \simeq \varepsilon_{ai} = \frac{1}{2}\left(\frac{\lambda_i^2 - 1}{\lambda_i^2}\right). \tag{5.69}$$

Hence replacing $E_{1-3}$ with $\varepsilon_{a1-3}$ in (5.68) leads to

$$\boldsymbol{\sigma} \simeq \mathbf{Q}(\mathbf{n}) \begin{bmatrix} f_1(\varepsilon_{a1-3}) & & \\ & f_2(\varepsilon_{a1-3}) & \\ & & f_3(\varepsilon_{a1-3}) \end{bmatrix} \mathbf{Q}(\mathbf{n})^{\mathrm{T}} \tag{5.70}$$

which, again assuming that the directions of principal stress and strain coincide, leads via (4.154) to (5.64). Hence, effectively the same results would be obtained from using $\mathbf{S} = \mathbf{CE}$ as $\boldsymbol{\sigma} = \mathbf{C}\boldsymbol{\varepsilon}_{\mathrm{a}}$ without introducing any transformations to the material modular matrices.

### 5.4.3 An approximate incremental formulation

We can, in an approximate manner, extend this procedure to an incremental formulation so that we again use the cheaper (5.57) and (5.58) instead of (5.19) and (5.26) and also avoid the use of the transformation (5.59). To this end, we replace (5.63) with

$$\boldsymbol{\sigma} = \boldsymbol{\sigma}_{\mathrm{o}} + \Delta\boldsymbol{\sigma}(\boldsymbol{\varepsilon}_{\mathrm{an}} - \boldsymbol{\varepsilon}_{\mathrm{ao}}) = \boldsymbol{\sigma}_{\mathrm{o}} + \Delta\boldsymbol{\sigma}(\Delta\boldsymbol{\varepsilon}_{\mathrm{a}}). \tag{5.71}$$

Here $\boldsymbol{\varepsilon}_{\mathrm{ao}}$ are the old (probably stored) Almansi strains at the last converged configuration, $\boldsymbol{\sigma}_o$ is the old (stored) Cauchy stress related to this configuration while $\boldsymbol{\varepsilon}_{an}$ is the current Almansi strain.

However, for both this procedure and that of the previous section, neither the internal force vector nor the tangent stiffness matrix would be fully consistent with the stress updating. Hence the iterative performance might possibly suffer a little in comparison with the 'purer' total Lagrangian formulation of Section 5.1.

## 5.5 SPECIAL NOTATION

$\mathbf{A}(\theta)$ = matrix containing displacement derivatives (see (5.10) or (5.34))
$\bar{\mathbf{A}}$ = special matrix (see (5.40)) (Section 5.1.4)
$\mathbf{b}_{nl}$ = vector connecting $\delta\mathbf{E}_1$ to $\delta\mathbf{p}$ (Section 5.1.4)
$\mathbf{B}_l$ = matrix connecting $\delta\mathbf{E}_l$ to $\delta\mathbf{p}$
$\mathbf{B}_{nl}$ = matrix connecting $\delta\mathbf{E}$ to $\delta\mathbf{p}$
$\mathbf{d}$ = vector relating linear part of $\mathbf{E}_2$ to $\delta\mathbf{p}$ (Section 5.1.4)
$\mathbf{D}$ = displacement-derivative matrix (see (4.72))
$\mathbf{E}_1$ = Green's strain along axisymmetric membrane (Section 5.1.4)
$\mathbf{E}_2$ = hoop Green's strain for axisymmetric membrane (Section 5.1.4)
$\mathbf{E}$ = Green's strain (as vector or tensor; the latter is sometimes $\mathbf{E}_2$)
$\mathbf{E}_l$ = linear part of Green's strain
$\mathbf{E}_{nl}$ = non-linear part of Green's strain
$\mathbf{F}$ = deformation gradient (see (4.71))
$\mathbf{G}$ = matrix connecting $\boldsymbol{\theta}$ to $\mathbf{p}$ (see (5.8))
$\mathbf{h}_\xi, \mathbf{h}_\eta$ = vectors containing derivatives with respect to $\xi$ and $\eta$ of shape function vector, $\mathbf{h}$
$\mathbf{H}$ = Boolean matrix (see (5.12) or (5.33))
$\mathbf{I}$ = unit matrix
$\mathbf{J}$ = Jacobian matrix (see (5.6))
$l_o, l_n$ = old and new lengths (Section 5.1.4)
$\mathbf{n}_1, \mathbf{n}_2, \mathbf{n}_3$ = unit vectors defining directions of principal stretch in final configuration (defines the Eulerian triad)
$\mathbf{N}_1, \mathbf{N}_2, \mathbf{N}_3$ = unit vectors defining directions of principal stretch in initial configuration (defines the Lagrangian triad)
$\mathbf{p}$ = nodal displacements
ordering of 2-D continuum element is $\mathbf{p}^T = (\mathbf{u}^T, \mathbf{v}^T)$
ordering of 3-D continuum element is $\mathbf{p}^T = (\mathbf{u}^T, \mathbf{v}^T, \mathbf{w}^T)$
ordering of axisymmetric element is $\mathbf{p}^T = (\mathbf{u}^T, \mathbf{w}^T)$
$\mathbf{Q}$ = orthogonal matrix containing principal directions, $\mathbf{N}$s or $\mathbf{n}$s
$R$ = radius (Section 5.1.4)
$S_1, S_2$ = second Piola–Kirchhoff stresses corresponding to strains $E_1$ and $E_2$ (Section 5.1.4)

$\mathbf{S}$ = second Piola–Kirchhoff stresses (as vector or tensor; the latter is sometimes $\mathbf{S}_2$)
$\hat{\mathbf{S}}$ = matrix containing second Piola–Kirchhoff stresses (see (5.24) or (5.36))
$\mathbf{u}$ = displacements, sometimes specifically $x$-direction nodal displacements
$\mathbf{v}$ = $y$-direction nodal displacements
$\mathbf{w}$ = $z$-direction nodal displacements
$\mathbf{x}$ (with components $x, y, z$) = initial coordinates in Sections 5.1 and 5.2 (sometimes used for nodal values)
$\mathbf{x}$ (with components $x, y, z$) = final coordinates ($\mathbf{x} = \mathbf{X} + \mathbf{u}$) in Sections 5.3 and 5.4 (sometimes used for nodal values)
$\mathbf{X}$ (with components $X, Y, Z$) = initial coordinates (in Sections 5.3 and 5.4)
$\alpha_o$ = length parameter relating to initial configuration ($l_o/2$) (Section 5.1.4)
$\varepsilon_a$ = Almansi strain
$\boldsymbol{\theta}$ = displacement derivatives in vector form (see (5.8) or (5.29))
$\lambda$ = stretch scalar (see (4.131))
$\boldsymbol{\sigma}$ = Cauchy stress (as vector or tensor; the latter is sometimes $\boldsymbol{\sigma}_2$)

### Subscripts

n = new
o = old

## 5.6 REFERENCES

[B1] Bathe, K. J., Ramm, E. & Wilson, E., Finite element formulations for large deformation dynamic analysis, *Int. J. Num. Meth. Engng.*, **9**, 353–386 (1975).
[B2] Bathe, K. J., *Finite Element Procedures in Engineering Analysis*, Prentice Hall, Englewood Cliffs (1982).
[Z1] Zienkiewicz, O. C., *The Finite Element Method*, 3rd edition, McGraw-Hill, London (1977).

# 6 Basic plasticity

## 6.1 INTRODUCTION

Books on plasticity can be found elsewhere [H2, M7, M10, J2, P1, H5, N3, C2, I1, S1] and a review of recent developments in [D4]. The main objective of the present chapter is to concentrate on those aspects that relate to a 'numerical solution'. In particular, we will be mainly thinking of the finite element method but many of the concepts also apply to other discretisation procedures such as finite differences. Books and manuals in this category can be found in [O4, S5, C2, D2]. The last two references relate primarily to geomechanical materials. In contrast, the present chapter will concentrate on the von Mises yield criterion [V1, V2], although much of the work will be general and applicable to other yield functions. (Other yield criteria will be treate in more detail in Volume 2.) Reviews on numerical work on plasticity can be found in [A3, W1, W3, D3] while the workshop proceedings [N3], although now a little out-of-date, gives a range of interesting papers and discussions.

In the present chapter, only isotropic hardening will be treated in any depth (with kinematic and mixed hardening being considered in Volume 2) while the flow rules will generally be assumed to be associative. Because the plastic flow rules are incremental in nature [H2], elasto-plastic problems, should strictly be solved using small equilibrium steps. For, no matter how accurately we may, within an increment, satisfy the flow rules and keep to the yield surface, the solution is only in equilibrium at the end of each increment after the equilibrium iterations (Chapters 1–3, 5) (see Section 6.2 for further discussion). Nonetheless, very acceptable solutions have often been obtained with large steps.

In keeping with the main, static, theme of the book, we will not consider the time-dependent viscoplasticity [O4, W1, Z3] but should note that viscoplastic approaches have been used with a 'pseudo-time', to analyse time-independent elasto-plasticity [Z3, O4]. In relation to such elasto-plasticity, having reached equilibrium at point A (Figure 6.1) on the effective stress/strain curve, the next step may continue to flow plastically to point B or else to unload elastically to C. Clearly, the two paths have very different stiffnesses. If it is known that the loads are to be reversed at point A, the elasto-plastic tangent stiffness matrix should not be used for the next increment and the elastic stiffness matrix should be used instead. But, in the absence of prior knowledge on load reversals, it will generally be assumed that plastic flow will continue and that the tangent stiffness relates to AB. However, even for monotonically increasing loads, certain areas of the structure can 'unload'. In such

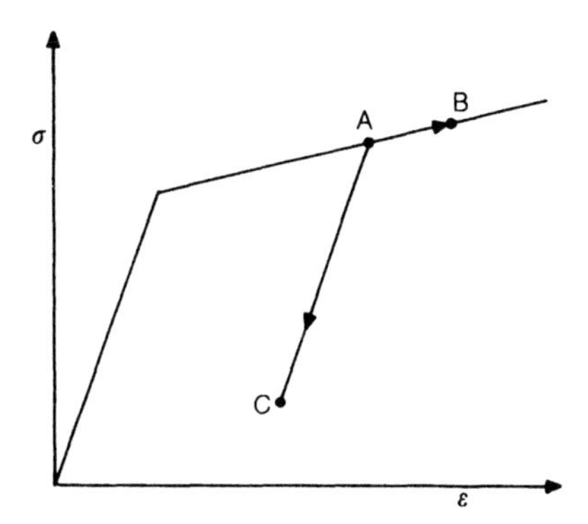


**Figure 6.1** One-dimensional stress–strain relationship.

circumstances, it is generally left to the iterative correction procedure (Chapters 1–3, 5) to discover those areas that are 'unloading'. However, if it is suspected that much unloading is occurring, it may be beneficial to revert to the elastic stiffness matrix and adopt a solution procedure based on the 'initial stress' method (see Section 1.2.3 and Figure 1.9). For problems involving combined geometric and material non-linearity, no data is then provided on the stability of the solution (Chapter 9 and Volume 2).

In general, there are three separate roles for the plasticity algorithms of a finite element code. These roles are:

(1) the formation of the standard tangent modular matrix [P1.1, M4.1, Y1.1, Z1.1, Z2.1] for use in the incremental tangent stiffness matrix of the structure or for use with the integration of the the stress/strain laws (see (3) below);
(2) the formation of a 'consistent' tangent modular matrix for use with Newton–Raphson iterations;
(3) the integration of the stress/strain laws to update the stresses.

As already briefly discussed in Sections 4.12 and 5.2.1, with material non-linearity, the structural tangent stiffness matrix takes the form

$$\mathbf{K}_t = \int \mathbf{B}^T\mathbf{C}_t\mathbf{B}\,\mathrm{d}V + \text{initial stress matrix} \tag{6.1}$$

where $\mathbf{C}_t$ is the standard tangential modular matrix, which is given by

$$\frac{\partial \boldsymbol{\sigma}}{\partial \boldsymbol{\varepsilon}} = \mathbf{C}_t. \tag{6.2}$$

The initial stress matrix in (6.1) only exists if geometric non-linearity is included. As previously discussed in Chapters 3 and 5, with such geometric non-linearity, we must specify the type of stress measure being used. However, for the present chapter, this issue will be avoided.

If certain forms of stress updating are adopted, it is possible to derive a 'consistent' tangent modular matrix, $\mathbf{C}_{tc}$, that is consistent with the numerical technique used for the stress updating. In general terms, the concept related to a 'consistent linearisation' were discussed in [H7]. The ideas appear to have been first applied to plasticity by Simo and Taylor [S6] and Runesson & Samuelsson [R3] (with subsequent work in [S6, S7, S5, A1, J1, H1, B3, C4, M8]). In comparison with the use of the 'standard' tangential modular matrix, the consistent tangent leads to a significantly faster convergence rate when the Newton–Raphson algorithm is used for the equilibrium iterations.

Most of the work in this chapter will be presented from an 'engineering approach', starting from the early tangential modular matrices [P1.1, M4.1, Y1.1, Z1.1, Z2.1] and leading on, via the 'implicit' integration procedures, to the 'consistent tangent matrices'. Some of this work can also be approached [M6, S5, S1, R3] from a more rigorous mathematical programming basis [M1, M10, S9] (often with identical results). These issues will be briefly discussed in Section 6.10 which can be considered as an Appendix to this chapter.

Before detailing the derivation of the 'standard' tangent modular matrix, $\mathbf{C}_t$, we will briefly discuss an important aspect of stress updating that has been omitted from many books on finite elements and plasticity.

## 6.2 STRESS UPDATING: INCREMENTAL OR ITERATIVE STRAINS?

Because of the incremental (or rate) nature of the flow rules [H1, M7], it is almost inevitable that solution procedures based on incremental predictor/corrector approaches (Chapters 1–3 and 9) will lead to some error [A2, C3]. This error will not relate to a lack of equilibrium but rather will be caused by errors in the integration of the flow rules (Section 6.6) and their relation to the complete incremental/iterative solution procedure. Most analysts assume a linear strain path within an increment (see [M2] for more on the 'loading path' and plasticity). Even if equilibrium is exactly satisfied at both the beginning and end of an increment, and sub-increments (Section 6.6.4) are used to help integrate the strain rules accurately, the solution will not correspond exactly with a solution in which the increment was itself cut into a number of smaller increments for each of which equilibrium was exactly ensured. To limit these errors, Tracey and Freese [T1] have developed an adaptive scheme that examines the local curvature of the yield surface and direction of the strain rate vector to select the load step sizes.

The errors will be strongly related to the adopted procedure for updating the stresses and strains. In relation to the incremental/iterative procedure, two distinct algorithms would seem possible:

(*A*) *Using iterative strains*

(1) Compute the iterative displacements, $\delta\mathbf{p}$, using, for example, $\delta\mathbf{p} = -\mathbf{K}_t^{-1}\mathbf{g}$.

(2) Compute the iterative strains, $\delta\boldsymbol{\varepsilon}$, from the iterative displacements, $\delta\mathbf{p}$, using $\delta\boldsymbol{\varepsilon} = \text{fn}(\delta\mathbf{p})$.

(3) Compute the iterative stresses using, $\delta\boldsymbol{\sigma} = \mathbf{C}_t(\boldsymbol{\sigma})\delta\boldsymbol{\varepsilon}$ or, preferably, by 'integrating the rate equations' (Section 6.6) possibly with the aid of sub-incrementation (Section 6.6.4).

(4) Update the stresses using, $\boldsymbol{\sigma}_n = \boldsymbol{\sigma}_o + \delta\boldsymbol{\sigma}$ where $\boldsymbol{\sigma}_o$ are the old stresses before the current iteration.

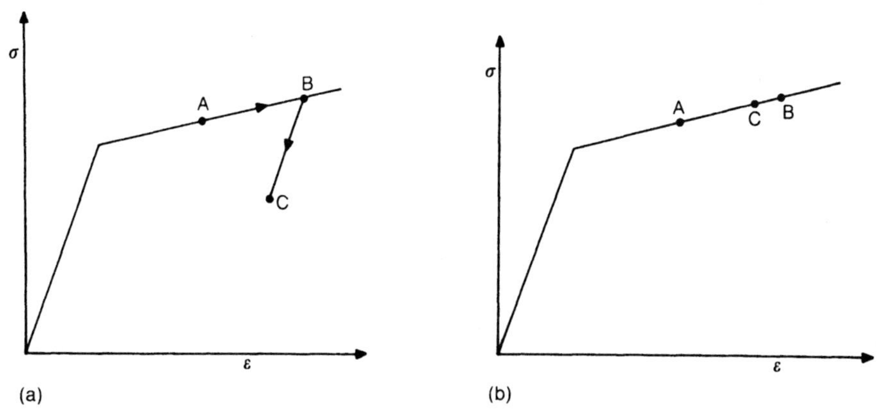


**Figure 6.2** One-dimensional illustration of the alternative updating stategies: (a) updating strategy, A; (b) updating strategy, B.

(*B*) *Using the incremental strains*

(1) Compute the iterative dsplacements, $\delta\mathbf{p}$, using, for example, $\delta\mathbf{p} = -\mathbf{K}_t^{-1}\mathbf{g}$.

(2) Update the incremental displacements (from the last converged equilibrium state) using $\Delta\mathbf{p}_n = \Delta\mathbf{p}_0 + \delta\mathbf{p}$, where $\Delta\mathbf{p}_o$ are the incremental displacements at the end of the last iteration.

(3) Compute the incremental strains, $\Delta\boldsymbol{\varepsilon}$, from the incremental displacements, $\Delta\mathbf{p}$, using $\Delta\boldsymbol{\varepsilon} = \mathrm{fn}(\Delta\mathbf{p})$.

(4) Compute the incremental stresses using, $\Delta\boldsymbol{\sigma} = \mathbf{C}_t(\boldsymbol{\sigma})\Delta\boldsymbol{\varepsilon}$ or, preferably, by integrating the rate equations.

(5) Update the stresses using, $\boldsymbol{\sigma}_n = \boldsymbol{\sigma}_o + \Delta\boldsymbol{\sigma}$ where $\boldsymbol{\sigma}_o$ are the old stresses at the end of the last increment.

Strategy A is not recommended as it may lead to 'spurious unloading' during the iterations. This phenomenon is illustrated in Figure 6.2(a) in which point A represents a converged equilibrium state. The tangential incremental solution then takes the stress to point B. At this stage, the iteative process produces a negative iterative displacement and hence a negative iterative strain. As a consequence, the stress will spuriously unload to point C (Figure 6.2(a)). These issues were discussed in [K1, C3] and the two different schemes have been compared by Marques [M4].

In the context of combined material and geometric non-linearity, Bushnell [B4] divorced the geometric non-linearity from the elasto-plastic formulation by performing equilibrium iterations on the former with fixed $\mathbf{C}_t$ matrices before changing them to conform with the new converged stresses. This procedure was repeated until the complete system converged. Little [L1] adopted a simpler but similar strategy to remove the material effects from the geometric effects. These procedures could become rather expensive for large problems. In addition, the 'true' equilibrium path will not be followed during the iterations. This 'true' path can be more closely (but still not exactly) followed if strategy B is adopted.

Using this procedure, the incremental stress is simply re-computed from the new incremental strain which, in relation to Figure 6.2(b), is still positive and hence the stress/strain configuration moves from point A to point B. The main advantages of

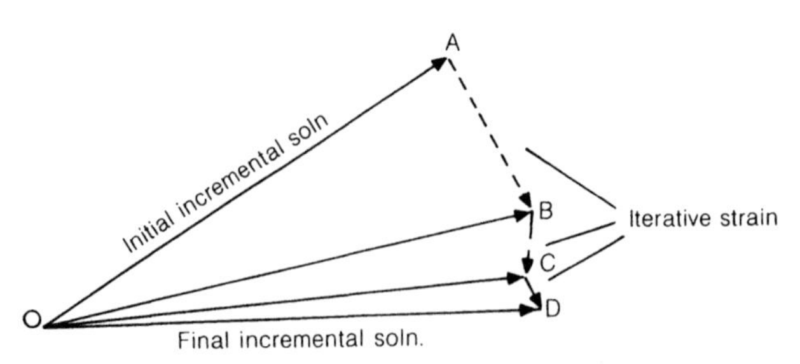


**Figure 6.3** Incremental and iterative strains.

strategy B are gained because the stresses are always updated from the stresses at the end of the last increment. These stresses are in equilibrium. This point is further illustrated in Figure 6.3. The true final strain increment is OD and by always working with incremental strains, the intermediate steps OA, OB, OC all lie reasonably close to this strain. In contrast, with strategy A, the steps AB, BC, CD, etc., may lie in entirely different directions.

Nyssen [N4] advocates a modified form of strategy A because he argues that strategy B uses too much work in integrating the rate equations (Section 6.6). Certainly, if sub-incrementation is adopted (Section 6.6.4), strategy B will require the same number of sub-increments for the later as for the earlier iterations even although the iterative strains will be considerably smaller. The modification to strategy A, proposed by Nyssen [N4], involves 'incremental reversibility' and allows plastic unloading within an increment so that an 'unloaded point' is only defined as elastic for the next increment. Hence, the rule of no plastic unloading is only applied in a piecewise incremental manner. Nyssen further modified this strategy by adding the proviso that such unloading was only allowed until the plastic work done becomes again equal to its value at the beginning of the considered increment.

While there may be some justification for a modified strategy A when sub-increments are used, there would appear to be none with a 'backward Euler integration scheme' (Section 6.6.6) coupled with a 'consistent tangent' (Section 6.7). In summary, the author strongly recommends strategy B (see also Dodds [D3]).

## 6.3 THE STANDARD ELASTO-PLASTIC MODULAR MATRIX FOR AN ELASTIC/PERFECTLY PLASTIC VON MISES MATERIAL UNDER PLANE STRESS

In some senses, plane stress is one of the more difficult stress states. The complexities will be discussed in Section 6.8.2. In the meantime, plane stress will be used to introduce plasticity calculations simply because it involves fewer components. However, the prime aim is to develop the general form of the matrix, vector and tensor equations which will also apply to more general stress states.

We will start with the simple plane-stress version of the von Mises [V1, V2] yield function:

$$f = (\sigma_x^2 + \sigma_y^2 - \sigma_x\sigma_y + 3\tau_{xy}^2)^{1/2} - \sigma_0 = \sigma_e - \sigma_0 \tag{6.3}$$

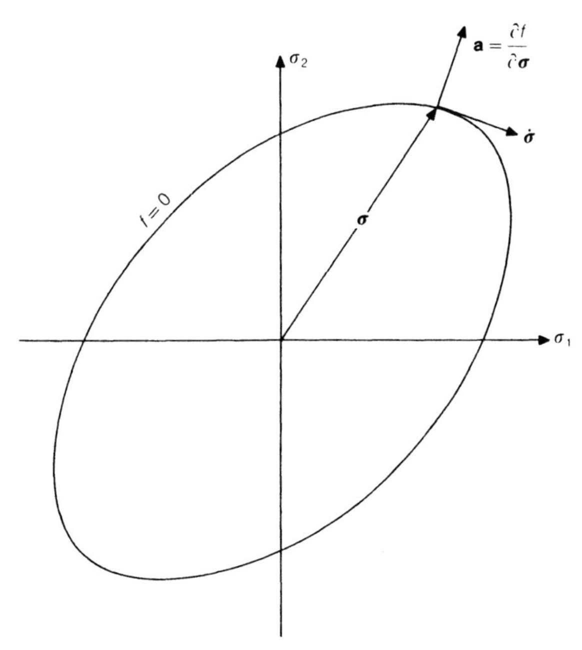


**Figure 6.4** The von Mises yield criterion under principal stress ($\sigma_1 = \sigma_x$, $\sigma_2 = \sigma_y$) and plane-stress conditions.

where $\sigma_e$ is the effective stress and $\sigma_o$ the yield stress. In conjunction with (6.3), the Prandtl–Reuss flow rules are:

$$\dot{\boldsymbol{\varepsilon}}_p = \dot{\lambda}\left(\frac{\partial f}{\partial \boldsymbol{\sigma}}\right) = \dot{\lambda}\mathbf{a} = \begin{pmatrix} \dot{\varepsilon}_{px} \\ \dot{\varepsilon}_{py} \\ \dot{\varepsilon}_{pxy} \end{pmatrix} = \frac{\dot{\lambda}}{2\sigma_e}\begin{pmatrix} 2\sigma_x - \sigma_y \\ 2\sigma_y - \sigma_x \\ 6\tau_{xy} \end{pmatrix} \tag{6.4}$$

where (Figure 6.4), the vector **a** is normal to the yield surface and $\dot{\lambda}$ is a positive constant usually referred to as the 'plastic strain-rate multiplier'. (*Note* that with the *present* notation, $\partial f/\partial \boldsymbol{\sigma}$ is a column vector.) In equation (6.4) and generally during this chapter, we are using the 'rate' form denoted by a dot. However, as discussed in Section 4.6, we are not considering dynamic effects so that we have a 'pseudo-time' and indeed the dotted quantities can be simply considered as small changes which we have in the past often designated via $\delta$s (strictly, $\delta\boldsymbol{\varepsilon} = \dot{\boldsymbol{\varepsilon}}\,\delta t$, but we will often loosely refer to $\dot{\boldsymbol{\varepsilon}}$ as a small change). In addition to (6.4), the stress changes are related to the strain changes via

$$\dot{\boldsymbol{\sigma}} = \begin{pmatrix} \dot{\sigma}_x \\ \dot{\sigma}_y \\ \dot{\sigma}_{xy} \end{pmatrix} = \begin{bmatrix} \mathbf{C} \end{bmatrix}\left(\begin{pmatrix} \dot{\varepsilon}_x \\ \dot{\varepsilon}_y \\ \dot{\varepsilon}_{xy} \end{pmatrix} - \begin{pmatrix} \dot{\varepsilon}_{px} \\ \dot{\varepsilon}_{py} \\ \dot{\varepsilon}_{pxy} \end{pmatrix}\right) = \mathbf{C}(\dot{\boldsymbol{\varepsilon}}_t - \dot{\boldsymbol{\varepsilon}}_p) = \mathbf{C}(\dot{\boldsymbol{\varepsilon}} - \dot{\lambda}\mathbf{a}) \tag{6.5}$$

where, assuming isotropic elasticity,

$$\mathbf{C} = \frac{E}{1-\nu^2}\begin{bmatrix} 1 & \nu & 0 \\ \nu & 1 & 0 \\ 0 & 0 & (1-\nu)/2 \end{bmatrix}. \tag{6.6}$$

Equation (6.5) relates the small changes in stress (or more strictly stress rates) to the small changes in elastic strain (or more strictly elastic strain rates), $\dot{\boldsymbol{\varepsilon}}_e = \dot{\boldsymbol{\varepsilon}}_t - \dot{\boldsymbol{\varepsilon}}_p$. In this equation and throughout the chapter, the subscript t will sometimes be dropped from the total strain changes (or total strain rates).

A negative 'plastic strain-rate multiplier', $\dot{\lambda}$, would imply plastic unloading from the yield surface. The latter cannot occur and, consequently, any negative, $\dot{\lambda}$s should be replaced by zero so that elastic unloading occurs (see also Section 6.10).

For plastic flow to occur, the stresses must remain on the yield surface and hence

$$\dot{f} = \frac{\partial f^{\mathrm{T}}}{\partial \boldsymbol{\sigma}} \dot{\boldsymbol{\sigma}} = \mathbf{a}^{\mathrm{T}} \dot{\boldsymbol{\sigma}} = \mathbf{a} : \dot{\boldsymbol{\sigma}} = 0. \tag{6.7}$$

In equation (6.7), we have adopted an approach that we will use often in the chapter, and have given both the vector and tensor forms although only the former relates directly to the precise forms in (6.4)–(6.6). Generally, we will not use (as in Chapters 4 and 5) subscripts such as 2 to indicate the order of the tensor. It should be obvious from the context which form is being used. In particular the use of the contraction symbol: will indicate the tensor form (see discussion below equation (4.6)) while the use of the symbol 'T' for transpose will imply the use of a vector.

The situation described by (6.7) is illustrated in Figure 6.4 and shows that, for plastic flow, the stress changes, $\dot{\boldsymbol{\sigma}}$, are instantaneously moving tangentially to the surface with $\dot{\boldsymbol{\sigma}}$ being orthogonal to the vector $\mathbf{a}$. Hence $\mathbf{a}$ is normal to the surface and the flow rules (equation (6.4)) invoke 'normality'.

In order to find the plastic strain-rate multiplier, $\dot{\lambda}$, equation (6.5) is premultiplied by the flow vector $\mathbf{a}^{\mathrm{T}}$ and, using equation (6.7)

$$\dot{\lambda} = \frac{\mathbf{a}^{\mathrm{T}} \mathbf{C} \dot{\boldsymbol{\varepsilon}}}{\mathbf{a}^{\mathrm{T}} \mathbf{C} \mathbf{a}} = \frac{\mathbf{a} : \mathbf{C} : \dot{\boldsymbol{\varepsilon}}}{\mathbf{a} : \mathbf{C} : \mathbf{a}}. \tag{6.8}$$

Consequently, substitution into equation (6.5) gives

$$\dot{\boldsymbol{\sigma}} = \mathbf{C}_t \dot{\boldsymbol{\varepsilon}} = \mathbf{C}\left(\mathbf{I} - \frac{\mathbf{a}\mathbf{a}^{\mathrm{T}}\mathbf{C}}{\mathbf{a}^{\mathrm{T}}\mathbf{C}\mathbf{a}}\right)\dot{\boldsymbol{\varepsilon}} = \left(\mathbf{C} - \frac{1}{\mathbf{a}:\mathbf{C}:\mathbf{a}}(\mathbf{C}:\mathbf{a}) \otimes (\mathbf{C}:\mathbf{a})\right) : \dot{\boldsymbol{\varepsilon}} \tag{6.9}$$

where $\mathbf{C}_t$ is the tangential modular matrix (or fourth-order tensor in the final form in (6.9)) which is not only a function of $E$ and $\nu$ but also, via $\mathbf{a}$, a function of the current stresses, $\boldsymbol{\sigma}$. This matrix can now be used in finite element expressions such as (6.1) to form the element and hence the structure tangent stiffness matrix.

A numerical example involving the computation of the elasto-plastic modular matrix of (6.9) is given in Section 6.9.2.

### 6.3.1 Non-associative plasticity

Before introducing hardening, we should make a brief mention of non-associative plasticity, which is mainly relevant to geomechanical materials such as soils. For such materials, experiments show that the flow direction is not usually normal to the yield surface, $f$. However, it can be considered as normal to some second function, $g$, known as the plastic potential. It then follows that in place of (6.4), $\dot{\boldsymbol{\varepsilon}}_p = \dot{\lambda}(\partial g / \partial \boldsymbol{\sigma}) = \dot{\lambda}\mathbf{b}$

rather than $\dot{\lambda}\mathbf{a}$ where, unless the plasticity is associative, $\mathbf{b} \neq \mathbf{a}$. With this difference, the basic formulation follows that in the previous section, but in place of (6.8),

$$\dot{\lambda} = \frac{\mathbf{a}^{\mathrm{T}}\mathbf{C}\dot{\boldsymbol{\varepsilon}}}{\mathbf{a}^{\mathrm{T}}\mathbf{C}\mathbf{b}},$$

while in place of the relationship in (6.9),

$$\mathbf{C}_{\mathrm{t}} = \mathbf{C}\left(\mathbf{I} - \frac{\mathbf{b}\mathbf{a}^{\mathrm{T}}\mathbf{C}}{\mathbf{a}^{\mathrm{T}}\mathbf{C}\mathbf{b}}\right).$$

It follows that, for non-associative plasticity, $\mathbf{C}_{\mathrm{t}}$ is generally non-symmetric.

## 6.4 INTRODUCING HARDENING

Although we will introduce the concepts of hardening with specific reference to the plane-stress plasticity that we have just introduced, the concepts are equally valid in relation to general stress states. Indeed the equations remain valid for such general states (such as the three-dimensional plasticity to be introduced in Section 6.5). The only item specific to plane-stress plasticity is the precise form of the equivalent plastic strain.

### 6.4.1 Isotropic strain hardening

Hardening can be introduced by changing the fixed yield stress, $\sigma_0$, in equation (6.3) to a variable stress, $\sigma_{\mathrm{o}}(\varepsilon_{\mathrm{ps}})$, so that

$$f = \sigma_{\mathrm{e}} - \sigma_{\mathrm{o}}(\varepsilon_{\mathrm{ps}}). \tag{6.10}$$

The variable yield stress is now a function of the equivalent plastic strain:

$$\varepsilon_{\mathrm{ps}} = \Sigma\delta\varepsilon_{\mathrm{ps}} = \int \dot{\varepsilon}_{\mathrm{ps}} \tag{6.11}$$

which is accumulated from the equivalent plastic strain rates,

$$\dot{\varepsilon}_{\mathrm{ps}} = \frac{2}{\sqrt{3}}(\dot{\varepsilon}_{\mathrm{p}x}^2 + \dot{\varepsilon}_{\mathrm{p}y}^2 + \dot{\varepsilon}_{\mathrm{p}x}\dot{\varepsilon}_{\mathrm{p}y} + \tfrac{1}{4}\dot{\gamma}_{\mathrm{p}xy})^{1/2}. \tag{6.12}$$

Under uniaxial tension, $\sigma_x$, $\dot{\varepsilon}_{\mathrm{p}y} = \dot{\varepsilon}_{\mathrm{p}z} = -\frac{1}{2}\dot{\varepsilon}_{\mathrm{p}x}$ so that there is no plastic volume change and $\dot{\varepsilon}_{\mathrm{ps}} = \dot{\varepsilon}_{\mathrm{p}x}$ while $\sigma_{\mathrm{e}} = \sigma_{\mathrm{o}} = \sigma_x$. Consequently, the relationship between $\sigma_{\mathrm{o}}$ and $\varepsilon_{\mathrm{ps}}$ can be taken from the uniaxial stress/plastic strain relationship. In particular, we will require $\partial\sigma_{\mathrm{o}}/\partial\varepsilon_{\mathrm{ps}}$ which, from Figure 6.5, is given by

$$H' = \frac{\partial\sigma_{\mathrm{o}}}{\partial\varepsilon_{\mathrm{ps}}} = \frac{\partial\sigma_{\mathrm{x}}}{\partial\varepsilon_{\mathrm{p}x}} = \frac{E_{\mathrm{t}}}{1 - E_{\mathrm{t}}/E}. \tag{6.13}$$

Once hardening is introduced, the tangency condition of equation (6.7) is modified to

$$\dot{f} = \frac{\partial f^{\mathrm{T}}}{\partial\boldsymbol{\sigma}}\dot{\boldsymbol{\sigma}} + \frac{\partial f}{\partial\sigma_{\mathrm{o}}}\frac{\partial\sigma_{\mathrm{o}}}{\partial\varepsilon_{\mathrm{ps}}}\dot{\varepsilon}_{\mathrm{ps}} = \mathbf{a}^{\mathrm{T}}\dot{\boldsymbol{\sigma}} - H'\dot{\varepsilon}_{\mathrm{ps}} = 0. \tag{6.14}$$

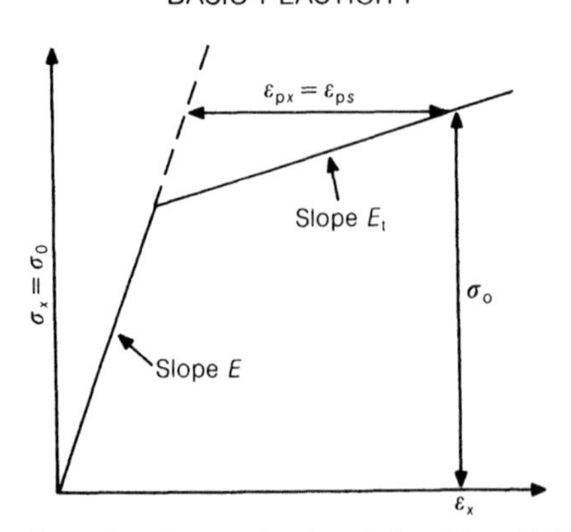


**Figure 6.5** One-dimensional stress/strain relationship with linear hardening.

Substitution from equation (6.4) into equation (6.12) gives

$$\dot{\varepsilon}_{ps} = \dot{\lambda} = B(\boldsymbol{\sigma})\dot{\lambda}. \tag{6.15}$$

For the present von Mises yield criterion, $B(\boldsymbol{\sigma}) = 1$, but for other criteria, this may not be so. Substituting from (6.15) into (6.14) gives

$$\dot{f} = \mathbf{a}^{\mathrm{T}}\dot{\boldsymbol{\sigma}} - H'B\dot{\lambda} = \mathbf{a}^{\mathrm{T}}\dot{\boldsymbol{\sigma}} - A'\dot{\lambda} = 0. \tag{6.16}$$

Hence, premultiplying equation (6.5) by $\mathbf{a}^{\mathrm{T}}$ and substituting into (6.16) give

$$\dot{\lambda} = \frac{\mathbf{a}^{\mathrm{T}}\mathbf{C}\dot{\boldsymbol{\varepsilon}}}{\mathbf{a}^{\mathrm{T}}\mathbf{C}\mathbf{a} + A'} = \frac{\mathbf{a}:\mathbf{C}:\dot{\boldsymbol{\varepsilon}}}{\mathbf{a}:\mathbf{C}:\mathbf{a} + A'} \tag{6.17}$$

while equation (6.9) is replaced by

$$\dot{\boldsymbol{\sigma}} = \mathbf{C}_t\dot{\boldsymbol{\varepsilon}} = \mathbf{C}\left(\mathbf{I} - \frac{\mathbf{a}\mathbf{a}^{\mathrm{T}}\mathbf{C}}{\mathbf{a}^{\mathrm{T}}\mathbf{C}\mathbf{a} + A'}\right)\dot{\boldsymbol{\varepsilon}} = \left(\mathbf{C} - \frac{1}{\mathbf{a}:\mathbf{C}:\mathbf{a} + A'}(\mathbf{C}:\mathbf{a})\otimes(\mathbf{C}:\mathbf{a})\right):\dot{\boldsymbol{\varepsilon}}. \tag{6.18}$$

For linear hardening, $A'$ is (via $H'$ in (6.13)) a single measurable constant. For non-linear hardening, $A'$ will vary with $\varepsilon_{ps}$ (and possibly other quantities)—indeed, more generally, $\sigma_0$ (or $A$) will vary with $\varepsilon_{ps}$.

The equivalent plastic strain can be considered as an 'internal variable' as it is 'internal' to the response [Z1, M3, M6]. Using this terminology, only the directly measurable total stresses and total strains are external. However, the plastic strains are required in order to define the response of the body. Following the introduction of the flow rules and the hardening hypothesis, in the previous developments one was left with one internal variable, $\varepsilon_{ps}$, to define the behaviour. The hardening behaviour was a function of this 'internal variable'. For a more complex material, one may require more 'internal variables' [M6].

### 6.4.2 Isotropic work hardening

Work hardening [H2] is more generally applicable than strain hardening. With work hardening, equation (6.10) is replaced by

$$f = \sigma_e - \sigma_o(W_p) \tag{6.19}$$

where $W_p$ is the plastic work, which is given by

$$W_p = \int \sigma_o \dot{\varepsilon}_{po} = \int \boldsymbol{\sigma}^T \dot{\boldsymbol{\varepsilon}}_p = \int \dot{\lambda} \boldsymbol{\sigma}^T \mathbf{a} \tag{6.20}$$

where $\dot{\varepsilon}_{po}$ is the one-dimensional plastic strain rate and the plastic work rate is

$$\dot{W}_p = \sigma_o \dot{\varepsilon}_{po} = \boldsymbol{\sigma}^T \boldsymbol{\varepsilon}_p = \dot{\lambda} \boldsymbol{\sigma}^T \mathbf{a}. \tag{6.21}$$

Instead of equations (6.14) or (6.16),

$$\begin{aligned} \dot{f} &= \mathbf{a}^T \dot{\boldsymbol{\sigma}} + \frac{\partial f}{\partial \sigma_o} \frac{\partial \sigma_o}{\partial W_p} \dot{W}_p = \mathbf{a}^T \dot{\boldsymbol{\sigma}} - \dot{\lambda} \frac{\partial \sigma_o}{\partial W_p} \mathbf{a}^T \boldsymbol{\sigma} \\ &= \mathbf{a}^T \dot{\boldsymbol{\sigma}} - \dot{\lambda} \frac{\partial \sigma_o}{\partial \varepsilon_{po}} \frac{\partial \varepsilon_{po}}{\partial W_p} \mathbf{a}^T \boldsymbol{\sigma} \\ &= \mathbf{a}^T \dot{\boldsymbol{\sigma}} - \dot{\lambda} \frac{H'}{\sigma_o} \mathbf{a}^T \boldsymbol{\sigma} = \mathbf{a}^T \dot{\boldsymbol{\sigma}} - A' \dot{\lambda} = 0 \end{aligned} \tag{6.22}$$

where

$$A' = \frac{H' \boldsymbol{\sigma}^T \mathbf{a}}{\sigma_o} = \frac{H' \boldsymbol{\sigma} : \mathbf{a}}{\sigma_o} \tag{6.23}$$

is the hardening constant for use in equations (6.17) and (6.18). If the yield function can be written in a similar form to equation (6.3), so that $f$ is a homogeneous function of order one, it is easily shown (Euler's theorem [H2]) that

$$\frac{\partial f^T}{\partial \boldsymbol{\sigma}} \boldsymbol{\sigma} = \mathbf{a}^T \boldsymbol{\sigma} = \mathbf{a} : \boldsymbol{\sigma} = \boldsymbol{\sigma}_o. \tag{6.24}$$

Hence $A' = H'$ (equation (6.13)) and for von Mises' yield function (with $B$ (equation (6.15) $= 1$), strain hardening and work hardening formulations coincide. This is not always the case.

### 6.4.3 Kinematic hardening

For seismic problems or low-cycle fatigue, the induced cyclic loading may involve relatively small plastic strains. In these circumstances, the Bauschinger effect [H2] may be significant. Assuming a linear hardening this effect is illustrated for a one-dimensional problem in Figure 6.6. Here, the yielding in tension has lowered the compressive strength, so that

$$(\sigma - \alpha) = \pm \sigma_o \tag{6.25}$$

where $\alpha$ is the 'kinematic shift' of the centre of the yield surface. As a result of this shift, with $\sigma_o$ being fixed (see Figure 6.6), the uniaxial stress $\sigma$ 'hardens'.

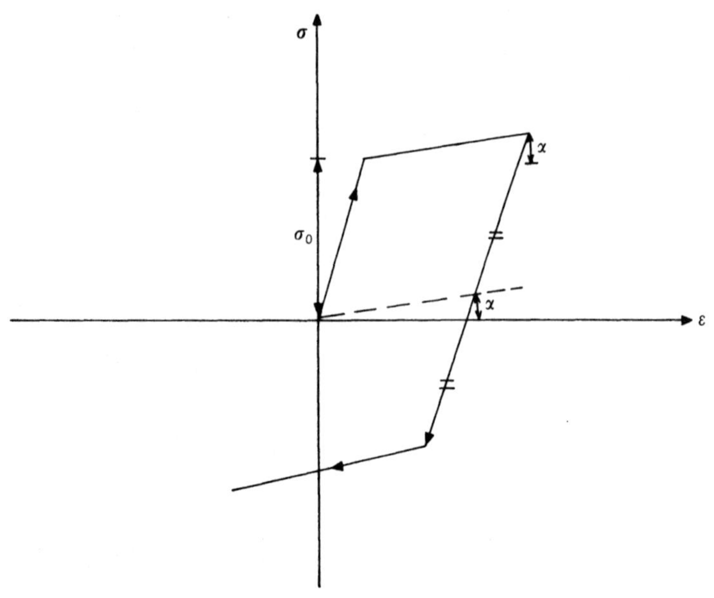


**Figure 6.6** One-dimensional illustration of kinematic hardening.

The Bauschinger effect [H2] cannot be treated by the methods of Sections 6.4.1 and 6.4.2 which involve isotropic hardening and one must introduce kinematic hardening [P2, Z2] or an 'overlay' [B1, O5] or an equivalent [M9] model. These issues are discussed in Volume 2.

## 6.5 VON MISES PLASTICITY IN THREE DIMENSIONS

For the general three-dimensional case, the von Mises yield criterion is

$$\begin{aligned} f &= \sigma_e - \sigma_0 = \sqrt{3}J_2^{1/2} - \sigma_0 \\ &= \frac{1}{\sqrt{2}}[(\sigma_x - \sigma_y)^2 + (\sigma_y - \sigma_z)^2 + (\sigma_z - \sigma_x)^2 + 6(\tau_{xy}^2 + \tau_{yz}^2 + \tau_{zx}^2)]^{1/2} - \sigma_o \\ &= \sqrt{3}[\tfrac{1}{2}(s_x^2 + s_y^2 + s_z^2) + \tau_{xy}^2 + \tau_{yz}^2 + \tau_{zx}^2]^{1/2} - \sigma_0 \\ &= \sqrt{\tfrac{3}{2}}(\mathbf{s}^{\mathrm{T}}\mathbf{L}\mathbf{s})^{1/2} - \sigma_0 = \sqrt{\tfrac{3}{2}}(\mathbf{s}:\mathbf{s})^{1/2} - \sigma_0 \end{aligned} \tag{6.26}$$

where

$$\mathbf{L} = \begin{bmatrix} 1 & & & & & \\ & 1 & & & & \\ & & 1 & & & \\ & & & 2 & & \\ & & & & 2 & \\ & & & & & 2 \end{bmatrix} \tag{6.27}$$

and

$$\mathbf{s}^{\mathrm{T}} = \{s_x, s_y, s_z, \tau_{xy}, \tau_{xz}, \tau_{yz}\} \tag{6.28}$$

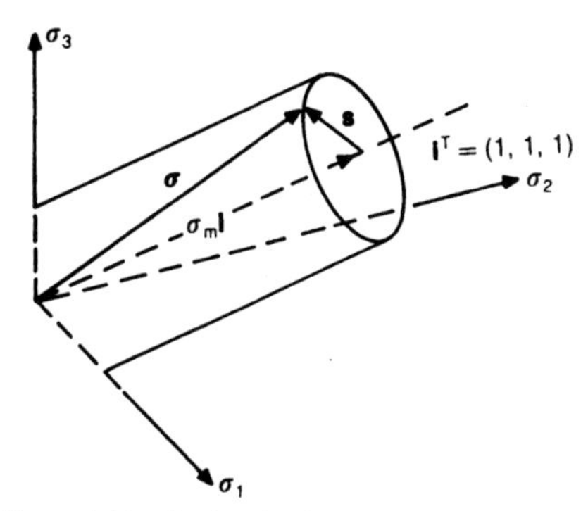


**Figure 6.7** Von Mises yield criterion in three-dimensional principal stress space.

are the deviatoric stresses previously defined in Section 4.2.2. They may also be written in tensor form (see (4.18) and in this case (see the last form in (6.26)) there is no need for the **L** matrix.

The three-dimensional von Mises yield criterion is plotted in principal stress space in Figure 6.7 where the stress vector $\boldsymbol{\sigma}$ is decomposed into a volumetric component (along the axis $\mathbf{i}^T = (1, 1, 1)$) and a deviatoric component, $\mathbf{s}$. From (6.26), the radius of the von Mises cylinder is clearly $\sqrt{\frac{2}{3}}\sigma_o$.

For three-dimensional plasticity, the equivalent plastic strain rate is given by*

$$\dot{\varepsilon}_{ps} = \sqrt{\tfrac{2}{3}}[\dot{\varepsilon}_{px}^2 + \dot{\varepsilon}_{py}^2 + \dot{\varepsilon}_{pz}^2 + \tfrac{1}{2}(\dot{\gamma}_{xy}^2 + \dot{\gamma}_{yz}^2 + \dot{\gamma}_{zx}^2)]^{1/2} = \sqrt{\tfrac{2}{3}}(\dot{\boldsymbol{\varepsilon}}_p : \dot{\boldsymbol{\varepsilon}}_p)^{1/2} = \sqrt{\tfrac{2}{3}}(\dot{\mathbf{e}}_p : \dot{\mathbf{e}}_p)^{1/2} \quad (6.29)$$

where $\dot{\mathbf{e}}_p$ are the deviatoric plastic strains (in tensor form—see (4.19)). The elastic stresses and strains are connected (see (4.10)) by

$$\begin{pmatrix} \sigma_x \\ \sigma_y \\ \sigma_z \\ \tau_{xy} \\ \tau_{xz} \\ \tau_{yz} \end{pmatrix} = \frac{E}{(1+\nu)(1-2\nu)} \times \begin{bmatrix} (1-\nu) & \nu & \nu & & & \\ \nu & (1-\nu) & \nu & & & \\ \nu & \nu & (1-\nu) & & & \\ & & & \frac{1}{2}(1-2\nu) & & \\ & & & & \frac{1}{2}(1-2\nu) & \\ & & & & & \frac{1}{2}(1-2\nu) \end{bmatrix} \begin{pmatrix} \varepsilon_x \\ \varepsilon_y \\ \varepsilon_z \\ \gamma_{xy} \\ \gamma_{xz} \\ \gamma_{yz} \end{pmatrix} \quad (6.30)$$

or

$$\boldsymbol{\sigma} = \mathbf{C}\boldsymbol{\varepsilon} \quad \text{or} \quad \boldsymbol{\sigma} = \mathbf{C}:\boldsymbol{\varepsilon}. \quad (6.31)$$

*With $\dot{\varepsilon}_{py} = \dot{\varepsilon}_{pz} = -\frac{1}{2}\dot{\varepsilon}_{px}$, $\dot{\varepsilon}_{ps}$ again degenerates to $\dot{\varepsilon}_{px}$ for the uniaxial case.

Differentiating equation (6.26) gives

$$\mathbf{a}^{\mathrm{T}} = \frac{\partial f^{\mathrm{T}}}{\partial \boldsymbol{\sigma}} = \frac{1}{2\sigma_{\mathrm{e}}}\{(2\sigma_x - \sigma_y - \sigma_z), (2\sigma_y - \sigma_x - \sigma_z), (2\sigma_z - \sigma_x - \sigma_y), 6\tau_{xy}, 6\tau_{yz}, 6\tau_{zx}\}$$

$$= \frac{3}{2\sigma_{\mathrm{e}}}\{s_x, s_y, s_z, 2\tau_{xy}, 2\tau_{yz}, 2\tau_{zx}\} = \frac{3}{2\sigma_{\mathrm{e}}}(\mathbf{Ls})^{\mathrm{T}} = \frac{\partial f^{\mathrm{T}}}{\partial \mathbf{s}} \tag{6.32}$$

or, using the tensor form,

$$\mathbf{a} = \frac{\partial f}{\partial \boldsymbol{\sigma}} = \frac{\partial f}{\partial \mathbf{s}} = \frac{3}{2\sigma_{\mathrm{e}}}\mathbf{s}. \tag{6.33}$$

As in (6.4), $\dot{\boldsymbol{\varepsilon}}_{\mathrm{p}} = \dot{\lambda}\mathbf{a}$ so that in (6.29)

$$\dot{\varepsilon}_{\mathrm{ps}} = \sqrt{\tfrac{2}{3}}\dot{\lambda}(\mathbf{a}^{\mathrm{T}}\mathbf{L}^{-1}\mathbf{a})^{1/2} = \sqrt{\tfrac{2}{3}}\dot{\lambda}(\mathbf{a}:\mathbf{a})^{1/2} = \sqrt{\frac{3}{2}\frac{\dot{\lambda}}{\sigma_{\mathrm{e}}}}(\mathbf{s}^{\mathrm{T}}\mathbf{Ls})^{1/2} = \dot{\lambda}. \tag{6.34}$$

With these new definitions of $\boldsymbol{\sigma}$, $\boldsymbol{\varepsilon}$, $\mathbf{C}$ and $\mathbf{a}$, an identical formulation to that of Sections 6.2–6.4 produces equation (6.17) for $\dot{\lambda}$ and (6.18) for $\mathbf{C}_{\mathrm{t}}$.

### 6.5.1 Splitting the update into volumetric and deviatoric parts

With a view to later developments on the 'radial return' method (Section 6.6.7), it is useful to split the stress update into volumetric and deviatoric components. (Background information was given in Sections 4.2.2 and 4.2.3.) To this end, (6.30) and (6.32) can be used to show (see also [D1]) that (using the matrix and vector forms)

$$\mathbf{Ca} = \frac{\sqrt{3}\mu}{\sqrt{\mathrm{J}_2}}\mathbf{s} = \frac{3\mu}{\sigma_{\mathrm{e}}}\mathbf{s} = 2\mu L^{-1}\mathbf{a} \tag{6.35}$$

and

$$\mathbf{a}^{\mathrm{T}}\mathbf{Ca} = \mathbf{a}:\mathrm{C}:\mathbf{a} = 3\mu. \tag{6.36}$$

Hence substitution into (6.18) gives

$$\dot{\boldsymbol{\sigma}} = \left(\mathbf{C} - \frac{3\mu}{\sigma_{\mathrm{e}}^2\left(1 + \frac{A'}{3\mu}\right)}\mathbf{s}\mathbf{s}^{\mathrm{T}}\right)\dot{\boldsymbol{\varepsilon}} = \left(\mathrm{C} - \frac{3\mu}{\sigma_{\mathrm{e}}^2\left(1 + \frac{A'}{3\mu}\right)}\mathbf{s}\otimes\mathbf{s}\right):\dot{\boldsymbol{\varepsilon}}. \tag{6.37}$$

In addition, the total strain rate, $\dot{\boldsymbol{\varepsilon}}$, can be split (see also Section 4.2.2) into

$$\dot{\boldsymbol{\varepsilon}} = \dot{\varepsilon}_{\mathrm{m}}\begin{pmatrix}1\\1\\1\\0\\0\\0\end{pmatrix} + \dot{\mathbf{e}} = \dot{\varepsilon}_{\mathrm{m}}\mathbf{j} + \dot{\mathbf{e}} \tag{6.38}$$

where $\dot{\varepsilon}_m$ is the mean strain rate,

$$\dot{\varepsilon}_m = (\dot{\varepsilon}_x + \dot{\varepsilon}_y + \dot{\varepsilon}_z)/3 = \tfrac{1}{3}\mathbf{j}^T\dot{\boldsymbol{\varepsilon}} \tag{6.39}$$

and $\dot{\mathbf{e}}$ are the deviatoric strains.

### 6.5.2 Using tensor notation

Using tensor forms, the equivalent of (6.38) is (see (4.19))

$$\dot{\boldsymbol{\varepsilon}} = \dot{\varepsilon}_m \mathbf{1} + \dot{\mathbf{e}} \tag{6.40}$$

where $\mathbf{1}$ is the second-order unit tensor (or identity matrix) which is often written as $\mathbf{I}$.

Equation (6.30) can be used to show that, using the matrix and vector forms,

$$\mathbf{C}\dot{\boldsymbol{\varepsilon}} = 3k\dot{\varepsilon}_m\mathbf{j} + 2\mu\mathbf{L}^{-1}\dot{\mathbf{e}} \tag{6.41a}$$

or using the equivalent tensor forms (see also (4.21) and (4.22)),

$$\mathbf{C}:\dot{\boldsymbol{\varepsilon}} = 3\mathrm{k}\mathbf{1} + 2\mu\dot{\mathbf{e}} \tag{6.41b}$$

where k is the bulk modulus (Section 4.2.2). In addition, from the definition of $\mathbf{s}$ in (6.32) and of $\mathbf{j}$ and $\dot{\mathbf{e}}$ in (6.38),

$$\mathbf{s}^T\mathbf{j} = 0. \tag{6.42}$$

Hence the first, matrix and vector, form in (6.37) can be modified to

$$\dot{\boldsymbol{\sigma}} = \dot{\sigma}_m\mathbf{j} + \dot{\mathbf{s}} = 3k\dot{\varepsilon}_m\mathbf{j} + 2\mu\left(\mathbf{L}^{-1} - \frac{3}{2\sigma_e^2\left(1 + \dfrac{A'}{3\mu}\right)}\mathbf{s}\mathbf{s}^T\right)\dot{\mathbf{e}} = \mathbf{C}_t\dot{\boldsymbol{\varepsilon}} \tag{6.43}$$

while the tensor form is

$$\dot{\boldsymbol{\sigma}} = \dot{\sigma}_m\mathbf{1} + \dot{\mathbf{s}} = 3k\dot{\varepsilon}_m\mathbf{1} + 2\mu\left(\mathbf{I} - \frac{3}{2\sigma_e^2\left(1 + \dfrac{A'}{3\mu}\right)}\mathbf{s}\otimes\mathbf{s}\right):\dot{\mathbf{e}} = \mathbf{C}_t:\dot{\boldsymbol{\varepsilon}}. \tag{6.44}$$

In (6.44), $\mathbf{1}$ is the second-order unit tensor while $\mathbf{I}$ is the fourth-order unit tensor—see (4.31).

Using the notation and procedure of Section 4.2.2, the $\mathbf{C}_t$ tensor in (6.44) can be written as

$$\mathbf{C}_t = \left(k - \frac{2\mu}{3}\right)(\mathbf{1}\otimes\mathbf{1}) + 2\mu\left(\mathbf{I} - \frac{3}{2\sigma_e^2\left(1 + \dfrac{A'}{3\mu}\right)}\mathbf{s}\otimes\mathbf{s}\right) \tag{6.45a}$$

while, taking account of (6.42), from (6.43), the matrix and vector form is

$$\mathbf{C}_t = \left(k - \frac{2\mu}{3}\right)\mathbf{j}\mathbf{j}^T + 2\mu\left(\mathbf{L}^{-1} - \frac{3}{2\sigma_e^2\left(1 + \dfrac{A'}{3\mu}\right)}\mathbf{s}\mathbf{s}^T\right). \tag{6.45b}$$

The latter is easily confirmed, with the aid of (6.42), by multiplying $\mathbf{C}_t$ from (6.45b) by $\dot{\boldsymbol{\varepsilon}}$ from (6.40).

## 6.6 INTEGRATING THE RATE EQUATIONS

Some issues relating to the 'integration of the flow rules' have already been discussed in Section 6.2. If the stress and strain increments were very small, we could effectively proceed by applying the previous tangential formulae with terms like $\dot{\boldsymbol{\varepsilon}}$ being replaced by terms like $\delta\boldsymbol{\varepsilon}$ and use the strain updating scheme of Strategy B as discussed in Section 6.2. However, the strain and subsequent stress changes will not be infinitesimally small and, as a consequence, errors would accumulate just as they would under a pure 'incremental' or 'forward-Euler scheme' at the structural level (see Chapter 1). Consequently, an uncorrected forward-Euler procedure at the Gauss-point level would lead to an unsafe drift from the 'yield surface'. In the same way as a pure incremental (tangential) procedure leads to a violation of equilibrium at the structural level (Chapter 1) so an equivalent tangential (forward-Euler) procedure leads to a violation of the yield criterion at the Gauss-point level.

Before addressing the problem of finite increment sizes, we should note that, even if the increments were infinitesimally small, it would be computationally inefficient to use the tangent modular matrix, $\mathbf{C}_t$, to compute stress rates $\dot{\boldsymbol{\sigma}}$. Instead of using equation (6.18), it would be more efficient to separately use (6.17) to compute $\dot{\lambda}$ and hence knowing $\mathbf{a} = \partial f/\partial\boldsymbol{\sigma}$, to compute $\dot{\boldsymbol{\sigma}}$ from the general form in (6.5) (This is illustrated in the numerical example of Section 6.9.2.) With infinitesimal strain increments it would then only be necessary to update the equivalent plastic strain, $\varepsilon_{ps}$ using equations (6.11) and (6.12) (or (6.15)), before proceeding to the next increment.

However, the strain increments will not be infinitesimal and as a consequence, we cannot replace terms like $\dot{\boldsymbol{\varepsilon}}$ with terms like $\Delta\boldsymbol{\varepsilon}$ although we can use $\delta\boldsymbol{\varepsilon}$ where $\delta\boldsymbol{\varepsilon}$ is infinitesimally small. For the von Mises yield criterion, we can however add a higher-order term and replace (6.7) by

$$\Delta f = \mathbf{a}^{\mathrm{T}}\Delta\boldsymbol{\sigma} + \tfrac{1}{2}\Delta\boldsymbol{\sigma}^{\mathrm{T}}\frac{\partial\mathbf{a}}{\partial\boldsymbol{\sigma}}\Delta\boldsymbol{\sigma} = \mathbf{a}:\Delta\boldsymbol{\sigma} + \tfrac{1}{2}\Delta\boldsymbol{\sigma}:\frac{\partial\mathbf{a}}{\partial\boldsymbol{\sigma}}:\Delta\boldsymbol{\sigma} \tag{6.46}$$

where differentiation of (6.32) gives

$$\frac{\partial\mathbf{a}}{\partial\boldsymbol{\sigma}} = \frac{1}{2\sigma_e}\begin{bmatrix} 2 & -1 & -1 & & & \\ -1 & 2 & -1 & & & \\ -1 & -1 & 2 & & & \\ & & & 6 & & \\ & & & & 6 & \\ & & & & & 6 \end{bmatrix} - \frac{1}{\sigma_e}\mathbf{a}\mathbf{a}^{\mathrm{T}} = \frac{1}{2\sigma_e}\mathbf{A} - \frac{1}{\sigma_e}\mathbf{a}\mathbf{a}^{\mathrm{T}}. \tag{6.47}$$

It is clear that the omission of the second-order terms in (6.46) will lead to error. We will later discuss methods which directly employ the 'second-order' information.

If we simply calculate $\mathbf{a} = \partial f/\partial\boldsymbol{\sigma}$ at the beginning of the increment and use equation (6.17) to compute $\Delta\lambda$, we adopt a 'forward-Euler scheme' which is bound to lead to stresses that lie outside the yield surface at the end of the increment (see Figure 6.8

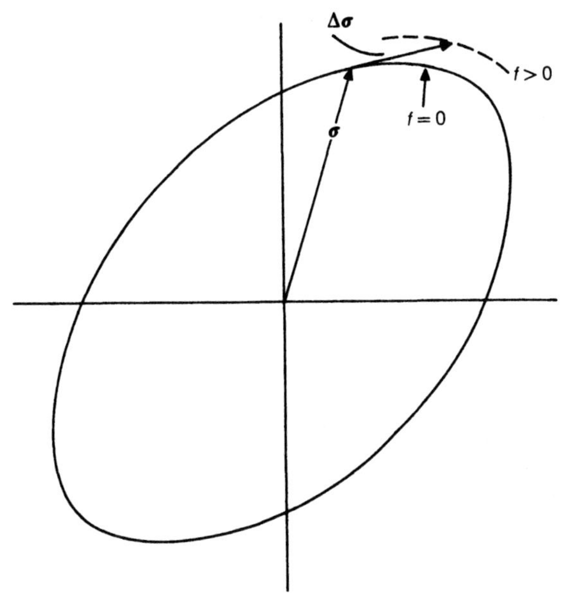


**Figure 6.8** Potential accumulation of error with the forward-Euler procedure.

and the numerical example in Section 6.9.2). Unless steps are taken to return the stresses to the yield surface or in some other way to ensure that the stresses remain at least very close to the surface, errors are bound to accumulate and the computed collapse load will generally be overpredicted.

There would appear to be three alternative procedures which can be used, either individually or in combinations to overcome this problem. They are:

(1) Add a return to the yield surface to the 'forward-Euler' scheme.
(2) Use sub-increments [H6, N1.1, Z2.1, S8, O4, M4, B4, N4).
(3) Use some form of backward or mid-point Euler scheme [W2, K2, K3, O2, S3, B3].

In each case, the aim is to update the stresses at a Gauss point given (a) the old stresses, strains and equivalent plastic strains and (b) the new strains. For all procedures, the first step is to use an elastic relationship to update the stresses. If these updated stresses are found to lie within the yield surface, the material at the Gauss point is assumed to have either remained elastic or to have unloaded elastically from the yield surface. In these circumstances, there is no need to 'integrate the rate equations'. However, if the elastic stresses are outside the yield surface, we need to adopt one of the 'integration' procedures.

Recent work has seen increasing use of the backward-Euler scheme without sub-incrementation (Sections 6.6.6 and 6.6.7). This method is popular because, for the von Mises yield criterion, it takes a particularly simple form and, in addition, it allows the generation of a 'consistent tangent modular matrix' (Section 6.7) which ensures quadratic convergence (see Section 1.2.3) for the overall structural iterations when the full Newton–Raphson method is adopted. Nonetheless, for some complicated yield criteria when coupled with complex hardening laws, the backward-Euler procedure is difficult to implement and hence techniques such as sub-incrementations may still be relevant.

### 6.6.1 Crossing the yield surface

A number, but not all, of the integration procedures require the location of the intersection [B2] of the elastic stress vector with the yield surface (Figure 6.9(a)). In such circumstances, we require

$$f(\boldsymbol{\sigma}_X + \alpha\Delta\boldsymbol{\sigma}_e) = 0 \tag{6.48}$$

where the original stresses, $\boldsymbol{\sigma}_X$ are such that

$$f(\boldsymbol{\sigma}_X) = f_X < 0 \tag{6.49}$$

while, with $\alpha = 1$, the elastic stresses $\boldsymbol{\sigma}_X + \Delta\boldsymbol{\sigma}_e$ give

$$f(\boldsymbol{\sigma}_B) = f(\boldsymbol{\sigma}_X + \Delta\boldsymbol{\sigma}_e) > 0. \tag{6.50}$$

For some yield surfaces, this problem can be solved exactly. For example, with the von Mises yield function, we can use the $\mathbf{A}$ matrix in (6.47) to re-express the yield function (6.26) in squared form as

$$f_2 = \sigma_e^2 - \sigma_o^2 = \tfrac{1}{2}\boldsymbol{\sigma}^T\mathbf{A}\boldsymbol{\sigma} - \sigma_o^2 = 0. \tag{6.51}$$

Substituting the stresses, $\boldsymbol{\sigma}_X + \alpha\Delta\boldsymbol{\sigma}_e$ into (6.51) gives

$$f_2 = \alpha^2\sigma_e(\Delta\boldsymbol{\sigma}_e)^2 + \alpha\Delta\boldsymbol{\sigma}_e^T\mathbf{A}\boldsymbol{\sigma}_X + \sigma_e(\sigma_X)^2 - \sigma_o^2 = 0 \tag{6.52}$$

where the $\sigma_e$ terms are simply the 'equivalent stress' terms of (6.26). We require the positive root of (6.52). A numerical example is given in Section 6.9.1.

Alternatively, for a general yield function we can use a truncated Taylor series with $\alpha$ as the only variable to set up an iterative scheme. Such a scheme might start with an initial estimate:

$$\alpha_0 = \frac{-f_X}{f_B - f_X} \tag{6.53}$$

and then use the truncated Taylor series:

$$f_n = f_o + \frac{\partial f}{\partial \boldsymbol{\sigma}}\frac{\partial \boldsymbol{\sigma}}{\partial \alpha}\delta\alpha = f_o + \mathbf{a}^T\Delta\boldsymbol{\sigma}_e\delta\alpha = 0 \tag{6.54}$$

to give a first change in $\alpha$, $\delta\alpha_o$. In applying (6.54), the 'old' yield function value, $f_o$, would for the iteration be computed from the stresses $\boldsymbol{\sigma} = \boldsymbol{\sigma}_X + \alpha_o\Delta\boldsymbol{\sigma}_e$ with $f_o$ being computed from these same stresses. The scalar $\alpha$ would then be updated using $\alpha_1 = \alpha_0 + \delta\alpha_0$ while a second iteration would involve

$$\delta\alpha_1 = \frac{-1}{\mathbf{a}^T\Delta\boldsymbol{\sigma}_e} f_1 \tag{6.55}$$

where $\mathbf{a}$ and $f_1$ would be computed at $\alpha_1$. Having computed the intersection point

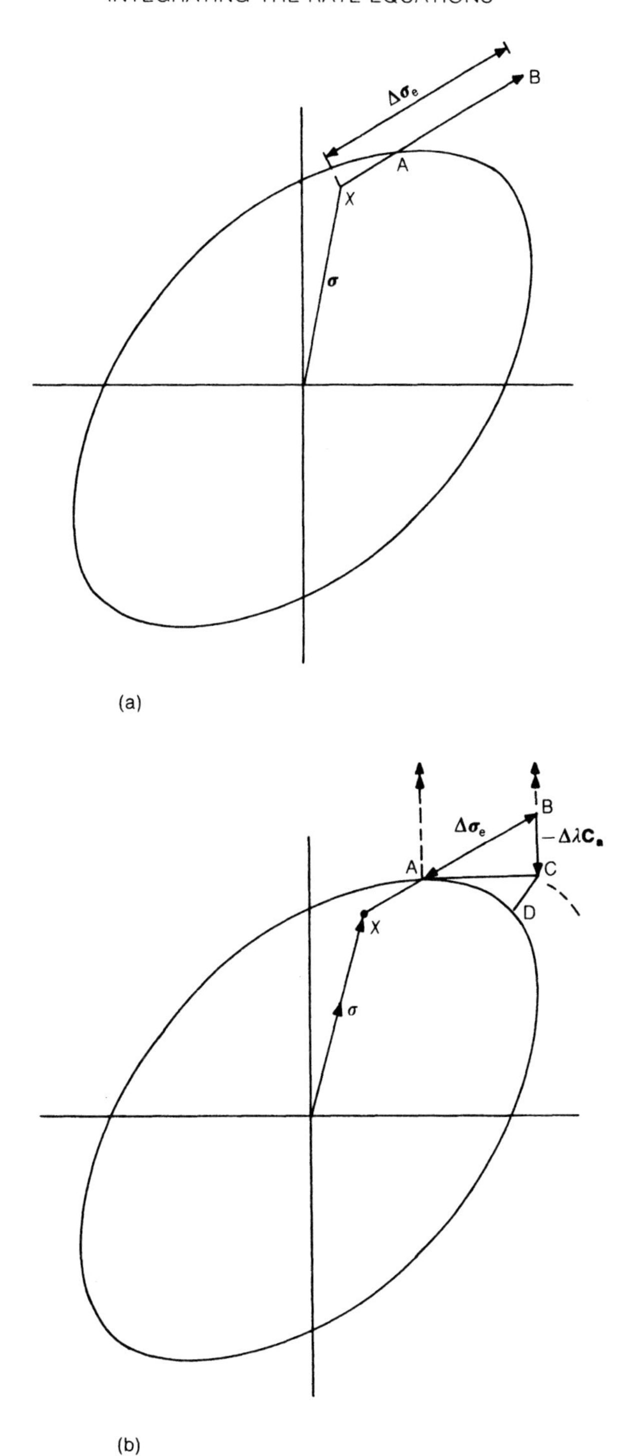


**Figure 6.9** The forward-Euler procedure: (a) Locating the intersection print, A; (b) Moving tangentialy from A to C (and (later) correcting to D).

$\boldsymbol{\sigma}_X + \alpha\Delta\boldsymbol{\sigma}_e$, the remaining portion of the strain increment, which is $(1-\alpha)\Delta\boldsymbol{\sigma}$, can be treated in an elasto-plastic manner.

### 6.6.2 Two alternative 'predictors'

We have already indicated that we may need some scheme to return the stresses to the yield surface following an initial 'predictor'. The standard predictor [O4] is the forward-Euler procedure which follows from (6.5) by replacing the rates with $\Delta$s so that:

$$\Delta\boldsymbol{\sigma} = \mathbf{C}\,\Delta\boldsymbol{\varepsilon} - \Delta\lambda\,\mathbf{Ca} = \Delta\boldsymbol{\sigma}_e - \Delta\lambda\,\mathbf{Ca} \tag{6.56}$$

where we are now moving from the intersection point A (Figure 6.9(b)) so that $\Delta\boldsymbol{\sigma}_e$ is now the elastic increment after reaching the yield surface (i.e. $(1-\alpha)$ times the $\Delta\boldsymbol{\sigma}_e$ in (6.48) or Figure 6.9(a)). In relation to Figure 6.9(b))

$$\boldsymbol{\sigma}_C = \boldsymbol{\sigma}_A + \Delta\boldsymbol{\sigma}_e - \Delta\lambda\,\mathbf{Ca} = \boldsymbol{\sigma}_B - \Delta\lambda\,\mathbf{Ca} \tag{6.57}$$

and the step can be interpreted as giving an elastic step from the intersection point A to B followed by a plastic return that is orthogonal to the yield surface at A. (To fully justify the pictorial representation in figures such as 6.9, the **C** matrix must be thought of as an identity matrix.)

An alternative predictor (Figure 6.10) uses the normal at the 'elastic trial point', B and hence avoids the necessity of computing the intersection point, A. A first-order Taylor exapansion about point B gives

$$f = f_B + \frac{\partial f^T}{\partial \boldsymbol{\sigma}}\Delta\boldsymbol{\sigma} + \frac{\partial f}{\partial \varepsilon_{ps}}\Delta\varepsilon_{ps} = f_B - \Delta\lambda\,\mathbf{a}_B^T\mathbf{C}\mathbf{a}_B - \Delta\lambda\,A' \tag{6.58}$$

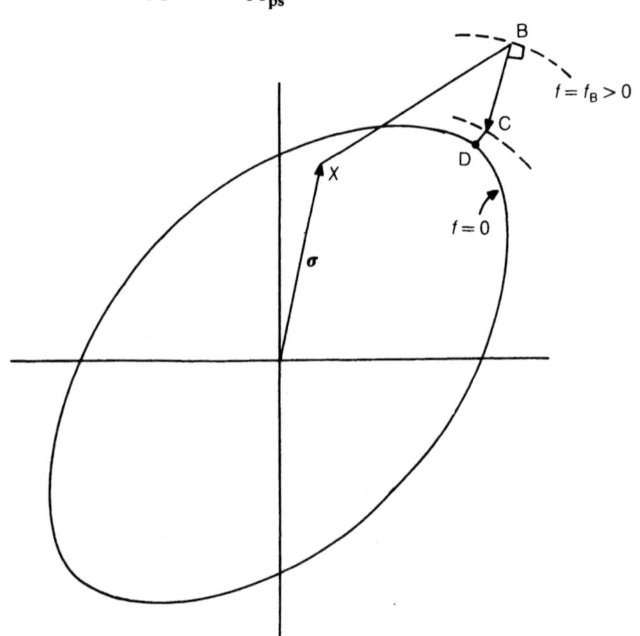


**Figure 6.10** An alternative 'predictor' (with later correction to D).

where the hardening parameter A′ has been defined in equations (6.13) and (6.16) (with $B = 1$). Equation (6.58) has used the incremental form of equation (6.5) with $\Delta\boldsymbol{\varepsilon} = \mathbf{0}$ because the total strain $\Delta\boldsymbol{\varepsilon}$ has already been applied in moving from point X to point B (Figure 6.10). If the new yield-function value $f$, is to be zero, equation (6.58) gives

$$\Delta\lambda = \frac{f_{\mathrm{B}}}{\mathbf{a}_{\mathrm{B}}^{\mathrm{T}}\mathbf{C}\mathbf{a}_{\mathrm{B}} + A'_{\mathrm{B}}} = \frac{f_{\mathrm{B}}}{\mathbf{a}_{\mathrm{B}}:\mathbf{C}:\mathbf{a}_{\mathrm{B}} + A'_{\mathrm{B}}} \tag{6.59}$$

where $\mathbf{a}$ and $A'$ are computed at B (Figure 6.10) and the final stresses, $\boldsymbol{\sigma}_{\mathrm{C}}$, are given by

$$\boldsymbol{\sigma}_{\mathrm{C}} = \boldsymbol{\sigma}_{\mathrm{B}} - \Delta\lambda\, \mathbf{C}\mathbf{a}_{\mathrm{B}}. \tag{6.60}$$

This method (for which a numerical example is given in Section 6.9.5.1) can be viewed as a form of backward-Euler scheme although, unlike the full backward-Euler procedure (Section 6.6.6), the final stresses at C will not always lie on the yield surface (see Figure 6.10). However, for the three-dimensional von Mises criterion (with linear hardening), the present method coincides with the well-known 'radial return algorithm' [W1, K2, K3, O2] which is a special form of the backward-Euler procedure. This relationship will be explored in Section 6.6.7.

### 6.6.3 Returning to the yield surface

In general, both of the previous methods produce stresses that lie outside the yield surface. It is now possible to simply scale the stresses at C (Figure 6.9(b) or 6.10) by a factor $r$ until the yield surface $f$ becomes zero [O2]. However, this technique, which should not be confused with the 'radial return method' (Section 6.6.7) which gives a radial return in *deviatoric* space, will generally involve an elastic component and is not recommended. An alternative technique [O1, C3], can be viewed as an extension of the previous backward-Euler predictor and has been related to 'operator splitting' by Ortiz and Simo [O1].

Using this approach, the total strains are kept fixed while additional plastic strains are introduced in order to 'relax' the stresses on to the yield surface. To this end, equation (6.60) can be repeated at point C (Figure 6.9(b) or 6.10) so that

$$\boldsymbol{\sigma}_{\mathrm{D}} = \boldsymbol{\sigma}_{\mathrm{C}} - \delta\lambda_{\mathrm{C}}\mathbf{C}\mathbf{a}_{\mathrm{C}} \tag{6.61}$$

where

$$\delta\lambda_{\mathrm{C}} = \left.\frac{f_{\mathrm{C}}}{\mathbf{a}^{\mathrm{T}}\mathbf{C}\mathbf{a} + A}\right|_{\mathrm{C}}. \tag{6.62}$$

If the resulting yield function at D (Figure 6.9(b) or 6.10) is insufficiently small, further relaxation can be applied. The final process leads to

$$\Delta\boldsymbol{\sigma} = \mathbf{C}\,\Delta\boldsymbol{\varepsilon} - \Delta\lambda_0\mathbf{C}\mathbf{a}_0 - \delta\lambda_{\mathrm{B}}\mathbf{C}\mathbf{a}_{\mathrm{B}} - \delta\lambda_{\mathrm{C}}\mathbf{C}\mathbf{a}_{\mathrm{C}} \tag{6.63}$$

where, for the forward-Euler procedure (Figure 6.9(b)), $\mathbf{a}_0$ is the normal at the intersection A while, for the backward-Euler predictor, $\mathbf{a}_0$ is the normal at B (Figure 6.10). The method is illustrated via a numerical example in Section 6.9.4.

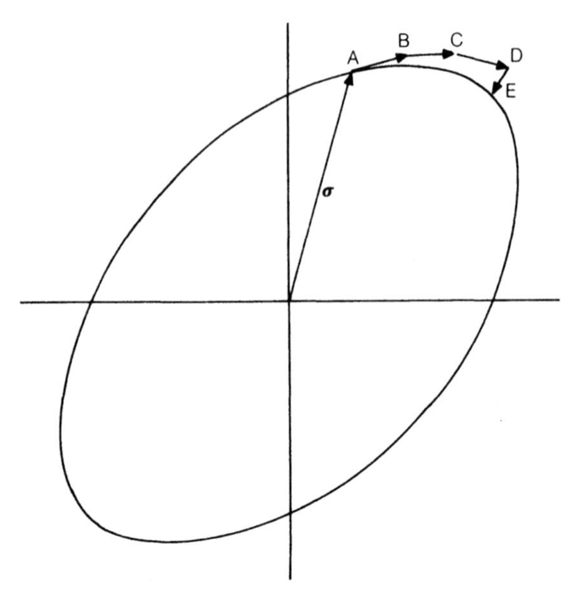


**Figure 6.11** Reducing the drift via sub-increments (with later correction to E).

### 6.6.4 Sub-incrementation

Instead of introducing some artificial return to the yield surface, the errors that are introduced by the forward-Euler tangential scheme can be significantly reduced by sub-incrementation [H6, N4, S3, S8, B4, M4]. Using such a technique (Figure 6.11), the incremental strain $\Delta\boldsymbol{\varepsilon}$ is divided into $m$ sub-steps each of $q\,\Delta\boldsymbol{\varepsilon}$, where $q = 1/m$ and the standard forward-Euler, tangential procedure is applied (at the Gauss-point level) at each step.

Some workers [N4, S8] use a two-step Euler procedure to estimate the error produced by the standard Euler technique and hence to compute the required number of steps. Such a two-step scheme starts with a standard step:

$$\boldsymbol{\sigma}_{B1} = \boldsymbol{\sigma}_A + \mathbf{C}_{tA}\Delta\boldsymbol{\varepsilon} = \boldsymbol{\sigma}_A + \Delta\boldsymbol{\sigma}_1 \tag{6.64}$$

and then recomputes the step using the average tangential modular matrix, so that

$$\boldsymbol{\sigma}_{B2} = \boldsymbol{\sigma}_A + \tfrac{1}{2}(\mathbf{C}_{tA} + \mathbf{C}_{tB1})\Delta\boldsymbol{\varepsilon} = \boldsymbol{\sigma}_A + \tfrac{1}{2}(\Delta\boldsymbol{\sigma}_1 + \Delta\boldsymbol{\sigma}_2) \tag{6.65}$$

where

$$\Delta\boldsymbol{\sigma}_2 = \mathbf{C}_{tB1}\Delta\boldsymbol{\varepsilon}. \tag{6.66}$$

Consequently, an estimate of the error is

$$\delta\boldsymbol{\sigma} = \boldsymbol{\sigma}_{B2} - \boldsymbol{\sigma}_{B1} = \tfrac{1}{2}(\Delta\boldsymbol{\sigma}_2 - \Delta\boldsymbol{\sigma}_1). \tag{6.67}$$

Nyssen [N4] uses this error estimate, $\delta\boldsymbol{\sigma}$, to propose the following measure for the truncation error in $f$ for one step:

$$e = 2\sigma_e(\Delta\boldsymbol{\sigma})/\sigma_0. \tag{6.68}$$

He then argues that the total error will be roughly $1/m$ times the error for a single step if $m$ sub-increments are used. Hence the required number of substeps to give a tolerance of $\beta$ in $f$ is

$$m = \frac{2\sigma_e(\Delta\boldsymbol{\sigma})}{\beta\sigma_0} \tag{6.69}$$

where he suggests a value of 0.05 for $\beta$.

Equation (6.46) would indicate that the truncation error in the standard Euler scheme is proportional to the square of the length of the stress increment $\Delta\boldsymbol{\sigma}$. Hence, it could be argued that

$$m \propto \left(\frac{\sigma_e(\Delta\boldsymbol{\sigma})}{\sigma_0}\right)^{1/2} \tag{6.70}$$

would be more appropriate. Sloan's work [S8] suggests such a scheme, although he works with the Euclidean norm rather than with $\sigma_e$ and computes the required reduction factor, $q$, as:

$$q = 0.8\left(\frac{\beta\|\boldsymbol{\sigma}\|}{\|\Delta\boldsymbol{\sigma}\|}\right)^{1/2} \tag{6.71}$$

where he advocates $10^{-3}$–$10^{-4}$ for $\beta$. Equations (6.70) and (6.71) are of the same form because $q$ is inversely proportional to $m$.

In practice, Sloan takes his procedure beyond the simple computation of a number of sub-increments. Instead, at each sub-increment, he uses equation (6.71) to indicate whether or not the current substep needs to be reduced further. In addition, he uses the two-step Euler scheme (equations (6.64)–(6.66)) at each substep. In contrast to most other workers, Sloan does not combine his substepping with a technique, such as that of Section 6.6.3, to return the stresses to the yield surface. Even if sub-incrementation is used, it is probably wise to introduce such a 'correction' either at the end of each substep or at the end of the increment (Figure 6.11—from D to E).

Other schemes have been proposed [B4, N4, S3] for estimating the number of sub-increments. For instance, Marques [M4], proposes a Fortan-type algorithm whereby

$$m = \mathsf{NSTEP} = \mathsf{INT(AMAX(STEP\,1, STEP\,2))} + 1 \tag{6.72a}$$

where

$$m_1 = \mathsf{STEP\,1} = \beta_1(\sigma_e(\boldsymbol{\sigma}_B) - \boldsymbol{\sigma}_e(\boldsymbol{\sigma}_A))/\sigma_e(\boldsymbol{\sigma}_A) \tag{6.72b}$$

$$m_2 = \mathsf{STEP\,2} = \beta_2\sigma_e(\boldsymbol{\sigma}_B - \boldsymbol{\sigma}_A)/\sigma_e(\boldsymbol{\sigma}_A). \tag{6.72c}$$

Equation (6.72b) is designed to limit the radial movement from the yield surface while equation (6.72c) is designed to limit the tangential movement.

The simplest form of sub-incrementation is illustrated by way of a numerical example in Section 6.9.3.

### 6.6.5 Generalised trapezoidal or mid-point algorithms

Ortiz and Popov [O2] have shown that a number of different integration algorithms can be included in the generalised algorithm:

$$\boldsymbol{\sigma}_C = \boldsymbol{\sigma}_A + \mathbf{C}(\Delta\boldsymbol{\varepsilon} - \Delta\boldsymbol{\varepsilon}_p) = \boldsymbol{\sigma}_B - \mathbf{C}\,\Delta\boldsymbol{\varepsilon}_p \tag{6.73}$$

$$\Delta\boldsymbol{\varepsilon}_p = \Delta\lambda[(1-\eta)\mathbf{a}_A + \eta\mathbf{a}_C)] \tag{6.74a}$$

or

$$\Delta\boldsymbol{\varepsilon}_p = \Delta\lambda[\mathbf{a}((1-\eta)\boldsymbol{\sigma}_A + \eta\boldsymbol{\sigma}_C)] \tag{6.74b}$$

$$f_C = \sigma_{eC}(\boldsymbol{\sigma}_C) - \sigma_{0C}(\varepsilon_{psC}) = \sigma_{eC}(\boldsymbol{\sigma}_C) - \sigma_{0C}(\varepsilon_{psB} + \Delta\varepsilon_{ps}(\Delta\boldsymbol{\varepsilon}_p)) \tag{6.75}$$

where (Figure 6.12(a)) A is the starting point and C the final point on the yield surface. If $\eta = 0$, equations (6.74a) and (6.74b) coincide and we obtain the previous 'explicit', forward-Euler, tangential algorithm. However, as we have already discussed, this algorithm does not directly lead to stresses that satisfy the yield criterion and hence equation (6.75) is not satisfied.

If $\eta = 1$, we produce a 'backward-Euler' or 'closest point' algorithm (see Sections 6.6.6 and 6.6.10). A slightly modified version of this algorithm was discussed in Section 6.6.2. However, in contrast to the situation of Figure 6.9(b), the full backward-Euler scheme involves a vector $\mathbf{a}_C$ that is normal to the yield surface at the *final* position C (Figure 6.12(b)) for which the stresses, $\boldsymbol{\sigma}_C$, satisfy (6.75). Except in special circumstances (see Section 6.6.7), $\mathbf{a}_C$ cannot be directly computed from data at A or B (Figure 6.12(a)). Hence an iterative procedure must be used at the Gauss-point level to solve the non-linear equations (6.73)–(6.75). This process will be described in more detail in Section 6.6.6.

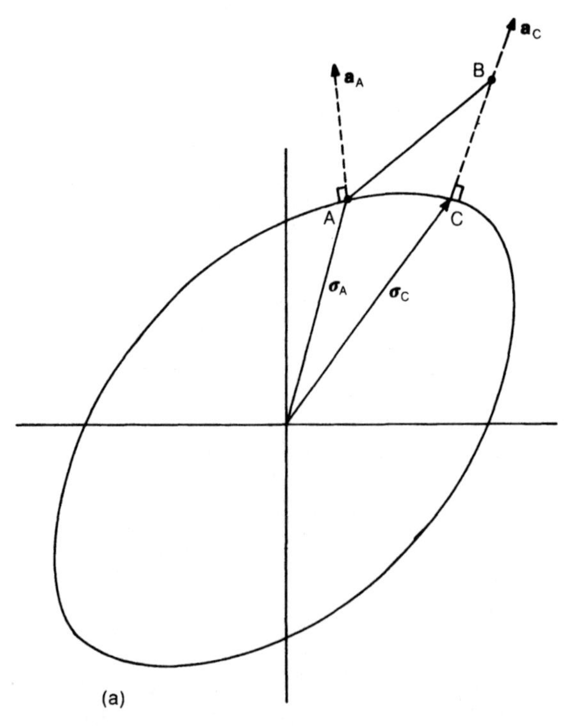


**Figure 6.12** General and backward-Euler returns: (a) the flow vectors $\mathbf{a}_A$ and $\mathbf{a}_C$; (b) backward-Euler return from inside the yield surface; (c) three-dimensional backward-Euler return.

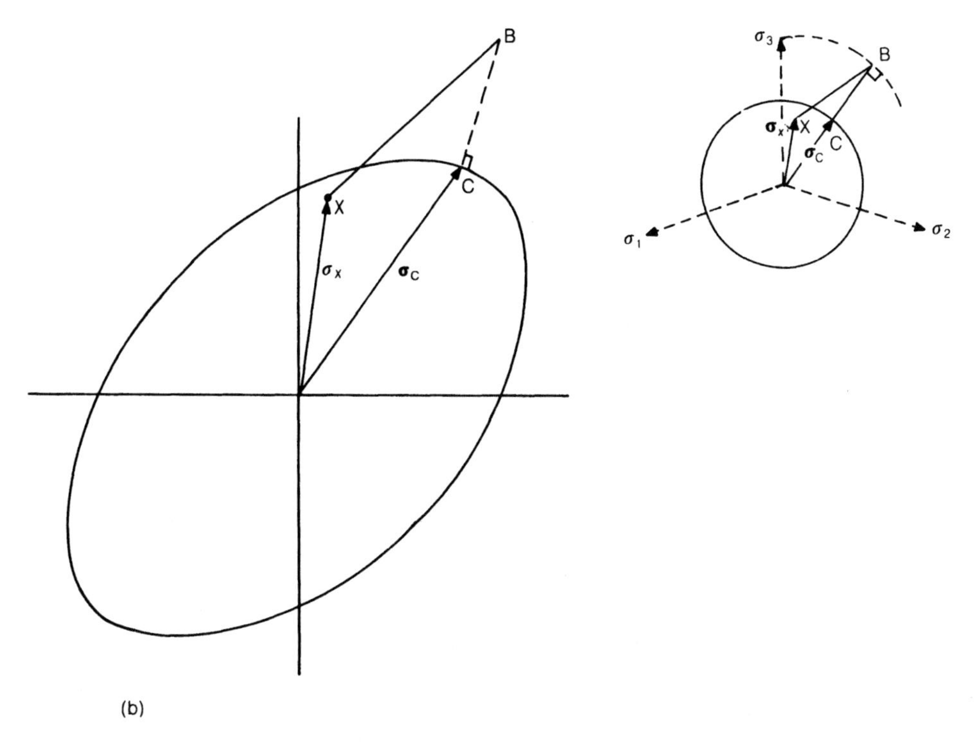


**Figure 6.12** *(continued)*

For $0 < \eta < 1$, either equations (6.74(a)) or (6.74(b)) may be used for the plastic flow. In the former instance, a generalised trapezoidal rule is produced while the latter gives a generalised mid-point rule. For $\eta = \frac{1}{2}$, the former procedure coincides with Rice and Tracey's mean-normal procedure [R2] which was devised for the three-dimensional von Mises yield criterion with perfect plasticity. This method involves a simple modification to the basic forward-Euler procedure. Specifically, it takes

$$\Delta\boldsymbol{\varepsilon}_p = \tfrac{1}{2}\Delta\lambda(\mathbf{a}_A + \mathbf{a}_B) \tag{6.76}$$

where $\mathbf{a}_B$ is normal to the 'enlarged' yield surface at the 'trial elastic position', B (Figure 6.10) in conjunction with the condition

$$\tfrac{1}{2}(\mathbf{a}_A + \mathbf{a}_B)^T\Delta\boldsymbol{\sigma} = 0 \tag{6.77}$$

in place of the incremental form of (6.7). For the three-dimensional von Mises yield criterion with perfect plasticity, this procedure ensures that the final stresses lie on the yield surface and no Gauss-point level iterations are required.

For the von Mises yield criterion with linear hardening, the generalised trapezoidal (equation (6.74a)) and mid-point (equation (6.74b)) rules coincide [O2]. More generally, Ortiz and Popov [O2] preferred the generalised mid-point rule and, in particular, $\eta = \frac{1}{2}$. However, they showed that for large steps, the ($\eta = 1$), backward-Euler scheme is better. It should also be noted that only with $\eta = 1$, is it unnecessary to compute the intersection point of the elastic predictor with the yield surface, i.e.

for $\eta < 1$, $\boldsymbol{\sigma}_A$ in equation (6.73) must be on the yield surface whereas for $\eta = 1, \boldsymbol{\sigma}_A$ is not required (see Figure 6.12(b)).

### 6.6.6 A backward-Euler return

The backward-Euler return is based on the equation

$$\boldsymbol{\sigma}_C = \boldsymbol{\sigma}_B - \Delta\lambda\, \mathbf{C}\mathbf{a}_C \tag{6.78}$$

which can be obtained from (6.73) and (6.74) with $\eta = 1$. A starting estimate for $\boldsymbol{\sigma}_C$ can be obtained from the second method of Section 6.6.2. Generally this starting estimate will not satisfy the yield function and further iterations will be required because the normal at the trial position B (Figure 6.10) will not generally equal the final normal. In order to derive such an iterative loop, a vector, $\mathbf{r}$, can be set up to represent the difference between the current stresses and the backward-Euler stresses, i.e.

$$\mathbf{r} = \boldsymbol{\sigma} - (\boldsymbol{\sigma}_B - \Delta\lambda\, \mathbf{C}\mathbf{a}_C) \tag{6.79}$$

and iterations are introduced to reduce $\mathbf{r}$ to (almost) zero while the final stresses should satisfy the yield criterion, $f = 0$.

With the trial elastic stresses, $\boldsymbol{\sigma}_B$ being kept fixed, a truncated Taylor expansion can be applied to equation (6.79) so as to produce a new residual, $\mathbf{r}_n$, where

$$\mathbf{r}_n = \mathbf{r}_o + \dot{\boldsymbol{\sigma}} + \dot{\lambda}\, \mathbf{C}\mathbf{a} + \Delta\lambda\, \mathbf{C}\frac{\partial \mathbf{a}}{\partial \boldsymbol{\sigma}}\dot{\boldsymbol{\sigma}} \tag{6.80}$$

$\dot{\boldsymbol{\sigma}}$ is the change in $\boldsymbol{\sigma}$ and $\dot{\lambda}$ is the change in $\Delta\lambda$. Setting $\mathbf{r}_n$ to zero gives

$$\dot{\boldsymbol{\sigma}} = -\left(\mathbf{I} + \Delta\lambda\, \mathbf{C}\frac{\partial \mathbf{a}}{\partial \boldsymbol{\sigma}}\right)^{-1}(\mathbf{r}_o + \dot{\lambda}\, \mathbf{C}\mathbf{a}) = -\mathbf{Q}^{-1}\mathbf{r}_o - \dot{\lambda}\mathbf{Q}^{-1}\mathbf{C}\mathbf{a}. \tag{6.81}$$

Also, a truncated Taylor series on the yield function (6.75) gives (in a similar fashion to (6.58)):

$$f_{Cn} = f_{Co} + \frac{\partial f^T}{\partial \boldsymbol{\sigma}}\dot{\boldsymbol{\sigma}} + \frac{\partial f}{\partial \varepsilon_{ps}}\dot{\varepsilon}_{ps} = f_{Co} + \mathbf{a}_C^T\dot{\boldsymbol{\sigma}} + A'_C\dot{\lambda} = 0 \tag{6.82}$$

so that (dropping the subscript C):

$$\dot{\lambda} = \frac{f_o - \mathbf{a}^T\mathbf{Q}^{-1}\mathbf{r}_o}{\mathbf{a}^T\mathbf{Q}^{-1}\mathbf{C}\mathbf{a} + A'}. \tag{6.83}$$

Consequently, (6.81) can be solved to obtain the iterative stress change, $\dot{\boldsymbol{\sigma}}$. Also, from (6.15), the iterative change in the equivalent plastic strain is

$$\dot{\varepsilon}_{ps} = B(\boldsymbol{\sigma})\dot{\lambda} \tag{6.84}$$

where, for many yield functions, $B(\boldsymbol{\sigma}) = 1$.

A numerical example involving this general backward-Euler return is given in Section 6.9.5.1.

### 6.6.7 The radial-return algorithm, a special form of backward-Euler procedure

It will now be shown that the radial return method is a special form of backward-Euler (or fully implicit integration) procedure. The radial return method was apparently first proposed by Wilkins [W2] and subsequently refined in [K2, K3, N1, S3].

For the von Mises yield criterion with linear hardening, no iterations are required for the backward-Euler procedure and the second method of Section 6.6.2 gives an exact solution. This can best be demonstrated by splitting the update of Section 6.6.2 into volumetric and deviatoric parts with the aid of some of the relationships that were developed in Section 6.5.1. Using the notation of (6.43) for the split into deviatoric and volumetric components, equation (6.57) can be re-expressed as

$$\boldsymbol{\sigma}_C = \sigma_{mB}\mathbf{j} + \mathbf{s}_B - \Delta\lambda\,\mathbf{C}\mathbf{a}_B = \sigma_{mB}\mathbf{j} + \mathbf{s}_C \tag{6.85}$$

where the elastic stresses, $\boldsymbol{\sigma}_B$, at B (Figure 6.12) have been split into volumetric ($\sigma_{mB}\mathbf{j}$) and deviatoric ($\mathbf{s}_B$) components (see also (6.38) for $\mathbf{j}$). With the aid of (6.35), (6.85) becomes

$$\boldsymbol{\sigma}_C = \sigma_{mC}\mathbf{j} + \mathbf{s}_C = \sigma_{mB}\mathbf{j} + \left(1 - \frac{3\mu\Delta\lambda}{\sigma_{eB}}\right)\mathbf{s}_B. \tag{6.86}$$

Because $\mathbf{s}_B$ has no component in the direction $\mathbf{j}$ (no volumetric component because of (6.42)), it follows that $\boldsymbol{\sigma}_{mC} = \boldsymbol{\sigma}_{mB}$ and

$$\mathbf{s}_c = \alpha\mathbf{s}_B = \left(1 - \frac{3\mu\Delta\lambda}{\sigma_{eB}}\right)\mathbf{s}_B. \tag{6.87}$$

These deviatoric stresses must satisfy the yield criterion of (6.75) so that (using (6.26) in either vector or tensor form for the yield function):

$$f_C = \sigma_{eC}(\mathbf{s}_C) - \sigma_{oC}(\varepsilon_{psC}) = \alpha\sigma_{eB} - \sigma_{oC}(\varepsilon_{psC}). \tag{6.88}$$

Using (6.87) for $\alpha$, with linear hardening (with fixed $A'$), (6.88) simplifies to

$$f_C = \sigma_{eB} - 3\mu\,\Delta\lambda - (\sigma_{oB} + A'\,\Delta\varepsilon_{ps}) = f_B - (3\mu + A')\,\Delta\lambda = 0 \tag{6.89}$$

(recalling—see (6.15)—that $\Delta\varepsilon_{ps} = \Delta\lambda$). From (6.89),

$$\Delta\lambda = \frac{f_B}{(3\mu + A')}. \tag{6.90}$$

Substituting from (6.36) into (6.59) gives the same relationship as above. Hence the current procedure which was designed to satisfy the yield function at the final position (via (6.88)) coincides (for the von Mises criterion) with the predictor in Section 6.6.2 based on a truncated Taylor series of the yield function at the 'trial position', B. The reason is that (6.87) defines a 'radial return' in deviatoric space (see Figure 6.12(c) which shows a cross-section of the von Mises cylinder of Figure 6.7). If $\mathbf{s}_C = \alpha\mathbf{s}_B$, then from the yield function of (6.26), $\sigma_{eC} = \alpha\sigma_{eB}$ and hence, from (6.32),

$$\mathbf{a}_C = \mathbf{a}_B. \tag{6.91}$$

From (6.85), (6.87) and (6.90), the complete update is

$$\boldsymbol{\sigma}_{C} = \sigma_{mB}\mathbf{j} + \alpha \mathbf{s}_{B}, \qquad \alpha = 1 - \frac{3\mu f_{B}}{(3\mu + A')\sigma_{eB}} \tag{6.92}$$

which takes an even simpler form without hardening. For the tensor form, one need simply replace **j** in (6.92) by **1**.

With non-linear hardening, the return is still radial in deviatoric space and again relates to (6.87). However, (6.90) can be longer be used for $\Delta\lambda$ although with $A' = A'_{B} = A'_{A}$ it can give a starting value for a scalar Newton–Raphson iteration. This scalar iteration relates to the satisfaction of the yield function (6.88) with $\Delta\lambda$ as the only variable. A truncated Taylor series then leads to

$$f_{Cn} = f_{Co} + \frac{\partial f}{\partial \Delta\lambda}\dot{\lambda} = f_{Co} + (3\mu + A'_{Co})\dot{\lambda} = 0 \tag{6.93}$$

where the subscript n means 'new' and o means 'old'. The term $A'_{Co} = H'_{Co}$ (see (6.16) with $B = 1$) is the slope of the uniaxial stress/plastic strain relationship (see (6.13)) at the old trial value of the equivalent plastic strain, $\varepsilon_{psCo}$.

## 6.7 THE CONSISTENT TANGENT MODULAR MATRIX

Simo and Taylor [S4] and Runesson and Samuelsson [R3] derived a tangent modular matrix that is fully consistent with the backward-Euler integration algorithm of Section 6.6.7. (Other work on the consistent tangent approach can be found in [S5-S7, A1, J1, H1, B3, C4, R1, M8].) As a consequence of the 'consistency', the use of the consistent tangent modular matrix significantly improves the convergence characteristics of the overall equilibrium iterations if a Newton–Raphson scheme is used for the latter. Standard techniques would use the modular matrix of (6.18) which is 'inconsistent' with the backward-Euler integrations scheme (or any other effective integration schemes unless the increment sizes are infinitesimal) and hence destroys the 'quadratic convergence' inherent in the Newton–Raphson method.

We will now give two derivations for consistent tangent relationships; one based on the general backward-Euler return of Section 6.6.6 and the other based on the specialised radial return of Section 6.6.7. For the von Mises yield criterion, the two techniques lead to exactly equivalent formulations.

### 6.7.1 Splitting the deviatoric from the volumetric components

We will firstly follow on from the radial return of Section 6.6.7. To this end, it is most convenient to work with the tensor forms for stress and strain but we will also give (sometimes on the same line) the matrix and vector forms. The former will involve the contraction symbol: while the latter will involve inner products designated via the 'T' symbol for transpose.

From (6.87), the basic return was

$$\mathbf{s}_{C} = \alpha \mathbf{s}_{B} = \left(1 - \frac{3\mu\Delta\lambda}{\sigma_{eB}}\right)\mathbf{s}_{B} \tag{6.94}$$

where with linear hardening $\Delta\lambda$ is given by (6.90) while for non-linear hardening, it would be obtained from the iterative procedure described at the end of the last section. To obtain a consistent tangent, we differentiate (6.94) to obtain

$$\dot{\mathbf{s}}_C = \alpha\dot{\mathbf{s}}_B + \dot{\alpha}\mathbf{s}_B = 2\mu\alpha\dot{\mathbf{e}}_B + \dot{\alpha}\mathbf{s}_B = 2\mu\alpha\dot{\mathbf{e}}_C + \dot{\alpha}\mathbf{s}_B \tag{6.95a}$$

or, with matrices and vectors,

$$\dot{\mathbf{s}}_C = 2\mu\alpha\mathbf{L}^{-1}\dot{\mathbf{e}}_C + \dot{\alpha}\mathbf{s}_B. \tag{6.95b}$$

In (6.95), we have used the linear elastic relationship of (6.41) (see also (4.21) and (4.22)) and have also used the basic return algorithm of (6.78) for the relationship $\dot{\mathbf{e}}_C = \dot{\mathbf{e}}_B$ (the movement from B to C being entirely governed by $\Delta\lambda$ and $\mathbf{a}_C = \mathbf{a}_B$).

From (6.94),

$$\dot{\alpha} = \frac{3\mu\dot{\lambda}}{\sigma_{eB}} + \frac{3\mu\Delta\lambda}{\sigma_{eB}^2}\dot{\sigma}_{eB} = \frac{(1-\alpha)}{\Delta\lambda}\dot{\lambda} + \frac{(1-\alpha)}{\sigma_{eB}}\dot{\sigma}_{eB} \tag{6.96}$$

while, from (6.26),

$$\dot{\sigma}_{eB} = \sqrt{\frac{3}{2}}\frac{\mathbf{s}_B:\dot{\mathbf{s}}_B}{\|\mathbf{s}_B\|} = \frac{3}{2\sigma_{eB}}\mathbf{s}_B:\dot{\mathbf{s}}_B = \frac{3\mu}{\sigma_{eB}}\mathbf{s}_B:\dot{\mathbf{e}}_B = \frac{3\mu}{\sigma_{eB}}\mathbf{s}_B:\dot{\mathbf{e}}_C = \frac{3\mu}{\sigma_{eB}}\mathbf{s}_B^T\dot{\mathbf{e}}_C. \tag{6.97}$$

We must now ensure that we remain on the yield surface at C, by differentiating (6.88) to obtain:

$$\dot{f}_C = \dot{\alpha}\sigma_{eB} + \alpha\dot{\sigma}_{eB} - A'_C\dot{\lambda} = 0 \tag{6.98}$$

where $A'_C$ is the tangential hardening parameter at C. Substituting from (6.96) into (6.98) gives

$$\dot{\sigma}_{eB} - (3\mu + A'_C)\dot{\lambda} = 0 \tag{6.99}$$

while substitution from (6.99) for $\dot{\lambda}$ and from (6.97) for $\dot{\sigma}_{eB}$ into (6.96) gives

$$\dot{\alpha} = 2\mu\beta\mathbf{s}_B:\dot{\mathbf{e}}_B = 2\mu\beta\mathbf{s}_B^T\dot{\mathbf{e}}_B \tag{6.100}$$

where

$$\beta = \frac{3}{2\sigma_{eB}^2}(1-\alpha)\left(1 - \frac{\sigma_{eB}}{\Delta\lambda(3\mu + A'_C)}\right) = \frac{3}{2\sigma_{eB}^2}\left(\frac{(1-\alpha)(3\mu + A'_C) - 3\mu}{(3\mu + A'_C)}\right). \tag{6.101}$$

Substituting from (6.101) into (6.95) gives

$$\dot{\mathbf{s}}_C = 2\mu(\alpha\mathbf{I} + \beta\mathbf{s}_B\otimes\mathbf{s}_B):\dot{\mathbf{e}}_C = 2\mu(\alpha\mathbf{L}^{-1} + \beta\mathbf{s}_B\mathbf{s}_B^T)\dot{\mathbf{e}}_C. \tag{6.102}$$

Knowing, from (6.94), that $\mathbf{s}_C = \alpha\mathbf{s}_B$ and hence $\sigma_{eC} = \alpha\sigma_{eB}$, equations (6.101) and (6.102) can easily be re-expressed to contain $\mathbf{s}_C$ and $\sigma_{eC}$ instead of $\mathbf{s}_B$ and $\sigma_{eB}$.

Combining (6.102) with the volumetric contribution (as in (6.43) and (6.44)) and using the notation of Section 4.2.2 (with **1** as a second-order unit tensor and **I** as a fourth-order unit tensor), leads to a consistent tangent modular tensor:

$$\mathbf{C}_t = \left(k - \frac{2\mu\alpha}{3}\right)(\mathbf{1}\otimes\mathbf{1}) + 2\mu(\alpha\mathbf{I} - \beta\mathbf{s}_B\otimes\mathbf{s}_B) \tag{6.103a}$$

or, in matrix and vector form, as

$$\mathbf{C}_t = \left(k - \frac{2\mu\alpha}{3}\right)(\mathbf{j}\mathbf{j}^T) + 2\mu(\alpha\mathbf{L}^{-1} - \beta\mathbf{s}_B\mathbf{s}_B^T). \tag{6.103b}$$

It is instructive to compare (6.102) with the equivalent 'inconsistent' relationship in (6.43) or (6.44). Considering, for simplicity, the case with no hardening so that $A' = 0$, the inconsistent form is (from (6.44))

$$\dot{\mathbf{s}} = 2\mu\left(\mathbf{I} - \frac{3}{2\sigma_e^2}\mathbf{s}\otimes\mathbf{s}\right):\dot{\mathbf{e}} = 2\mu\left(\mathbf{I} - \frac{\mathbf{s}\otimes\mathbf{s}}{\mathbf{s}:\mathbf{s}}\right):\dot{\mathbf{e}} \tag{6.104}$$

while, from (6.102), the 'consistent' form is

$$\dot{\mathbf{s}} = 2\mu\alpha\left(\mathbf{I} - \frac{\mathbf{s}\otimes\mathbf{s}}{\mathbf{s}:\mathbf{s}}\right):\dot{\mathbf{e}}. \tag{6.105}$$

For the radial return of (6.87), the scalar $\alpha$ can be significantly less than unity and hence the inconsistent relationship can differ appreciably from the consistent form and hence destroy the favourable quadratic convergence characteristics of the full Newton–Raphson method. From the discussions of Chapter 1, this inconsistency would not affect the final answers (provided Strategy B of Section 6.2 is used) but will affect the convergence rate. In order to gain the potential benefits of the 'consistent tangent', the author believes it is important to use 'line searches' (see Section 9.2) for the early iterations before the iterative procedure reaches the bowl of Newton convergence. During these early iterations, the structural model is deciding which Gauss points are elastic, which unload, which remain plastic and which become plastic.

### 6.7.2 A combined formulation

A consistent tangent modular matrix can be derived without splitting the stresses and strains into volumetric and deviatoric components. Such a matrix can be derived even when the backward-Euler algorithm does not degenerate to a radial return in deviatoric space. The derivation of a more general form of the consistent tangent modular matrix is therefore relevant to a wider range of yield criteria.

Returning to the conventional matrix notation, the standard backward-Euler algorithm can be expressed (see (6.78)) as

$$\boldsymbol{\sigma} = \boldsymbol{\sigma}_B - \Delta\lambda\mathbf{C}\mathbf{a} \tag{6.106}$$

where we are dropping the suffix C relating to the current configuration following the return (see Figure 6.12(b)) so that if a variable has no suffix it is assumed to relate to this configuration. The suffix B in (6.106) shows that $\boldsymbol{\sigma}_B$ are the elastic 'trial' stresses (Figure 6.12(b)). Differentiation of (6.106) gives

$$\dot{\boldsymbol{\sigma}} = \mathbf{C}\dot{\boldsymbol{\varepsilon}} - \dot{\lambda}\mathbf{C}\mathbf{a} - \Delta\lambda\mathbf{C}\frac{\partial\mathbf{a}}{\partial\boldsymbol{\sigma}}\dot{\boldsymbol{\sigma}} \tag{6.107}$$

where the last term in equation (6.107) is omitted from the derivation of the standard tangent modular matrix. From equation (6.107),

$$\dot{\boldsymbol{\sigma}} = \left(\mathbf{I} + \Delta\lambda\mathbf{C}\frac{\partial \mathbf{a}}{\partial \boldsymbol{\sigma}}\right)^{-1}\mathbf{C}(\dot{\boldsymbol{\varepsilon}} - \dot{\lambda}\mathbf{a}) = \mathbf{Q}^{-1}\mathbf{C}(\dot{\boldsymbol{\varepsilon}} - \dot{\lambda}\mathbf{a}) = \mathbf{R}(\dot{\boldsymbol{\varepsilon}} - \dot{\lambda}\mathbf{a}) \tag{6.108}$$

where the $\mathbf{Q}$ matrix has appeared before in relation to the backward-Euler return (see (6.81)).

To remain on the yield surface, $\dot{f}$ should be zero, and hence from (6.16)

$$\mathbf{a}^{\mathrm{T}}\dot{\boldsymbol{\sigma}} = \mathbf{a}^{\mathrm{T}}\mathbf{R}\dot{\boldsymbol{\varepsilon}} - \dot{\lambda}\mathbf{a}^{\mathrm{T}}\mathbf{R}\mathbf{a} - A'\dot{\lambda} = 0 \tag{6.109}$$

and hence

$$\dot{\boldsymbol{\sigma}} = \mathbf{C}_{\mathrm{ct}}\dot{\boldsymbol{\varepsilon}} = \left(\mathbf{R} - \frac{\mathbf{R}\mathbf{a}\mathbf{a}^{\mathrm{T}}\mathbf{R}^{\mathrm{T}}}{\mathbf{a}^{\mathrm{T}}\mathbf{R}\mathbf{a}A'}\right)\dot{\boldsymbol{\varepsilon}}. \tag{6.110}$$

In deriving equation (6.110), account has been taken of the symmetry of $\mathbf{R}$. This symmetry can be proved with the aid of the symmetries of $\mathbf{C}$ and $\partial\mathbf{a}/\partial\mathbf{a}$ and the relationship

$$\mathbf{A}^{-1}\mathbf{B} = (\mathbf{B}^{-1}\mathbf{A})^{-1}. \tag{6.111}$$

In contrast to the consistent tangent matrix, the standard tangent modular matrix is derived by setting $\Delta\lambda$ to zero in equation (6.107). In this situation, the matrix $\partial\mathbf{a}/\partial\boldsymbol{\sigma}$ is unused.

A numerical example involving the consistent tangent modular matrix of (6.110) is given in Section 6.9.6.1.

## 6.8 SPECIAL TWO-DIMENSIONAL SITUATIONS

### 6.8.1 Plane strain and axial symmetry

The three-dimensional formulations of Sections 6.5–6.7 can be simply reduced to two dimensions by setting $\tau_{xz} = \tau_{yz} = \gamma_{xz} = \gamma_{yz} = 0$ so that equations (6.30) reduce to

$$\boldsymbol{\sigma}_4 = \begin{bmatrix} \sigma_x \\ \sigma_y \\ \sigma_z \\ \tau_{xy} \end{bmatrix} = \frac{E}{(1+\nu)(1-2\nu)}\begin{bmatrix} (1-\nu) & & & \\ \nu & (1-\nu) & & \\ \nu & \nu & (1-\nu) & \\ & & & \frac{1}{2}(1-2\nu) \end{bmatrix}\begin{bmatrix} \varepsilon_x \\ \varepsilon_y \\ \varepsilon_z \\ \gamma_{xy} \end{bmatrix} = \mathbf{C}_4\boldsymbol{\varepsilon}_4 \tag{6.112}$$

where the subscript 4 relates to the four stress and strain components. For axial symmetry (Figure 4.3), $\sigma_z$ can be taken as the hoop stress, while for plane strain, $\varepsilon_z$ is set to zero.

### 6.8.2 Plane stress

We have already described (Section 6.3) a forward-Euler formulation for plane-stress problems in which $\sigma_z = 0$. It has been argued that, for such problems, a backward-Euler scheme is impossible, or at least very difficult. However, it can be achieved [J1, S6, S7] without too much difficulty. Before describing such a formulation, we will redrive the forward-Euler relationships (Section 6.3) starting from the four stresses and strains of equation (6.112). In the previous developments (Section 6.3), we assumed that the response could be related to the three-component forms of (6.3) and (6.5). We will now prove the validity of these assumptions. To this end, with fixed elastic properties, we can, from (6.112) write the four-parameter equivalent of (6.5) as

$$\dot{\boldsymbol{\sigma}}_4 = \mathbf{C}_4(\dot{\boldsymbol{\varepsilon}}_4 - \dot{\boldsymbol{\varepsilon}}_{p4}) = \mathbf{C}_4(\dot{\boldsymbol{\varepsilon}}_4 - \dot{\lambda}\mathbf{a}_4) \tag{6.113}$$

where $\mathbf{a}_4$ contains the first four terms from (6.32) and $\mathbf{C}_4$ comes from (6.112).

From (6.36)

$$\mathbf{a}_4^{\mathrm{T}}\mathbf{C}_4\mathbf{a}_4 = 3\mu \tag{6.114}$$

so that (6.17) gives

$$\dot{\lambda} = \frac{\mathbf{a}_4^{\mathrm{T}}\mathbf{C}_4\dot{\boldsymbol{\varepsilon}}_4}{3\mu + A'}. \tag{6.115}$$

Because $\dot{\sigma}_z = 0$, the third row of equation (6.113) (with $\mathbf{C}_4$ from (6.112)) gives

$$\dot{\varepsilon}_z = \frac{-\nu}{(1-\nu)}(\dot{\varepsilon}_x + \dot{\varepsilon}_y) - \frac{\dot{\lambda}(1-2\nu)(\sigma_x + \sigma_y)}{2\sigma_e(1-\nu)}. \tag{6.116}$$

After some algebra, if (6.116) is substituted into (6.115),

$$\dot{\lambda} = \frac{\mathbf{a}_3^{\mathrm{T}}\mathbf{C}_3\dot{\boldsymbol{\varepsilon}}_3}{\mathbf{a}_3^{\mathrm{T}}\mathbf{C}_3\mathbf{a}_3 + A'} \tag{6.117}$$

where $\mathbf{C}_3$ is the three-parameter $\mathbf{C}$ matrix of (6.6) and $\mathbf{a}_3$ is the three-dimensional normal vector of (6.4). Further substitution for $\dot{\lambda}$ from (6.117) and $\dot{\varepsilon}_z$ from (6.116) can be applied to (6.113) to give (6.18) with $\mathbf{C}$ as $\mathbf{C}_3$ and $\mathbf{a}$ as $\mathbf{a}_3$.

Apart from showing that the two schemes produce identical results, this derivation illustrates that the elastic rate $\dot{\sigma}_{ze}$, which comes from the third row of (6.113) (see also (6.112) when the plastic strain rate, $\dot{\varepsilon}_{zp}$, is set to zero), is non-zero while the complete $\dot{\sigma}_z$ is zero. Hence, there would appear to be difficulties in applying a backward-Euler integration scheme. Using such an approach and starting from (6.78) with $\mathbf{C}$ from (6.112) leads to

$$\sigma_{x\mathrm{C}} = \sigma_{x\mathrm{A}} + E'\{(1-\nu)\Delta\bar{\varepsilon}_x + \nu\Delta\bar{\varepsilon}_y\} + \nu E'\Delta\varepsilon_z - \frac{\Delta\lambda\mu}{\sigma_{e\mathrm{C}}}(2\sigma_x - \sigma_y)_{\mathrm{C}} \tag{6.118}$$

$$\sigma_{y\mathrm{C}} = \sigma_{y\mathrm{A}} + E'\{(1-\nu)\Delta\bar{\varepsilon}_y + \nu\Delta\bar{\varepsilon}_x\} + \nu E'\Delta\varepsilon_z - \frac{\Delta\lambda\mu}{\sigma_{e\mathrm{C}}}(2\sigma_y - \sigma_x)_{\mathrm{C}} \tag{6.119}$$

$$\tau_{xy\mathrm{C}} = \tau_{xy\mathrm{A}} + \mu\Delta\bar{\gamma}_{xy} \tag{6.120}$$

$$\sigma_{z\mathrm{C}} = 0 = \nu E'(\Delta\bar{\varepsilon}_x + \Delta\bar{\varepsilon}_y) + \nu E'\Delta\varepsilon_z + \frac{\Delta\lambda\mu}{\sigma_{e\mathrm{C}}}(\sigma_x + \sigma_y)_{\mathrm{C}} \tag{6.121}$$

where

$$E' = \frac{E}{(1+\nu)(1-2\nu)} \tag{6.122}$$

and the bars over $\Delta\varepsilon_x$, $\Delta\varepsilon_y$ and $\Delta\gamma_{xy}$ indicate that these are known quantities while, operating within a normal three-parameter plane-stress environment, $\Delta\varepsilon_z$ is unknown. However, the latter term can be computed from (6.121). After some manipulation, substitution into (6.118) and (6.119) gives, in conjunction with (6.120),

$$\boldsymbol{\sigma}_C = \boldsymbol{\sigma}_A + \mathbf{C}\,\Delta\boldsymbol{\varepsilon} - \Delta\lambda\mathbf{C}\mathbf{a}_C = \boldsymbol{\sigma}_B - \Delta\lambda\mathbf{C}\mathbf{a}_C \tag{6.123}$$

where all vectors and matrices involve three components (see (6.6) for $\mathbf{C}$ and (6.4) for $\mathbf{a}$). Hence there is no need to be concerned that the algorithm returns from four stresses at B to three stresses at points A and C (Figure 6.12). Instead, we can simply operate on the three-term equations (6.123). This could be achieved using the return of Section 6.6.6 and the consistent tangent of Section 6.7.2. However, as shown by Jetteur [J1] and Simo *et al.* [S6, S7], a rather simpler return can be produced. To this end, equations (6.123) can be expanded and manipulated to give

$$\left(1 + \frac{\Delta\lambda' E}{2(1-\nu)}\right)(\sigma_x + \sigma_y)_C = A_1(\sigma_x + \sigma_y)_C = (\sigma_x + \sigma_y)_B \tag{6.124}$$

$$(1 + 3\Delta\lambda'\mu)(\sigma_x - \sigma_y)_C = A_2(\sigma_x - \sigma_y)_C = (\sigma_x - \sigma_y)_B \tag{6.125}$$

$$(1 + 3\Delta\lambda'\mu)\tau_{xyC} = A_2\tau_{xyC} = \tau_{xyB} \tag{6.126}$$

where $\sigma_{xB}$, $\sigma_{yB}$ and $\tau_{xyB}$ are known, and for convenience we write

$$\Delta\lambda' = \Delta\lambda/\sigma_{eC}. \tag{6.127}$$

For plane-stress conditions, the effective stress of (6.3) can be re-expressed via

$$\sigma_e^2 = \tfrac{1}{4}((\sigma_x + \sigma_y)^2 + 3(\sigma_x - \sigma_y)^2 + 12\tau_{xy}^2). \tag{6.128}$$

For the following, it is simplest to work with the squared form of the yield function, $f_2 = \sigma_e^2 - \sigma_o^2$. Substitution from (6.124)–(6.126) for $(\sigma_x + \sigma_y)_C$, etc., gives

$$f_2 = \frac{1}{4}\left(\frac{C_1}{(1+\Delta\lambda'\mu r)^2} + \frac{C_2}{(1+3\Delta\lambda'\mu)^2}\right) - \sigma_{0C}^2 = 0 \tag{6.129}$$

where

$$r = \frac{(1+\nu)}{(1-\nu)} \tag{6.130}$$

and

$$C_1 = (\sigma_x + \sigma_y)_B^2, \qquad C_2 = 3(\sigma_x - \sigma_y)_B^2 + 12\tau_{xyB}^2. \tag{6.131}$$

Equation (6.129) is non-linear in $\Delta\lambda'$ and may be solved with the aid of a scalar Newton–Raphson iteration derived from the truncated Taylor series:

$$f_{2n} = f_{2o} + \left.\frac{\partial f_2}{\partial \Delta\lambda'}\right|_o \dot{\lambda}' = 0 \tag{6.132}$$

where, as before, the subscript o means old while the subscript n means new. Without hardening, the $\partial f/\partial\Delta\lambda'$ term is obtained from (6.129) as

$$\frac{\partial f}{\partial \Delta\lambda'} = -\frac{\mu}{2}\left(\frac{C_1 r}{(1+\Delta\lambda'\mu r)^3} + \frac{3C_2}{(1+3\Delta\lambda'\mu)^3}\right). \tag{6.133}$$

With hardening, we can obtain $\dot{\lambda}' = (\dot{\lambda} - \Delta\lambda'\sigma_{ec})/\sigma_{ec}$ from (6.127) and can supplement (6.129) with $g = \sigma_{ec}^2 - \sigma_{0c}^2 = 0$. The iterative process is then applied to the latter in conjunction with (6.129) and up-dates both $\Delta\lambda$ and $\sigma_{ec}$.

#### 6.8.2.1 A consistent tangent modular matrix for plane stress

Following the return of the previous section, a consistent tangent modular matrix could be computed using the general procedure of Section 6.7.2. However, a number of simplifications can be made [J1, S6].

Differentiation of (6.124)–(6.126) gives

$$\dot{\sigma}_{xC} + \dot{\sigma}_{yC} = \frac{1}{A_1}(\dot{\sigma}_{xB} + \dot{\sigma}_{yB}) - \frac{E\dot{\lambda}'}{2(1-\nu)A_1}(\sigma_{xC} + \sigma_{yC})$$

$$= \frac{E}{(1-\nu)A_1}(\dot{\varepsilon}_{xC} + \dot{\varepsilon}_{yC}) - \frac{E\dot{\lambda}'}{2(1-\nu)A_1}(\sigma_{xC} + \sigma_{yC}) \tag{6.134}$$

$$\dot{\sigma}_{xC} - \dot{\sigma}_{yC} = \frac{1}{A_2}(\dot{\sigma}_{xB} - \dot{\sigma}_{yB}) - \frac{3\mu\dot{\lambda}'}{A_2}(\sigma_{xC} - \sigma_{yC})$$

$$= \frac{2\mu}{A_2}(\dot{\varepsilon}_{xC} - \dot{\varepsilon}_{yC}) - \frac{3\mu\dot{\lambda}'}{A_2}(\sigma_{xC} - \sigma_{yC}) \tag{6.135}$$

$$\dot{\tau}_{xyC} = \frac{1}{A_2}\dot{\tau}_{xyB} - \frac{3\mu}{A_2}\dot{\lambda}'\tau_{xyC} = \frac{\mu}{A_2}\dot{\gamma}_{xyC} - \frac{3\mu}{A_2}\dot{\lambda}'\tau_{xyC} \tag{6.136}$$

from which

$$\dot{\boldsymbol{\sigma}} = \mathbf{R}(\dot{\boldsymbol{\varepsilon}} - \dot{\lambda}'\mathbf{a}') \tag{6.137}$$

where

$$\mathbf{a}^{\mathrm{T}} = \frac{1}{\sigma_e}\mathbf{a}'^{\mathrm{T}} = \frac{1}{2\sigma_e}(2\sigma_x - \sigma_y, 2\sigma_y - \sigma_x, 6\tau_{xy}) \tag{6.138}$$

and

$$\mathbf{R} = \begin{bmatrix} \dfrac{E}{2(1-\nu)A_1} + \dfrac{\mu}{A_2} & \dfrac{E}{2(1-\nu)A_1} - \dfrac{\mu}{A_2} & 0 \\ \dfrac{E}{2(1-\nu)A_1} - \dfrac{\mu}{A_2} & \dfrac{E}{2(1-\nu)A_1} + \dfrac{\mu}{A_2} & 0 \\ 0 & 0 & \mu/A_2 \end{bmatrix}. \tag{6.139}$$

Equation (6.137) is of the same form as (6.108) and hence, assuming no hardening, the constituent tangent modular matrix is given by (6.110) with $\mathbf{R}$ from (6.139) but with $\mathbf{a}'$ from (6.138) replacing $\mathbf{a}$. With hardening we can use (6.110) with $R$ from (6.139) provided $A'$ is replaced by $A'(1 - \Delta\lambda' A')$. The method is illustrated via a numerical example in Section 6.9.6.2. Extensions of these concepts to shell analysis have been given by Ramm and Matzenmiller [R1].

## 6.9 NUMERICAL EXAMPLES

A range of benchmark tests for plasticity has been given in [H3, H4]. In this section, we will provide hand-based numerical computations using the various techniques developed in the previous sections. These numerical examples will all be related to the plane-stress yield function of Figure 6.13. Assuming that we are working in principle stress space with $\gamma_{xy} = \tau_{xy} = 0$, from (6.3), with $\sigma_1 = \sigma_x$ and $\sigma_2 = \sigma_y$, the yield function is

$$f = \sigma_e - \sigma_o = (\sigma_1^2 + \sigma_2^2 - \sigma_1\sigma_2)^{1/2} - \sigma_o = 0. \tag{6.140}$$

The adopted material properties are

$$E = 200\,000\ \mathrm{N/mm^2}, \qquad \nu = 0, \qquad \sigma_o = 200\ \mathrm{N/mm^2}. \tag{6.141}$$

### 6.9.1 Intersection point

**Question** Starting from point X (Fig. 6.13(a)) with

$$\boldsymbol{\sigma}_X^T = (\sigma_1, \sigma_2) = (120, -80) \tag{6.142}$$

apply an elastic increment $\Delta\sigma_e$ relating to a strain increment of

$$\Delta\boldsymbol{\varepsilon}^T = (0.0009, 0.0009) \tag{6.143}$$

and compute the intersection point A in Figure 6.13 and hence the ratio $\alpha$ of equation (6.48) which ensures that $f(\boldsymbol{\sigma}_X + \alpha\Delta\boldsymbol{\sigma}_e) = 0$.

**Solution** From (6.140), the initial stresses, $\boldsymbol{\sigma}_X$ in (6.142) give $f_X = -25.64$ and, from (6.140), $\sigma_e(\boldsymbol{\sigma}_o) = 174.4$. From (6.143), the elastic incremental stresses are $\Delta\boldsymbol{\sigma}_e^T = (180, 180)$ so that $\sigma_e(\Delta\boldsymbol{\sigma}_e) = 180$. Equation (6.52) then provides the quadratic equation

$$32\,400\alpha^2 + 7200\alpha - 9600 = 0 \tag{6.144}$$

for the intersection point $\alpha$. The roots of (6.144) are

$$\alpha_1 = 0.444, \qquad \alpha_2 = -0.667 \tag{6.145}$$

which give the intersection points A and A′ on Figure 6.13(a). The required intersection point is at A for which the stresses are $\boldsymbol{\sigma}_A^T = (200, 0)$.

A forward Euler integration scheme would then proceed to apply the ratio $(1 - \alpha) = 0.667$ of the strain increment of (6.143) in an elasto-plastic manner.

The reader might like to try using the iterative procedure of Section 6.6.1 instead of the direct solution given above.

### 6.9.2 A forward-Euler integration

**Question** Assume that we have computed the intersection point, A, with stresses

$$\boldsymbol{\sigma}_A^T = (200, 0). \tag{6.146}$$

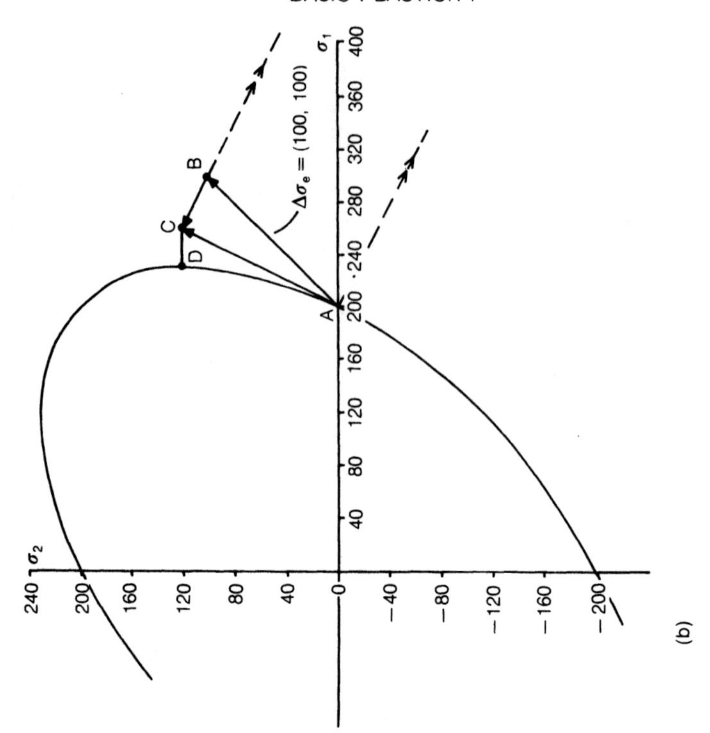

$\sigma_2$
$\sigma_1$
A
B
C
D
$\Delta\sigma_e = (100, 100)$
240
200
160
120
80
40
0
−40
−80
−120
−160
−200
40
80
120
160
200
240
280
320
360
400


(b)

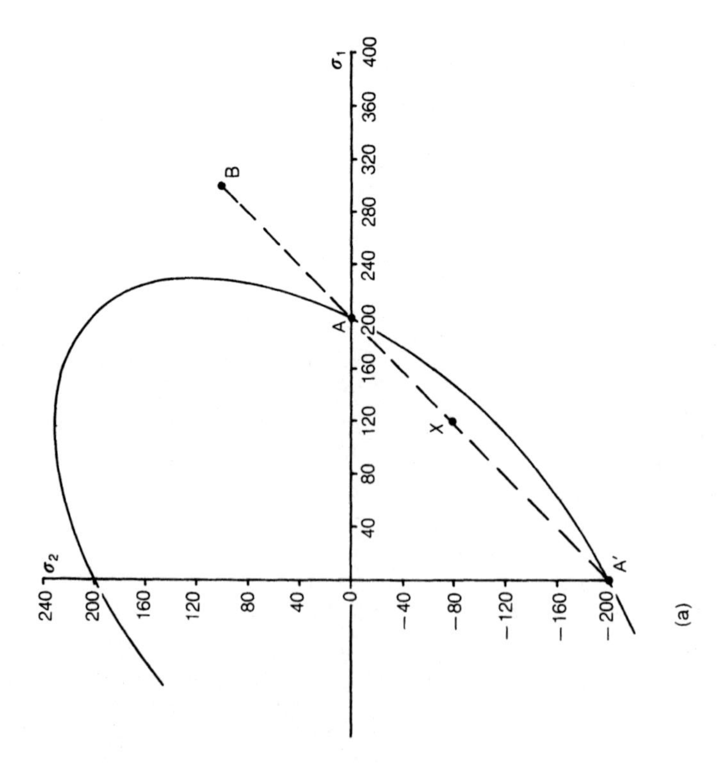

$\sigma_2$
$\sigma_1$
A
A′
B
X
240
200
160
120
80
40
0
−40
−80
−120
−160
−200
40
80
120
160
200
240
280
320
360
400


(a)

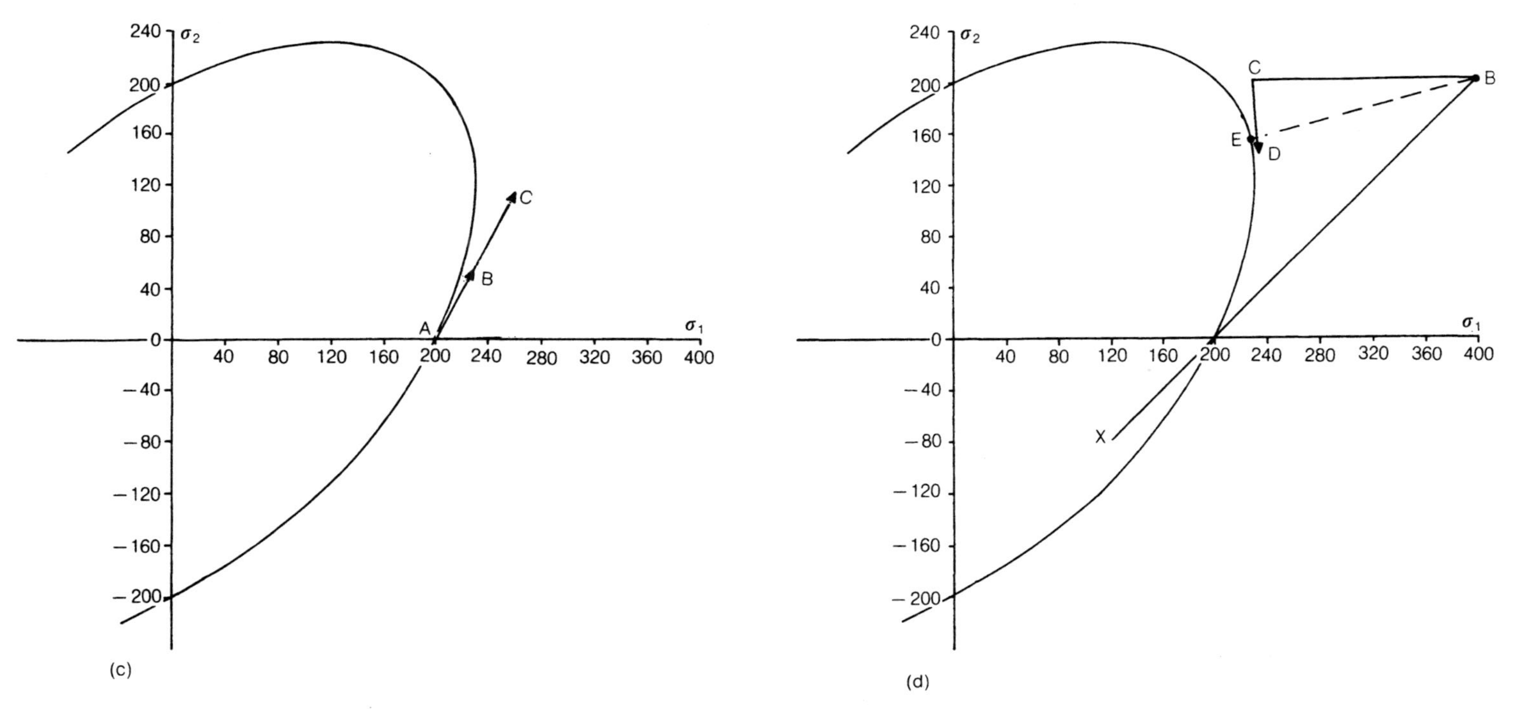


**Figure 6.13** Numerical examples: (a) Computing the intersection print; (b) a forward-Euler step from A to C and (later) a correction to D; (c) using two sub-increments; (d) a backward-Euler return.

Apply a forward Euler step relating to a strain increment of

$$\Delta\boldsymbol{\varepsilon}^{\mathrm{T}} = (0.0005,\ 0.0005) \tag{6.147}$$

and plot the point graphically.

**Solution** We will firstly obtain a solution by computing the tangent modular matrix at point A (Figure 6.13(b)). From (6.4),

$$\mathbf{a}_{\mathrm{A}}^{\mathrm{T}} = (1.0,\ -0.5) \tag{6.148}$$

so that in (6.9),

$$\mathbf{C}_{\mathrm{tA}} = \mathbf{C}\left(\mathbf{I} - \frac{\mathbf{a}\mathbf{a}^{\mathrm{T}}\mathbf{C}}{\mathbf{a}^{\mathrm{T}}\mathbf{C}\mathbf{a}}\right)$$

$$= 200\,000\begin{bmatrix} 1 & 0 \\ 0 & 1 \end{bmatrix} - \frac{1}{250\,000}\begin{bmatrix} 160\,000 & -80\,000 \\ -80\,000 & 160\,000 \end{bmatrix} = \begin{bmatrix} 4000 & 80\,000 \\ 8000 & 160\,000 \end{bmatrix}. \tag{6.149}$$

Hence a forward Euler step leads to

$$\boldsymbol{\sigma}_{\mathrm{C}} = \boldsymbol{\sigma}_{\mathrm{A}} + \Delta\boldsymbol{\sigma}_{\mathrm{A}} = \boldsymbol{\sigma}_{\mathrm{A}} + \mathbf{C}_{\mathrm{tA}}\Delta\boldsymbol{\varepsilon} = \begin{pmatrix} 200 \\ 0 \end{pmatrix} + \begin{pmatrix} 60 \\ 120 \end{pmatrix} = \begin{pmatrix} 260 \\ 120 \end{pmatrix} \tag{6.150}$$

which as shown in Figure 6.13(b) lies outside the yield surface.
A more computationally efficient solution would involve:

Compute $\Delta\lambda$ from (6.8) (but with $\Delta$'s instead of dots) so that with $\mathbf{a}$ from (6.148) and $\Delta\boldsymbol{\varepsilon}$ from (6.147),

$$\Delta\lambda = \frac{\mathbf{a}^{\mathrm{T}}\mathbf{C}\,\Delta\boldsymbol{\varepsilon}}{\mathbf{a}^{\mathrm{T}}\mathbf{C}\mathbf{a}} = \frac{50}{250\,000} = 0.0002. \tag{6.151}$$

Compute $\Delta\boldsymbol{\sigma}_{\mathrm{A}}$ from (6.5) (again with $\Delta$'s instead of dots) to give

$$\Delta\boldsymbol{\sigma}_{\mathrm{A}} = \mathbf{C}\,\Delta\boldsymbol{\varepsilon} - \Delta\lambda\,\mathbf{C}\mathbf{a} = \Delta\boldsymbol{\sigma}_{\mathrm{e}} - \Delta\lambda\,\mathbf{C}\mathbf{a} = \begin{pmatrix} 100 \\ 100 \end{pmatrix} - 0.0002\begin{pmatrix} 200\,000 \\ -100\,000 \end{pmatrix} = \begin{pmatrix} 60 \\ 120 \end{pmatrix}. \tag{6.152}$$

### 6.9.3 Sub-increments

**Question** Repeat the solution of 6.9.2 but use two sub-increments and plot the solution.
**Solution** From (6.147), the sub-incremental steps are, $\Delta\boldsymbol{\varepsilon} = (0.000\,25, 0.0025)$. For the first sub-increment, using a similar procedure to that in (6.151) but with half the step size gives $\Delta\lambda = 0.0001$. Hence the equivalent of (6.152) gives $\Delta\boldsymbol{\sigma}_{\mathrm{A}} = (30, 60)$. Graphically, this takes the solution to point B in Figure 6.13(c) with $\boldsymbol{\sigma}_{\mathrm{B}} = (230, 60)$. At this stage, for the second sub-increment, we have from (6.4),

$$\mathbf{a}^{\mathrm{T}} = (0.9679,\ -0.2662) \tag{6.153}$$

so that the incremental form of (6.8) gives $\Delta\lambda = 0.000\,174$ and, using the incremental form of (6.5),

$$\Delta\boldsymbol{\sigma}_B = \mathbf{C}\,\Delta\boldsymbol{\varepsilon} - \Delta\lambda\,\mathbf{Ca} = \begin{pmatrix} 50 \\ 50 \end{pmatrix} - 0.000174\begin{pmatrix} 200\,000 \\ -100\,000 \end{pmatrix} = \begin{pmatrix} 16.30 \\ 59.27 \end{pmatrix} \quad (6.154)$$

and the final stress at C of Figure (6.13(c)) is (246.3, 119.2). Clearly, in comparison with the procedure of Section 6.9.2 (Figure 6.13(b)), the use of sub-increments has reduced the drift from the yield surface.

### 6.9.4 Correction or return to the yield surface

**Question** Using the procedure of Section 6.6.3, return the final stresses obtained in Section 6.9.2, i.e. $\boldsymbol{\sigma}^T = (260, 120)$ to the yield surface.
**Solution** In relation to Figure 6.13(b), from (6.140), the yield function at C is $f_C =$ 25.39, while $\mathbf{a}_C^T = (0.8874, -0.044\,37)$. Hence in (6.62), $\delta\lambda_C = 25.39/157\,900 = 0.000\,161$ and from (6.61), an improved solution is

$$\boldsymbol{\sigma}_D = \boldsymbol{\sigma}_C - \delta\lambda_C\,\mathbf{Ca}_C = \begin{pmatrix} 260 \\ 120 \end{pmatrix} - 0.000\,161\begin{pmatrix} 177\,480 \\ 8\,874 \end{pmatrix} = \begin{pmatrix} 260 \\ 120 \end{pmatrix} - \begin{pmatrix} 28.54 \\ -1.43 \end{pmatrix} = \begin{pmatrix} 231.5 \\ 121.4 \end{pmatrix} \quad (6.155)$$

which is illustrated in Figure 6.13(b). As a result of this correction the yield function has been reduced from 25.39 to 0.531. The reader might try applying one further iteration.

### 6.9.5 Backward-Euler return

**Question** Starting from point X in Figure 6.13(d) (see equation (6.142)), apply a backward-Euler return, appropriate to an elastic increment of

$$\Delta\boldsymbol{\varepsilon}^T = (0.0014, 0.0014). \quad (6.156)$$

Firstly use the general procedure of Section 6.6.6 and then the special plane stress form of Section 6.8.2.

#### *6.9.5.1 General method*

**Solution** We start with the second predictor in Section 6.6.2. To this end (see Figure 6.13(d)), we firstly obtain the elastic increments $\Delta\boldsymbol{\sigma}_e^T = (280, 280)$ so that the stresses at point B in Figure 6.13(d) are

$$\boldsymbol{\sigma}_B = (400, 200) \quad (6.157)$$

while the $\mathbf{a}$ vector is $\mathbf{a}_B = (0.866, 0)$ and the yield function is $f_B = 146.4$. Hence, from (6.59), $\Delta\lambda = 146.4/150\,000 = 0.000\,976$ and hence, from (6.60),

$$\boldsymbol{\sigma}_C = \boldsymbol{\sigma}_B - \Delta\lambda\,\mathbf{Ca}_B = \begin{pmatrix} 400 \\ 200 \end{pmatrix} - 0.000\,976\begin{pmatrix} 173\,200 \\ 0 \end{pmatrix} = \begin{pmatrix} 230.9 \\ 200 \end{pmatrix}. \quad (6.158)$$

We now apply the backward-Euler corrector of Section 6.6.6 for which we require $\mathbf{a}_C = (0.6031, 0.3893)$ and $f_C = 17.13$. From (6.79), we now have

$$\mathbf{r} = \boldsymbol{\sigma}_C - (\boldsymbol{\sigma}_B - \Delta\lambda\,\mathbf{C}\mathbf{a}_C) = \begin{pmatrix} 230.9 \\ 200 \end{pmatrix} - \left( \begin{pmatrix} 400 \\ 200 \end{pmatrix} - 0.000\,976 \begin{pmatrix} 120\,620 \\ 77\,860 \end{pmatrix} \right) = \begin{pmatrix} -51.34 \\ 76.00 \end{pmatrix}. \tag{6.159}$$

Using (6.81),

$$\mathbf{Q} = \left( \mathbf{I} + \Delta\lambda\,\mathbf{C} \frac{\partial \mathbf{a}}{\partial \boldsymbol{\sigma}}\bigg|_C \right)$$

$$= \left[ \begin{bmatrix} 1 & 0 \\ 0 & 1 \end{bmatrix} + 0.000\,976 \times 200\,000 \begin{bmatrix} 0.002\,93 & -0.003\,39 \\ -0.003\,39 & 0.003\,91 \end{bmatrix} \right]$$

$$= \begin{bmatrix} 1.572 & -0.661 \\ -0.661 & 1.572 \end{bmatrix} \tag{6.160}$$

and from (6.83),

$$\dot{\lambda} = \frac{f_C - \mathbf{a}_C^T \mathbf{Q}^{-1} \mathbf{r}_o}{\mathbf{a}_C^T \mathbf{Q}^{-1} \mathbf{C} \mathbf{a}_C} = \frac{17.13 - 3.86}{101\,900} = 0.000\,1302 \tag{6.161}$$

while in (6.81)

$$\dot{\boldsymbol{\sigma}} = -\mathbf{Q}^{-1}\mathbf{r}_o - \dot{\lambda}\mathbf{Q}^{-1}\mathbf{C}\mathbf{a} = -\begin{pmatrix} -17.26 \\ 36.65 \end{pmatrix} - 0.000\,1302 \begin{pmatrix} 113\,100 \\ 86\,550 \end{pmatrix} = \begin{pmatrix} 2.53 \\ -47.92 \end{pmatrix} \tag{6.162}$$

so that, at the end of the iteration, the *new* point C is given by $\boldsymbol{\sigma}_C^T = (233.5, 152.1)$, $f_C = 5.256$, $\Delta\lambda = 0.001\,06$. This point is plotted on Figure 6.13(d) as point D.

The reader might like to try a further iteration which leads to

$$\boldsymbol{\sigma}_C^T = (226.3, 153.6), \qquad f_C = 0.0927, \qquad \Delta\lambda = 0.001\,160 \tag{6.163}$$

which is plotted as point E in Figure 6.13(d).

#### 6.9.5.2 *Specific plane-stress method*

A more efficient return can be made by applying the special plane-stress return of Section 6.8.2 from the trial solution of (6.158). To this end, with $r = 1$ from (6.130), from (6.131) and (6.157), we obtain

$$C_1 = 360\,000 \qquad C_2 = 120\,000. \tag{6.164}$$

To start the iterative procedure, we will compute $\Delta\lambda'$ from (6.127) using the $\Delta\lambda$ value in (6.158) (0.000 976) and the $\sigma_{eC}$ value relating to the stresses, $\boldsymbol{\sigma}_C$, in (6.158) so that

$$\Delta\lambda' = 0.000\,976/217.1 = 0.4495\mathrm{e} - 5. \tag{6.165}$$

The yield function value, $f_2$ in (6.129) is then

$$f_2 = \frac{1}{4}\left( \frac{360\,000}{(1.45)^2} + \frac{120\,000}{(2.349)^2} \right) - (200)^2 = 48\,273 - 40\,000 = 8273. \tag{6.166}$$

In order to iteratively change $\Delta\lambda'$, we require, from (6.133),

$$\frac{\partial f}{\partial \Delta\lambda'} = \frac{-100\,000}{2}\left(\frac{360\,000}{(1.45)^3} + \frac{3 \times 120\,000}{(2.349)^3}\right) = -0.7299\text{e}10 \tag{6.167}$$

so that from (6.132), the change in $\Delta\lambda'$ is

$$\dot{\lambda}' = \frac{-8273}{-0.7299\text{e}10} = 0.1133\text{e}-5. \tag{6.168}$$

Hence the updated $\Delta\lambda'$ value is $0.4495\text{e}-5 + 0.1133\text{e}-5 = 0.5629\text{e}-5$ and, from (6.129),

$$f_2 = 40\,997 - 40\,000 = 997. \tag{6.169}$$

This corresponds to a yield function value in the standard form of (6.140) of $f = 2.477$.

Another iteration is hardly necessary but leads to

$$\dot{\lambda}' = \frac{-997}{-0.5432\text{e}10} = 0.180\text{e}-6 \tag{6.170}$$

so that the final $\Delta\lambda'$ value is $0.5809\text{e}-5$ which, from (6.127), corresponds to an unscaled $\Delta\lambda$ value of

$$\Delta\lambda' = 0.5809\text{e}-5, \qquad \Delta\lambda = 0.5809\text{e}-5 \times 200 = 0.001\,162. \tag{6.171}$$

Substituting for $\Delta\lambda'$ from (6.171) into (6.124)–(6.126) leads to the final stresses as

$$\boldsymbol{\sigma}_C^T = (226.2, 153.3) \tag{6.172}$$

for which $f = 0.17\text{e}-4$ and $f_2 = 0.678\text{e}-2$. The stresses in (6.172) are very close to those obtained in (6.163) and are plotted as point E in Figure 6.13(d).

### 6.9.6 Consistent and inconsistent tangents

**Question** From the point obtained at the end of the previous section with stresses given by (6.172) and plastic multipliers given by (6.171), compute (i) the inconsistent tangent and (ii) the consistent tangent ready for use in a structural Newton–Raphson iteration.

From (6.4), we have

$$\mathbf{a}^T = (0.7478, 0.2010) \tag{6.173}$$

so that in (6.9), the inconsistent tangent modular matrix is given by

$$\mathbf{C}_t = \mathbf{C}\left(\mathbf{I} - \frac{\mathbf{a}\mathbf{a}^T\mathbf{C}}{\mathbf{a}^T\mathbf{C}\mathbf{a}}\right) = \begin{bmatrix} 0.1348\text{e}5 & -0.5014\text{e}5 \\ -0.5014\text{e}5 & 0.1865\text{e}6 \end{bmatrix}. \tag{6.174}$$

#### *6.9.6.1 Solution using the general method*

Using the general method of Section 6.7.2 for the consistent approach, we firstly use (6.108) with $\Delta\lambda$ from (6.171) and $\partial\mathbf{a}/\partial\boldsymbol{\sigma}$ from (6.47) to compute

$$\mathbf{Q} = \left(\mathbf{I} + \Delta\lambda \mathbf{C} \left.\frac{\partial \mathbf{a}}{\partial \boldsymbol{\sigma}}\right|_{\mathrm{C}}\right)$$

$$= \begin{bmatrix} 1 & 0 \\ 0 & 1 \end{bmatrix} + 0.001\,162 \times 200\,000 \begin{bmatrix} 1 & 0 \\ 0 & 1 \end{bmatrix} \begin{bmatrix} 0.002\,204 & -0.003\,252 \\ -0.003\,252 & 0.004\,798 \end{bmatrix}$$

$$= \begin{bmatrix} 1.512 & -0.7558 \\ -0.7558 & 2.115 \end{bmatrix} \tag{6.175}$$

and, from (6.108),

$$\mathbf{R} = \mathbf{Q}^{-1}\mathbf{C} = \begin{bmatrix} 0.8051 & 0.2877 \\ 0.2877 & 0.5757 \end{bmatrix} \begin{bmatrix} 200\,000 & 0 \\ 0 & 200\,000 \end{bmatrix} = \begin{bmatrix} 0.1610\mathrm{e}6 & 0.5753\mathrm{e}5 \\ 0.5753\mathrm{e}5 & 0.1151\mathrm{e}6 \end{bmatrix} \tag{6.176}$$

so that in (6.110),

$$\mathbf{C}_{\mathrm{tC}} = \mathbf{R}\left(\mathbf{I} - \frac{\mathbf{a}\mathbf{a}^{\mathrm{T}}\mathbf{R}}{\mathbf{a}^{\mathrm{T}}\mathbf{R}\mathbf{a}}\right) = \begin{bmatrix} 0.1610\mathrm{e}6 & 0.5753\mathrm{e}5 \\ 0.5753\mathrm{e}5 & 0.1151\mathrm{e}6 \end{bmatrix} - \frac{1}{0.1120\mathrm{e}6} \begin{bmatrix} 0.1742\mathrm{e}11 & 0.8732\mathrm{e}10 \\ 0.8732\mathrm{e}10 & 0.4378\mathrm{e}10 \end{bmatrix}$$

$$= \begin{bmatrix} 5493 & -0.2044\mathrm{e}5 \\ -0.2044\mathrm{e}5 & 0.7602\mathrm{e}5 \end{bmatrix}, \tag{6.177}$$

which should be contrasted with the 'inconsistent' solution in (6.174).

#### 6.9.6.2 *Solution using the specific plane-stress method*

As already pointed out in Section 6.8.2.1, one can devise a more economical way of computing the consistent tangent for the plane stress case [J1, S6, S7]. To this end, we compute $\mathbf{a}'$ from (6.138) as

$$\mathbf{a}'^{\mathrm{T}} = \tfrac{1}{2}(2\sigma_x - \sigma_y, 2\sigma_y - \sigma_x) = (149.6, 40.20). \tag{6.178}$$

In order to compute $\mathbf{R}$, we require $A_1$ and $A_2$ from (6.124) and (6.125), which with $\Delta\lambda'$ from (6.171) are given by

$$A_1 = 1. + 0.5811\mathrm{e}-5 \times 200\,000 = 1.581,$$
$$A_2 = 1. + 3 \times 0.5511\mathrm{e}-5 \times 100\,000 = 2.743 \tag{6.179}$$

and, from (6.139), $\mathbf{R}$ is given by

$$\mathbf{R} = \begin{bmatrix} 0.9970\mathrm{e}5 & 0.2679\mathrm{e}5 \\ 0.2679\mathrm{e}5 & 0.9970\mathrm{e}5 \end{bmatrix}$$

so that using (6.110) but with $\mathbf{a}'$ instead of $\mathbf{a}$ (see Section 6.8.2.1),

$$\mathbf{C}_{\mathrm{tC}} = \mathbf{R}\left(\mathbf{I} - \frac{\mathbf{a}\mathbf{a}^{\mathrm{T}}\mathbf{R}}{\mathbf{a}^{\mathrm{T}}\mathbf{R}\mathbf{a}}\right) = \begin{bmatrix} 0.9970\mathrm{e}5 & 0.2679\mathrm{e}5 \\ 0.2679\mathrm{e}5 & 0.9970\mathrm{e}5 \end{bmatrix} - \frac{1}{0.2713\mathrm{e}10} \begin{bmatrix} 0.2556\mathrm{e}15 & 0.1281\mathrm{e}15 \\ 0.1281\mathrm{e}15 & 0.6424\mathrm{e}14 \end{bmatrix}$$

$$= \begin{bmatrix} 5493 & -0.2044\mathrm{e}5 \\ -0.2044\mathrm{e}5 & 0.7602\mathrm{e}5 \end{bmatrix} \tag{6.180}$$

which corresponds with the solution in (6.177).

## 6.10 PLASTICITY AND MATHEMATICAL PROGRAMMING

The links between plasticity and mathematical programming can be found in [S5, M1, S1, R3, J3, M5, M10, S9]. The present developments follow closely those in [S5, M5, R3]. For simplicity, the work will be related to perfect plasticity. Extensions to include hardening can be found in the previous references. An essential prerequisite to the understanding of this section is some basic knowledge on constrained optimisation with inequalities; in particular the use of Lagrangian functions, Lagrangian multipliers and the Kuhn–Tucker conditions. Good books covering these topics are due to Fletcher [F1] and Luenberger [L2].

We will start with the principle of maximum plastic work which Hill [H2] attributes to von Mises [V1]. This principle firstly requires that the stresses must be restricted by the yield surface and, secondly, requires that they should be such as to maximise the increment (or rate) of plastic work, i.e.

$$f(\boldsymbol{\sigma}) \leqslant 0. \tag{6.181a}$$

$$\max\{\dot{W} = \boldsymbol{\sigma}^{\mathrm{T}}\dot{\boldsymbol{\varepsilon}}_{\mathrm{p}}\}. \tag{6.181b}$$

Using standard techniques of mathematical programming [F1, L2], we firstly turn the maximum into a minimum by changing $\dot{W}$ to $-\dot{W}$ and, secondly, create a Lagrangian function by adding a Lagrangian multiplier $\dot{\lambda}$ times the constraint of (6.181a), so that

$$L(\boldsymbol{\sigma}, \dot{\lambda}) = -\boldsymbol{\sigma}^{\mathrm{T}}\dot{\boldsymbol{\varepsilon}}_{\mathrm{p}} + \dot{\lambda} f(\boldsymbol{\sigma}). \tag{6.182}$$

We now make $L$ stationary with respect to variations on $\boldsymbol{\sigma}$ and $\dot{\lambda}$. This leads to the Kuhn–Tucker conditions:

$$\frac{\partial L}{\partial \boldsymbol{\sigma}} = -\dot{\boldsymbol{\varepsilon}}_{\mathrm{p}} + \dot{\lambda}\frac{\partial f}{\partial \boldsymbol{\sigma}} = \mathbf{0} \tag{6.183}$$

$$f(\boldsymbol{\sigma}) \leqslant 0 \tag{6.184a}$$

$$\dot{\lambda} \geqslant 0 \tag{6.184b}$$

$$\dot{\lambda} f(\boldsymbol{\sigma}) = 0. \tag{6.184c}$$

The 'complementarity condition' (6.184c) requires that either the yield function is zero or $\dot{\lambda}$ is zero and there is no plastic flow.

Equation (6.183) is the flow rule, (6.184a) the yield criterion, and (6.184b) the condition for a 'positive plastic strain-rate multiplier'. All of these conditions have been considered before in the earlier developments of Section 6.3.

There is one further condition that can be derived from the principle of maximum plastic work (equation (6.181)). This is the essential 'convexity' [F1, L2] of the yield surface. To prove this, we must re-write (6.181b) as

$$\boldsymbol{\sigma}^{*\mathrm{T}}\dot{\boldsymbol{\varepsilon}}_{\mathrm{p}} \geqslant \boldsymbol{\sigma}^{\mathrm{T}}\dot{\boldsymbol{\varepsilon}}_{\mathrm{p}} \tag{6.185a}$$

or

$$(\boldsymbol{\sigma} - \boldsymbol{\sigma}^*)^{\mathrm{T}}\dot{\boldsymbol{\varepsilon}}_{\mathrm{p}} \leqslant 0, \tag{6.185b}$$

where $\boldsymbol{\sigma}^*$ are the actual stresses that maximise the plastic work rate, $\dot{W}$, and $\boldsymbol{\sigma}$ are any other admissible stresses (satisfying (6.181a)). (Note that many workers [S5, R3, S1, M5] use $\boldsymbol{\tau}$ and $\boldsymbol{\sigma}$ in place of the current $\boldsymbol{\sigma}$ and $\boldsymbol{\sigma}^*$.) Because of the flow rule (6.183) and the condition of non-negative $\dot{\lambda}$ (6.184b), (6.185b) can be re-expressed as

$$(\boldsymbol{\sigma}-\boldsymbol{\sigma}^*)^{\mathrm{T}}\frac{\partial f}{\partial \boldsymbol{\sigma}} \leqslant 0 \tag{6.186}$$

which is illustrated in Figure 6.14 and ensures that the region contained by $f$ is 'convex' [F1, L2].

By writing the plastic strain rate, $\dot{\boldsymbol{\varepsilon}}_p$, as the difference between the total strain rate, $\dot{\boldsymbol{\varepsilon}}$, and the elastic rate, $\dot{\boldsymbol{\varepsilon}}_e = \mathbf{C}^{-1}\dot{\boldsymbol{\sigma}}$, we can rewrite (6.185b) in the form

$$(\boldsymbol{\sigma}-\boldsymbol{\sigma}^*)^{\mathrm{T}}(\dot{\boldsymbol{\varepsilon}}-\mathbf{C}^{-1}\dot{\boldsymbol{\sigma}}) \leqslant 0. \tag{6.187}$$

Integrating (6.187) over the volume leads to

$$\int (\boldsymbol{\sigma}-\boldsymbol{\sigma}^*)^{\mathrm{T}}(\dot{\boldsymbol{\varepsilon}}-\mathbf{C}^{-1}\dot{\boldsymbol{\sigma}})\,\mathrm{d}V \leqslant 0 \tag{6.188}$$

which is a 'variational inequality' which has been used as the starting point of some numerical developments in plasticity.

Yet another alternative to the principle of maximum plastic work (6.181), is provided by adopting a complementary energy form [M5] with

$$\min\{\tfrac{1}{2}\dot{\boldsymbol{\sigma}}^{\mathrm{T}}\mathbf{C}^{-1}\dot{\boldsymbol{\sigma}}-\dot{\boldsymbol{\sigma}}^{\mathrm{T}}\dot{\boldsymbol{\varepsilon}}\} \tag{6.189}$$

in place of (6.181b). In (6.189), the total strain rate, $\dot{\boldsymbol{\varepsilon}}$, is fixed and the stress rates, $\dot{\boldsymbol{\sigma}}$, are variables. Adding the constraint of (6.181a) leads, instead of (6.182), to a Lagrangian:

$$L(\dot{\boldsymbol{\sigma}},\dot{\lambda}) = \tfrac{1}{2}\dot{\boldsymbol{\sigma}}^{\mathrm{T}}\mathbf{C}^{-1}\dot{\boldsymbol{\sigma}}-\dot{\boldsymbol{\sigma}}^{\mathrm{T}}\dot{\boldsymbol{\varepsilon}}+\dot{\lambda}f(\boldsymbol{\sigma}) \tag{6.190}$$

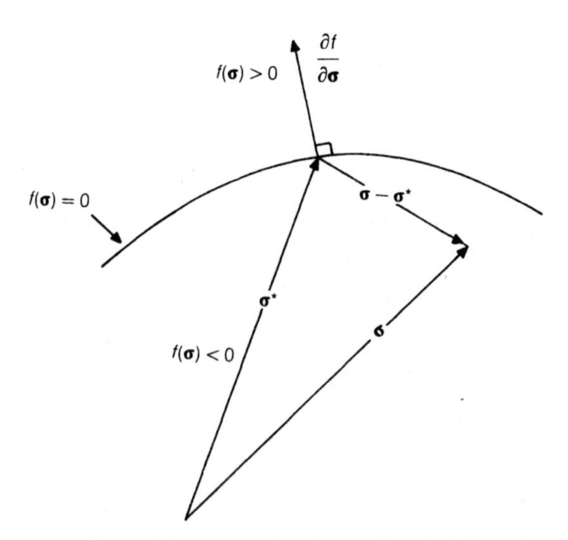


**Figure 6.14** Illustration of equation (6.186).

from which in place of (6.183) we have

$$\frac{\partial L}{\partial \dot{\boldsymbol{\sigma}}} = \mathbf{C}^{-1}\dot{\boldsymbol{\sigma}} - \dot{\boldsymbol{\varepsilon}} + \dot{\lambda}\frac{\partial f}{\partial \boldsymbol{\sigma}} = \mathbf{C}^{-1}\dot{\boldsymbol{\sigma}} - \dot{\boldsymbol{\varepsilon}} + \dot{\boldsymbol{\varepsilon}}_p = 0 \tag{6.191}$$

along with the other Kuhn–Tucker conditions of (6.184). In equation (6.191) we have used the flow-rule relationship of (6.183) which then leads in (6.191) to the standard additive decomposition into elastic and plastic strains. The converse is equally true.

### 6.10.1 A backward-Euler or implicit formulation

We will now derive the backward-Euler procedure of Section 6.6.6, by starting from an incremental form of (6.189) to which we add the constant ($\dot{\boldsymbol{\varepsilon}}$ or $\Delta\boldsymbol{\varepsilon}$ are fixed), $\frac{1}{2}\Delta\boldsymbol{\varepsilon}^{\mathrm{T}}\mathbf{C}\,\Delta\boldsymbol{\varepsilon}$, which does not affect the minimisation process [M5]. It follows that the function to minimise is $F$, where

$$F = \tfrac{1}{2}\Delta\boldsymbol{\sigma}^{\mathrm{T}}\mathbf{C}^{-1}\Delta\boldsymbol{\sigma} - \Delta\boldsymbol{\varepsilon}^{\mathrm{T}}\Delta\boldsymbol{\sigma} + \tfrac{1}{2}\Delta\boldsymbol{\varepsilon}^{\mathrm{T}}\mathbf{C}\,\Delta\boldsymbol{\varepsilon} \tag{6.192a}$$

$$F = \tfrac{1}{2}(\Delta\boldsymbol{\varepsilon} - \mathbf{C}^{-1}\Delta\boldsymbol{\sigma})^{\mathrm{T}}(\mathbf{C}\,\Delta\boldsymbol{\varepsilon} - \Delta\boldsymbol{\sigma}) \tag{6.192b}$$

$$F = \tfrac{1}{2}[\boldsymbol{\varepsilon}_e + \Delta\boldsymbol{\varepsilon} - \mathbf{C}^{-1}(\boldsymbol{\sigma} + \Delta\boldsymbol{\sigma})]^{\mathrm{T}}[\mathbf{C}(\boldsymbol{\varepsilon}_e + \Delta\boldsymbol{\varepsilon}) - \boldsymbol{\sigma} - \Delta\boldsymbol{\sigma}] \tag{6.192c}$$

$$F = \tfrac{1}{2}[\boldsymbol{\sigma}_{\mathrm{elt}} - (\boldsymbol{\sigma} + \Delta\boldsymbol{\sigma})]^{\mathrm{T}}\mathbf{C}^{-1}[\boldsymbol{\sigma}_{\mathrm{elt}} - (\boldsymbol{\sigma} + \Delta\boldsymbol{\sigma})] \tag{6.192d}$$

$$F = \tfrac{1}{2}(\boldsymbol{\sigma}_{\mathrm{B}} - \boldsymbol{\sigma}_{\mathrm{C}})^{\mathrm{T}}\mathbf{C}^{-1}(\boldsymbol{\sigma}_{\mathrm{B}} - \boldsymbol{\sigma}_{\mathrm{C}}). \tag{6.192e}$$

In the step from (6.192b) to (6.192c), we add and subtract terms $\boldsymbol{\varepsilon}_e = \mathbf{C}^{-1}\boldsymbol{\sigma}$ which involve the elastic strains, $\boldsymbol{\varepsilon}_e$, and total stresses at the beginning of the increment. The stresses $\boldsymbol{\sigma}_{\mathrm{elt}}$ are the elastic 'trial stresses' given by

$$\boldsymbol{\sigma}_{\mathrm{elt}} = \boldsymbol{\sigma}_{\mathrm{B}} = \boldsymbol{\sigma} + \mathbf{C}\,\Delta\boldsymbol{\varepsilon} = \mathbf{C}\boldsymbol{\varepsilon}_e + \mathbf{C}\,\Delta\boldsymbol{\varepsilon}. \tag{6.193}$$

In the final step in (6.192) (and in (6.193)) we have also introduced the subscripts B and C which relate to Figure 6.12 and the earlier developments of Sections 6.6.5 and 6.6.6. Maintaining this notation, we now have to minimise (6.192e) with the final stresses, $\boldsymbol{\sigma}_{\mathrm{C}}$, as the variables and with the yield surface constraint (now with an equality because we are assuming plastic flow) being applied at the end via $f(\boldsymbol{\sigma}_{\mathrm{C}}) = 0$. Hence the Lagrangian equivalent to the minimisation of (6.192e) is

$$L(\boldsymbol{\sigma}_{\mathrm{C}}, \Delta\lambda) = \tfrac{1}{2}(\boldsymbol{\sigma}_{\mathrm{B}} - \boldsymbol{\sigma}_{\mathrm{C}})^{\mathrm{T}}\mathbf{C}^{-1}(\boldsymbol{\sigma}_{\mathrm{B}} - \boldsymbol{\sigma}_{\mathrm{C}}) + \Delta\lambda f(\boldsymbol{\sigma}_{\mathrm{C}}). \tag{6.194}$$

The equivalent to (6.191) is now

$$\frac{\partial L}{\partial \boldsymbol{\sigma}_{\mathrm{C}}} = -\mathbf{C}^{-1}\boldsymbol{\sigma}_{\mathrm{B}} + \mathbf{C}^{-1}\boldsymbol{\sigma}_{\mathrm{C}} + \Delta\lambda\left.\frac{\partial f}{\partial \boldsymbol{\sigma}}\right|_{\mathrm{C}} = -\mathbf{C}^{-1}\boldsymbol{\sigma}_{\mathrm{B}} + \mathbf{C}^{-1}\boldsymbol{\sigma}_{\mathrm{C}} + \Delta\lambda\mathbf{a}_{\mathrm{C}} = 0 \tag{6.195}$$

where we are introducing more of the notation of Sections 6.6.5 and 6.6.6 with $\mathbf{a} = \partial f/\partial\boldsymbol{\sigma}$. Equation (6.195) provides the backward-Euler relationship:

$$\boldsymbol{\sigma}_{\mathrm{C}} = \boldsymbol{\sigma}_{\mathrm{B}} - \Delta\lambda\,\mathbf{C}\mathbf{a}_{\mathrm{C}} \tag{6.196}$$

previously given in (6.78).

From (6.192e), $F$ can be identified as the scaled, squared length between point B (the elastic trial point) and C (the final point satisfying the yield function). It follows that $\boldsymbol{\sigma}_C$ (as given by (6.196)) is the closest-point projection onto the yield surface in the energy norm,

$$E = \sqrt{[(\boldsymbol{\sigma}_B - \boldsymbol{\sigma}_C)\mathbf{C}^{-1}(\boldsymbol{\sigma}_B - \boldsymbol{\sigma}_C)]} \tag{6.197}$$

induced by the metric, $\mathbf{C}^{-1}$ [S5].

## 6.11 SPECIAL NOTATION

$\mathbf{1}$ (or $\mathbf{1}_2$) = unit second-order tensor (see (4.30) and (4.31))
$\mathbf{a} = \partial f/\partial \boldsymbol{\sigma}$ which is defined here as a *column* vector
$A'$ = hardening parameter (= H′B)
$A_1, A_2$ = scalars for plane-stress analysis (see (6.124) and (6.125))
$\mathbf{A}$ = special matrix within $\partial \mathbf{a}/\partial \boldsymbol{\sigma}$ (see (6.47))
$B(\boldsymbol{\sigma})$ = stress parameter (see (6.15))
$C_1, C_2$ = stress parameters for plane-stress analysis (see (6.131))
$\mathbf{C}_t$ = tangential constitutive tensor (or matrix),
$\mathbf{C}_{tc}$ = consistent tangential constitutive tensor (or matrix)
$\mathbf{C}_3$ = constitutive matrix with three stress components; $\mathbf{C}_4$ = constitutive matrix with four stress components (Section 6.8)
$\mathbf{e}$ = deviatoric strains

$$E' = \frac{E}{(1+\nu)(1-2\nu)}$$

$f$ = yield function
$\gamma_{xy}$, etc. = engineering shear strain = $\varepsilon_{xy}/2$
$H'$ = hardening parameter
$\mathbf{I}$ = unit fourth-order tensor (see (4.30) and (4.31)) or (sometimes) unit matrix (or second-order tensor)
$\mathbf{j}^T = (1, 1, 1, 0, 0, 0)$
$J_2$ = second stress deviator invariant
$k$ = bulk modulus
$L$ = Lagrangian function (Section 6.10)
$\mathbf{L}$ = special matrix (see (6.27)) required for use with vector stress and strain forms

$$r = \frac{1+\nu}{1-\nu} \text{ (Section 6.8.2)}$$

$\mathbf{r}$ = residual vector (see (6.79))
$\mathbf{R}$ = special matrix (see (6.108) and (6.139))
$\mathbf{s}$ = deviatoric stresses (see (6.28) for vector form)
$W_p$ = plastic work
$\alpha$ = scalar for crossing the yield surface (Section 6.6.1)
$\alpha$ = scalar for radial return (see (6.94))
$\beta$ = scalar for consistent tangent (see (6.101))
$\varepsilon_{ps}$ = equivalent plastic strain

$\dot{\varepsilon}_m$ = mean strain rate
$\boldsymbol{\varepsilon}$ = vector or tensor of strains (the latter is sometimes written as $\boldsymbol{\varepsilon}_2$)
$\dot{\boldsymbol{\varepsilon}}$ = strain rate
$\dot{\boldsymbol{\varepsilon}}_t$ = total strain rate (the subscript t is often dropped)
$\dot{\boldsymbol{\varepsilon}}_p$ = plastic strain rate
$\boldsymbol{\varepsilon}_d$ (or $\boldsymbol{\varepsilon}_{2d}$ or $\mathbf{e}$) = deviatoric strains
$\dot{\lambda}$ = plastic strain-rate multiplier
$\mu$ = shear modulus
$\sigma_e$ = effective stress
$\sigma_o$ = yield stress
$\sigma_m$ = mean stress
$\boldsymbol{\sigma}$ = stress (as vector or tensor; the latter is sometimes written as $\boldsymbol{\sigma}_2$)
$\dot{\boldsymbol{\sigma}}$ = stress rate
$\tau_{xy}$, etc. = shear stress

### Subscripts

e = effective, elastic
$n$ = new
o = old
p = plastic

## 6.12 REFERENCES


[A1] *ABAQUS—Theory manual*, Vers. 4.6, Hibbit, Karlsson & Sorensen, providence, Rhode Island, USA (1984).

[A2] Argyris, J. H., Vaz, L. E. & Willam, K. J., Improved solution methods for rate problems, *Comp. Math. Appl. Mech. & Engng.*, **16**, 231–277 (1978).

[A3] Armen, H., Assumptions, models and computational methods for plasticity, *Computers & Structures*, **10**, 161–174 (1979).

[B1] Besseling, J. F., A theory of elastic, plastic, and creep deformations of an initially isotropic material showing anisotropic strain-hardening, creep recovery and secondary creep, *JAM*, **24**, 529–536 (1958).

[B2] Bicanic, N. P., Exact evaluation of contact stress state in computational elasto-plasticity, *Engineering Comp.*, **6**, 67–73 (1989).

[B3] Braudel, H. J., Abouaf, M. & Chenot, J. L., An implicit and incremental formulation for the solution of elastoplastic problems by the finite element method, *Computers and Structures*, **22** (5), 801–814 (1986).

[B4] Bushnell, D., A strategy for the solution of problems involving large deflections, plasticity and creep, *Int. J. Num. Meth. Engng.*, **10**, 1343–1356 (1976).

[C1] Caddemi, S. & Martin, J. B., Convergence of the Newton–Raphson algorithm in incremental elastic-plastic finite element analysis, *Computational Plasticity: Models, Software & Applications*, Vol. 1, ed. D. R. J. Owen *et al.*, Pineridge, Swansea, pp. 27–48 (1989).

[C2] Chen, W. F., *Plasticity in Reinforced Concrete*, McGraw-Hill, New York (1984).

[C3] Crisfield, M. A., Numerical analysis of structures, *Developments in Thin-walled Structures*—1, ed. J. Rhodes *et al.*, Applied Science, pp. 235–284 (1981).

[C4] Crisfield, M. A., Consistent schemes for plasticity computation with the Newton–Raphson method, *Computational Plasticity: Models, Software and Applications*, Part 1, Pineridge, Swansea, pp. 133–259 (1987).

[D1] De Borst, R., Non-linear analysis of frictional materials, Ph.D. Thesis, Inst. TNO for Building Materials and Structures, Delft (1986).

[D2] Desai, C. S. & Siriwardane, H. J., *Constitutive Laws for Engineering Materials*, Prentice Hall (1984).

[D3] Dodds, R. H., Numerical techniques for plasticity computations in finite element analysis, *Computers & Struct.*, **26**(5), 767–779 (1987).

[D4] Drucker, D. C., Conventional and unconventional plastic response and representation, *Appl. Mech. Rev.*, **41**(4), 151–167 (1988).

[F1] Fletcher, R., *Practical Methods of Optimisation*, 2nd edition, Wiley (1987).

[H1] Hibbitt, H. D., Some issues in numerical simulation of the nonlinear response of metal shells, *Proc. FEMSA '86 Symp. on Finite Element Methods in South Africa*, University of Witwatersrand, Johannesburg (February 1986).

[H2] Hill, R., *The Mathematical Theory of Plasticity*, Oxford University Press (1950).

[H3] Hinton, E., Hellen, T. K. & Lyons, L. P. R., On elasto-plastic benchmark philosophies, *Computational Plasticity: Models, Software and Applications*, ed. D. R. J. Owen *et al.*, Pineridge, Swansea, pp. 389–408 (1989).

[H4] Hinton, E, & Ezzat, M. H., Fundamental tests for two- and three-dimensional, small strain, elastoplastic finite element analysis, National Agency for Finite Element Methods and Standards Report SSEPT, (1987).

[H5] Hodge, P. G., *Plastic Analysis of Structures*, McGraw-Hill (1959).

[H6] Huffington, N. G., Numerical analysis of elastoplastic stress, Memorandum Report No. 2006, Ballistic Res. Labs., Aberdeen Proving Ground, Maryland (1969).

[H7] Hughes, T. J. R. & Pister, K. S., Consistent linearisation in mechanics of solids and structures, *Comp. & Struct.*, **8**, 391–397 (1978).

[I1] Ilyushin, A. A., *Plasticité—deformations elasto-plastiques* (transated from Russian in French), Editions Eyrolles, Paris, (1956).

[J1] Jetteur, P., Implicit integration algorithm for elastioplasticity in plane stress analysis, *Engineering Computations*, **3**(3), 251–253 (1986).

[J2] Johnson, W. & Mellor, P. B., *Engineering Plasticity*, Ellis Horwood, Chichester (1983)

[J3] Johnson, C., A mixed finite element method for plasticity problems with hardening, *SIAM J. Num. Anal.*, **14**, 575–584 (1977).

[K1] Key, S. W., Stone, C. M. & Krieg, R. D., A solution strategy for the quasi-static large-deformation, inelastic response of axisymmetric solids, *Europe/US Workshop on Nonlinear Finite Element Analysis in Structural Mechanics*, Bochum (1980).

[K2] Krieg, R. D. & Krieg, D. B., Accuracies of numerical solution methods for the elastic-perfectly platic model, *Trans. ASME*, 510–515 (November 1977).

[K3] Kreig, R. D. & Key, S. M., Implementation of a time independent plasticity theory into structural computer programs, *Constitutive Equations in Viscoplasticity: Computational and Engineering Aspects*, AMD-20, ed. J. A. Stricklin *et al.*, ASME, New York, pp. 125–138 (1976).

[L1] Little, G. H., Rapid analysis of plate collapse by live-energy minimisation, *Int. J. Mech. Sci.*, **19**, 725–743 (1977).

[L2] Luenberger, D. G., *Linear and Nonlinear Programming*, 2nd edition, Addison-Wesley, Reading, Mass. (1984).

[M1] Maeir, G. & Nappi, A., On the unified framework provided by mathematical programming to plasticity, *Mechanics of Materials Behaviour*, ed. G. J. Dvorak *et al.*, Elsevier, Amsterdam, pp. 253–273 (1983).

[M2] Martin, J. B. & Bird, W. W., Integration along the path of loading in elastic–plastic problems, *Engineering Computations*, **5**, 217–223 (1988).

[M3] Martin, J. B., An internal variable approach to the formulation of finite element problems in plasticity, *Physical Nonlinearities in Structural Analysis*, ed. J. Holt *et al.*, Springer-Verlag, pp. 165–176 (1981).

[M4] Marques, J. M. M. C., Stress computation in elastoplasticity, *Engineering Computations*, **1**, 42–51 (1984).

[M5] Matthies, H. G., A decomposition method for the integration of the elastic–plastic rate problem, *Int. J. Num. Meth. Engng.*, **28**, 1–11 (1989).

[M6] Matthies, H., The rate problem for complex material behaviour with internal variables, *Computational Plasticity: Models, Software & Applications*, Vol. 1, ed. D. R. J. Owen *et al.*, Pineridge, Swansea, pp. 27–48 (1989).
[M7] Mendelson, A., *Plasticity: Theory and Application*, McMillan, New York (1968).
[M8] Mitchell, G. P. & Owen, D. R. J., Numerical solutions for elastic–plastic problems, *Engineering Computations*, **5**(4), 274–284 (1988).
[M9] Mroz, Z., An attempt to describe the behaviour of metals under cyclic loads using a more general work-hardening model, *Acta Mechanica*, **7**(2–3), 199–212 (1969).
[M10] Martin, J. B., *Plasticity—Fundamentals and General Results*, The MIT Press, Cambridge, Mass., London (1975).
[N1] Nagtegal, J. C., On the implementation of inelastic constitutive equations with special reference to large deformation problems, *Comp. Meth. Appl. Mech. & Engng.*, **33**, 469–484 (1982).
[N2] Neal, B. G., *The Plastic Methods of Structural Analysis*, Chapman Hall, London (1963).
[N3] Nemat-Nasser, S. (ed.), Theoretical foundation for large-scale computations for nonlinear material behaviour, *Proc. Workshop*, Evanston, Illinois, 1983, Martinus Nijhoff, Dordrecht (1984).
[N4] Nyssen, C., An efficient and accurate iterative method allowing large incremental steps to solve elasto-plastic problems, *Computers & Structures*, **13**, 63–71 (1981).
[O1] Ortiz, M. & Simo, J. C., An analysis of a new class of integration algorithms for elastoplastic constitutive relations, *Int. J. Num. Meth. Engng.*, **23**, 353–366 (1986).
[O2] Ortiz, M. & Popov, E. P., Accuracy and stability of integration algorithms for elastoplastic constitutive relations, *Int. J. Num. Meth. Engng.*, **21**(9), 1561–1576 (1985).
[O3] Ortiz, M., Pinsky, P. M. & Taylor, R. L., Operator split methods for the numerical solution of the elastoplastic dynamic problem, *Comp. Meth. Appl. Mech. & Engng.*, **39**, 137–157 (1983).
[O4] Owen, D. R. J. & Hinton, E., *Finite Elements in Plasticity—Theory and Practice*, Pineridge Press, Swansea (1980).
[O5] Owen, D. R. J., Prakahs, A. & Zienkiewicz, O. C., Finite element analysis of non-linear composite materials by use of overlay systems, *Int. J. Num. Meth. Engng.*, **4**, 1251–1267 (1974).
[P1] Prager, W., *Introduction to Plasticity*, Addison-Wesley (1959).
[P2] Prager, W., A new method of analysing stress and strains in work-hardening plastic solids, *J. Applied Mechanics*, **23**, 493–496 (1956).
[R1] Ramm, E. & Matzenmiller, A., Consistent linearization in elasto-plastic shell analysis, *Engineering Computations*, **5**(4), 289–299 (1988).
[R2] Rice, J. R. & Tracey, D. M., Computation fracture mechanics, *Proc. Symp. Num. Meth. Struct. Mech.*, ed. S. J. Fenves, Academic Press, p. 585 (1973).
[R3] Runesson, K. & Samuelsson, A., Aspects on numerical techniques in small deformation plasticity, NUMETA 85, *Numerical Methods in Engineering: Theory and Applications*, ed. J. Middleton *et al.*, A. A. Balkema, Rotterdam, Vol. 1, pp. 337–348 (1985).
[R4] Runesson, K., Samuelsson, A. & Bernspang, L., Numerical techniques in plasticity including solution advancement control, *Int. J. Num. Meth. Engng.*, **22**, 769–788 (1986).
[S1] Samuelsson, A. & Froier, M., Finite elements in plasticity—a variational inequality approach, MAFELAP III, ed. J. R. Whiteman, Academic Press, New York (1979).
[S2] Save, M. A. & Massonet, C., *Plastic Analysis and Design of Plates, Shells and Discs*, North-Holland, Amsterdam (1972).
[S3] Schreyer, H. L., Kulak, R. F. & Kramer, J. M., Accurate numerical solutions for elasto-plastic models, *ASME Journal of Pressure Vessel Technology*, **101**, 226–234 (1979).
[S4] Simo, J. C. & Taylor, R. J., Consistent tangent operators for rate-independent elasto-plasticity, *Comp. Meth. Applied Mechanics and Engng.*, **48**, 101–118 (1985).
[S5] Simo, J. C. & Hughes, T. J. R., *Elasto-plasticity and Viscoplasticity—Computational Aspects*, to be published.
[S6] Simo, J. C. & Taylor, R. L., A return mapping algorithm for plane stress elastoplasticity, *Int. J. Num. Meth. Engng.*, **22**, 649–670 (1986).

[S7] Simo, J. C. & Govindjee, S., Exact closed-form solution of the return mapping algorithm in plane stress elasto-viscoplasticity, *Engineering Computations*, **5**(3), 254–258 (1988).
[S8] Sloan, S. W., Substepping schemes for the numerical integration of elastoplastic stress-strain relations, *Int. J. Num. Meth. Engng.*, **24**, 893–912 (1987).
[S9] Strang, G., Matthies, H. & Temam, R., Mathematical and computational methods in plasticity, *Variational Methods in the Mechanics of Solids*, ed. S. Nemat-Nasser, Pergamon Press, Oxford (1980).
[T1] Tracey, D. M. & Freese, C. E., Adaptive load incrementation in elasto-plastic finite element analysis, *Comp. & Struct.*, **13**, 45–53 (1981).
[V1] Von Mises, R., Mechanik der plastischen Formanderung von Kristallen, *Z. Angew. Math. Mech.*, **8**(3), 161–185 (1928).
[V2] Von Mises, R., Mechanik der festen Körper im plastisch deformablen Zustand, Götting Nachr. Math. Phys. K1, 582–592 (1913).
[W1] Waszczyszyn, Z., Computational methods and plasticity, Report LR-583, Tech. Univ. Delft, Faculty of Aerospace (February 1989).
[W2] Wilkins, M. L., Calculation of elastic-plastic flow, *Methods of Computational Physics*, Vol. 3, ed. B. Alder *et al.*, Academic Press (1964).
[W3] Willam, K. J., Recent issues in computational plasticity, *Computational Plasticity; Models, Software & Applications*, ed. D. R. J. Owen *et al.*, Part 2, Pineridge, Swansea, pp. 1353–1378 (1989).
[Z1] Ziegler, H., *An Introduction to Thermodynamics*, North-Holland, Amsterdam (1983).
[Z2] Ziegler, H., A modification of Prager's rule, *Quarterly of Appl. Math.*, **17**, 55–65 (1959).
[Z3] Zienkiewicz, O. C. & Cormeau, I. C., Viscoplasticity, plasticity and creep in elastic solids—a unified numerical solution approach, *Int. J. Num. Meth. Engng.*, **8**, 821–845 (1974).

# 7 Two-dimensional formulations for beams and rods

Much work has been devoted to finite element methods for beam and rod elements acting in a two-dimensional plane [B2–B4,C3–C5,E1,H1,H2,K1,M2,O1,W3,W4]. A significant proportion of this work has involved the total and updated Lagrangian methods [B2.5,B1,B4,C4,E1,F1,H2,K1,M2,W3,W4] which have already been considered in Chapter 3 for trusses and Chapter 5 for continuum elements. Beam elements for two-dimensional analysis are not only of interest in their own right but also in a didactic role. With this in mind, we will begin this chapter by following on from the work of Chapters 1 and 2 and will adopt a shallow-arch formulation [C4,C5] which is a degenerate form of total Lagrangian technique. We will later introduce a corotational approach which can be considered as an extension of the procedure adopted for trusses in Section 3.6 which used a 'rotated engineering strain'. Finally, in Section 7.5, we will consider a degenerate-continuum approach using the total Lagrangian formulation [B2.5,B1,S4,W4]. This work can be considered as a natural extension of the work in Chapter 5 on continua and a precursor to the work in Chapter 8 on degenerate-continuum shells.

We should note that some work on arch and ring elements has involved interpolations for tangential and transverse displacements (in relation to the curved arch) [P1]. We will not consider these approaches in the present chapter because in relation to shell analysis (Chapter 8), most successful elements use interpolations in relation to fixed orthogonal axes.

## 7.1 A SHALLOW-ARCH FORMULATION

We will firstly consider an initially flat element as shown in Figure 7.1. From Chapter 4, equation (4.84), the axial strain in the $x$-direction can, using a degenerated form of the Green strain, be expressed (see also (2.3)) as

$$\varepsilon_x = \frac{\mathrm{d}u}{\mathrm{d}x} + \frac{1}{2}\left(\frac{\mathrm{d}w}{\mathrm{d}x}\right)^2. \tag{7.1}$$

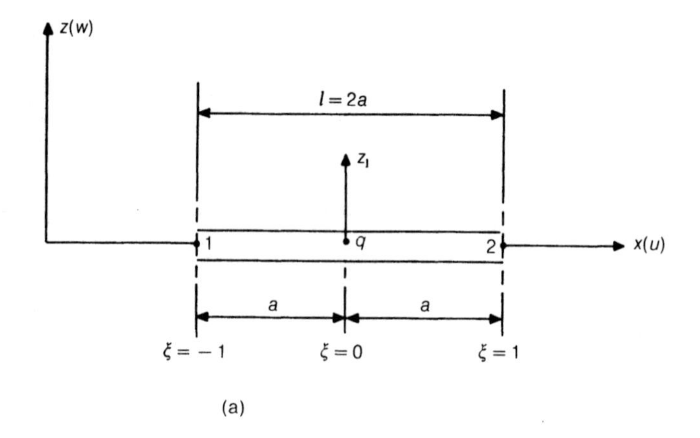


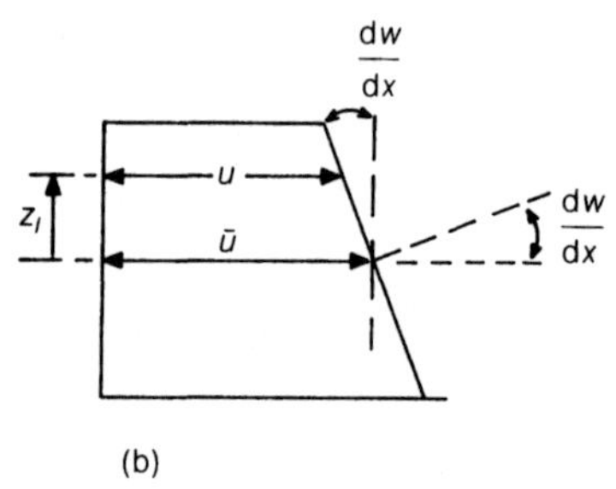


**Figure 7.1** An initially flat shallow-arch element: (a) coordinates and nodes; (b) detail.

Assuming plane sections remain plane, the displacement in the $x$-direction, $u$, at distance $z_l$ from the centroid is given (see Figure 7.1(b)) by

$$u = \bar{u} - z_l \frac{\mathrm{d}w}{\mathrm{d}x}. \tag{7.2}$$

Combining (7.1) and (7.2), gives

$$\varepsilon_x = \frac{\mathrm{d}\bar{u}}{\mathrm{d}x} + \frac{1}{2}\left(\frac{\mathrm{d}w}{\mathrm{d}x}\right)^2 - z_l \frac{\mathrm{d}^2 w}{\mathrm{d}x^2} = \bar{\varepsilon} + z_l \chi \tag{7.3}$$

where $\chi$ is the curvature. For an initially curved element (Figure 7.2), equation (7.3) must be modified to

$$\varepsilon_x = \frac{\mathrm{d}\bar{u}}{\mathrm{d}x} + \frac{1}{2}\left(\left(\frac{\mathrm{d}(z+w)}{\mathrm{d}x}\right)^2 - \left(\frac{\mathrm{d}z}{\mathrm{d}x}\right)^2\right) - z_l \frac{\mathrm{d}^2 w}{\mathrm{d}x^2} = \bar{\varepsilon} + z_l \chi. \tag{7.4}$$

The virtual work equation can be expressed as

$$V = \int\left(\int \sigma_x \delta\varepsilon_{xv}\, \mathrm{d}z_l\right)\mathrm{d}x - \mathbf{q}_e^{\mathrm{T}} \delta \mathbf{p}_v \tag{7.5}$$

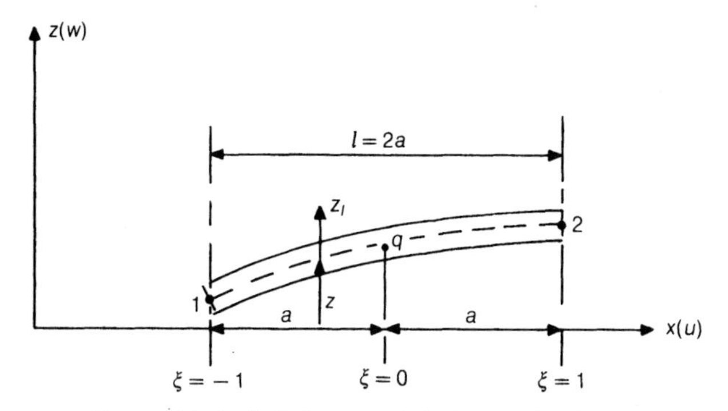


**Figure 7.2** An initially curved shallow-arch element.

where $\delta\varepsilon_{xv}$ is obtained by differentiation of (7.4) to give

$$\delta\varepsilon_{xv} = \delta\bar{\varepsilon}_v + z_l\delta\chi_v = \frac{d\delta\bar{u}_v}{dx} + \frac{d(z+w)}{dx}\frac{d\delta w_v}{dx} + z_l\delta\chi_v. \tag{7.6}$$

Substitution into (7.5), followed by integration through the depth, $z_l$, leads to

$$V = \int\left(N\left(\frac{d\delta\bar{u}_v}{dx} + \frac{dw'}{dx}\frac{d\delta w_v}{dx}\right) + M\,\delta\chi_v\right)dx - \mathbf{q}_e^T\delta\mathbf{p}_v \tag{7.7}$$

where, for convenience, we are writing $w' = w + z$ and the stress resultants in (7.7) are

$$N = \int\sigma_x\,dz_l, \qquad M = \int\sigma_x z_l\,dz_l. \tag{7.8}$$

At this stage, the finite element shape functions can be introduced so that

$$\bar{u} = \mathbf{h}_u^T\mathbf{u}, \qquad w = \mathbf{h}_w^T\mathbf{w}, \qquad z = \mathbf{h}_w^T\mathbf{z}. \tag{7.9}$$

Using a thin-beam, Kirchhoff assumption (as here), the lowest-order function we can use for w is a cubic [C2.2]. For the present, we will adopt a quadratic, hierarchical function for $\bar{u}$ [C2.2] but will later (Section 7.1.3) consider in more detail the issue of appropriate, matching, functions. With the chosen function, we have (see Figures 7.1 and 7.2),

$$\mathbf{u} = (u_1, u_2, \Delta u_q) \tag{7.10}$$

$$\mathbf{w}^T = (w_1, \theta_{x1}, w_2, \theta_{x2}), \qquad \theta_x = \frac{dw}{dx} \tag{7.11a}$$

$$\mathbf{z}^T = (z_1, \alpha_{x1}, z_2, \alpha_{x2}), \qquad \alpha_x = \frac{dz}{dx} \tag{7.11b}$$

$$\mathbf{h}_u^T = \tfrac{1}{2}(1-\xi, 1+\xi, 2(1-\xi^2)) \tag{7.12}$$

$$\mathbf{h}_w^T = \tfrac{1}{8}(4 - 6\xi + 2\xi^3, l(\xi^2-1)(\xi-1), 4 + 6\xi - 2\xi^3, l(\xi^2-1)(\xi+1)). \tag{7.13}$$

In relation to these equations nodes 1 and 2 are the end-nodes (Figures 7.1 and 7.2) while node q is a central node at which the hierarchical mid-side displacement $\Delta u_q$

(q for quadratic) acts. (The superscript will often be omitted from the nodal values of $\bar{u}$, but it should be noted that all nodal $u$-values refer to displacement at the reference plane.) The more basic linear displacement function for $\bar{u}$ can be obtained by setting $\Delta u_q$ to zero. The nodal values $\theta_{x1}$ and $\theta_{x2}$ are the nodal values of $dw/dx$ and are variables, while the nodal quantities $\alpha_{x1}$ and $\alpha_{x2}$ are fixed nodal values of $dz/dx$ which, along with the fixed nodal quantities $z_1$ and $z_2$, define the initial configuration of the element.

With a view to the computation of the strains, differentiation of (7.9) leads to

$$\frac{d\bar{u}}{dx} = \frac{1}{l}\begin{pmatrix} -1 \\ 1 \\ -4\xi \end{pmatrix}^{T} \mathbf{u} = \mathbf{b}_u^T \mathbf{u} \tag{7.14}$$

$$\frac{dw}{dx} = \frac{1}{4l}\begin{pmatrix} 6(\xi^2 - 1) \\ l(3\xi^2 - 2\xi - 1) \\ 6(1 - \xi^2) \\ l(3\xi^2 + 2\xi - 1) \end{pmatrix}^{T}, \qquad \mathbf{w} = \mathbf{b}_w^T \mathbf{w}, \qquad \frac{dz}{dx} = \mathbf{b}_w^T \mathbf{z} \tag{7.15}$$

$$\chi = -\frac{d^2 w}{dx^2} = \frac{1}{l^2}\begin{pmatrix} 6\xi \\ l(3\xi - 1) \\ -6\xi \\ l(3\xi + 1) \end{pmatrix} \mathbf{w} = \mathbf{c}^T \mathbf{w} \tag{7.16}$$

so that, in (7.4),

$$\bar{\varepsilon} = \mathbf{b}_u^T \mathbf{u} + \tfrac{1}{2}(\mathbf{b}_w^T \mathbf{w}')^2 - \tfrac{1}{2}(\mathbf{b}_w^T \mathbf{z})^2 \tag{7.17}$$

where $\mathbf{w}' = \mathbf{w} + \mathbf{z}$. Hence, knowing the nodal displacements, $\bar{\mathbf{u}}$ and $\mathbf{w}$, the strain $\varepsilon_x$ of (7.4) can be computed at any depth, $z_l$. Assuming, for the present elastic properties, $N$ and $M$ in (7.8) can be re-expressed as

$$N = EA\bar{\varepsilon}, \qquad M = EI\chi. \tag{7.18}$$

The shape-function relationships of (7.9) relate to total displacements but identical expressions apply for the virtual displacements and hence substitution into (7.6) gives

$$\delta\varepsilon_{xv} = \delta\bar{\varepsilon}_v + z_l \delta\chi_v = \mathbf{b}_u^T \delta\mathbf{u}_v + (\mathbf{b}_w^T \mathbf{w}')\mathbf{b}_w^T \delta\mathbf{w}_v + z_l \mathbf{c}^T \delta\mathbf{w}_v. \tag{7.19}$$

Further substitution into the virtual work of (7.7) gives

$$V = \int (N(\mathbf{b}_u^T \delta\mathbf{u}_v + (\mathbf{b}_w^T \mathbf{w}')\mathbf{b}_w^T \delta\mathbf{w}_v) + M\mathbf{c}^T \delta\mathbf{w}_v)\, dx - \mathbf{U}_e^T \delta\mathbf{u}_v - \mathbf{W}_e^T \delta\mathbf{w}_w$$

$$= \mathbf{U}_i^T \delta\mathbf{u} + \mathbf{W}_i^T \delta\mathbf{w} - \mathbf{U}_e^T \delta\mathbf{u} - \mathbf{W}_e^T \delta\mathbf{w} \tag{7.20}$$

where the usual internal force vector, $\mathbf{q}_i$, can be written as

$$\mathbf{q}_i^T = (\mathbf{U}_i^T, \mathbf{W}_i^T) \tag{7.21}$$

with

$$\mathbf{U}_i = \int N \mathbf{b}_u \, dx \tag{7.22a}$$

$$\mathbf{W}_i = \int (N(\mathbf{b}_w^T \mathbf{w}')\mathbf{b}_w + M\mathbf{c})\, dx. \tag{7.22b}$$

In (7.20)–(7.22), the internal forces $U_i$ correspond to the nodal displacements, $\bar{\mathbf{u}}$, in (7.10) and the 'forces', $\mathbf{W}_i$ to the nodal 'displacements', $\mathbf{w}$ in (7.11a).

### 7.1.1 The tangent stiffness matrix

The tangent stiffness matrix is obtained in the usual manner by differentiation of the internal force vector. To this end, it is most convenient to work in terms of submatrices, so that

$$\mathbf{K}_t = \begin{bmatrix} \mathbf{K}_{uu} & \mathbf{K}_{uw} \\ \mathbf{K}_{uw}^T & \mathbf{K}_{ww} \end{bmatrix} = \begin{bmatrix} \dfrac{\partial \mathbf{U}_i}{\partial \mathbf{u}} & \dfrac{\partial \mathbf{U}_i}{\partial \mathbf{w}} \\ \dfrac{\partial \mathbf{W}_i}{\partial \mathbf{u}} & \dfrac{\partial \mathbf{W}_i}{\partial \mathbf{w}} \end{bmatrix}. \quad (7.23)$$

Then, from (7.21) and (7.22) with the aid of (7.18) and the non-virtual form of (7.19),

$$\mathbf{K}_{uu} = \frac{\partial \mathbf{U}_i}{\partial \mathbf{u}} = \int b_u \frac{\partial N}{\partial u} \mathrm{d}x = \int EA\mathbf{b}_u\mathbf{b}_u^T \mathrm{d}x \quad (7.24a)$$

$$\mathbf{K}_{uw} = \frac{\partial \mathbf{U}_i}{\partial w} = \int \mathbf{b}_u \frac{\partial N}{\partial w} \mathrm{d}x = \int EA(\mathbf{b}_w^T\mathbf{w}')\mathbf{b}_u\mathbf{b}_w^T \mathrm{d}x \quad (7.24b)$$

$$\mathbf{K}_{ww} = \frac{\partial \mathbf{W}_i}{\partial \mathbf{w}} = \int \left( \mathbf{c}\frac{\partial M}{\partial \mathbf{w}} + \mathbf{b}_w(\mathbf{b}_w^T\mathbf{w}')\frac{\partial N}{\partial \mathbf{w}} + N\mathbf{b}_w\mathbf{b}_w^T \right) \mathrm{d}x$$

$$= \int (EI\mathbf{c}\mathbf{c}^T + EA(\mathbf{b}_w^T w')^2\mathbf{b}_w\mathbf{b}_w^T + N\mathbf{b}_w\mathbf{b}_w^T)\,\mathrm{d}x \quad (7.24c)$$

where the last term in (7.24c) can be identified as the initial stress matrix.

### 7.1.2 Introduction of material non-linearity or eccentricity

In deriving (7.24), it has been assumed that

$$\delta N = EA\,\delta\bar{\varepsilon}_x, \qquad \delta M = EI\,\delta\chi. \quad (7.25)$$

More generally, the tangent $E$-value at depth $z_l$ will not be $E$ but, say, $\hat{E}$ so that, using (7.8) and (7.3).

$$\delta N = \int \delta\sigma\,\mathrm{d}z_l = \int \hat{E}(\delta\bar{\varepsilon} + z_l\,\delta\chi)\,\mathrm{d}z_l = \int \hat{E}\,\mathrm{d}z_l\,\delta\bar{\varepsilon} + \int \hat{E}z_l\,\mathrm{d}z_l\,\delta\chi = \text{'}EA\text{'}\,\delta\bar{\varepsilon} + \text{'}EX\text{'}\,\delta\chi \quad (7.26)$$

$$\delta M = \int \delta\sigma z_l\,\mathrm{d}z_l = \int \hat{E}z_l(\delta\bar{\varepsilon} + z_l\,\delta\chi)\,\mathrm{d}z_l = \int \hat{E}z_l\,\mathrm{d}z_l\,\delta\bar{\varepsilon} + \int \hat{E}z_l^2\,\mathrm{d}z_l\,\delta\chi = \text{'}EX\text{'}\,\delta\bar{\varepsilon} + \text{'}EI\text{'}\,\delta\chi. \quad (7.27)$$

Even if the material is elastic so that $\hat{E} = E$, the cross-coupling term, '$EX$', is zero only if the reference plane from which $z_l$ (Figures 7.1 and 7.2) is measured is at the centroid.

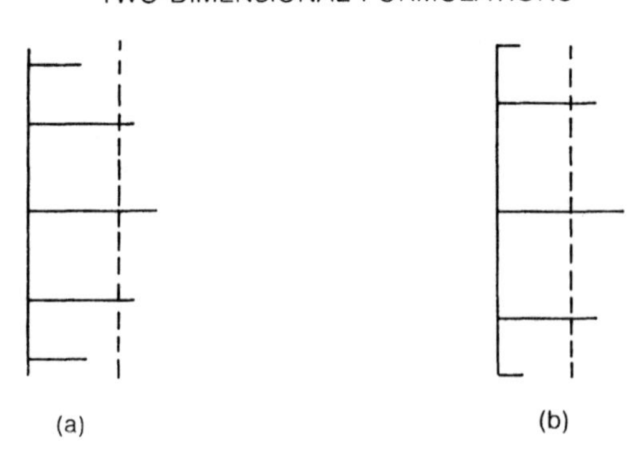


**Figure 7.3** Schemes for numerical integration through the depth (five-point): (a) Gaussian; (b) Lobatto.

As a consequence of (7.26) and (7.27), the stiffness matrices of (7.24) are modified to

$$\mathbf{K}_{\mathrm{uu}} = \int \text{‘}EA\text{’}\mathbf{b}_u\mathbf{b}_u^{\mathrm{T}}\,\mathrm{d}x \tag{7.28}$$

$$\mathbf{K}_{\mathrm{uw}} = \int (\text{‘}EA\text{’}(\mathbf{b}_w^{\mathrm{T}}\mathbf{w}')\mathbf{b}_u\mathbf{b}_w + \text{‘}EX\text{’}\mathbf{b}_u\mathbf{c}^{\mathrm{T}})\,\mathrm{d}x \tag{7.29}$$

$$\mathbf{K}_{\mathrm{ww}} = \int (\text{‘}EI\text{’}\mathbf{c}\mathbf{c}^{\mathrm{T}} + (\text{‘}EA\text{’}(\mathbf{b}_w^{\mathrm{T}}\mathbf{w}')^2 + N)\mathbf{b}_w\mathbf{b}_w^{\mathrm{T}} + \text{‘}EX\text{’}(\mathbf{b}_w^{\mathrm{T}}\mathbf{w}')(\mathbf{b}_w\mathbf{c}^{\mathrm{T}} + \mathbf{c}\mathbf{b}_w^{\mathrm{T}})\,\mathrm{d}x. \tag{7.30}$$

For non-linear materials, in obtaining ‘$EA$’, ‘$EX$’ and ‘$EI$’ from (7.26) and (7.27), one must integrate through the depth of the beam. Numerical integration is usually adopted with the stresses and other relevant variables (such as the plastic strain) being stored at integration points through the depth. Hence at the integration points (see Figure 7.3) one may compute the tangent $E$-values, $\hat{E}$, from the stored stresses, and plastic strains. With plane sections remaining plane, it is unnecessary to store the strains at the integration points through the depth since the strains are completely defined, from (7.4), by $\bar{\varepsilon}$ and $\chi$. This approach can be easily extended to shells (see Section 8.1.2).

Burgoyne and the author [B5] believe that the best procedure for the through-thickness integration is either standard Gaussian integration or the Lobatto rule, using, say, five points, if the material is non-linear. Two Gaussian points are adequate for a linear response although, of course, the integration can then be performed explicitly. In contrast to the Gaussian rule (Figure 7.3(a)), the Lobatto rule has points on the surface (7.3(b)). While the positioning and weights for Gaussian integration are given in many text-books the Lobatto values are not. Hence the latter are given in the Appendix.

### 7.1.3 Numerical integration and specific shape functions

As already discussed, with the adopted Kirchhoff bending theory, we cannot use a lower-order function for $w$ than the cubic adopted in (7.9) and (7.13). In contrast, we could, with respect to the continuity requirements, adopt any functions, from linear upwards, for $\bar{u}$. Strictly, from (7.1), with a cubic $w$, we require a quintic $\bar{u}$ [D1, C2.2] in order to balance the functions and thus ensure that we can represent a constant membrane strain and, in particular, the zero membrane strain associated with 'inextensional bending'. Such a quintic element would be extremely cumbersome, particularly when extended to shells.

Instead of using such an element, we can use a number of techniques to remove or, at least ameliorate, the 'membrane locking' [S3, C4, C2.2, M2]. If we adopt the lowest possible, linear, function for $\bar{u}$, then one solution is to use a single point, reduced integration for the membrane strain, $\bar{\varepsilon}$, in (7.4) [M2]. Alternatively, one could adopt linear functions for $w$ and $z$ with respect to the membrane strain, $\bar{\varepsilon}$, although a cubic $w$ for the curvature, $\chi$ [G2.1, M2].

These devices might appear a little *ad hoc*. Certainly, they will lead to deformation modes involving $w$ that lead to no membrane strain. However, the bending energy should ensure that there are no mechanisms. Also these methods can be put on a more rigorous footing via the Hu–Washizu variational principle [W1, C2.2] with a constant membrane strain. (More discussion on this theme will be given in Section 7.1.5.) The latter would effectively involve replacing the membrane strain, $\bar{\varepsilon}$ from (7.4), with an effective membrane strain, $\bar{\varepsilon}_{\text{eff}}$, so that

$$\bar{\varepsilon}_{\text{eff}} = \frac{1}{l}\int \bar{\varepsilon}\,\mathrm{d}x. \tag{7.31}$$

Using the basic linear function for $\bar{u}$ (with $\Delta\bar{u}_{\text{q}} = 0$), with the shape function of (7.10)–(7.13), this would lead to

$$\bar{\varepsilon} = \frac{\bar{u}_{21}}{l} + \frac{1}{2l}\mathbf{w}'^{\mathrm{T}}\int \mathbf{b}_w\mathbf{b}_w^{\mathrm{T}}\,\mathrm{d}x\,\mathbf{w}' - \frac{1}{2l}\mathbf{z}^{\mathrm{T}}\int \mathbf{b}_w\mathbf{b}_w^{\mathrm{T}}\,\mathrm{d}x\,\mathbf{z} \tag{7.32}$$

where we are using the subscript '21' to denote the difference between the variables at nodes 1 and 2, i.e. $\bar{u}_{21} = \bar{u}_2 - \bar{u}_1$.

One problem with the adoption of a linear $\bar{u}$ is that, in association with the quadratic $\mathrm{d}w/\mathrm{d}x$ stemming from the cubic $w$, the terms in (7.2) do not match. Consequently (ignoring the non-linear $(\mathrm{d}w/\mathrm{d}x)^2$ term), we have a solution that even for 'bending-dominant problems' [C7] depends on the chosen reference plane (at which $\bar{u}$ acts). Hence, with eccentricity (so that '$EX$' in (7.26) and (7.27) are non-zero), we can produce over-stiff solutions [G1, C2.2, C7]. As indicated in Section 7.1.2, effective eccentricity can be induced by material non-linearity [C7].

Returning to 'membrane locking', if instead of using (7.31) or (7.32), we adopt the full functions, it is essential to use (at least) a quadratic function for $\bar{u}$ and include the $\Delta\bar{u}_{\text{q}}$ term to limit the self-straining (see also the more detailed discussion in Section 7.1.5). If this quadratic term is not included, the solutions will be dramatically over-stiff [C4]. With this quadratic term included, if the integration of the internal force and tangent stiffness matrices is performed using two-point Gaussian integration, very reasonable solutions are obtained. The stresses must, of course, also be sampled at these reduced integration stations [C2.2].

### 7.1.4 Introducing shear deformation

Instead of directly introducing the Kirchhoff hypothesis, we can adopt a Timoshenko beam formulation which includes shear deformation [T1, C2.2]. As a consequence, the rotation of the normal, $\theta$ becomes a separate variable (Figure 7.4) and the curvature is given by $\chi = \mathrm{d}\theta/\mathrm{d}x$, while the shearing virtual work:

$$V_{\mathrm{s}} = \int Q\,\delta\gamma_{\mathrm{v}}\,\mathrm{d}x = \int Q\left(\delta\theta_{\mathrm{v}} + \frac{\mathrm{d}\delta w_{\mathrm{v}}}{\mathrm{d}x}\right)\mathrm{d}x \tag{7.33}$$

must be added to (7.7) (with $Q$ as the transverse shear force).

We will adopt quadratic hierarchical functions for both $u$, $w$ and $\theta$, so that

$$\bar{u} = \mathbf{h}_{\mathrm{u}}^{\mathrm{T}}\mathbf{u}, \qquad w = \mathbf{h}_{\mathrm{w}}^{\mathrm{T}}\mathbf{w}, \qquad \theta = \mathbf{h}_{\theta}^{\mathrm{T}}\boldsymbol{\theta} \tag{7.34}$$

where $\mathbf{h}_{\mathrm{w}} = \mathbf{h}_{\theta} = \mathbf{h}_{\mathrm{u}}$ (Equation (7.12)) and, in addition to (7.10),

$$\mathbf{w}^{\mathrm{T}} = (w_1, w_2, \Delta w_{\mathrm{q}}), \qquad \boldsymbol{\theta}^{\mathrm{T}} = (\theta_1, \theta_2, \Delta\theta_{\mathrm{q}}). \tag{7.35}$$

It should be noted that the nodal rotational variables are now of the opposite sign to those adopted for the Kirchhoff formulation (currently they are the rotations of the normal (Figure 7.4) while before (see (7.11) and Figure 7.1) they were $(\mathrm{d}w/\mathrm{d}x)$s).

Differentiation of $\bar{u}$ (7.34) leads to (7.14) while differentiation of $w$ and $\theta$ in (7.34) gives

$$\frac{\mathrm{d}w}{\mathrm{d}x} = \mathbf{b}_{\mathrm{w}}^{\mathrm{T}}\mathbf{w}, \qquad \frac{\mathrm{d}\theta}{\mathrm{d}x} = \mathbf{b}_{\theta}^{\mathrm{T}}\boldsymbol{\theta} \tag{7.36}$$

where $\mathbf{b}_w = \mathbf{b}_\theta = \mathbf{b}_u$ (7.14). (Although we are here using the same shape functions for all of the variables, we will keep the subscripts to help with an understanding of the equations.) Using (7.36), the shear strain and curvature are given by

$$\gamma = \theta + \frac{\mathrm{d}w}{\mathrm{d}x} = \mathbf{b}_{\mathrm{w}}^{\mathrm{T}}\mathbf{w} + \mathbf{h}_{\theta}^{\mathrm{T}}\boldsymbol{\theta}, \qquad \chi = \frac{\mathrm{d}\theta}{\mathrm{d}x} = \mathbf{b}_{\theta}^{\mathrm{T}}\boldsymbol{\theta}. \tag{7.37}$$

Consequently, from (7.7) and (7.33) (compare (7.20)) the internal virtual work becomes

$$V_{\mathrm{i}} = \int N(\mathbf{b}_u^{\mathrm{T}}\delta\mathbf{u}_{\mathrm{v}} + (\mathbf{b}_w^{\mathrm{T}}\mathbf{w}')\mathbf{b}_w^{\mathrm{T}}\delta\mathbf{w}_{\mathrm{v}} + M\mathbf{b}_\theta^{\mathrm{T}}\delta\boldsymbol{\theta}_{\mathrm{v}} + Q(\mathbf{h}_\theta^{\mathrm{T}}\delta\boldsymbol{\theta}_{\mathrm{v}} + \mathbf{b}_w^{\mathrm{T}}\delta\mathbf{w}_{\mathrm{v}})\,\mathrm{d}x \tag{7.38}$$

and the internal force vector is

$$\mathbf{q}_{\mathrm{i}}^{\mathrm{T}} = (\mathbf{U}_{\mathrm{i}}^{\mathrm{T}}, \mathbf{W}_{\mathrm{i}}^{\mathrm{T}}, \mathbf{T}_{\mathrm{i}}^{\mathrm{T}}) \tag{7.39}$$

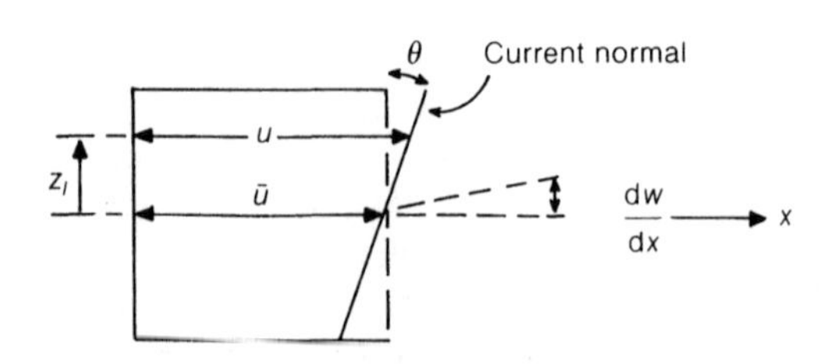


**Figure 7.4** Detail for element with shear deformation.

where the $\mathbf{T}_i$ terms are work-conjugate to the nodal variables, $\boldsymbol{\theta}$. The components of $\mathbf{q}_i$ are given by

$$\mathbf{U}_i = \int N\mathbf{b}_u \,dx \tag{7.40}$$

$$\mathbf{W}_i = \int (N(\mathbf{b}_w^T\mathbf{w}')\mathbf{b}_w + Q\mathbf{b}_w)\,dx \tag{7.41}$$

$$\mathbf{T}_i = \int (M\mathbf{b}_\theta + Q\mathbf{h}_\theta)\,dx. \tag{7.42}$$

It is convenient to subdivide the tangent stiffness matrix so that

$$\mathbf{K}_t = \begin{bmatrix} \mathbf{K}_{uu} & \mathbf{K}_{uw} & \mathbf{K}_{u\theta} \\ \mathbf{K}_{uw}^T & \mathbf{K}_{ww} & \mathbf{K}_{w\theta} \\ \mathbf{K}_{u\theta}^T & \mathbf{K}_{w\theta}^T & \mathbf{K}_{\theta\theta} \end{bmatrix} \tag{7.43}$$

where, from (7.40)–(7.42) and with the aid of (7.17), (7.28), (7.29) and (7.37), the sub-matrices are given by

$$\mathbf{K}_{uu} = \frac{\partial \mathbf{U}_i}{\partial \mathbf{u}} = \int \text{‘}EA\text{’}\mathbf{b}_u \frac{\partial \bar{\varepsilon}}{\partial \mathbf{u}} = \int \text{‘}EA\text{’}\mathbf{b}_u\mathbf{b}_u^T\,dx \tag{7.44a}$$

$$\mathbf{K}_{uw} = \frac{\partial \mathbf{U}_i}{\partial \mathbf{w}} = \int \text{‘}EA\text{’}\mathbf{b}_u \frac{\partial \bar{\varepsilon}}{\partial \mathbf{w}} = \int \text{‘}EA\text{’}(\mathbf{b}_w^T\mathbf{w}')\mathbf{b}_u\mathbf{b}_w^T\,dx \tag{7.44b}$$

$$\mathbf{K}_{u\theta} = \frac{\partial \mathbf{U}_i}{\partial \boldsymbol{\theta}} = \int \text{‘}EX\text{’}\mathbf{b}_u \frac{\partial \chi}{\partial \boldsymbol{\theta}}\,dx = \int \text{‘}EX\text{’}\mathbf{b}_u\mathbf{b}_\theta^T\,dx \tag{7.44c}$$

$$\mathbf{K}_{ww} = \frac{\partial \mathbf{W}_i}{\partial \mathbf{w}} = \int \left( \text{‘}EA\text{’}(\mathbf{b}_w^T\mathbf{w}')\mathbf{b}_w \frac{\partial \bar{\varepsilon}}{\partial \mathbf{w}} + \text{‘}GA\text{’}\mathbf{b}_w \frac{\partial \chi}{\partial \mathbf{w}} + N\mathbf{b}_w\mathbf{b}_w^T \right) dx$$

$$= \int (\text{‘}EA\text{’}(\mathbf{b}_w^T\mathbf{w}')^2\mathbf{b}_w\mathbf{b}_w^T + \text{‘}GA\text{’}\mathbf{b}_w\mathbf{b}_w^T + N\mathbf{b}_w\mathbf{b}_w^T)\,dx \tag{7.44d}$$

$$\mathbf{K}_{w\theta} = \int \left( \text{‘}EX\text{’}(\mathbf{b}_w^T\mathbf{w}')\mathbf{b}_w \frac{\partial \chi}{\partial \boldsymbol{\theta}} + \text{‘}GA\text{’}\mathbf{b}_w \frac{\partial \gamma}{\partial \mathbf{w}} \right) dx$$

$$= \int (\text{‘}EX\text{’}(\mathbf{b}_w^T\mathbf{w}')\mathbf{b}_w\mathbf{b}_\theta^T + \text{‘}GA\text{’}\mathbf{b}_w\mathbf{h}_\theta)\,dx \tag{7.44e}$$

$$\mathbf{K}_{\theta\theta} = \int \left( \text{‘}EI\text{’}\mathbf{b}_\theta \frac{\partial \chi}{\partial \boldsymbol{\theta}} + \text{‘}GA\text{’}\mathbf{h}_\theta \frac{\partial \gamma}{\partial \boldsymbol{\theta}} \right) dx$$

$$= \int (\text{‘}EI\text{’}\mathbf{b}_\theta\mathbf{b}_\theta^T\,dx + \text{‘}GA\text{’}\mathbf{h}_\theta\mathbf{h}_\theta^T)\,dx. \tag{7.44f}$$

The $N\mathbf{b}_w\mathbf{b}_w^T$ term in (7.44d) is the ‘initial stress’ or ‘geometric’ stiffness matrix. In (7.44d)–(7.44f), the ‘$GA$’ terms should include a shape-function factor for shear ($\frac{5}{6}$ for a rectangular section) [C2].

### 7.1.5 Specific shape functions, order of integration and shear-locking

Very acceptable solutions are achieved when the proposed quadratic shape functions are used in conjunction with two-point Gaussian integration. However, if the length-to-thickness ratio becomes very high, shear-locking can occur [C2.2]. This limitation can be overcome by forcing the shear strain to be effectively constant (with $x$) and constraining out the $\Delta w_q$ term so that [C2.2]

$$\Delta w_q = \frac{l}{8}(\theta_2 - \theta_1). \tag{7.45}$$

In these circumstances, the final element has $u$, $w$ and $\theta$ at the end nodes and $\Delta u$ and $\Delta\theta$ at the mid-element node, q. The constraining-out can be effected once the full stiffness matrix is formed but it is better to modify the $\mathbf{b}_w$ function.

We could eliminate all mid-element variables, by setting $\Delta\theta_q$ to zero (thus giving a linear bending moment) and constraining out $\Delta\bar{u}_q$ in order to provide a constant membrane strain. This would lead to a very similar solution to that obtained with only linear functions if one-point integration for all of the terms. This relationship was first raised in Section 7.1.3 and will now be amplified.

For simplicity, we will start with the membrane strain–displacement relationship in (7.3) rather than that of (7.4) which relates to an initially curved member. (The theory is easily modified to cover the latter.) Assuming a Timoshenko beam formulation as in the previous section, $w$ may be taken as a quadratic. In these circumstances, from (7.3), to be able to reproduce a constant $\bar{\varepsilon}$, $\bar{u}$ must strictly be a cubic. Using hierarchical displacement functions [C2.2], this can be provided via

$$\bar{u} = \frac{1}{2}\begin{pmatrix} 1-\xi \\ 1+\xi \end{pmatrix}^{\mathrm{T}} \begin{pmatrix} \bar{u}_1 \\ \bar{u}_2 \end{pmatrix} + (1-\xi^2)\Delta\bar{u}_q + \tfrac{8}{3}\xi(1-\xi^2)\Delta\bar{u}_c \tag{7.46}$$

where we have added to (7.9), (7.10) and (7.12) a hierarchical cubic function which is zero at $\xi = 0$ and $\Delta\bar{u}_c$ at $\xi = \frac{1}{2}$. We can now substitute (7.46) and the quadratic function of (7.34) and (7.35) for $w$ into (7.3) to find an expression for $\bar{\varepsilon}$. The expression is simplified by using non-dimensional displacements, $\hat{u} = \bar{u}/l$ and $\hat{w} = w/l$ and can be expressed as

$$\bar{\varepsilon} = (\hat{u}_{21} + \tfrac{1}{2}\hat{w}_{21}^2 + \tfrac{16}{3}\Delta\hat{u}_c) - 4\xi(\Delta\hat{u}_q + \hat{w}_{21}\Delta\hat{w}_q) + 8\xi^2(\Delta\hat{w}_q^2 - 2\Delta\hat{u}_c). \tag{7.47}$$

Hence, for a constant strain, we require

$$\Delta\hat{u}_q = -\hat{w}_{21}\,\Delta\hat{w}_q = [\tfrac{1}{8}\hat{w}_{21}\theta_{21}] \tag{7.48}$$

and

$$\Delta\hat{u}_c = \tfrac{1}{2}\Delta\hat{w}_q^2 = [\tfrac{1}{128}\theta_{21}^2]. \tag{7.49}$$

The square-bracketed values at the end of equations (7.48) and (7.49) stem from the introduction of (7.45). If we actually impose (7.48) and (7.49) as constraints and eliminate $\Delta\hat{u}_q$ and $\Delta\hat{u}_c$, we are left with

$$\bar{\varepsilon} = \hat{u}_{21} + \tfrac{1}{2}\hat{w}_{21}^2 + \tfrac{8}{3}\Delta\hat{w}_q^2 = [\hat{u}_{21} + \tfrac{1}{2}\hat{w}_{21}^2 + \tfrac{1}{24}\theta_{21}^2]. \tag{7.50}$$

An identical expression would be obtained (more simply) via the use of (7.31) using the quadratic $w$ function and linear $\bar{u}$ function (with no introduction of $\Delta\bar{u}_q$ and $\Delta\bar{u}_c$). In contrast, the use of a single-point integration would (see (7.47) with $\Delta\hat{u}_q = \Delta\hat{u}_c = 0$) lead to

$$\bar{\varepsilon} = \hat{u}_{21} + \tfrac{1}{2}\hat{w}_{21}^2. \tag{7.51}$$

The latter expression could also be obtained by using a linear function for $w$ in the membrane strain (see earlier discussion in Section 7.1.3). The difference between (7.50) and (7.51) will vanish as the mesh is refined and $\Delta w_q$ tends to zero. Under constant curvature, equation (7.32) for the Kirchhoff element of Sections 7.1–7.1.3 would again lead to (7.50).

If we do not apply any averaging on the membrane strain, it is essential to include a quadratic membrane displacement, $\bar{u}$. From (7.47), with $\Delta\bar{u}_q$ included, the shape functions will allow a membrane strain field with no $\xi$ term. Without a cubic ($\Delta\bar{u}_c$) term, we are strictly left with an unmatched quadratic term due to $\Delta w_q$ (see (7.47)) and hence cannot fully recover a constant membrane state. However, an 'unmatched' quadratic term is far less serious than an 'unmatched' linear term (as would arise, without averaging, if we did not include a quadratic, $\Delta\bar{u}_q$). Consider, firstly, a solution starting with a cubic $\bar{u}$, followed by the imposition of the constraint of (7.49) to eliminate $\Delta\hat{u}_c$. As an alternative, consider a solution obtained with only a quadratic $\bar{u}$. It is not difficult to show (using (7.47)) that both solutions give exactly the same membrane strain, $\bar{\varepsilon}$, at the two-point Gaussian integrations stations, $\xi = \pm\sqrt{\frac{1}{3}}$. Hence, if we adopt two-point Gaussian integration and use a quadratic $\bar{u}$, we can effectively recover a constant membrane strain. In the limit, as the element reaches a state of constant curvature, the cubic $w$ function of the Kirchhoff element of Sections 7.1–7.1.3 will become a quadratic and, in these, circumstances, the previous arguments will apply equally to that element.

As previously discussed in Section 7.1.3, the 'eccentricity issue' [C7] is also relevant to the choice of matching shape functions.

## 7.2 A SIMPLE COROTATIONAL ELEMENT USING KIRCHHOFF THEORY

A form of corotational technique has already been introduced in Section 3.6 for truss elements. The corotational technique [B2, B3, C3, C6, H1, M1, N1, O1, O2] was initially introduced by Wempner [W2] and Belytschko and co-workers [B2, B3] and has much in common with the 'natural approach' of Argyris *et al.* [A1]. Belytschko and co-workers [B2, B3] mainly applied the method to dynamic analysis using an 'explicit formulation' and hence the issue of the tangent stiffness matrix was not directly addressed although Belytschko did schematically outline the procedure for the generation of a consistent tangent matrix [B2]. He showed that, contrary to some arguments [T2], it was possible to derive a tangent stiffness matrix using corotational procedures (which, as already indicated in Chapters 3 and 5, are different from updated Lagrangian techniques).

The term 'corotational' has been used in a number of different contexts but will be taken here to relate to the provision of a single element frame that continuously rotates with the element and with respect to which standard, small-strain, small-displacement (or engineering) relationship can be applied (Locally shallow strain terms can be added—see Section 7.2.7.) Although corotational procedures have often been applied, they have not always been approached in a fully consistent manner particularly with regard to the generation of the tangent stiffness matrix. The key element in such a consistent derivation is the introduction of the variation of the local–global transformation matrices. This was recognised by Oran [O1, O2] who derived some elegant corotational formulations for beams and rods. (He also included 'beam–column' terms which may have somewhat obscured the corotational basis.)

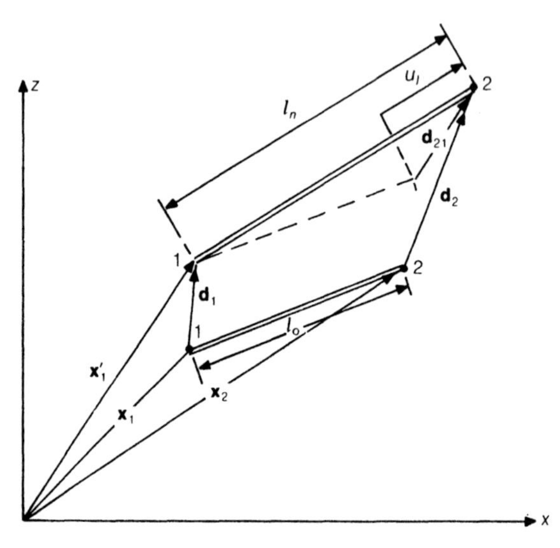


**Figure 7.5** Corotational stretch, $u_l$.

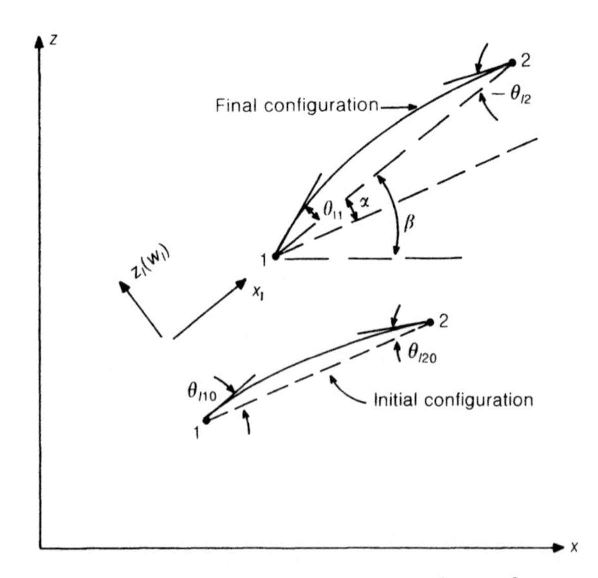


**Figure 7.6** Local corotational slopes, $\theta_{l1}$ and $\theta_{l2}$.

There has recently been a resurgence of interest in corotational formulations for both three-dimensional beams and shells [R1, C6, N1]. These issues will be addressed in Volume 2.

In this section, we will derive a two-dimensional corotational beam element using Kirchhoff theory. The element will be firstly described using simple engineering concepts without resource to shape functions. Throughout, the subscript, $l$, will be used for 'local' coordinates (see Figures 7.5 and 7.6) while if no subscript is given, the variable can be taken to be 'global' and relating to the fixed orthogonal axes. The formulation has much in common with that of Oran [O1, O2].

### 7.2.1 Stretching 'stresses' and 'strains'

The stretching side of the element formulation follows that given for the truss element of Sections 3.4 and (more directly) 3.6, which was derived using a 'rotated engineering strain'. Using this approach (see Figure 7.5), the local 'strain-inducing' extension is

$$u_l = l_\mathrm{n} - l_\mathrm{o} = ((\mathbf{x}_{21} + \mathbf{d}_{21})^\mathrm{T}((\mathbf{x}_{21} + \mathbf{d}_{21}))^{1/2} - (\mathbf{x}_{21}^\mathrm{T}\mathbf{x}_{21})^{1/2} \tag{7.52}$$

where the subscript 21 has the previous meaning with $\mathbf{x}_{21} = \mathbf{x}_2 - \mathbf{x}_1$, where $\mathbf{x}_1$ and $\mathbf{x}_2$ are the initial position vectors of nodes 1 and 2. In practice, (7.52) is badly conditioned and it is better to adopt

$$l_\mathrm{n} - l_\mathrm{o} = \frac{(l_\mathrm{n} - l_\mathrm{o})}{(l_\mathrm{n} + l_\mathrm{o})}(l_\mathrm{n} + l_\mathrm{o}) = \frac{l_\mathrm{n}^2 - l_\mathrm{o}^2}{l_\mathrm{n} + l_\mathrm{o}} \tag{7.53}$$

from which can be derived (see also Section 3.8), the mid-point formula [B2]

$$u_l = \frac{2}{(l_\mathrm{n} + l_\mathrm{o})}(\mathbf{x}_{21} + \tfrac{1}{2}\mathbf{d}_{21})^\mathrm{T}\mathbf{d}_{21}. \tag{7.54}$$

Using the initial basic form for the element, the axial strain is assumed constant as $u_l/l_\mathrm{o}$ and the axial force is given by

$$N = EAu_l/l_\mathrm{o}. \tag{7.55}$$

### 7.2.2 Bending 'stresses' and 'strains'

The standard engineering beam-theory relationships are assumed to apply in the local system so that with the local transverse displacements being zero (Figure 7.6):

$$\begin{pmatrix} \bar{M}_1 \\ \bar{M}_2 \end{pmatrix} = \frac{2EI}{l_o}\begin{bmatrix} 2 & 1 \\ 1 & 2 \end{bmatrix}\begin{pmatrix} \theta_{l1} \\ \theta_{l2} \end{pmatrix} \tag{7.56}$$

where $\theta_{l1}$ and $\theta_{l2}$ are the local slopes, $(\mathrm{d}w/\mathrm{d}x)_l$ (see Figure 7.6) in the corotating frame. These slopes are given by

$$\theta_{l1} = \theta_1 - \alpha - \theta_{l1\mathrm{o}}, \qquad \theta_{l2} = \theta_2 - \alpha - \theta_{l2\mathrm{o}} \tag{7.57}$$

where $\alpha$ defines the rigid rotation of the bar (Figure 7.6) and $\theta_{l1\mathrm{o}}$ and $\theta_{l2\mathrm{o}}$ are the initial local slopes. The rigid rotation, $\alpha$, can be found using the cross-product of the unit vectors along the bar in its initial and final configuration via

$$s = \sin\alpha = \frac{1}{l_o l_n}(\mathbf{x}_{21} \times (\mathbf{x}_{21} + \mathbf{d}_{21})) = \frac{1}{l_o l_n}(x_{21}w_{21} - z_{21}u_{21}). \tag{7.58}$$

Or, alternatively, using the inner product relationship:

$$c = \cos\alpha = \frac{1}{l_o l_n}(\mathbf{x}_{21}^{\mathrm{T}}(\mathbf{x}_{21} + \mathbf{d}_{21})). \tag{7.59}$$

Provided $|\alpha| < \pi$, one can decide which formula to use by deciding in which quadrant the beam lies, i.e.

$$\begin{aligned} \alpha &= \sin^{-1} s \text{ if } (s \geqslant 0 \text{ and } c \geqslant 0) \text{ or } (s \leqslant 0 \text{ and } c \geqslant 0) \\ \alpha &= \cos^{-1} c \text{ if } (s \geqslant 0 \text{ and } c \leqslant 0) \\ \alpha &= -\cos^{-1} c \text{ if } (s \leqslant 0 \text{ and } c \leqslant 0). \end{aligned} \tag{7.60}$$

The previous relationships fail if $|\alpha| > \pi$. However, in most circumstances, it is possible to extend the range to $2\pi$. To this end, we need to know the direction (clockwise or anticlockwise) in which the element has rotated. Then, knowing both c and s, we can uniquely find the quadrant in which the element lies. The direction of the rotation can be obtained from the sign of the (total) nodal rotations. More generally, in three dimensions, one can use quarternions to uniquely update a nodal triad—see [C6, S2] and Volume 2.

### 7.2.3 The virtual local displacements

In order to apply the principal of virtual work, we need to differentiate (7.52) (or (7.54)). An alternative, but completely equivalent, geometric approach follows from

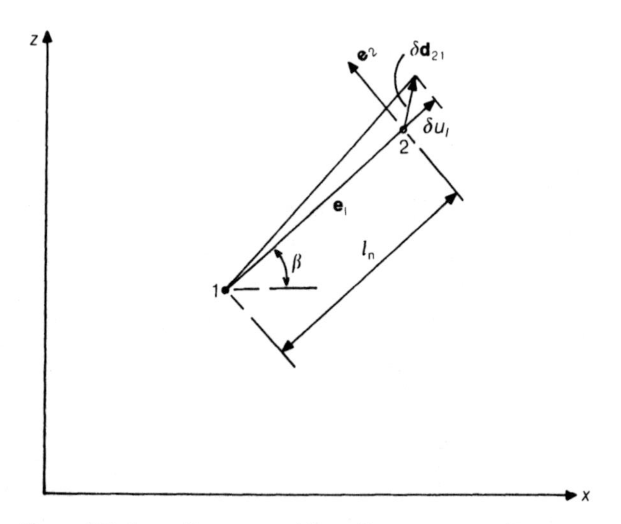


**Figure 7.7** A small movement from the current configuration.

Figure 7.7 and involves

$$\delta u_l = \mathbf{e}_1^{\mathrm{T}} \delta \mathbf{d}_{21} = \begin{pmatrix} \cos\beta \\ \sin\beta \end{pmatrix}^{\mathrm{T}} \delta \mathbf{d}_{21} = \frac{1}{l_{\mathrm{n}}} \begin{pmatrix} x'_{21} \\ z'_{21} \end{pmatrix}^{\mathrm{T}} \delta \mathbf{d}_{21} \tag{7.61}$$

where $\mathbf{e}_1$ is the unit vector lying along the current configuration of the bar (Figure 7.7) and $\mathbf{x}'_{21} = \mathbf{x}_{21} + \mathbf{d}_{21}$. In future we will refer to $\cos\beta$ as $c$ and $\sin\beta$ as $s$.

Equation (7.61) can be re-expressed as

$$\delta u_l = (-c, -s, 0, c, s, 0)\delta \mathbf{p} = \mathbf{r}^{\mathrm{T}} \delta \mathbf{p} \tag{7.62}$$

where the components of $\delta \mathbf{p}$ are the infinitesimal displacement changes at the nodes with $\mathbf{p}$ ordered as

$$\mathbf{p}^{\mathrm{T}} = (u_1, w_1, \theta_1, u_2, w_2, \theta_2). \tag{7.63}$$

From Figure 7.7, the unit vector orthogonal to $\mathbf{e}_1$ is given by

$$\mathbf{e}_2^{\mathrm{T}} = (-s, c) \tag{7.64}$$

and hence (see Figure 7.7), a small rigid rotation from the current configuration is given by

$$\delta\alpha = \frac{1}{l_{\mathrm{n}}} \mathbf{e}_2^{\mathrm{T}} \delta \mathbf{d}_{21} \tag{7.65}$$

or, in terms of the nodal displacements, $\delta \mathbf{p}$,

$$\delta\alpha = \frac{1}{l_{\mathrm{n}}}(s, -c, 0, -s, c, 0)\delta \mathbf{p} = \frac{1}{l_{\mathrm{n}}} \mathbf{z}^{\mathrm{T}} \delta \mathbf{p}. \tag{7.66}$$

From (7.57) and (7.66), we can write

$$\delta \boldsymbol{\theta}_l = \begin{pmatrix} \theta_{l1} \\ \theta_{l2} \end{pmatrix} = \left[ \begin{bmatrix} 0 & 0 & 1 & 0 & 0 & 0 \\ 0 & 0 & 0 & 0 & 0 & 1 \end{bmatrix} - \frac{1}{l_{\mathrm{n}}} \begin{bmatrix} \mathbf{z}^{\mathrm{T}} \\ \mathbf{z}^{\mathrm{T}} \end{bmatrix} \right] \delta \mathbf{p} = \mathbf{A}^{\mathrm{T}} \delta \mathbf{p}. \tag{7.67}$$

Hence the complete vector of local 'strain producing' displacement changes is given by

$$\delta \mathbf{p}_{\mathrm{l}} = \begin{bmatrix} \delta u_{\mathrm{l}} \\ \delta\theta_{11} \\ \delta\theta_{12} \end{bmatrix} = \begin{bmatrix} \mathbf{r}^{\mathrm{T}} \\ \mathbf{A}^{\mathrm{T}} \end{bmatrix} \delta \mathbf{p} = \mathbf{B} \delta \mathbf{p}. \tag{7.68}$$

### 7.2.4 The virtual work

Equating the internal virtual work in both the local and global systems and making use of the virtual form of (7.68), we can write

$$V_{\mathrm{i}} = \delta \mathbf{p}_{\mathrm{v}}^{\mathrm{T}} \mathbf{q}_{\mathrm{i}} = N\,\delta u_{l\mathrm{v}} + \bar{M}_1\,\delta\theta_{l1\mathrm{v}} + \bar{M}_2 \delta\theta_{l2\mathrm{v}} = \delta \mathbf{p}_{l\mathrm{v}}^{\mathrm{T}} \mathbf{q}_{l\mathrm{i}} = \delta \mathbf{p}_{\mathrm{v}}^{\mathrm{T}} \mathbf{B}^{\mathrm{T}} \mathbf{q}_{l\mathrm{i}} \tag{7.69}$$

where the local 'internal forces' are

$$\mathbf{q}_{l\mathrm{i}}^{\mathrm{T}} = (N, \bar{M}_1, \bar{M}_2). \tag{7.70}$$

Equation (7.69) must apply for any arbitrary $\delta \mathbf{p}_{\mathrm{v}}$ and hence the global internal force vector, $\mathbf{q}_i$, is given by

$$\mathbf{q}_i = \mathbf{B}^T \mathbf{q}_{li} \tag{7.71}$$

which is readily computed once the stress resultants, $\mathbf{q}_{li}$, are known from (7.55) and (7.56).

### 7.2.5 The tangent stiffness matrix

Differentiation of (7.71) leads to

$$\delta \mathbf{q}_i = \mathbf{B}^T \delta \mathbf{q}_{li} + N\,\delta \mathbf{B}_1 + \bar{M}_1\,\delta \mathbf{B}_2 + \bar{M}_2\,\delta \mathbf{B}_3 = \mathbf{K}_{t1}\,\delta \mathbf{p} + \mathbf{K}_{t\sigma}\delta \mathbf{p}, \tag{7.72}$$

where $\mathbf{B}_2$, for instance, denotes the second column of $\mathbf{B}^T$ (from (7.68)). Assuming linear material behaviour, differentiation of (7.55) and (7.56) leads to

$$\begin{bmatrix} \delta N \\ \delta \bar{M}_1 \\ \delta \bar{M}_2 \end{bmatrix} = \frac{EA}{l_o} \begin{bmatrix} 1 & 0 & 0 \\ 0 & 4r^2 & 2r^2 \\ 0 & 2r^2 & 4r^2 \end{bmatrix} \delta \mathbf{p}_l = \mathbf{C}_l \delta \mathbf{p}_l \tag{7.73}$$

where $r$ is the radius of gyration. Using (7.73), the first term on the right-hand side of (7.72) is easily computed as

$$\mathbf{K}_{t1} = \mathbf{B}^T \mathbf{C}_l \mathbf{B} \tag{7.74}$$

which is the standard tangent stiffness matrix. The geometric stiffness matrix comes from the last three terms in (7.72). From (7.66) and (7.62), differentiation of the first column of $\mathbf{B}^T$ (see (7.68)) leads to

$$\delta \mathbf{B}_1 = \delta \mathbf{r} = \delta \beta \mathbf{z} \tag{7.75}$$

with $\mathbf{z}$ from (7.66). But from Figure 7.6, $\delta\beta = \delta\alpha$ and hence, from (7.75) and (7.66),

$$\delta \mathbf{B}_1 = \frac{1}{l_n} \mathbf{z}\mathbf{z}^T \delta \mathbf{p}. \tag{7.76}$$

From (7.66)–(7.68),

$$\delta \mathbf{B}_2 = \frac{1}{l_n} \delta \mathbf{z} + \frac{1}{l_n^2} \mathbf{z}\,\delta u_l. \tag{7.77}$$

Therefore, from (7.61), (7.62) and (7.66),

$$\delta \mathbf{B}_2 = \frac{1}{l_n^2} (\mathbf{r}\mathbf{z}^T + \mathbf{z}\mathbf{r}^T) \delta \mathbf{p}. \tag{7.78}$$

From (7.67) and (7.68), $\delta \mathbf{B}_3 = \delta \mathbf{B}_2$ and hence, from (7.72)–(7.78), the complete tangent stiffness matrix is given by

$$\mathbf{K}_t = \mathbf{B}^T \mathbf{C}_l \mathbf{B}^T + \frac{\mathbf{N}}{l_n} \mathbf{z}\mathbf{z}^T + \frac{(\bar{M}_1 + \bar{M}_2)}{l_n^2} (\mathbf{r}\mathbf{z}^T + \mathbf{z}\mathbf{r}^T). \tag{7.79}$$

### 7.2.6 Using shape functions

A more conventional finite element formulation would introduce shape functions. To this end, the local displacement, $u_l(\xi)$ would be expressed as

$$u_l(\xi) = \tfrac{1}{2}(1 + \xi)u_l \tag{7.80a}$$

with

$$x_l = \tfrac{1}{2}(1 + \xi)l_o \tag{7.80b}$$

so that the local strain is obtained by differentiation to give

$$\varepsilon_{xl} = \frac{du_l}{dx_l} = \frac{du_l}{d\xi}\frac{d\xi}{dx_l} = u_l/l_o \tag{7.81}$$

while the local transverse displacement would be given by the conventional cubic (but with $w_l$ being zero at the two ends—Figure 7.6). Hence

$$w_l = \frac{l_o}{8}\begin{pmatrix}(\xi^2 - 1)(\xi - 1)\\(\xi^2 - 1)(\xi + 1)\end{pmatrix}^{\mathrm{T}}\begin{pmatrix}\theta_{l1}\\\theta_{l2}\end{pmatrix}. \tag{7.82}$$

Differentiation of (7.82) leads to

$$\theta_l = \frac{dw_l}{dx_l} = \frac{1}{4}\begin{pmatrix}-1 - 2\xi + 3\xi^2\\-1 + 2\xi + 3\xi^2\end{pmatrix}^{\mathrm{T}}\boldsymbol{\theta}_l = \mathbf{s}^{\mathrm{T}}\boldsymbol{\theta}_l \tag{7.83}$$

and further differentiation to

$$\chi = \frac{d\theta_l}{dx_l} = \frac{1}{l_o}\begin{pmatrix}-1 + 3\xi\\1 + 3\xi\end{pmatrix}^{\mathrm{T}}\boldsymbol{\theta}_l = \mathbf{b}^{\mathrm{T}}\boldsymbol{\theta}_l. \tag{7.84}$$

From (7.81), the axial strain is constant and hence the axial force is given by (7.55). The (local) bending moment is obtained from (7.84) as

$$M = EI\chi = \mathbf{EIb}^{\mathrm{T}}\boldsymbol{\theta}_l \tag{7.85}$$

and hence the internal virtual work follows from (7.81) and (7.84) as

$$V_{\mathrm{i}} = \mathbf{q}_{\mathrm{i}}^{\mathrm{T}}\delta\mathbf{p}_{\mathrm{v}} = \int_0^{l_o}\left(M\,\delta\chi_{\mathrm{v}} + N\frac{\delta u_{l\mathrm{v}}}{l_o}\right)dx_l = N\,\delta u_{l\mathrm{v}} + \delta\boldsymbol{\theta}_{l\mathrm{v}}^{\mathrm{T}}\int_0^{l_o} EI\mathbf{b}\mathbf{b}^{\mathrm{T}}\boldsymbol{\theta}_l dx_l \tag{7.86}$$

from which

$$V_{\mathrm{i}} = N\,\delta u_l + \frac{EI}{l_o}\delta\boldsymbol{\theta}_{l\mathrm{v}}^{\mathrm{T}}\begin{bmatrix}4 & 2\\2 & 4\end{bmatrix}\boldsymbol{\theta}_l. \tag{7.87}$$

On account of the definitions of $\bar{M}_1$ and $\bar{M}_2$ in (7.56), (7.87) coincides with (7.69) and hence, as anticipated, the internal force vector, $\mathbf{q}_{\mathrm{i}}$ of (7.71), is the same under the two formulations. It follows that the stiffness matrix, (7.79), will also coincide.

### 7.2.7 Including higher-order axial terms

Equation (7.55) was based on the approximation that the axial strain in the bar is equal to the relative axial deformation of the two ends divided by the original axial length. Such an assumption does not allow for any straining caused by the beam shape departing from a straight line (see Figure 7.6). Such an effect can be simply introduced by introducing the *local* shallow-arch terms (Section 7.1) [B3] or by invoking Green's strain (see (4.75)), relative to the corotating system. The latter is more general and will be developed here. Assuming for the present that the bar is initially straight, with the aid of (4.75), (7.81) and (7.83), we have

$$\varepsilon_{xl}(\xi) = \frac{\mathrm{d}u_l}{\mathrm{d}x_l} + \frac{1}{2}\left(\frac{\mathrm{d}u_l}{\mathrm{d}x_l}\right)^2 + \tfrac{1}{2}\theta_l^2 = \frac{u_l}{l_o} + \frac{1}{2}\left(\frac{u_l}{l_o}\right)^2 + \tfrac{1}{2}\boldsymbol{\theta}_l^{\mathrm{T}}\mathbf{s}\mathbf{s}^{\mathrm{T}}\boldsymbol{\theta}_l. \tag{7.88}$$

If the shallow arch equations are used, the $\frac{1}{2}(u_l/l_o)^2$ term in (7.88) would be neglected. As already discussed in Sections 7.1.3 and 7.1.5, unless extra variables are provided for $u_l$, only a constant axial strain can be accommodated and hence one should adopt (7.31) to modity (7.88) to give

$$\varepsilon_{xl} = \frac{u_l}{l_o} + \frac{1}{2}\left(\frac{u_l}{l_o}\right)^2 + \frac{1}{2l_o}\boldsymbol{\theta}_l^{\mathrm{T}}\int \mathbf{s}\mathbf{s}^{\mathrm{T}}\,\mathrm{d}x_l\boldsymbol{\theta}_l \tag{7.89}$$

so that the last term in (7.88) is changed to take its average value. On performing the integration, (7.89) becomes

$$\varepsilon_{xl} = \frac{u_l}{l_o} + \frac{1}{2}\left(\frac{u_l}{l_o}\right)^2 + \tfrac{1}{60}\boldsymbol{\theta}_l^{\mathrm{T}}\begin{bmatrix} 4 & -1 \\ -1 & 4 \end{bmatrix}\boldsymbol{\theta}_l. \tag{7.90}$$

Alternatively, it might be better to provide shallow terms relating to the state under constant curvature. From (7.50) we would then have

$$\varepsilon_{xl} = \frac{u_l}{l_o} + \tfrac{1}{24}\boldsymbol{\theta}_l^{\mathrm{T}}\begin{bmatrix} 1 & -1 \\ -1 & 1 \end{bmatrix}\boldsymbol{\theta}_l. \tag{7.90a}$$

It should be noted that, in the limit, as the mesh is refined, any of these additional terms (above $u_l/l_o$) will vanish. For the present developments, we will work from (7.90), but there should be no difficulty in adapting the equations to relate to (7.90a).

With a view to the virtual work, (7.90) can be differentiated to give (with the aid of (7.61) and (7.67)),

$$\delta\varepsilon_{xl} = \frac{1}{l_o}\left(1 + \frac{u_l}{l_o}\right)\mathbf{r}^{\mathrm{T}}\delta\mathbf{p} + \tfrac{1}{30}\boldsymbol{\theta}_l^{\mathrm{T}}\begin{bmatrix} 4 & -1 \\ -1 & 4 \end{bmatrix}\mathbf{A}^{\mathrm{T}}\delta\mathbf{p}. \tag{7.91}$$

Hence, in place of (7.68), we have

$$\delta\mathbf{p}_l = \begin{bmatrix} l_o\delta\varepsilon_{xl} \\ \delta\theta_{l1} \\ \delta\theta_{l2} \end{bmatrix} = \begin{bmatrix} \left(1 + \frac{u_l}{l_o}\right)\mathbf{r}^{\mathrm{T}} + \frac{l_o}{30}\boldsymbol{\theta}_l^{\mathrm{T}}\begin{bmatrix} 4 & -1 \\ -1 & 4 \end{bmatrix}\mathbf{A}^{\mathrm{T}} \\ \mathbf{A}^{\mathrm{T}} \end{bmatrix}\delta\mathbf{p} = \mathbf{B}\,\delta\mathbf{p} \tag{7.92}$$

and following (7.69)–(7.71), the internal force vector, $\mathbf{q}_i$, is given by (7.71) with $\mathbf{q}_{li}$ defined by (7.70) and $\mathbf{B}$ by (7.92).

The tangent stiffness matrix is obtained by differentiating (7.71) so that, in place of (7.79), we have

$$\mathbf{K}_t = \mathbf{K}_{t1} + \mathbf{K}_{t\sigma} = \mathbf{B}^{\mathrm{T}}\mathbf{C}_l\mathbf{B}^{\mathrm{T}} + \mathbf{K}_{t\sigma}, \tag{7.93}$$

where $\mathbf{C}_l$ was given in (7.73) and the 'initial stress' matrix is given by

$$\mathbf{K}_{t\sigma} = \mathbf{B}^{\mathrm{T}}\mathbf{C}^*\mathbf{B} + \frac{N(1 + u_l/l_{\mathrm{o}})}{l_{\mathrm{n}}}\mathbf{z}\mathbf{z}^{\mathrm{T}} + N\frac{\mathbf{r}\mathbf{r}^{\mathrm{T}}}{l_{\mathrm{o}}} + \frac{(\bar{M}_1 + \bar{M}_2 + \frac{1}{10}Nl_{\mathrm{o}}(\theta_{l1} + \theta_{l2})}{l_{\mathrm{n}}^2}(\mathbf{r}\mathbf{z}^{\mathrm{T}} + \mathbf{z}\mathbf{r}^{\mathrm{T}}) \tag{7.94}$$

where

$$\mathbf{C}^* = \frac{Nl_{\mathrm{o}}}{30}\begin{bmatrix} 0 & 0 & 0 \\ 0 & 4 & -1 \\ 0 & -1 & 4 \end{bmatrix}. \tag{7.95}$$

A formulation involving the local shallow-arch equations would involve setting $u_l$ to zero in (7.92) and (7.94).

### 7.2.8 Some observations

The workings of the previous section have assumed that the bar was initially straight rather than curved. If the latter applies, one may simply modify (7.88) to give

$$\varepsilon_{xl}(\xi) = \frac{\mathrm{d}u_l}{\mathrm{d}x_l} + \frac{1}{2}\left(\frac{\mathrm{d}u_l}{\mathrm{d}x_l}\right)^2 + \tfrac{1}{2}\theta_l^2 - \tfrac{1}{2}\theta_{lo}^2 \tag{7.96}$$

where the last term in (7.96) includes the effects of the initial local slopes ($\theta_{lo}$).

The formulation including the higher-order terms gives, as expected, more accurate results than the basic co-rotational formulation [C1]. However, this is achieved at some cost. In particular, once the higher-order terms are included, the formulation seems to be more badly conditioned and some convergence difficulties occur [C1]. (It is possible that these difficulties may be ameliorated by using a development based on (7.90a) rather than (7.90).) However, as the mesh is refined, there should be little difference between the two formulations, both in terms of accuracy and convergence characteristics. For linear buckling problems with coarse meshes, the extra terms would be required.

## 7.3 A SIMPLE COROTATIONAL ELEMENT USING TIMOSHENKO BEAM THEORY

If we include shear deformation, with the aid of Timoshenko beam theory, the stretching strain is exactly as in Section 7.2.1 but, see Figure 7.8, the bending and shear 'stresses' and 'strains' are given by

$$M = EI\chi_l = -\frac{EI}{l_{\mathrm{o}}}(\theta_{l2} - \theta_{l1}) = -\frac{EI}{l_{\mathrm{o}}}(\theta_2 - \theta_1) \tag{7.97}$$

$$Q = \text{`}GA\text{'}\gamma = -\text{`}GA\text{'}\left(\frac{\theta_{l2} + \theta_{l1}}{2}\right) = -\text{`}GA\text{'}\left(\frac{\theta_1 + \theta_2}{2} - \alpha\right) \tag{7.98}$$

where $\alpha$ is, as before, the rigid rotation given by (7.58)–(7.60). The minus signs in

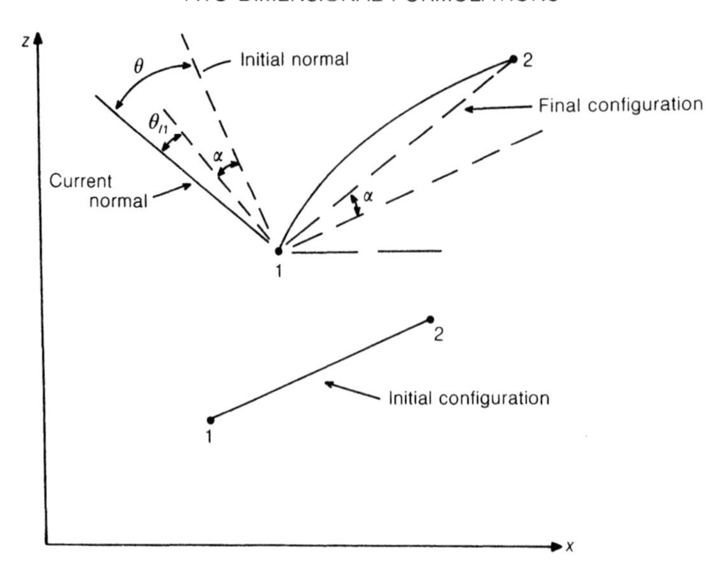


**Figure 7.8** Local corotational rotation of normal at node 1, $\theta_{l1}$.

(7.97) and (7.98) are required in order to maintain the sign convention adopted for the shallow-arch formulation of Section 7.1.4. (Note that the $\theta_l$s in Figure 7.4 and 7.8 are of opposite sign.) For simplicity, in (7.97) and (7.98), we have ignored the influence of any initial $\theta_l$ terms although the latter are easily introduced.

From (7.97) and (7.98) and using (7.62) and (7.66), in place of (7.68), we have

$$l_o \delta \boldsymbol{\varepsilon}_l = \begin{bmatrix} \delta u_l \\ l_o \delta \chi \\ l_o \delta \gamma \end{bmatrix} = \left[ \begin{bmatrix} & & \mathbf{r}^T & & & \\ 0 & 0 & 1 & 0 & 0 & -1 \\ 0 & 0 & -l_o/2 & 0 & 0 & -l_o/2 \end{bmatrix} + \frac{l_o}{l_n} \begin{bmatrix} \mathbf{0}^T \\ \mathbf{0}^T \\ \mathbf{z}^T \end{bmatrix} \right] \delta \mathbf{p} = \mathbf{B} \delta \mathbf{p} \tag{7.99}$$

where $\mathbf{r}$ and $\mathbf{z}$ have been given in (7.62) and (7.66). Consequently, the internal virtual work is given by

$$V_i = \delta \mathbf{p}_v^T \mathbf{q}_i = l_o \mathbf{q}_{li}^T \delta \boldsymbol{\varepsilon}_{lv} = \delta \mathbf{p}_v^T \mathbf{B}^T \mathbf{q}_{li} \tag{7.100}$$

where the local internal forces are

$$\mathbf{q}_{li}^T = (N, M, Q) \tag{7.101}$$

and the (global) internal force vector is given by

$$\mathbf{q}_i = \mathbf{B}^T \mathbf{q}_{li} \tag{7.102}$$

with $\mathbf{B}$ from (7.99). Differentiation of (7.102) leads to

$$\delta \mathbf{q}_i = \mathbf{B}^T \delta \mathbf{q}_{li} + (N\, \delta \mathbf{B}_1 + M\, \delta \mathbf{B}_2 + Q\, \delta \mathbf{B}_3) \tag{7.103}$$

where $\delta \mathbf{B}_2$, for example, is the differential of the second column of $\mathbf{B}^T$, given in (7.99).

In place of (7.73), we now have, using (7.99),

$$\delta \mathbf{q}_{li} = \begin{bmatrix} \delta N \\ \delta M \\ \delta Q \end{bmatrix} = \begin{bmatrix} EA & 0 & 0 \\ 0 & EI & 0 \\ 0 & 0 & GA \end{bmatrix} \delta \boldsymbol{\varepsilon}_l = \mathbf{C}_l \delta \boldsymbol{\varepsilon}_l = \frac{1}{l_o} \mathbf{C}_l \mathbf{B}^T \delta \mathbf{p} \tag{7.104}$$

so that the first term of (7.103) gives the standard tangent stiffness matrix as

$$\mathbf{K}_{t1} = \frac{1}{l_o} \mathbf{B}^T \mathbf{C}_l \mathbf{B}. \tag{7.105}$$

To form the geometric stiffness matrix, differentiation of the columns of $\mathbf{B}^T$ (given in (7.99)) leads to

$$\delta \mathbf{B}_1 = \delta \mathbf{B}_1 (\text{equation (7.76)}), \qquad \delta \mathbf{B}_2 = 0, \qquad \delta \mathbf{B}_3 = -l_o \delta \mathbf{B}_2 (\text{equation (7.78)}). \tag{7.106}$$

Substitution into (7.103) leads to

$$\mathbf{K}_t = \frac{1}{l_o} \mathbf{B}^T \mathbf{C}_l \mathbf{B} + \frac{N}{l_n} \mathbf{z}\mathbf{z}^T - \frac{Q l_o}{l_n^2} (\mathbf{r}\mathbf{z}^T + \mathbf{z}\mathbf{r}^T). \tag{7.107}$$

## 7.4 AN ALTERNATIVE ELEMENT USING REISSNER'S BEAM THEORY

With a view to later work (in Volume 2) involving three-dimensional beam elements, we will now describe an alternative two-noded element based on Reissner's beam theory [R2]. The element is closely related to a three-dimensional beam due to Vu–Quoc and Simo [S1, S2] (which will be described in Volume 2). This element is not, within the previous definitions, corotational. In other words, it does not use a single corotating frame for the element and then use standard small-strain engineering terms with respect to that rotating frame.

For this element, the bending moment and curvature are effectively the same (see (7.97)) as those of the previous element and are given by

$$M = EI\chi = \frac{EI}{l_o} (\theta_2 - \theta_1) \tag{7.108}$$

where the rotations of the nodal normals, $\theta_1$ and $\theta_2$ are shown in Figure 7.9.

In order to define the axial and shear strains, we require the orthogonal unit vectors, $\mathbf{t}_1$ and $\mathbf{t}_2$, at the central Gauss point (Figure 7.9). These are related to the average normal rotation

$$\theta_{av} = 0.5(\theta_1 + \theta_2) \tag{7.109}$$

via

$$\mathbf{t}_1^T = (\cos\theta_{av}, \sin\theta_{av}), \qquad \mathbf{t}_2^T = (-\sin\theta_{av}, \cos\theta_{av}). \tag{7.110}$$

In addition, we require (see Figure 7.9) the vector,

$$\mathbf{x}'_{21} = \mathbf{x}_{21} + \mathbf{d}_{21} \tag{7.111}$$

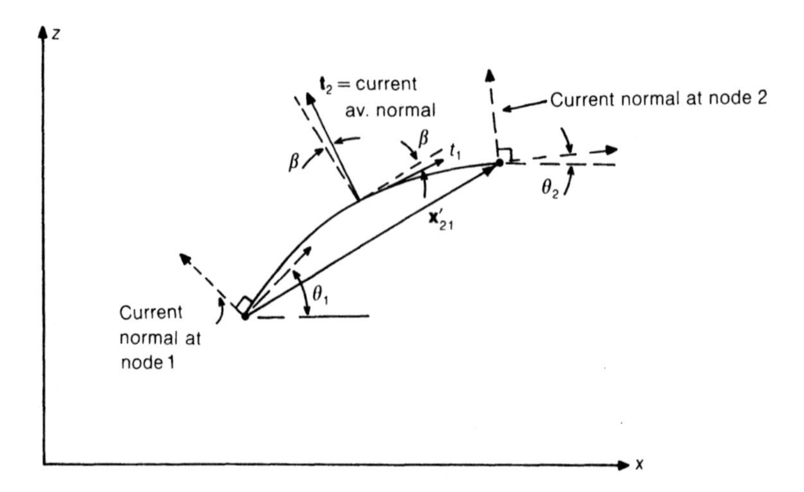


**Figure 7.9** Current configuration for arch element using Reissner's theory.

where, as before, we use the notation $\mathbf{x}_{21} = \mathbf{x}_2 - \mathbf{x}_1$, etc., and $\mathbf{x}$ describes the initial coordinates while $\mathbf{d}$ describes the total nodal displacements and (see Figure 7.5), $\mathbf{x}' = \mathbf{x} + \mathbf{d}$ describes the current coordinates. Using these terms, the strains can be expressed as

$$\varepsilon_l = \frac{1}{l_o}\mathbf{t}_1^T\mathbf{x}'_{21} - 1 = \frac{1}{l_o}\mathbf{t}_1^T\mathbf{x}_{21} + \frac{1}{l_o}\mathbf{r}^T\mathbf{p} - 1 \tag{7.112}$$

where $\mathbf{r}$ is as in (7.62) but with

$$s = \sin\theta_{av}, \qquad c = \cos\theta_{av}. \tag{7.113}$$

Also, the shear strain, $\gamma$, is given by

$$\gamma = \frac{1}{l_o}\mathbf{t}_2^T\mathbf{x}'_{21} = \frac{1}{l_o}\mathbf{t}_2^T\mathbf{x}'_{21} + \frac{1}{l_o}\mathbf{z}^T\mathbf{p} \tag{7.114}$$

where $\mathbf{z}^T = (s, -c, 0, -s, c, 0)$ with $s$ and $c$ being given by (7.113).

An explanation of these strain measures can be found from Figure 7.9. In particular,

$$\mathbf{x}'_{21} = l_n\mathbf{e}_1 \tag{7.115}$$

where $\mathbf{e}_1$ is the unit vector along the tangent. Hence, from (7.112),

$$\varepsilon_l = \frac{1}{l_o}(l_n(\mathbf{e}_1^T\mathbf{t}_1) - l_o) = \frac{1}{l_o}(l_n\cos\beta - l_o) \tag{7.116}$$

(with $\beta$ as in Figure 7.9), while from (7.114),

$$\gamma = \frac{l_n}{l_o}\mathbf{e}_1^T\mathbf{t}_2 = \frac{l_n\sin\beta}{l_o}. \tag{7.117}$$

As $\beta \to 0$, equation (7.116) coincides with the previous 'rotated engineering strain' while, again with $\beta \to 0$ and $l_n \to l_o$, equation (7.117) gives $\gamma \to \beta$ which again corresponds with the previous measure for the shear strain.

To obtain the infinitesimal 'strain' changes, we differentiate (7.108), (7.112) and (7.114) to give

$$l_o \delta \boldsymbol{\varepsilon}_l = \begin{bmatrix} \delta u_l \\ l_o \delta \chi \\ l_o \delta \gamma \end{bmatrix} = \begin{bmatrix} & \mathbf{r}^T + c_2 \mathbf{s}^T & \\ 0 \quad 0 & -1 \quad 0 \quad 0 & 1 \\ & \mathbf{z}^T + c_1 \mathbf{s}^T & \end{bmatrix} = \mathbf{B}^T \delta \mathbf{p} \tag{7.118}$$

where $s$ and $c_1$ and $c_2$ are given by

$$\mathbf{s}^T = (0, 0, 1, 0, 0, 1) \tag{7.119}$$

$$c_1 = -0.5\mathbf{t}_1^T \mathbf{x}'_{21} = -0.5(l_o + u_l) = -0.5 l_n \tag{7.120}$$

$$c_2 = 0.5\mathbf{t}_2^T \mathbf{x}'_{21} = 0.5 \gamma l_o. \tag{7.121}$$

As usual, to compute the internal force vector, we start with the (internal) virtual work:

$$V_i = \delta \mathbf{p}_v^T \mathbf{q}_i = l_o \mathbf{q}_{li}^T \delta \boldsymbol{\varepsilon}_{lv} = \delta \mathbf{p}_v^T \mathbf{B}^T \mathbf{q}_{li} \tag{7.122}$$

where

$$\mathbf{q}_{li}^T = (N, M, Q) \tag{7.123}$$

and we have used (7.118) for $\delta \boldsymbol{\varepsilon}_l$. For arbitrary $\delta p_v$, from (7.122), the internal force vector is given by

$$\mathbf{q}_i = \mathbf{B}^T \mathbf{q}_{li} \tag{7.124}$$

with $\mathbf{B}$ from (7.118). To obtain the tangent stiffness matrix, differentiation of (7.124) leads to

$$\delta \mathbf{q}_i = \mathbf{B}^T \delta \mathbf{q}_{li} + (N \delta \mathbf{B}_1 + M \delta \mathbf{B}_2 + Q \delta \mathbf{B}_3) = \mathbf{K}_t \delta \mathbf{p} \tag{7.125}$$

where $\delta \mathbf{B}_2$, for example, is the differential of the second column of $\mathbf{B}^T$ (given in (7.118)). Using (7.104), the first term in (7.125) leads to (7.105) with $\mathbf{B}$ from (7.118). To form the geometric stiffness matrix, differentiation of the columns of $\mathbf{B}^T$ (given in (7.118)) leads to

$$\delta \mathbf{B}_1 = \tfrac{1}{2}(\mathbf{s}\mathbf{z}^T \delta \mathbf{p} + \mathbf{z}\mathbf{s}^T \delta \mathbf{p}) + \tfrac{1}{2} c_1 \mathbf{s}\mathbf{s}^T \delta \mathbf{p} \tag{7.126}$$

$\delta \mathbf{B}_2 = \mathbf{0}$ and

$$\delta \mathbf{B}_3 = -\tfrac{1}{2}(\mathbf{s}\mathbf{r}^T \delta \mathbf{p} + \mathbf{r}\mathbf{s}^T \delta \mathbf{p}) - \tfrac{1}{2} c_2 \mathbf{s}\mathbf{s}^T \delta \mathbf{p}. \tag{7.127}$$

Substitution into (7.125) gives

$$\mathbf{K}_t = \frac{1}{l_o} \mathbf{B}^T \mathbf{C}_l \mathbf{B} + \frac{N}{2}(\mathbf{s}\mathbf{z}^T + \mathbf{z}\mathbf{s}^T) + \frac{N}{2} c_1 \mathbf{s}\mathbf{s}^T - \frac{Q}{2}(\mathbf{s}\mathbf{r}^T + \mathbf{r}\mathbf{s}^T) - \frac{Q}{2} c_2 \mathbf{s}\mathbf{s}^T. \tag{7.128}$$

### 7.4.1 The introduction of shape functions and extension to a general isoparametric element

The previous work can be extended to cover a general isoparametric formulation using the same functions ($\mathbf{h}$) for the geometry ($x$ and $z$) and each of the 'displacements' ($u$, $w$ and $\theta$). In these circumstances, we require the 'length parameter', $\alpha$ (see Section 3.3.4), at the Gauss point. This parameter, which relates to the initial geometry (and hence remains fixed at a particular Gauss point), is given by

$$\alpha^2 = \left(\frac{dx}{d\xi}\right)^2 + \left(\frac{dz}{d\xi}\right)^2 \tag{7.129}$$

where

$$x = \mathbf{h}^T\mathbf{x}, \qquad z = \mathbf{h}^T\mathbf{z}, \qquad \frac{dx}{d\xi} = \mathbf{h}_\xi^T\mathbf{x}, \qquad \frac{dz}{d\xi} = \mathbf{h}_\xi^T\mathbf{z} \tag{7.130}$$

and $\mathbf{x}$ and $\mathbf{z}$ have the nodal values of $x$ and $z$ respectively.

In place of equations (7.108), (7.112) and (7.114), the 'strains' become

$$\chi = \frac{1}{\alpha}\mathbf{h}_\xi^T\boldsymbol{\theta} \tag{7.131}$$

$$\varepsilon = \frac{1}{\alpha}\mathbf{t}_1^T\frac{d\mathbf{x}'}{d\xi} - 1 \tag{7.132}$$

and

$$\gamma = \frac{1}{\alpha}\mathbf{t}_2^T\frac{d\mathbf{x}'}{d\xi} \tag{7.133}$$

where $\mathbf{x}'$ relates to the current nodal geometry, i.e.

$$\mathbf{x}' = \mathbf{x} + \mathbf{d} \tag{7.134}$$

and the unit vectors $\mathbf{t}_1$ and $\mathbf{t}_2$ relate to the particular point (usually the Gauss point), $\xi$, and are obtained from (7.110) but using $\theta$ at $\xi$ (via the shape functions and the nodal $\theta$s) instead of $\theta_{av}$.

Differentiation of (7.131)–(7.133) gives the 'strain changes' at any point $\xi$ as

$$\alpha\delta\boldsymbol{\varepsilon}_l = \alpha\begin{bmatrix}\delta\varepsilon_l \\ \delta\chi \\ \delta\gamma\end{bmatrix} = \begin{bmatrix} 0 & 0 & \mathbf{r}^T + c_2\mathbf{s}^T & & & \\ 0 & 0 & \mathbf{h}_\xi(1) & 0 & 0 & \mathbf{h}_\xi(2) & \cdots \\ & & \mathbf{z}^T + c_1\mathbf{s}^T & & & \end{bmatrix} = \mathbf{B}\delta\mathbf{p} \tag{7.135}$$

where $\mathbf{h}_\xi$ is the differential of $\mathbf{h}$ with respect to $\xi$ and

$$\mathbf{r}^T = (c\mathbf{h}_\xi(1), s\mathbf{h}_\xi(1), 0, c\mathbf{h}_\xi(2), s\mathbf{h}_\xi(2), 0, \ldots) \tag{7.136}$$

$$\mathbf{z}^T = (-s\mathbf{h}_\xi(1), c\mathbf{h}_\xi(1), 0, -s\mathbf{h}_\xi(2), c\mathbf{h}_\xi(2), 0, \ldots), \tag{7.137}$$

where at the particular point $\xi$

$$c = \cos\theta, \qquad s = \sin\theta \qquad \text{and} \qquad \theta = \mathbf{h}^T\boldsymbol{\theta} \tag{7.138}$$

while

$$c_1 = -\mathbf{t}_1^T\frac{d\mathbf{x}'}{d\xi} = -\alpha(1+\varepsilon) \tag{7.139}$$

$$c_2 = \mathbf{t}_2^T\frac{d\mathbf{x}'}{d\xi} = \alpha\gamma \tag{7.140}$$

with

$$\mathbf{s}^T = (0, 0, \mathbf{h}(1), 0, 0, \mathbf{h}(2), 0, \ldots). \tag{7.141}$$

In (7.135) the ordering of the nodal variables would follow an extension from (7.63).

With the aid of (7.135), the (internal) virtual work is given by

$$V_{\mathrm{i}} = \mathbf{q}_{\mathrm{i}}^{\mathrm{T}} \delta \mathbf{p}_{l\mathrm{v}} = \int \mathbf{q}_{l\mathrm{i}}^{\mathrm{T}} \delta \boldsymbol{\varepsilon}_{l\mathrm{v}} \, \mathrm{d}x_l = \int \delta \mathbf{p}_{l\mathrm{v}}^{\mathrm{T}} \mathbf{B}^{\mathrm{T}} \mathbf{q}_{l\mathrm{i}} \, \mathrm{d}\xi \tag{7.142}$$

where $\mathbf{q}_{l\mathrm{i}}$ is given by (7.123) and relates to a particular point $\xi$. It follows that the internal force vector, $\mathbf{q}_{\mathrm{i}}$, is given by

$$\mathbf{q}_{\mathrm{i}} = \int \mathbf{B}^{\mathrm{T}} \mathbf{q}_{l\mathrm{i}} \, \mathrm{d}\xi. \tag{7.143}$$

Following closely the procedure of Section 7.4, the tangent stiffness matrix becomes

$$\mathbf{K}_{\mathrm{t}} = \int \left( \frac{1}{\alpha} \mathbf{B}^{\mathrm{T}} \mathbf{C}_l \mathbf{B} + N(\mathbf{s}\mathbf{z}^{\mathrm{T}} + \mathbf{z}\mathbf{s}^{\mathrm{T}}) + N c_1 \mathbf{s}\mathbf{s}^{\mathrm{T}} - Q(\mathbf{s}\mathbf{r}^{\mathrm{T}} + \mathbf{r}\mathbf{s}^{\mathrm{T}}) - Q c_2 \mathbf{s}\mathbf{s}^{\mathrm{T}} \right) \mathrm{d}\xi. \tag{7.144}$$

The theory can be applied to any order of element. For linear shape functions, it will lead to the same element as that of Section 7.4 provided one-point Gaussian integration is adopted.

## 7.5 AN ISOPARAMETRIC DEGENERATE-CONTINUUM APPROACH USING THE TOTAL LAGRANGIAN FORMULATION

Figure 7.10 shows a three-noded isoparametric degenerate continuum element for which the linear theory can be found in, for example, [C2.2]. The following non-linear theory will relate to a general non-linear isoparametric element [B2.5, B1, S4, W4]. Although the previous formulations in this chapter have involved a two-dimensional beam element in the $x$–$z$ plane, we are now (Figure 7.10) describing an element in the $x$–$y$ plane. This change is introduced in order to correspond with the two-dimensional continuum formulation of Section 5.1. The present degenerate-continuum formulation can be considered as a special case of the latter.

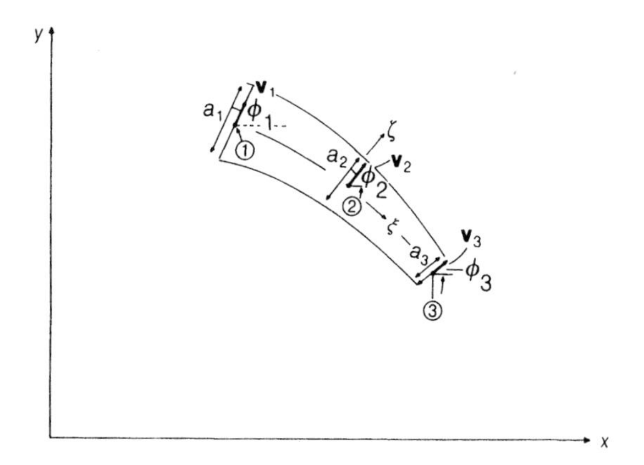


**Figure 7.10** Three-noded degenerate-continuum arch element.

Following standard degenerate-continuum techniques, the geometry can be interpolated (see Figure 7.10) via

$$x = \sum h_i(\xi)\bar{x}_i + \zeta \sum h_i(\xi) a_i \cos\varphi_i = \mathbf{h}^T\bar{\mathbf{x}} + \zeta \mathbf{h}^T \mathbf{v}_x(a\cos\varphi) \tag{7.145}$$

where

$$\mathbf{v}_x(a\cos\varphi)^T = \tfrac{1}{2}(a_1\cos\varphi_1, a_2\cos\varphi_2, \ldots) \tag{7.146}$$

contains the $x$-direction components (see Figure 7.10) of the director vectors, $\mathbf{v}$. In a similar fashion,

$$y = \sum h_i(\xi)\bar{y}_i + \tfrac{1}{2}\zeta \sum h_i(\xi) a_i \sin\varphi_i = \mathbf{h}^T\bar{\mathbf{y}} + \tfrac{1}{2}\zeta \mathbf{h}^T \mathbf{v}_y(a\sin\varphi) \tag{7.147}$$

where $\mathbf{v}_y(a\sin\varphi)$ is a similar vector to that in (7.146) but with sines instead of cosines. The vectors $\bar{\mathbf{x}}$ and $\bar{\mathbf{y}}$ in (7.145) and (7.147) contain the nodal $x$ and $y$ coordinates of the element centre-line.

To be consistent with (7.145) and (7.147), the displacements, $u$ and $v$ can be interpolated in a similar manner, so that

$$u = \sum h_i(\xi)\bar{u}_i + \tfrac{1}{2}\zeta \sum h_i(\xi) a_i(\cos\varphi_i - \cos\varphi_{oi}) \tag{7.148a}$$

$$v = \sum h_i(\xi)\bar{v}_i + \tfrac{1}{2}\zeta \sum h_i(\xi) a_i(\sin\varphi_i - \sin\varphi_{oi}) \tag{7.148b}$$

where $\varphi_{oi}$ are the initial values of $\varphi_i$ and $\bar{u}_i$ and $\bar{v}_i$ are the nodal displacements of the centre-line. At any Gauss point $(\xi, \zeta)$, we can obtain

$$\begin{bmatrix} \dfrac{\partial u}{\partial x} \\ \dfrac{\partial u}{\partial y} \end{bmatrix} = \mathbf{J}^{-1} \begin{bmatrix} \dfrac{\partial u}{\partial \xi} \\ \dfrac{\partial u}{\partial \zeta} \end{bmatrix}, \quad \begin{bmatrix} \dfrac{\partial v}{\partial x} \\ \dfrac{\partial v}{\partial y} \end{bmatrix} = \mathbf{J}^{-1} \begin{bmatrix} \dfrac{\partial v}{\partial \xi} \\ \dfrac{\partial v}{\partial \zeta} \end{bmatrix} \tag{7.149}$$

where $\mathbf{J}$ is the standard Jacobian (see (5.6)) relating to the initial geometry and, for example, from (7.148a),

$$\frac{\partial u}{\partial \xi} = \sum h_{\xi i}\bar{u}_i + \tfrac{1}{2}\zeta \sum h_{\xi i} a_i(\cos\varphi_i - \cos\varphi_{io}). \tag{7.150}$$

Having obtained the shape function and their derivatives, the procedure follows closely that already given in Section 5.1 for the two-dimensional continuum. In particular, from (5.10)–(5.12), the Green strains are given by

$$\mathbf{E} = \begin{bmatrix} \varepsilon_x \\ \varepsilon_y \\ \gamma_{xy} \end{bmatrix} = \begin{bmatrix} 1 & 0 & 0 & 0 \\ 0 & 0 & 0 & 1 \\ 0 & 1 & 1 & 0 \end{bmatrix} \begin{bmatrix} \dfrac{\partial u}{\partial x} \\ \dfrac{\partial u}{\partial y} \\ \dfrac{\partial v}{\partial x} \\ \dfrac{\partial v}{\partial y} \end{bmatrix} + \frac{1}{2} \begin{bmatrix} \dfrac{\partial u}{\partial x} & 0 & \dfrac{\partial v}{\partial x} & 0 \\ 0 & \dfrac{\partial u}{\partial y} & 0 & \dfrac{\partial v}{\partial y} \\ \dfrac{\partial u}{\partial y} & \dfrac{\partial u}{\partial x} & \dfrac{\partial v}{\partial y} & \dfrac{\partial v}{\partial x} \end{bmatrix} \begin{bmatrix} \dfrac{\partial u}{\partial x} \\ \dfrac{\partial u}{\partial y} \\ \dfrac{\partial v}{\partial x} \\ \dfrac{\partial v}{\partial y} \end{bmatrix}$$

$$= [\mathbf{H} + \tfrac{1}{2}\mathbf{A}(\boldsymbol{\theta})]\boldsymbol{\theta}. \tag{7.151}$$

In order to compute the internal forces, we also require (see (5.15))

$$\delta \mathbf{E} = \mathbf{B}_{nl}(\mathbf{p})\delta \mathbf{p} = [\mathbf{H} + \mathbf{A}(\boldsymbol{\theta})]\mathbf{G}\,\delta \mathbf{p} \tag{7.152}$$

where (see (7.151) for $\boldsymbol{\theta}$):

$$\delta \boldsymbol{\theta} = \mathbf{G}\,\delta \mathbf{p} \tag{7.153}$$

and for the full continuum case, $\mathbf{G}$ was given by (5.8). For the present degenerate continuum, the nodal variables in $\delta \mathbf{p}$ are ordered as

$$\delta \mathbf{p}^{\mathrm{T}} = (\delta \mathbf{u}^{\mathrm{T}}, \delta \mathbf{v}^{\mathrm{T}}, \delta \boldsymbol{\varphi}^{\mathrm{T}}) \tag{7.154}$$

where $\delta \mathbf{u}$ contains the nodal values of $\delta \bar{u}$, $\delta \mathbf{v}$ the nodal values of $\delta \bar{v}$ and $\delta \boldsymbol{\varphi}$ the nodal values of $\delta \varphi$ (see the footnote on page 25 regarding the ordering of the variables). In order to obtain $\delta \boldsymbol{\theta}$ (see (7.151) for $\boldsymbol{\theta}$), we require from (7.148),

$$\delta u = \mathbf{h}(\xi)^{\mathrm{T}}\delta \bar{\mathbf{u}} - \tfrac{1}{2}\zeta \mathbf{h}^{\mathrm{T}}\mathbf{D}(a \sin \varphi)\,\delta \boldsymbol{\varphi} \tag{7.155}$$

$$\delta v = \mathbf{h}(\xi)^{\mathrm{T}}\delta \bar{\mathbf{v}} + \tfrac{1}{2}\zeta \mathbf{h}^{\mathrm{T}}\mathbf{D}(a \cos \varphi)\delta \boldsymbol{\varphi} \tag{7.156}$$

where

$$\mathbf{D}(a \sin \varphi) = \begin{bmatrix} a_1 \sin \varphi_1 & & & \\ & a_2 \sin \varphi_2 & & \\ & & a_3 \sin \varphi_3 & \\ & & & \text{etc.} \end{bmatrix} \tag{7.157}$$

with a similar matrix for $\mathbf{D}(a \cos \varphi)$. In order to obtain the matrix $\mathbf{G}$ in (7.153), we must differentiate (7.155) and (7.156) so that, considering (7.155),

$$\frac{\partial \delta u}{\partial \xi} = \mathbf{h}_{\xi}^{\mathrm{T}}\delta \bar{\mathbf{u}} - \tfrac{1}{2}\zeta \mathbf{h}_{\xi}^{\mathrm{T}}\mathbf{D}(a \sin \varphi)\,\delta \boldsymbol{\varphi} \tag{7.158}$$

and

$$\frac{\partial \delta u}{\partial \zeta} = -\tfrac{1}{2}\mathbf{h}^{\mathrm{T}}\mathbf{D}(a \sin \varphi)\,\delta \boldsymbol{\varphi} \tag{7.159}$$

so that (compare the continuum form in (5.9)),

$$\delta \boldsymbol{\theta} = \begin{bmatrix} \dfrac{\partial \delta u}{\partial x} \\[2mm] \dfrac{\partial \delta u}{\partial y} \\[2mm] \dfrac{\partial \delta v}{\partial x} \\[2mm] \dfrac{\partial \delta v}{\partial y} \end{bmatrix} = \frac{1}{2}\begin{bmatrix} 2\mathbf{J}^{-1}(1,1)\mathbf{h}_{\xi}^{\mathrm{T}} & \mathbf{0}^{\mathrm{T}} & \mathbf{b}_1^{\mathrm{T}} \\ 2\mathbf{J}^{-1}(2,1)\mathbf{h}_{\xi}^{\mathrm{T}} & \mathbf{0}^{\mathrm{T}} & \mathbf{b}_2^{\mathrm{T}} \\ \mathbf{0}^{\mathrm{T}} & 2\mathbf{J}^{-1}(1,1)\mathbf{h}_{\xi}^{\mathrm{T}} & \mathbf{b}_3^{\mathrm{T}} \\ \mathbf{0}^{\mathrm{T}} & 2\mathbf{J}^{-1}(2,1)\mathbf{h}_{\xi}^{\mathrm{T}} & \mathbf{b}_4^{\mathrm{T}} \end{bmatrix}\begin{bmatrix} \delta \bar{\mathbf{u}} \\ \delta \bar{\mathbf{v}} \\ \delta \boldsymbol{\varphi} \end{bmatrix} = \mathbf{G}\,\delta \mathbf{p} \tag{7.160}$$

with

$$\mathbf{b}_1 = -\mathbf{D}(a \sin \varphi)(\zeta \mathbf{J}^{-1}(1,1)\mathbf{h}_\xi + \mathbf{J}^{-1}(1,2)\mathbf{h}) \tag{7.161}$$

$$\mathbf{b}_2 = -\mathbf{D}(a \sin \varphi)(\zeta \mathbf{J}^{-1}(2,1)\mathbf{h}_\xi + \mathbf{J}^{-1}(2,2)\mathbf{h}) \tag{7.162}$$

$$\mathbf{b}_3 = \mathbf{D}(a \cos \varphi)(\zeta \mathbf{J}^{-1}(1,1)\mathbf{h}_\xi + \mathbf{J}^{-1}(1,2)\mathbf{h}) \tag{7.163}$$

$$\mathbf{b}_4 = \mathbf{D}(a \cos \varphi)(\zeta \mathbf{J}^{-1}(2,1)\mathbf{h}_\xi + \mathbf{J}^{-1}(2,2)\mathbf{h}). \tag{7.164}$$

The principle of virtual work now leads, with the aid of (7.152), to

$$\mathbf{q}_\mathrm{i} = \int \mathbf{B}_{nl}(\mathbf{p})^\mathrm{T} S \,\mathrm{d}V_\mathrm{o} = \int \mathbf{G}^\mathrm{T}[\mathbf{H} + \mathbf{A}(\boldsymbol{\theta})]^\mathrm{T} \mathbf{S} \,\mathrm{d}V_\mathrm{o} \tag{7.165}$$

where

$$\mathrm{d}V_\mathrm{o} = \det(\mathbf{J})\,\mathrm{d}\xi\,\mathrm{d}\zeta \tag{7.166}$$

and $S$ are the second Piola–Kirchhoff stresses of (5.20). In Section 5.1.2, it was shown that the continuum tangent stiffness matrix of (5.26) could be obtained either from the continuum form of the variation of the virtual work (as in (5.4) from (4.83)) or by direct differentiation of the internal force vector. In the current degenerate case, it is important to adopt the latter procedure in order to be fully consistent with the kinematic assumptions inherent in the model. To this end, differentiation of (7.165) where

$$\delta \mathbf{q}_\mathrm{i} = \int (\mathbf{B}_{nl}^\mathrm{T}\, \delta \mathbf{S} + \delta \mathbf{B}_{nl}^\mathrm{T} \mathbf{S})\,\mathrm{d}V_\mathrm{o} = \mathbf{K}_{\mathrm{t}1}\, \delta \mathbf{p} + [\mathbf{K}_{\sigma 1} + \mathbf{K}_{\sigma 2}]\, \delta \mathbf{p} \tag{7.167}$$

where (see also (5.26))

$$\mathbf{K}_{\mathrm{t}1} = \int \mathbf{B}_{nl}^\mathrm{T} \mathbf{C}_\mathrm{t} \mathbf{B}_{nl}\,\mathrm{d}V_\mathrm{o} = \int \mathbf{G}^\mathrm{T}[\mathbf{H} + \mathbf{A}(\boldsymbol{\theta})]^\mathrm{T} \mathbf{C}_\mathrm{t} \mathbf{G}[\mathbf{H} + \mathbf{A}(\boldsymbol{\theta})]\,\mathrm{d}V_\mathrm{o} \tag{7.168}$$

where the $\mathbf{C}_\mathrm{t}$ matrix in (7.168) relates $\delta\mathbf{S}$ to $\delta\mathbf{E}$ and should reflect the plane-stress assumption. This issue is discussed further in Section 8.2.1 which deals with shells. The $\mathbf{K}_{\sigma 1}$ and $\mathbf{K}_{\sigma 2}$ terms in (7.167) are given by

$$\int \mathbf{G}^\mathrm{T}\, \delta \mathbf{A}(\boldsymbol{\theta})^\mathrm{T} \mathbf{S}\,\mathrm{d}V_\mathrm{o} = \mathbf{K}_{\sigma 1} \delta \mathbf{p} \tag{7.169}$$

while

$$\int \delta \mathbf{G}^\mathrm{T}[\mathbf{H} + \mathbf{A}(\boldsymbol{\theta})]^\mathrm{T} \mathbf{S}\,\mathrm{d}V_\mathrm{o} = \mathbf{K}_{\sigma 2} \delta \mathbf{p}. \tag{7.170}$$

Equation (7.169) leads to

$$\mathbf{K}_{\sigma 1} = \int \mathbf{G}^\mathrm{T} \hat{\mathbf{S}} \mathbf{G}\,\mathrm{d}V_\mathrm{o} \tag{7.171}$$

which is of the same form as $\mathbf{K}_{\mathrm{t}\sigma}$ in (5.26) with $\hat{\mathbf{S}}$ from (5.24).

There is no full continuum equivalent of $\mathbf{K}_{\sigma 2}$. Indeed, this matrix would be missed if one worked directly from the continuum form of the variation of the virtual work (see (4.83) and (5.4)). From (7.170), $\mathbf{K}_{\sigma 2}$ is given by

$$\mathbf{K}_{\sigma 2}\delta\mathbf{p} = \int (S'_1\delta\mathbf{G}_1 + S'_2\delta\mathbf{G}_2 + S'_3\delta\mathbf{G}_3 + S'_4\delta\mathbf{G}_4)\,\mathrm{d}V_o \tag{7.172}$$

where

$$S'_1 = S_x\left(1 + \frac{\partial u}{\partial x}\right) + S_{xy}\frac{\partial u}{\partial y} \tag{7.173}$$

$$S'_2 = S_{xy}\left(1 + \frac{\partial u}{\partial x}\right) + S_y\frac{\partial u}{\partial y} \tag{7.174}$$

$$S'_3 = S_{xy}\left(1 + \frac{\partial v}{\partial y}\right) + S_x\frac{\partial v}{\partial x} \tag{7.175}$$

$$S'_4 = S_y\left(1 + \frac{\partial v}{\partial y}\right) + S_{xy}\frac{\partial v}{\partial x} \tag{7.176}$$

and differentiation of the first column of $\mathbf{G}^{\mathrm{T}}$ in (7.160) gives $\delta\mathbf{G}_1$, etc. It is not difficult to show that, for $k = 1, 4$,

$$\delta\mathbf{G}_k = \begin{bmatrix} \mathbf{0} & \mathbf{0} & \mathbf{0} \\ \mathbf{0} & \mathbf{0} & \mathbf{0} \\ \mathbf{0} & \mathbf{0} & \mathbf{D}_k \end{bmatrix}\delta\mathbf{p} \tag{7.177}$$

where $\mathbf{D}_k$ are diagonal matrices with the $i$th term being given by

$$\mathbf{D}_1(i,i) = -\tfrac{1}{2}(\zeta\mathbf{J}^{-1}(1,1)\mathbf{h}_{\xi i} + \mathbf{J}^{-1}(1,2)\mathbf{h}_i)a_i\cos\varphi_i \tag{7.178}$$

$$\mathbf{D}_2(i,i) = -\tfrac{1}{2}(\zeta\mathbf{J}^{-1}(2,1)\mathbf{h}_{\xi i} + \mathbf{J}^{-1}(2,2)\mathbf{h}_i)a_i\cos\varphi_i \tag{7.179}$$

$$\mathbf{D}_3(i,i) = -\tfrac{1}{2}(\zeta\mathbf{J}^{-1}(1,1)\mathbf{h}_{\xi i} + \mathbf{J}^{-1}(1,2)\mathbf{h}_i)a_i\sin\varphi_i \tag{7.180}$$

$$\mathbf{D}_4(i,i) = -\tfrac{1}{2}(\zeta\mathbf{J}^{-1}(2,1)\mathbf{h}_{\xi i} + \mathbf{J}^{-1}(2,2)\mathbf{h}_i)a_i\sin\varphi_i \tag{7.181}$$

## 7.6 SPECIAL NOTATION

$a_i$ = thickness at node $i$ (Section 7.5)
$A$ = area of beam
$\mathbf{A}$ = matrix (7.67) connecting $\delta\boldsymbol{\theta}_l$ with $\delta\mathbf{p}$ (Section 7.2)
$\mathbf{A}(\boldsymbol{\theta})$ = matrix of displacement-derivatives ((7.151) in Section 7.5)
$\mathbf{b}$ = vector connecting curvature, $\chi$, to $\boldsymbol{\theta}_l$ (7.84)
$\mathbf{b}_u$ = differential of $\mathbf{h}_u$ with respect to $x$
$\mathbf{b}_w$ = differential of $\mathbf{h}_w$ with respect to $x$
$\mathbf{B}$ = matrix connecting $\delta\mathbf{p}_l$ or $l_o\delta\varepsilon_l$ to $\delta p$ ((7.68) or (7.92) for Kirchhoff theory; (7.99) for Timoshenko theory; (7.118) for Reissner theory)
$c = \cos\beta$ (Section 7.2.3), $= \cos\theta$ (Section 7.4.1)
$\mathbf{c}$ = vector relating curvature, $\chi$, to $\mathbf{w}$
$\mathbf{C}_l$ = local constitutive matrix
$\mathbf{d}$ = displacement vector

$\mathbf{d}_{21} = \mathbf{d}_2 - \mathbf{d}_1; \mathbf{d}_i$ = displacement vector at node $i$
$\mathbf{e}_1$ = unit vector between nodes of beam element
$\mathbf{e}_2$ = unit vector orthogonal to $\mathbf{e}_1$
$\mathbf{E}$ = Green's strain (Section 7.5)
$\mathbf{G}$ = matrix connecting $\delta\boldsymbol{\theta}$ (displacement derivatives) to $\delta\mathbf{p}$ (Section 7.5)
$\mathbf{h}_u$ = shape function vector for $\bar{u}$ displacement
$\mathbf{h}_w$ = shape function vector for $w$ displacement
$\mathbf{h}_\theta$ = shape function vector for $\theta$
$\mathbf{H}$ = Boolean matrix (Section 7.5)
$I$ = second moment of area
$l$ = initial 'length' of beam element (between nodes) in Section 7.1
$l_o$ = initial 'length' of beam element, $l_n$ = current 'length'
$M$ = bending moment in beam
$\bar{M}_1$ = internal bending moment in local system at node 1; $\bar{M}_2$ at node 2
$N$ = axial force in beam
$\mathbf{p}$ = vector of nodal 'displacements'
ordering such that $\mathbf{p}^T = (\mathbf{u}^T, \mathbf{w}^T)$ in Sections 7.1–7.1.3
ordering such that $\mathbf{p}^T = (\mathbf{u}^T, \mathbf{w}^T, \boldsymbol{\theta}^T)$ in Section 7.1.4
ordering as in (7.63) for Sections 7.2–4
ordering as in (63) for Section 7.5
$Q$ = transverse shear force
$r$ = radius of gyration
$\mathbf{r}$ = special geometry vector (7.62)
$s = \sin\beta$ (Section 7.2.3), $= \sin\theta$ (Section 7.4.1)
$\mathbf{s}$ = vector connecting $\theta_l$ to the nodal values, $\boldsymbol{\theta}_l$ (Section 7.2.6)
$\mathbf{s}$ = special vector (7.119) for Reissner theory
$\mathbf{S}$ = second Piola–Kirchhoff stresses (Section 7.5)
$\mathbf{t}_1, \mathbf{t}_2$ = unit vectors for Section 7.4
$\mathbf{T}$ = vector of nodal forces corresponding to $\boldsymbol{\theta}$ (Section 7.1.4)
$\bar{u}$ = axial displacement at reference plane
$\mathbf{u}$ = vector of in-plane nodal displacements at reference plane ($\bar{u}$)
$\mathbf{U}$ = nodal forces corresponding to $\mathbf{u}$
$\mathbf{v}_i$ = unit director vectors at node $i$ (Section 7.5)
$w' = w + z$
$\mathbf{w}$ = vector of out-of-plane nodel 'displacements' (7.11a) in Sections 7.1–7.1.3 ($w$s and $\theta$s)
$\mathbf{w}$ = vector of nodal $w$s in Section 7.1.4
$\mathbf{w}' = \mathbf{w} + \mathbf{z}$
$\mathbf{W}$ = nodal forces corresponding to $\mathbf{w}$
$x_{21} = x_2 - x_1$
$\mathbf{x}_1$ = initial position vector of node 1, $\mathbf{x}_2$ for node 2
$\mathbf{x}_{21} = \mathbf{x}_2 - \mathbf{x}_1$
$\mathbf{x}' = \mathbf{x} + \mathbf{d}$
$\bar{\mathbf{x}}$ = vector containing nodal values of $x$ at element centre-line (Section 7.5)
$\bar{\mathbf{y}}$ = vector containing nodal values of $y$ at element centre-line (Section 7.5)
$\mathbf{z}$ = vector of initial out-of-plane variables (7.11b) for Section 7.1
$\mathbf{z}$ = special geometry vector (7.66) for Sections 7.2 and 7.3

$\alpha$ = rigid-body rotation angle of beam (Sections 7.1 and 7.2)
$\alpha$ = 'length parameter' (7.129) in Section 7.4.1
$\alpha_x = dz/dx$
$\beta$ = final orientation angle for beam element (Section 7.2)
$\theta = dw/dx$ for Kirchhoff theory
$\theta$ = rotation of normal for Timoshenko theory
$\theta$ = rotation parameters (Figure 7.9) for Reissner theory
$\theta_l = \theta$ in local frame with $\theta_{l1}$ and $\theta_{l2}$ at nodes; $\theta_{l1_0}$ and $\theta_{l2_0}$ are initial values of $\theta_{l1}$ and $\theta_{l2}$ respectively
$\theta_x = dw/dx$
$\boldsymbol{\theta}$ = nodal values of $\theta$
$\boldsymbol{\theta}$ = vector form of displacement derivatives as in (7.151) (Section 7.5)
$\varphi_i$ = angular orientation of $\mathbf{v}_i$ (Section 7.5)
$\varphi_{io}$ = initial orientation of $\mathbf{v}_i$ (Section 7.5)
$\delta\boldsymbol{\varphi}$ = vector of nodal values of $\delta\varphi$ (Section 7.5)
$\chi$ = curvature (Section 7.1)
$\boldsymbol{\chi}$ = curvature vector (Section 7.2)

## Subscript

$l$ = local or linear
nl = non-linear

## Superscript

$\bar{}$ = 'at reference plane'
$\hat{}$ = quantity divided by $l$ (Section 7.1)

## 7.7 REFERENCES

[A1] Argyris, J. H., Balmer, H., Doltsinis, J. St., Dunne, P. C., Haase, M., Klieber, M., Malejannakis, G. A., Mlejenek, J. P., Muller, M. & Scharpf, D. W., Finite element method—the natural approach, *Comp. Meth. Appl. Mech. & Engng.*, **17/18**, 1–106 (1979).
[B1] Bathe, K. J. & Bolourchi, S., Large displacement analysis of three-dimensional beam structures, *Int. J. Num. Meth. Engng.*, **14**, 961–986 (1975).
[B2] Belytschko, T. & Hseih, B. J., Non-linear transient finite element analysis with convected co-ordinates, *Int. J. Num. Meth. Engng.*, **7**, 255–271 (1973).
[B3] Belytschko, T. & Glaum, L. W., Applications of higher order corotational stretch theories to nonlinear finite element analysis, *Computers & Structures*, **10**, 175–182 (1979).
[B4] Brink, K. & Kratzig, W. B., Geometrically correct formulations for curved finite bar elements under large deformations, *Nonlinear Finite Element Analysis in Structural Mechanics*, ed. W. Wunderlich *et al.*, Springer-Verlag, Berlin, pp. 236–256 (1981).
[B5] Burgoynne, C. & Crisfield, M. A., Numerical integration strategy for plates and shells, *Int. J. Num. Meth. Engng.*, 105–121 (1990).
[C1] Cole, G., Consistent co-rotational formulations for geometrically non-linear beam elements with special reference to large rotations, Ph.D. Thesis, Kingston Polytechnic, Surrey (1990).

[C2] Cowper, G. R., The shear coefficient in Timoshenko's beam theory, *J. Appl. Mech.*, **33**, 335–340 (1966).

[C3] Crisfield, M. A. & Cole, G., Co-rotational beam elements for two- and three-dimensional non-linear analysis, *Discretisation Methods in Structural Mechanics*, ed. G. Kuhn *et al.*, Springer-Verlag, pp. 115–124 (1989).

[C4] Crisfield, M. A. & Puthli, R. S., Approximations in the non-linear analysis of thin plated structures, *Finite Elements in Non-linear Mechanics*, Vol. I, ed. P. G. Bergan *et al.*, Tapir Press, Trondheim, Norway, pp. 373–392 (1978).

[C5] Crisfield, M. A. Large-deflection elasto-plastic buckling analysis of eccentrically stiffened plates using finite elements, Transport & Road Research Laboratory Report, LR 725, Crowthorne, Berks., England (1976).

[C6] Crisfield, M. A., A consistent co-rotational formulation for non-linear, three-dimensional beam elements, *Comp. Meth. Appl. Mech. & Engng.*, **81**, 131–150 (1990).

[C7] Crisfield, M. A., The 'eccentricity issue' in the design of beam, plate and shell elements, *Communications in Appl. Num. Meth.*, **7**, 47–56 (1991).

[D1] Dawe, D. J., Some higher order elements for arches and shells, *Finite Elements for Thin Shells and Curved Members*, ed. D. G. Ashwell *et al.*, Wiley, London, pp. 131–153 (1976).

[E1] Epstein, M. & Murray, D. W., Large deformation in-plane analysis of thin walled beams, *Computers and Structures*, **6**, 1–9 (1976).

[F1] Frey, F. & Cescotto, S., Some new aspects of the incremental total Lagrangian description in nonlinear analysis, *Finite Elements in Nonlinear Structural Mechanics*, Tapir, Trondheim (1978).

[G1] Gupta, A. K. & Ma, P. S., Error in eccentric beam formulation, *Int. J. Num. Meth. Engng.*, **11**, 1473–1477 (1977).

[H1] Hsiao, K. M. & Hou, F. Y., Nonlinear finite element analysis of elastic frames, *Computers and Structures*, **26**, 693–701 (1987).

[H2] Haefner, L. & Willam, K. J., Large deflection formulations of a simple beam element including shear deformations, *Engineering Computations*, **1**, 359–368 (1984).

[K1] Karamanlidis, D., Honecker, A. & Knothe, K., Large deflection finite element analysis of pre- and postcritical response of thin elastic frames, *Nonlinear Finite Element Analysis in Structural Mechanics*, ed. W. Wunderlich *et al.*, Springer-Verlag, Berlin, pp. 217–235 (1981).

[M1] Mattiason, K., On the co-rotational finite element formulation for large deformation problems, Pub. 83: 1, Dept. of Structural Mechanics, Chalmers Univ. of Technology (1983).

[M2] Moan, T. & Soreide, T., Analysis of stiffened plates considering nonlinear material and geometric behaviour, *Proc. World Cong. on Finite Element Methods in Struct. Mech.*, ed. J. Robinson, Bournemouth, England, Vol. II, pp. 14.1–14.28 (1975).

[N1] Nour-Omid, B. & Rankin, C. C., Finite rotation analysis and consistent linearisation using projectors, *Comp. Meth. Appl. Mech. & Engng.* (to be published).

[O1] Oran, C. & Kassimali, A., Large deformations of framed structures under static and dynamic loads, *Computers & Structs.*, **6**, 539–547 (1976).

[O2] Oran, C., Tangent stiffness in space frames, *Proc. ASCE, J. Struct. Div.*, **99**, ST6, 973–985 (1973).

[P1] Prathap, G., The curved beam/deep arch/finite ring element revisited, *Int. J. Num. Meth. Engng.*, **21**, 389–407 (1985).

[R1] Rankin, C. C. & Brogan, F. A., An element independent corotational procedure for the treatment of large rotations, *Collapse Analysis of Structures*, ed. L. H. Sobel & K. Thomas, *ASME*, New York, 85–100 (1984).

[R2] Reissner, E., On one-dimensional, large-displacement, finite-strain beam theory, *Stud. Appl. Math.*, **52**, 87–95 (1973).

[S1] Simo, J. C., A finite strain beam formulation. The three dimensional dynamic problem: Part 1, *Comp. Meth. Appl. Mech. & Engng.*, **49**, 55–70 (1985).

[S2] Simo, J. C. & Vu-Quoc, L., A three-dimensional finite strain rod model: Part 2: Computational aspects, *Comp. Meth. Appl. Mech. & Engng.*, **58**, 79–116 (1986).

[S3] Stolarski, H. & Belytschko, T., Membrane locking and reduced integration for curved elements, *J. Appl. Mech.*, **49**, 172–176 (1982).

[S4] Surana, K. S., Geometrically non-linear formulation for two-dimensional curved beam elements, *Computers & Structures*, **17**, 105–114 (1983).

[T1] Timoshenko, S., On the correction for shear of the differential equation for transverse vibration of prismatic bars, *Phil. Mag.*, **41**, 744–746 (1921).

[T2] Tang, S. C., Yeung, K. S. & Chon, C. T., On the tangent stiffness matrix in a convected coordinate system, *Computers & Structures*, **12**, 849–856 (1980).

[W1] Washizu, K., *Variational Methods in Elasticity and Plasticity*, 2nd edition, Pergamon Press, Oxford (1975).

[W2] Wempner, G., Finite elements, finite rotations and small strains of flexible shells, *Int. J. Solids & Structs.*, **5**, 117–153 (1969).

[W3] Wen, R. K. & Rahimzadeh, J., Nonlinear elastic frame analysis by finite element, *Proc. ASCE, J. Struct. Div.*, **109**(8), 1952–1971 (1983).

[W4] Wood, R. D. & Zienkiewicz, O. C., Geometrically nonlinear finite element analysis of beams, frames, arches and axisymmetric shells, *Computers & Structures*, **7**, 725–735 (1977).

# 8 Shells

Calladine [C1] gives an ineresting review of the developments of theories for shell structures. We are here specifically concerned with finite element analysis. For linear analysis, some of the earliest work involved facet formulations [C3, Z2]. This concept was extended to non-linear analysis by Horrigmoe and Bergan [H1] using a corotational approach (although in the paper, [H1], the term 'updated Lagrangian' was used). Another early facet formulation, using a corotational approach, was given by Backlund [B1] who used the Morley triangle [M4] which has mid-side rotational connectors. As discussed in the previous chapter, a corotational approach can be suppelemented by the addition of local shallow shell terms from Marguerre's theory [M2.4]. Such procedures have been applied to shells by, amongst others, Jetteur *et al.* [J2, J3] and Stolarski *et al.* [S7].

Probably the majority of work on non-linear shell elements has followed on from the linear work of Ahmad *et al.* [A1] using the degenerate-continuum approach with a total or updated Lagrangian formulation [W4.7, B2.5, B2, D1, F1, S4, S8, M2, P2, R1, R2, D1, H2] or using some 'incremental' (rate) form of strain measure [H3, L1, B3–B6, S5–S6] (often related to the corotational approach [B6]). As with linear analysis, problems occur with shear locking and, again as with linear analysis, these can be ameliorated by various forms of (selective) reduced integration. [Z1, P3, H3, H5, B3, B4, B6, S6]. Alternatively, substitute shape-functions can be used (possibly in reltion to the covariant strain components [D1, H2, J1]). Membrane locking (see Sections 7.1.3 and 7.1.5) can also occur and, again reduced integration [B3, B4, B6, S6] or substitute functions (possibly via 'stress projection' [B7]) [H2, B7, S6] can help. Alternatively (or additionally), some form of corotational (rate) strain measure may be used at the integration points [B6, S6]. Reduced integration can lead to problems with mechanisms. These can be overcome using some form of 'stabilisation technique' [B3, B5].

The isoparametric degenerate-continuum approach adopts shape functions for the components of displacement in a fixed rectangular cartesian system. Consequently, it allows the exact satisfaction of the rigid-body modes even when the plane-section constraint is applied in the through-thickness direction [C2.2]. However, this is achieved only when the continuum approach is adopted throughout and the various shape-function manipulations involving the Jacobian are applied at all of the integration points including those in the 'through-thickness direction'.

Considerable savings in computer time can be achieved by using the through-thickness integration only for the treatment of material non-linearity. In relation to a beam (although not, specifically, in relation to a degenerate-continuum approach), the ideas were discussed in Section 7.1.2. In order to introduce such a technique one must work with membrane strains, $\bar{\varepsilon}$, and curvatures, $\chi$, and the corresponding 'stress resultants', $\mathbf{N}$ and $\mathbf{M}$. In relation to linear analysis, a simple procedure was advocated by Zienkiewicz *et al.* [Z1] for divorcing the through-thickness integration from the main shape-function procedure. As noted in [I1, C2.2, C9, M2], the simple approach of [Z1] can lead to stressing under rigid-body movement (although this drawback should vanish in the limit as the mesh is refined). Milford and Schnobrich have proposed a solution [M2] involving a truncated Taylor series for the inverse Jacobian. Much further work has followed with an emphasis on non-linear analysis and the development of a 'continuum-based resultant form' [S5, S6, L1, B3].

As with linear analysis, an important issue in relation to the design of non-linear shell elements is that of the 'sixth degree of freedom' or 'drilling rotation'. In a conventional formulation for a membrane plate, we have locally five degrees of freedom including two (out-of-plane) rotations. If some form of coordinate transformation is adopted (say between local and global coordinates), we can then end up with six degrees of freedom at the structural level (now with three rotations). If all of the elements lie in the same plane, a mechanism results. To overcome this problem, Zienkiewicz suggested a fictitious spring stiffness relating to the (local) in-plane rotations [Z2.1]. (An alternative formulation using a form of integrated spring stiffness has recently been suggested by Providos [P4] and has been found to give good results.) While the fictitious spring approach may work adequately for linear problems, there are more dangers with non-linear analysis when, with material non-linearity, the stiffnesses of the elements can vary significantly.

It is possible for a smooth shell to work with only five variables by setting up some averaged system at the nodes with respect to which the out-of-plane rotations are taken. This approach was suggested, in relation to a facet analysis by Horrigmoe and Bergan [H1]. Also, as will be shown in Section 8.2, it is possible to work with only five variables for a total Lagrangian continuum formulation. In this work, which follows that of Ramm and Matzenmiller [R2], the two 'rotation variables' are not rotations in the true sense but rather define the orientation of the director (which is of unit length).

In order to encompass both smooth and non-smooth shells, some formulations use five degrees of freedom for the smooth regions and six for those involving junctions [C12]. This of course leads to complexities in the 'house-keeping'. An alternative that is currently receiving much research interest is to always use six variables, with the local in-plane rotation being included as a 'true' rather than a 'fictitious' connector [A3, C2, J2, J3, B8, N1, T1]. When rotations are used as variables, we also have to consider their non-commutativity when they become large [A4]. This issue is even more relevant to three-dimensional beams and rods and will be considered in Volume 2.

This brief review of previous work on non-linear shell analysis has inevitably been somewhat cursory. Some recent papers that have not been discussed can be found in [S1–S3, P1, W2]. The reader might also refer to the review paper by Wempner [W1] and the books edited by Hughes and Hinton [H4].

The present chapter will concentrate on two procedures for non-linear shell analysis;

firstly a shallow-shell formulation and secondly the degenerate-continuum approach. The former follows on directly from the work on beams in Section 7.1. An important class of problems that can be treated by the shallow-shell formulation is that of imperfect steel plates. Such plates (often stiffened) form the main components of a wide range of structures from bridges to dock gates and ship decks. Appropriate non-linear analysis usually requires a combination of geometric and material effects. The author has applied such analyses, using the shallow-shell formulation, to both stiffened and unstiffened imperfect plates with a view to the assessment of the strength of bridges [C3.6, C7, C8, C10, C11].

While the range of application of the shallow-shell approach is a little limited, along with the shallow-arch formulations of Section 7.1, they have an important teaching role. Also, as indicated in Section 8.1.7, their range of application can be extended.

The second area covered in this chapter is that of the degenerate-continuum approach. As already discussed, there has been a mass of research work in this area and there are still many problems to be overcome. In the present chapter, we will concentrate on one aspect only, that of 'consistency' in relation to a total Lagrangian approach. The earliest approaches [B2.5, B2, R1] discretised the problem after setting up the continuum expressions involving the tangent stiffness matrix. As a consequence, they did not produce a 'consistent tangent stiffness matrix'. This limitation was partly corrected by Surana [S8] and, later, in an unambiguous manner, by Ramm and Matzenmiller [R2]. The presentation in this chapter will be closely related to the latter.

## 8.1 A RANGE OF SHALLOW SHELLS

In this section, we will extend the shallow-arch formulation of Section 7.1 to a shallow shell. In contrast to Section 7.1, where we started with a 'Kirchhoff formulation', we will here directly introduce shear deformation at the very start and thus adopt a form of Mindlin–Reissner analysis [M2, R2, C2.2]. This formulation will provide a starting point even when we exclude shear deformation and adopt a form of Kirchhoff bending because, following the work in [C5, C6, C2.2], we will introduce the latter using a form of 'discrete Kirchhoff' hypothesis.

### 8.1.1 Strain–displacement relationships

From the assumption that plane sections remain plane, we can extend (7.3) (see also (4.84)) to give

$$\mathbf{E} = \bar{\mathbf{E}} + z_l \boldsymbol{\chi} = \bar{\mathbf{E}}_l + \mathbf{E}_{nl} + z_l \boldsymbol{\chi} \tag{8.1}$$

where the $\bar{\mathbf{E}}$ terms relate to the reference plane (possibly but not necessarily, the centre of the shell) and

$$\bar{\mathbf{E}}_l^{\mathrm{T}} = \left(\frac{\partial \bar{u}}{\partial x}, \frac{\partial \bar{v}}{\partial y}, \frac{\partial \bar{u}}{\partial y} + \frac{\partial \bar{v}}{\partial x}\right) \tag{8.2}$$

where $\bar{u}$ and $v$ are the in-plane displacements at the reference plane and (see (7.4) and (4.84)),

$$\bar{\mathbf{E}}_{nl} = \begin{bmatrix} \frac{1}{2}\left(\frac{\partial(w+z)}{\partial x}\right)^2 \\ \frac{1}{2}\left(\frac{\partial(w+z)}{\partial y}\right)^2 \\ \frac{\partial(w+z)}{\partial x}\frac{\partial(w+z)}{\partial y} \end{bmatrix} - \begin{bmatrix} \frac{1}{2}\left(\frac{\partial z}{\partial x}\right)^2 \\ \frac{1}{2}\left(\frac{\partial z}{\partial y}\right)^2 \\ \frac{\partial z}{\partial x}\frac{\partial z}{\partial y} \end{bmatrix} \tag{8.3}$$

where $z$ is the initial vertical coordinate of the shell reference plane and $w$ the subsequent deformation (see Figure 8.1 where the variables $\theta_x$ and $\theta_y$ refer to the rotations of the normal). Using the Mindlin–Reissner approach, the curvatures are given by

$$\boldsymbol{\chi}^{\mathrm{T}} = \left(\frac{\partial\theta_x}{\partial x}, \frac{\partial\theta_y}{\partial y}, \frac{\partial\theta_x}{\partial y} + \frac{\partial\theta_y}{\partial x}\right). \tag{8.4}$$

Equation (8.4) takes a very similar form to (8.2). For the virtual work, we also require the virtual form of (8.1) which is (see also (4.85)):

$$\delta\mathbf{E}_{\mathrm{v}} = \delta\bar{\mathbf{E}}_{\mathrm{v}} + z_l\delta\boldsymbol{\chi}_{\mathrm{v}}$$

$$= \delta\bar{\mathbf{E}}_{l\mathrm{v}} + \mathbf{T}\,\delta\mathbf{s}_{\mathrm{v}} + z_l\delta\boldsymbol{\chi}_{\mathrm{v}} = \delta\bar{\mathbf{E}}_{l\mathrm{v}} + \begin{bmatrix} \frac{\partial(w+z)}{\partial x} & 0 \\ 0 & \frac{\partial(w+z)}{\partial y} \\ \frac{\partial(w+z)}{\partial y} & \frac{\partial(w+z)}{\partial x} \end{bmatrix} \begin{bmatrix} \frac{\partial\delta w_{\mathrm{v}}}{\partial x} \\ \frac{\partial\delta w_{\mathrm{v}}}{\partial y} \end{bmatrix} + z_l\delta\boldsymbol{\chi}_{\mathrm{v}} \tag{8.5}$$

where $\delta\bar{\mathbf{E}}_{l\mathrm{v}}$ and $\delta\boldsymbol{\chi}_{\mathrm{v}}$ take similar forms to (8.2) and (8.4) respectively.

The vertical shear strains will be expressed in standard form as

$$\boldsymbol{\gamma} = \begin{pmatrix} \gamma_{xz} \\ \gamma_{yz} \end{pmatrix} = \begin{pmatrix} \theta_x \\ \theta_y \end{pmatrix} + \begin{bmatrix} \frac{\partial w}{\partial x} \\ \frac{\partial w}{\partial y} \end{bmatrix} = \boldsymbol{\theta} + \mathbf{s}. \tag{8.6}$$

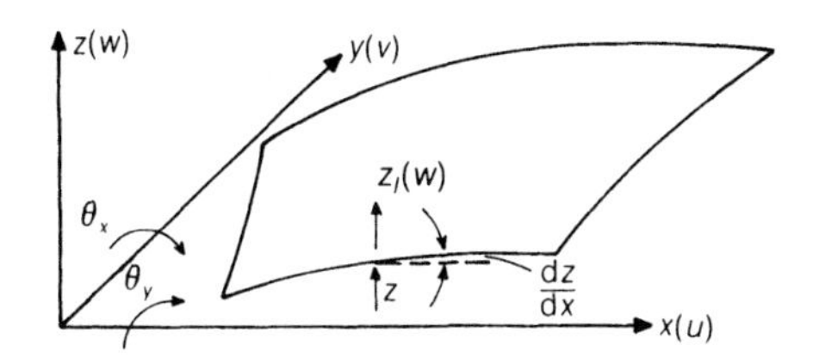


**Figure 8.1** Coordinate system for a shallow shell.

### 8.1.2 Stress–strain relationships

Using a similar procedure to that adopted for the arch (Section 7.1.2), we will not necessarily assume that the reference plane is at the centre of the section. We will also allow for the possibility of material non-linearity so that the stress–strain relationships at any depth $z_l$ are taken to be tangential, so that

$$\delta\mathbf{S} = \mathbf{C}_t\,\delta\mathbf{E} = \mathbf{C}_t(\delta\bar{\mathbf{E}} + z_l\,\delta\boldsymbol{\chi}) = \frac{E}{(1-\nu^2)}\begin{bmatrix} 1 & \nu & 0 \\ \nu & 1 & 0 \\ 0 & 0 & (1+\nu)/2 \end{bmatrix}(\delta\bar{\mathbf{E}} + z_l\,\delta\boldsymbol{\chi}) \tag{8.7}$$

where $\delta\mathbf{S}^{\mathrm{T}} = (\delta S_x, \delta S_y, \delta S_{xy})$ are the second Piola–Kirchhoff stresses at any depth $z_l$. In the last relationship in (8.7), we have written $\mathbf{C}_t$ using the linear-elastic, isotropic relationship for plane-stress conditions (Section 4.2.1). With plasticity, we would, instead, use the tangential elasto-plastic modular matrix (preferably 'consistent'—see Section 6.8.2.1) under conditions of 'plane stress'.

Using a very similar procedure to that adopted in one dimension in equations (7.26) and (7.27), we now integrate (8.7) through the thickness $z_l$ to obtain the relationships for the stress resultants as

$$\begin{aligned}\delta\mathbf{N} &= \int \delta\mathbf{S}\,\mathrm{d}z_l = \int \mathbf{C}_t(\delta\bar{\mathbf{E}} + z_l\,\delta\boldsymbol{\chi})\,\mathrm{d}z_l \\ &= \int \mathbf{C}_t\,\mathrm{d}z_l\,\delta\bar{E} + \int \mathbf{C}_t z_l\,\mathrm{d}z_l\,\delta\boldsymbol{\chi} = \mathbf{C}_{\mathrm{m}}\,\delta\bar{\mathbf{E}} + \mathbf{C}_{\mathrm{mb}}\delta\boldsymbol{\chi}\end{aligned} \tag{8.8}$$

$$\begin{aligned}\delta M &= \int \delta\mathbf{S} z_l\,\mathrm{d}z_l = \int \mathbf{C}_t z_l(\delta\bar{\mathbf{E}} + z_l\,\delta\boldsymbol{\chi})\,\mathrm{d}z_l \\ &= \int \mathbf{C}_t z_l\,\mathrm{d}z_l\,\delta\bar{\mathbf{E}} + \int \mathbf{C}_t z_l^2\,\mathrm{d}z_l\,\delta\boldsymbol{\chi} = \mathbf{C}_{\mathrm{mb}}^{\mathrm{T}}\,\delta\bar{\mathbf{E}} + \mathbf{C}_{\mathrm{b}}\delta\boldsymbol{\chi}.\end{aligned} \tag{8.9}$$

Equations (8.8) and (8.9) define the membrane constitutive matrix, $\mathbf{C}_{\mathrm{m}}$, the bending constitutive matrix, $\mathbf{C}_{\mathrm{b}}$ and the cross-coupling matrix, $\mathbf{C}_{\mathrm{mb}}$. All three of these matrices should be considered as 'tangential' although, to save space, the subscript t has been dropped. A discussion of the numerical integrations in (8.8) and (8.9) has already been given for the one-dimensional case in Section 7.1.2. The arguments remain valid for the present two-dimensional analyses.

To supplement (8.8) and (8.9), we will assume tangential vertical shear relationships of the form

$$\delta\mathbf{Q} = \int \begin{pmatrix} \delta\tau_{xz} \\ \delta\tau_{yz} \end{pmatrix} \mathrm{d}z_l = \mathbf{C}_y\delta\boldsymbol{\gamma} = \alpha G t \begin{bmatrix} 1 & 0 \\ 0 & 1 \end{bmatrix} \delta\boldsymbol{\gamma}. \tag{8.10}$$

In the last relationship in (8.10), we have assumed a linear-elastic, isotropic relationship with $\alpha$ as the 'shear factor' (usually taken as $\frac{5}{6}$ or $\pi^2/12$ [C2.7]). The author has applied (8.10) to the analysis of concrete bridges using an approximate non-linear relationship for the shear stiffnesses [C8].

### 8.1.3 Shape functions

The present shallow shell theory can be applied with a range of different elements with different shape functions. We will here concentrate on applications involving (a) standard isoparametric functions related to a standard Mindlin–Reissner formulation, and (b) a 'constrained Mindlin–Reissner' [C6, C2.2] and (c) a 'discrete Kirchhoff formulation [C5, C2.2] adopted by the author. In the latter cases ((b) and (c)), only an outline is given, the precise form of the constraints having already been described for linear applications in the references given previously.

For all of these analyses, one can assume that the five variables $\bar{u}, \bar{v}, w, \theta_x$ and $\theta_y$ are initially expanded using the same hierarchical shape functions (so would the initial geometry) so that

$$\bar{u} = \mathbf{h}^{\mathrm{T}}\bar{\mathbf{u}}, \quad \bar{v} = \mathbf{h}^{\mathrm{T}}\bar{\mathbf{v}}, \quad w = \mathbf{h}^{\mathrm{T}}\mathbf{w}; \qquad \theta_x = \mathbf{h}^{\mathrm{T}}\boldsymbol{\theta}_x, \quad \theta_y = \mathbf{h}^{\mathrm{T}}\boldsymbol{\theta}_y. \tag{8.11}$$

For the present purposes, one can consider these functions to be either serendipity (8-noded) or Lagrangian (9-noded)—see Figure 8.2. Also, it is most convenient to think of the shape functions as being of a hierarchical form [C2.2] so that the mid-side and central variables are 'relative' and can therefore more easily be constrained out at some later time. Collectively, the nodal variables $(\bar{\mathbf{u}}, \bar{\mathbf{v}}, \mathbf{w}, \boldsymbol{\theta}_x, \boldsymbol{\theta}_y)$ will, as usual, be referred to as $\mathbf{p}$. For some algebraic expressions it is also useful to refer to the membrane nodal displacements $\bar{\mathbf{u}}$ and $\bar{\mathbf{v}}$, collectively as $\mathbf{p}_{\mathrm{m}}$ and the normal rotations $\theta_x$ and $\theta_y$ collectively as $\mathbf{p}_\theta$.

Using these shape functions, we can use the Jacobian in the standard way to obtain

$$\bar{\mathbf{E}}_l = \mathbf{B}_{\mathrm{m}}\mathbf{p}_{\mathrm{m}}, \qquad \delta\bar{\mathbf{E}}_l = \mathbf{B}_{\mathrm{m}}\delta\mathbf{p}_{\mathrm{m}} \tag{8.12}$$

$$\boldsymbol{\chi} = \mathbf{B}_{\mathrm{b}}\mathbf{p}_\theta, \quad \delta\boldsymbol{\chi} = \mathbf{B}_{\mathrm{b}}\delta\mathbf{p}_\theta \tag{8.13}$$

where with the same shape functions for $\bar{u}, \bar{v}, \theta_x$ and $\theta_y$, $\mathbf{B}_{\mathrm{m}}$ and $\mathbf{B}_{\mathrm{b}}$ will take the same form. Also the slopes, $\mathbf{s}$ (see (8.6)) and slope changes, $\delta\mathbf{s}$, can be obtained as

$$\mathbf{s} = \mathbf{B}_{\mathrm{s}}\mathbf{w}, \qquad \delta\mathbf{s} = \mathbf{B}_{\mathrm{s}}\delta\mathbf{w} \tag{8.14}$$

so that from (8.6) and (8.14),

$$\boldsymbol{\gamma} = \boldsymbol{\theta} + \mathbf{s} = \mathbf{H}\,\delta\mathbf{p}_\theta + \mathbf{B}_{\mathrm{s}}\mathbf{w}, \qquad \delta\boldsymbol{\gamma} = \delta\boldsymbol{\theta} + \delta\mathbf{s} = \mathbf{H}\,\delta\mathbf{p}_\theta + \mathbf{B}_{\mathrm{s}}\,\delta\mathbf{w} \tag{8.15}$$

where the $\mathbf{H}$ matrix in (8.15) is simply composed of terms from the shape functions

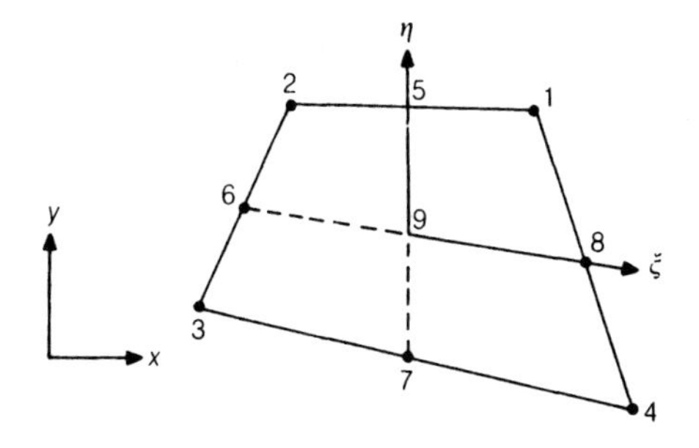


**Figure 8.2** 8- or 9-noded element.

**h** of (8.11). Finally, from (8.5), (8.12) and (8.14),

$$\delta\bar{\mathbf{E}} = \delta\bar{\mathbf{E}}_l + \delta\bar{\mathbf{E}}_{nl} = \mathbf{B}_{\mathrm{m}}\delta\mathbf{p}_{\mathrm{m}} + \mathbf{T}(\mathbf{w})\,\delta\mathbf{s} = \mathbf{B}_{\mathrm{m}}\delta\mathbf{p}_{\mathrm{m}} + \mathbf{T}(\mathbf{w})\mathbf{B}_{\mathrm{s}}\,\delta\mathbf{w}. \tag{8.16}$$

The only non-linearity in the relationships of (8.13), (8.15) and (8.16) is in (8.16) and comes from the **T** matrix which, from (8.5) and via (8.14), is a function of the current nodal displacements, **w**.

Further discussion on the issue of matching shape functions and numerical integration will be given in Section 8.1.5.

### 8.1.4 Virtual work and the internal force vector

The virtual work can be expressed as

$$V = \int (\mathbf{S}^{\mathrm{T}}\delta\mathbf{E} + \mathbf{Q}^{\mathrm{T}}\delta\boldsymbol{\gamma})\,\mathrm{d}V_{\mathrm{o}} - \mathbf{q}_{\mathrm{e}}^{\mathrm{T}}\delta\mathbf{p}_{\mathrm{v}} = \int (\mathbf{N}^{\mathrm{T}}\delta\bar{\mathbf{E}}_{\mathrm{v}} + \mathbf{M}^{\mathrm{T}}\delta\boldsymbol{\chi}_{\mathrm{v}} + \mathbf{Q}^{\mathrm{T}}\delta\boldsymbol{\gamma})\,\mathrm{d}A_{\mathrm{o}} - \mathbf{q}_{\mathrm{e}}^{\mathrm{T}}\delta\mathbf{p}_{\mathrm{v}}. \tag{8.17}$$

Substituting from the virtual forms of (8.13), (8.15) and (8.16) into (8.17) gives

$$V = \int (\mathbf{q}_{\mathrm{mi}}^{\mathrm{T}}\delta\mathbf{p}_{\mathrm{m}} + \mathbf{q}_{\mathrm{wi}}^{\mathrm{T}}\delta\mathbf{w} + \mathbf{q}_{\theta\mathrm{i}}^{\mathrm{T}}\delta\mathbf{p}_{\theta\mathrm{i}})\,\mathrm{d}A_{\mathrm{o}} - \mathbf{q}_{\mathrm{e}}^{\mathrm{T}}\delta\mathbf{p}_{\mathrm{v}} \tag{8.18}$$

where the subvectors of the internal force vector, $\mathbf{q}_{\mathrm{i}}$ are given by

$$\mathbf{q}_{\mathrm{mi}} = \int \mathbf{B}_{\mathrm{m}}^{\mathrm{T}}\mathbf{N}\,\mathrm{d}A_{\mathrm{o}} \tag{8.19a}$$

$$\mathbf{q}_{\mathrm{wi}} = \int \mathbf{B}_{\mathrm{s}}^{\mathrm{T}}(\mathbf{T}(\mathbf{w})^{\mathrm{T}}\mathbf{N} + \mathbf{Q})\,\mathrm{d}A_{\mathrm{o}} \tag{8.19b}$$

$$\mathbf{q}_{\theta\mathrm{i}} = \int (\mathbf{B}_{\mathrm{b}}^{\mathrm{T}}\mathbf{M} + \mathbf{H}^{\mathrm{T}}\mathbf{Q})\,\mathrm{d}A_{\mathrm{o}}. \tag{8.19c}$$

With a view to future developments, it is useful to re-express the strain change/nodal displacement relationships of (8.13), (8.15) and (8.16) as

$$\delta\boldsymbol{\chi} = \bar{\mathbf{B}}_{\chi}\,\delta\mathbf{p} \tag{8.20}$$

$$\delta\boldsymbol{\gamma} = \bar{\mathbf{B}}_{\gamma}\,\delta\mathbf{p} \tag{8.21}$$

$$\delta\bar{\mathbf{E}} = \bar{\mathbf{B}}_{\varepsilon}(\mathbf{p})\,\delta\mathbf{p}. \tag{8.22}$$

If a standard Mindlin–Reissner relationship were adopted the $\bar{\mathbf{B}}$ matrices in (8.20)–(8.22) would follow directly from (8.13), (8.15) and (8.16) and, for example, the $\mathbf{B}_{\chi}$ matrix obtained from (8.13) would imply only connections between $\delta\boldsymbol{\chi}$ and the $\delta\mathbf{p}_{\theta}$ terms. However, in moving from (8.13), (8.15) and (8.16) we could with either the constrained Mindlin–Reissner or discrete Kirchhoff formulations, constrain out the mid-side hierarchical nodal $\Delta w$s to be functions of the nodal $\theta$s [C5, C6, C2.2]. Hence, in these circumstances, there would be coupling between $\delta\boldsymbol{\chi}$ and the nodal $\delta\mathbf{p}_{\theta}$s while the $\Delta w$s would be dummy terms (see [C2.2]). For the discrete Kirchhoff element, we would also constrain out the $\Delta\theta$s [C2.2]. For the current formulation to be general, we will assume that any constraints have been applied in producing the $\bar{\mathbf{B}}$ matrices in (8.20)–(8.22) and will also adopt the general (possibly constrained) form

$$\delta \mathbf{s} = \bar{\mathbf{B}}_s \, \delta \mathbf{p} \tag{8.23}$$

in place of (8.14).

Using (8.20)–(8.23), the virtual work of (8.17) and (8.18) would be replaced by

$$V = \int (\mathbf{N}^T \bar{\mathbf{B}}_\varepsilon(\mathbf{p}) \, \delta \mathbf{p}_v + \mathbf{M}^T \bar{\mathbf{B}}_\chi \, \delta \mathbf{p}_v + \mathbf{Q}^T \bar{\mathbf{B}}_\gamma \, \delta \mathbf{p}_V) \, dA_o - \mathbf{q}_e^T \, \delta \mathbf{p}_v = \int \mathbf{q}_i^T \, \delta \mathbf{p}_v \, dA_o - \mathbf{q}_e^T \, \delta \mathbf{p}_v \tag{8.24}$$

while the subvectors of (8.19a)–(8.19c) would collectively become

$$\mathbf{q}_i = \int (\bar{\mathbf{B}}_\varepsilon(\mathbf{p})^T \mathbf{N} + \bar{\mathbf{B}}_\chi^T \mathbf{M} + \bar{\mathbf{B}}_\gamma^T \mathbf{Q}) \, dA_o. \tag{8.25}$$

### 8.1.5 The tangent stiffness matrix

The tangent stiffness matrix follows in the normal way by differentiation of (8.25), so that

$$\delta \mathbf{q}_i = \int (\bar{\mathbf{B}}_\varepsilon(\mathbf{p})^T \, \delta \mathbf{N} + \bar{\mathbf{B}}_\chi^T \, \delta \mathbf{M} + \bar{\mathbf{B}}_\gamma^T \, \delta \mathbf{Q}) \, dA_o + \int \delta \bar{\mathbf{B}}_\varepsilon(\mathbf{p})^T \mathbf{N} \, dA_o = \mathbf{K}_{t1} \, \delta \mathbf{p} + \mathbf{K}_{t\sigma} \, \delta \mathbf{p}. \tag{8.26}$$

Substituting from (8.8)–(8.10) and (8.20)–(8.22) into the first term in (8.26) gives

$$\mathbf{K}_{t1} = \int \begin{bmatrix} \bar{\mathbf{B}}_\varepsilon^T \\ \bar{\mathbf{B}}_\chi^T \\ \mathbf{B}_\gamma^T \end{bmatrix} \begin{bmatrix} \mathbf{C}_m & \mathbf{C}_{mb} & \mathbf{0} \\ \mathbf{C}_{mb}^T & \mathbf{C}_b & \mathbf{0} \\ \mathbf{0} & \mathbf{0} & \mathbf{C}_\gamma \end{bmatrix} \begin{bmatrix} \bar{\mathbf{B}}_\varepsilon \\ \bar{\mathbf{B}}_\chi \\ \mathbf{B}_\gamma \end{bmatrix} \cdot dA_o \tag{8.27}$$

The geometric stiffness matrix, $\mathbf{K}_{t\sigma}$ comes (see (8.26)) from the variation of $\bar{\mathbf{B}}_\varepsilon$ of (8.22). It is easier to derive this matrix by firstly reverting to the subvector forms of $\mathbf{q}_i$ in (8.19a)–(8.19c), from which the only term not stemming from variations of $\mathbf{N}, \mathbf{M}$ or $\mathbf{Q}$ (which lead to (8.27)) comes from (8.19b) as

$$\delta \mathbf{q}_{wi} = \int \mathbf{B}_s^T \, \delta \mathbf{T}(\mathbf{w})^T \mathbf{N} \, dA_o = \int \mathbf{B}_s^T \mathbf{N}_2 \delta \mathbf{s} \, dA_o = \int \mathbf{B}_s^T \mathbf{N}_2 \mathbf{B}_s \, dA_o \, \delta \mathbf{w} = \mathbf{K}_{t\sigma} \delta \mathbf{w} \tag{8.28}$$

where

$$\mathbf{N}_2 = \begin{bmatrix} N_x & N_{xy} \\ N_{xy} & N_y \end{bmatrix}. \tag{8.29}$$

In (8.28), use has been made of (8.5) for $\mathbf{T}$ and (8.14) for $\delta \mathbf{s}$. The $\mathbf{K}_{t\sigma}$ matrix in (8.28) could equally be derived directly from the continuum form as in (4.87) and (4.88) via

$$\delta V = \int \delta \mathbf{s}_v^T \mathbf{N}_2 \delta \mathbf{s} \, dA_o. \tag{8.30}$$

Allowing for any possible displacement constraints that may lead to (8.23) rather than (8.14), the complete tangent stiffness matrix becomes

$$\mathbf{K}_t = \mathbf{K}_{t1}(8.27) + \int \bar{\mathbf{B}}_s^T \mathbf{N}_2 \bar{\mathbf{B}}_s \, dA_o. \tag{8.31}$$

### 8.1.6 Numerical integration, matching shape functions and 'locking'

We have already considered the introduction of various techniques to avoid 'shear locking' (see also [C2.2] for more detail in relation to the present elements). As previously discussed in Sections 7.1.3 and 7.1.5, we must also consider 'membrane locking'. From these earlier discussions, we know that in order to avoid or, at the least, limit, such 'locking', we must use, at least, quadratic expansions (serendipity or Lagrangian) for the in-plane displacements, $\bar{u}$ and $\bar{v}$. (Allman [A2] has given polynomial expressions for $\bar{u}$ and $\bar{v}$ which are strictly required in order to recover all admissible states of constant strain with the von Karman equations [V1.4] when $w$ is expanded as a quadratic so as to represent the constant curvature states.) When coupled with two-by-two integration, the resulting configuration has been found, by the author, to give good results for a wide range of problems [C8, C11].

An alternative approach, using only linear shape functions for $\bar{u}$ and $\bar{v}$, could involve an extension of the methods suggested in Sections 7.1.3 and 7.1.5 for arch elements. One approach would simply involve the use of single-point integration for the membrane strain terms [M2.7]. Alternatively (see equations (7.31), (7.32) and (7.50)), the membrane strains could be replaced by constant, averaged values. For the two-dimensional shell, rather than the one-dimensional beam, it is not immediately obvious how to apply such a technique. For a constant strain triangle, Stolarski *et al.* [S7] (followed by Jetteur and Frey [J2, J3]) applied the previous one-dimensional relationships along the sides of the triangles. For a triangle with constant strain, the membrane strain vector, $\bar{\boldsymbol{\varepsilon}}$, can easily be related to these three axial strains. An extension of these ideas to quadrilaterals has been given by Jetteur and Frey [J2, J3].

Before leaving this section, we should note that the 'eccentricity issue' (see Section 7.1.3 and [C7.7]) is also relevant to the choice of matching shape functions. It is most relevant in relation to the provision of eccentric stiffeners, but also affects the attached plate or shell element, particularly in the presence of material non-linearity [C7.7].

### 8.1.7 Extensions to the shallow-shell formulation

The simplest way of extending the range of the shallow-shell formulation is to initially define a set of local flat surfaces with respect to which the shallow-shallow equations are defined. Prior to assembly of the overall systems, standard coordinate transformations can be applied. With triangular elements and with quadrilaterals for a cylindrical shell [C4] these facet systems are easily defined. For quadrilaterals with a general shell, some form of weighted averaging is required to define this surface [H1, J2, J3, N1]. In a non-linear environment, if the shallow-shell equations (and transformations) are always related to the initial locally flat system, the resulting formulation will be valid for deformations involving small rotations from the initial configuration. Morley [M5] and Providas [P4] have used a similar approach (but with the von Karman rather than the Marguerre relationships) in conjunction with Morley's triangular element [M4] which has mid-side rotational connectors.

The limitation to small rotations from the initial configuration can be removed by adopting a corotational formulation for the local flat surfaces (facets) in conjunction with a locally shallow formulation using the Marguerre equations. (These ideas have been discussed in relation to arches in Section 7.2.7.) Such an approach has been applied by Jetteur and Frey [J2, J3] and Stolarski *et al.* [S7].

## 8.2 A DEGENERATE-CONTINUUM ELEMENT USING A TOTAL LAGRANGIAN FORMULATION

In order to produce a non-linear degenerate shell element, we may, conceptually, simply extend the work of Section 7.5 into three dimensions. Early work applied the discretisation after the derivation of the tangent relationships at the continuum level [B2, R1] and consequently did not produce a consistent tangent stiffness matrix. The limitation was pointed out by Frey and Cescotto [F1] and later by Surana [S8]. In the present section, we will follow closely the formulation of Ramm and Matzenmiller [R2] which is described further in a paper by Stander *et al.* [S4]. The developments are closely related to both the three-dimensional continuum formulation of Section 5.1.3 and the two-dimensional 'arch' formulation of Section 7.5 and, to avoid repetition, some reference will be made to equations given in those sections.

As a starting point (Figure 8.3), the geometry is expressed as

$$\mathbf{r} = \bar{\mathbf{r}} + \tfrac{1}{2}\zeta\,\Delta\mathbf{r} = \begin{bmatrix} \bar{x} \\ \bar{y} \\ \bar{z} \end{bmatrix} + \tfrac{1}{2}\zeta \begin{bmatrix} \Delta\bar{x} \\ \Delta\bar{y} \\ \Delta\bar{z} \end{bmatrix} = \sum \mathbf{h}_i(\xi,\eta)\bar{\mathbf{r}}_i + \tfrac{1}{2}\zeta \sum \mathbf{h}_i(\xi,\eta) a_i \mathbf{v}_i \qquad (8.32)$$

where $\xi$ and $\eta$ are the non-dimensional coordinates in the plane of the shell and $\zeta$ the non-dimensional coordinate in the thickness direction. The position vector, $\bar{\mathbf{r}}$ in (8.32) relates to the centroidal or reference surface while $\Delta\mathbf{r}$ is in the direction of the

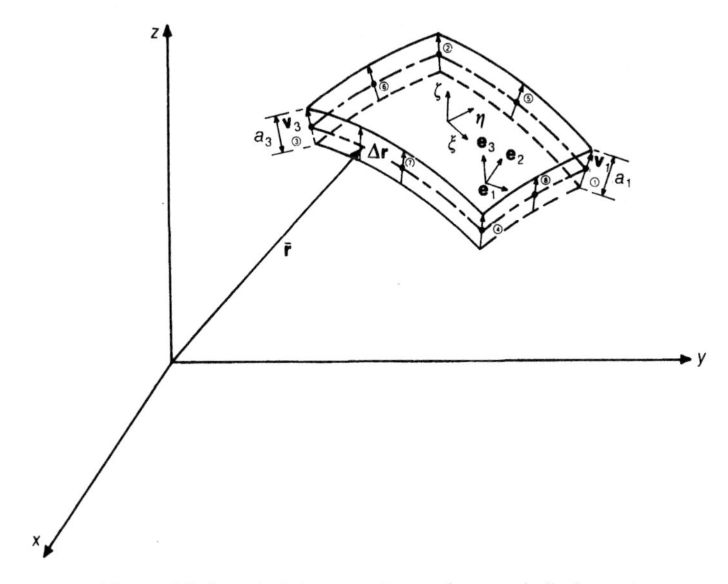


**Figure 8.3** 8-noded degenerate-continuum shell element.

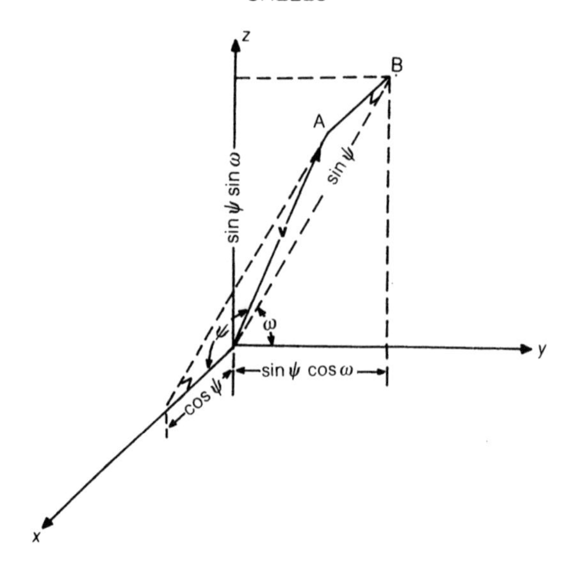


**Figure 8.4** 'Rotational' variables, $\psi$ and $\omega$ for degenerate-continuum shell element.

'director' (through the thickness). At the nodes, the coordinates of the reference surface are given by $\bar{\mathbf{r}}_i$, while the through thickness, director vectors are $\Delta\mathbf{r}_i = a_i\mathbf{v}_i$, where (Figure 8.3) $\mathbf{a}_i$ is the thickness and $\mathbf{v}_i$ the unit vector in the 'thickness direction'. The vector $\mathbf{v}_i$ is defined by *two* parameters $\psi_i$ and $\omega_i$ (Figure 8.4), where

$$\mathbf{v}_i = \begin{pmatrix} \cos\psi \\ \sin\psi\cos\omega \\ \sin\psi\sin\omega \end{pmatrix}_i . \tag{8.33}$$

In Figure 8.4, OA represents the unit vector $\mathbf{v}$ with $\psi$ as the angle between this vector and the $x$-axis. OB is the projection of $\mathbf{v}$ (OA) on to the $y$–$z$ plane while $\omega$ defines the angle between this projection and the $y$-axis. The relationship in Figure 8.4 and (8.33) is not a unique way of defining the unit vector, $\mathbf{v}_i$—other possibilities exist but with $\mathbf{v}_i$ of unit length they should involve two parameters.

For consistency with (8.32), the deflections,

$$\mathbf{d}^{\mathrm{T}} = (u, v, w) \tag{8.34}$$

follow as

$$\mathbf{d} = \bar{\mathbf{d}} + \Delta\mathbf{d} = \sum \mathbf{h}_i(\xi, \eta)\bar{\mathbf{d}}_i + \tfrac{1}{2}\zeta \sum \mathbf{h}_i a_i(\mathbf{v}_i - \mathbf{v}_{io}) \tag{8.35}$$

where

$$\mathbf{v}_{io}^{\mathrm{T}} = (\cos\psi_{io}, \sin\psi_{io}\cos\omega_{io}, \sin\psi_{io}\sin\omega_{io}) \tag{8.36}$$

and $\psi_{io}$, $\omega_{io}$ are the initial values of $\psi$ and $\omega$ at node i and $\mathbf{d}_i$ contains the nodal values of the deflections at the reference surface (see (8.34) for $\mathbf{d}$).

Assuming $\bar{\mathbf{d}}_i$ and $\mathbf{v}_i$ are known, the Green strain can be computed from (5.11) with $\mathbf{H}$ from (5.33) and $\mathbf{A}(\boldsymbol{\theta})$ from (5.34). The vector $\boldsymbol{\theta}$ (see (5.29)) of displacement gradients can be computed from the components of $\partial\mathbf{d}/\partial\mathbf{x}$, $\partial\mathbf{d}/\partial\mathbf{y}$, $\partial\mathbf{d}/\partial\mathbf{z}$ with the latter being computed using the usual Jacobian terms $\mathbf{J}$ (see [C2.2]) in conjunction with the partial differentials of (8.35):

$$\frac{\partial \mathbf{d}}{\partial \xi} = \sum h_{\xi i}\bar{\mathbf{d}}_i + \tfrac{1}{2}\zeta \sum h_{\xi i} a_i(\mathbf{v}_i - \mathbf{v}_{io}) \tag{8.37}$$

$$\frac{\partial \mathbf{d}}{\partial \eta} = \sum h_{\eta i}\bar{\mathbf{d}}_i + \tfrac{1}{2}\zeta \sum h_{\eta i} a_i(\mathbf{v}_i - \mathbf{v}_{io}) \tag{8.38}$$

$$\frac{\partial \mathbf{d}}{\partial \zeta} = \frac{1}{2}\sum h_i a_i(\mathbf{v}_i - \mathbf{v}_{io}) \tag{8.39}$$

where, as usual, $h_{\xi i}$ is the partial differential of $h_i$ with respect to $\xi$. Differentiaion of (8.35) leads to

$$\delta u = \mathbf{h}(\xi,\eta)^T\delta\bar{\mathbf{u}} - \tfrac{1}{2}\zeta\mathbf{h}^T\mathbf{D}(a\sin\psi)\,\delta\boldsymbol{\psi} \tag{8.40}$$

$$\delta v = \mathbf{h}(\xi,\eta)^T\delta\bar{\mathbf{v}} + \tfrac{1}{2}\zeta\mathbf{h}^T\mathbf{D}(a\cos\psi\cos\omega)\,\delta\boldsymbol{\psi} - \tfrac{1}{2}\zeta\mathbf{h}^T\mathbf{D}(a\sin\psi\sin\omega)\,\delta\boldsymbol{\omega} \tag{8.41}$$

$$\delta w = \mathbf{h}(\xi,\eta)^T\delta\bar{\mathbf{w}} + \tfrac{1}{2}\zeta\mathbf{h}^T\mathbf{D}(a\cos\psi\sin\omega)\,\delta\boldsymbol{\psi} + \tfrac{1}{2}\zeta\mathbf{h}^T\mathbf{D}(a\sin\psi\cos\omega)\,\delta\boldsymbol{\omega} \tag{8.42}$$

where $\delta\bar{\mathbf{u}}$ contain the nodal displacement changes of $\bar{u}, \delta\bar{\mathbf{v}}$ the nodal displacement changes in $\bar{v}$ and $\delta\bar{\mathbf{w}}$ the nodal displacement changes in $\bar{w}$, while $\delta\boldsymbol{\psi}$ contains the nodal changes in the $\boldsymbol{\psi}$s (see Figure 8.4) and $\delta\boldsymbol{\omega}$ contains he nodal changes in the $\omega$s (see Figure 8.4). Collectively, these nodal displacement changes can be combined as

$$\delta\mathbf{p}^T = (\delta\mathbf{u}^T, \delta\mathbf{v}^T, \delta\mathbf{w}^T, \delta\boldsymbol{\psi}^T, \delta\boldsymbol{\omega}^T) \tag{8.43}$$

(see the footnote on page 25 regarding the ordering of the variables). The matrices such as $\mathbf{D}(a\sin\psi)$ in (8.40)–(8.42) are diagonal and take the form previously discussed for the two-dimensional 'arch' in Section 7.5 (see equation (7.157)).

Using (8.37)–(8.42), the change in the displacement gradients $\delta\boldsymbol{\theta}$ (see (5.29),

$$\delta\boldsymbol{\theta}^T = \left(\frac{\partial\delta u}{\partial x}, \frac{\partial\delta u}{\partial y}, \frac{\partial\delta u}{\partial z}, \frac{\partial\delta v}{\partial x}, \frac{\partial\delta v}{\partial y}, \frac{\partial\delta v}{\partial z}, \frac{\partial\delta w}{\partial x}, \frac{\partial\delta w}{\partial y}, \frac{\partial\delta w}{\partial z}\right) \tag{8.44}$$

can be related to the nodal variables, $\delta\mathbf{p}$, via

$$\delta\boldsymbol{\theta} = \frac{1}{2}\begin{bmatrix} \mathbf{a}_1^T & \mathbf{0}^T & \mathbf{0}^T & \mathbf{b}_1^T & \mathbf{0}^T \\ \mathbf{a}_2^T & \mathbf{0}^T & \mathbf{0}^T & \mathbf{b}_2^T & \mathbf{0}^T \\ \mathbf{a}_3^T & \mathbf{0}^T & \mathbf{0}^T & \mathbf{b}_3^T & \mathbf{0}^T \\ \mathbf{0}^T & \mathbf{a}_1^T & \mathbf{0}^T & \mathbf{c}_1^T & \mathbf{e}_1^T \\ \mathbf{0}^T & \mathbf{a}_2^T & \mathbf{0}^T & \mathbf{c}_2^T & \mathbf{e}_2^T \\ \mathbf{0}^T & \mathbf{a}_3^T & \mathbf{0}^T & \mathbf{c}_3^T & \mathbf{e}_3^T \\ \mathbf{0}^T & \mathbf{0}^T & \mathbf{a}_1^T & \mathbf{d}_1^T & \mathbf{f}_1^T \\ \mathbf{0}^T & \mathbf{0}^T & \mathbf{a}_2^T & \mathbf{d}_2^T & \mathbf{f}_2^T \\ \mathbf{0}^T & \mathbf{0}^T & \mathbf{a}_3^T & \mathbf{d}_3^T & \mathbf{f}_3^T \end{bmatrix}\delta\mathbf{p} = \mathbf{G}\delta\mathbf{p}. \tag{8.45}$$

In equation (8.45),

$$\mathbf{a}_k = 2(\mathbf{J}^{-1}(k,1)\mathbf{h}_\xi + \mathbf{J}^{-1}(k,2)\mathbf{h}_\eta) \tag{8.46}$$

$$\mathbf{b}_k = -\mathbf{D}(\mathrm{a}\sin\psi)\mathbf{z}_k \tag{8.47}$$

with

$$\mathbf{z}_k = \zeta \mathbf{J}^{-1}(k,1)\mathbf{h}_\xi + \zeta \mathbf{J}^{-1}(k,2)\mathbf{h}_\eta + \mathbf{J}^{-1}(k,3)\mathbf{h}. \tag{8.48}$$

Also,

$$\mathbf{c}_k = \mathbf{D}(a\cos\psi\cos\omega)\mathbf{z}_k, \qquad \mathbf{d}_k = \mathbf{D}(a\cos\psi\sin\omega)\mathbf{z}_k \tag{8.49}$$

$$\mathbf{e}_k = -\mathbf{D}(a\sin\psi\sin\omega)\mathbf{z}_k, \qquad \mathbf{f}_k = \mathbf{D}(a\sin\psi\cos\omega)\mathbf{z}_k. \tag{8.50}$$

With the new definition of (8.45) for $\mathbf{G}$, (5.33) for $\mathbf{H}$, and (5.34) for $\mathbf{A}(\boldsymbol{\theta})$, equation (5.19) defines the internal force vector, $\mathbf{q}_i$.

### 8.2.1 The tangent stiffness matrix

As for the two-dimensional 'arch' formulation of Section 7.5, the tangent stiffness matrix can be expressed as

$$\mathbf{K}_t = \mathbf{K}_{t1} + \mathbf{K}_{\sigma 1} + \mathbf{K}_{\sigma 2} \tag{8.51}$$

with $\mathbf{K}_{t1}$ being defined by (7.168). Special treatment is required for the effective modular matrix $\mathbf{C}_t$ which will be discussed later.

The geometric stiffness matrix $\mathbf{K}_{\sigma 1}$ is given, as for the continuum formulation, by (5.25) (although with $\mathbf{G}$ from (8.45)) with $\hat{\mathbf{S}}$ being given by (5.36). For the second contribution to the geometric stiffness matrix, in place of the 'arch equation' of (7.172), we now have

$$\mathbf{K}_{\sigma 2}\,\delta\mathbf{p} = \iint \sum_{k=1,9} \mathbf{F}(k)\delta\mathbf{G}_k \,\mathrm{d}V_o \tag{8.52}$$

where $\mathbf{F}(k)$ is the $k$th component of the vector:

$$\mathbf{F} = [\mathbf{H} + \mathbf{A}(\boldsymbol{\theta})]^{\mathrm{T}}\mathbf{S} \tag{8.53}$$

with $\mathbf{S}$ from (5.35). Differentiation of the $k$th column of $\mathbf{G}^{\mathrm{T}}$ (from (8.45)) gives $\delta\mathbf{G}_k\,\delta\mathbf{p}$, where it can be shown that

$$\delta\mathbf{G}_k = \mathbf{P}_k\delta\mathbf{p} = \begin{bmatrix} \mathbf{0} & \mathbf{0} & \mathbf{0} & \mathbf{0} & \mathbf{0} \\ \mathbf{0} & \mathbf{0} & \mathbf{0} & \mathbf{0} & \mathbf{0} \\ \mathbf{0} & \mathbf{0} & \mathbf{0} & \mathbf{0} & \mathbf{0} \\ \mathbf{0} & \mathbf{0} & \mathbf{0} & \mathbf{D}_{1k} & \mathbf{D}_{2k} \\ \mathbf{0} & \mathbf{0} & \mathbf{0} & \mathbf{D}_{2k} & \mathbf{D}_{3k} \end{bmatrix} \delta\mathbf{p} \tag{8.54}$$

where $\mathbf{D}_{1k}, \mathbf{D}_{2k}$ and $\mathbf{D}_{3k}$ are diagonal matrices with $(i,i)^{\text{th}}$ terms which involve $\mathbf{z}_{1-3}$ from (8.48) and are given by the following equations:

for $k = 1, 3$:

$$\begin{aligned} \mathbf{D}_{1k}(i,i) &= -a_i\cos\psi_i\,\mathbf{z}_k(\mathrm{i}) \\ \mathbf{D}_{2k}(i,i) &= \mathbf{D}_{3k}(i,i) = 0 \end{aligned} \tag{8.55}$$

for $k = 4, 6$:

$$\begin{aligned} \mathbf{D}_{1k}(i,i) &= -a_i\sin\psi_i\cos\omega_i\,\mathbf{z}_{k-3}(i) \\ \mathbf{D}_{2k}(i,i) &= -a_i\cos\psi_i\sin\omega_i\,\mathbf{z}_{k-3}(i) \\ \mathbf{D}_{3k}(i,i) &= -a_i\sin\psi_i\cos\omega_i\,\mathbf{z}_{k-3}(i) \end{aligned} \tag{8.56}$$

for $k = 7, 9$:

$$\mathbf{D}_{1k}(i,i) = -a_i \sin\psi_i \sin\omega_i \mathbf{z}_{k-6}(i)$$
$$\mathbf{D}_{2k}(i,i) = a_i \cos\psi_i \cos\omega_i \mathbf{z}_{k-6}(i)$$
$$\mathbf{D}_{3k}(i,i) = -a_i \sin\psi_i \sin\omega_i \mathbf{z}_{k-6}(i). \tag{8.57}$$

In forming the standard tangent stiffness matrix, $\mathbf{K}_{t1}$ (see (7.168)), the constitutive matrix (or tensor), $\mathbf{C}_t$, should include the plane-stress constraint that there be no stress in the initial thickness direction. At a particular (Gauss) point $\xi, \eta$, the unit vector, $\mathbf{e}_3$, in this direction can be computed from

$$\mathbf{e}_3 = \frac{\Delta\mathbf{r}_o}{\|\Delta\mathbf{r}_o\|} \qquad \text{with } \Delta\mathbf{r}_o = \sum h_i a_i \mathbf{v}_{io} \tag{8.58}$$

(see (8.32) for $\Delta\mathbf{r}$). To find the unit vectors $\mathbf{e}_1$ and $\mathbf{e}_2$, orthogonal to $\mathbf{e}_3$ (Figure 8.3), one can use procedures similar to those used in linear analysis [C2.2, Z2.1]. Assuming a linear material response, the local material modular matrix is given by

$$\mathbf{C}_l = \frac{E}{(1-\nu^2)} \begin{bmatrix} 1 & \nu & 0 & 0 & 0 & 0 \\ \nu & 1 & 0 & 0 & 0 & 0 \\ 0 & 0 & 0 & 0 & 0 & 0 \\ 0 & 0 & 0 & A & 0 & 0 \\ 0 & 0 & 0 & 0 & \frac{5}{6}A & 0 \\ 0 & 0 & 0 & 0 & 0 & \frac{5}{6}A \end{bmatrix} \tag{8.59}$$

where $A = \frac{1}{2}(1-\nu)$. To transform to global coordinates (for use in 7.168), the tensor relationship of (4.55) can be used with $C^{o}_{abcd}$ relating to $\mathbf{C}_l$ and $C^{n}_{ijkl}$ to the global $\mathbf{C}_t$ in (7.168). The $\mathbf{T}$ tensor in (4.55) would be $\mathbf{T}_{gl} = \mathbf{T}_{lg}^{T}$, where the matrix form of $\mathbf{T}_{lg}$ is given by (see (4.35)):

$$\mathbf{T}_{lg} = \begin{pmatrix} \mathbf{e}_1^{T} \\ \mathbf{e}_2^{T} \\ \mathbf{e}_3^{T} \end{pmatrix}. \tag{8.60}$$

In order to produce an effective degenerate continuum shell element, the present theory must be supplemented by techniques to avoid 'shear' and 'membrane' locking. Also, for an efficient solution, the through-thickness integration should be divorced from the main calculations involving the Jacobians and shape functions. These concepts have been briefly discussed in the introduction and will be considered further in Volume 2.

## 8.3 SPECIAL NOTATION

$\mathbf{a}_i$ = nodal 'thicknesses' (Section 8.2)
$\mathbf{a}_k, \mathbf{b}_k, \mathbf{c}_k, \mathbf{d}_k, \mathbf{f}_k$ = vectors for defining $\mathbf{G}$ (see (8.45)) (Section 8.2)
$\mathbf{A}(\boldsymbol{\theta})$ = matrix containing displacement derivatives (see (5.34)) (Section 8.2)
$\mathbf{B}_m$ = matrix connecting $\bar{\mathbf{E}}_l$ to $\mathbf{p}_m$ (Section 8.1)
$\mathbf{B}_b$ = matrix connecting $\boldsymbol{\chi}$ to $\mathbf{p}_\theta$ (Section 8.1)
$\mathbf{B}_s$ = matrix connecting the slopes, $\mathbf{s}$, to $\mathbf{w}$ (Section 8.1)

$\bar{\mathbf{B}}_s$ = modified **B** matrices after applying constraints (Section 8.1)
$\bar{\mathbf{B}}_\varepsilon$ = $\bar{\mathbf{B}}$ matrix connecting $\delta\bar{\mathbf{E}}$ to $\delta\mathbf{p}$ (Section 8.1)
$\bar{\mathbf{B}}_\chi$ = $\bar{\mathbf{B}}$ matrix connecting $\delta\boldsymbol{\chi}$ to $\delta\mathbf{p}$ (Section 8.1)
$\bar{\mathbf{B}}_\gamma$ = $\bar{\mathbf{B}}$ matrix connecting $\delta\boldsymbol{\gamma}$ to $\delta\mathbf{p}$ (Section 8.1)
$\mathbf{C}_m$, $\mathbf{C}_b$ and $\mathbf{C}_{mb}$ = submatrices (membrane, bending and coupling) of tangential constitutive matrix, $\mathbf{C}_t$ (Section 8.1)
$\mathbf{d}$ = displacement vector (Section 8.2)
$\bar{\mathbf{d}}$ = displacement vector relating to reference surface (Section 8.2)
$\mathbf{D}$ = diagonal matrices (see (7.157)) (Section 8.2)
$\mathbf{e}_1, \mathbf{e}_2, \mathbf{e}_3$ = triad of orthogonal unit vectors at Gauss-point with $\mathbf{e}_3$ in 'thickness direction' (Section 8.2)
$\mathbf{E}$ = strains (Section 8.1)
$\bar{\mathbf{E}}$ = shallow-shell strains at reference plane (Section 8.1)
$\bar{\mathbf{E}}_l$ = linear part of $\mathbf{E}$ (Section 8.1)
$\bar{\mathbf{E}}_{nl}$ = non-linear part of $\bar{\mathbf{E}}$ (Section 8.1)
$G$ = shear modulus
$\mathbf{F} = \mathbf{H} + \mathbf{A}(\boldsymbol{\theta})$—see (8.53) (Section 8.2)
$\mathbf{G}$ = matrix connecting $\delta\boldsymbol{\theta}$ to $\delta\mathbf{p}$ (Section 8.2)
$\mathbf{h}_\xi, \mathbf{h}_\eta$ = derivatives w.r.t $\xi$ and $\eta$ of shape function vector, $\mathbf{h}$ (Section 8.2)
$\mathbf{H}$ = shape-function matrix (see 8.15) (Section 8.1)
$\mathbf{H}$ = Boolean matrix (see (5.33)) (Section 8.2)
$\mathbf{M}$ = vector of bending moments (Section 8.1)
$\mathbf{N}$ = vector of in-plane stress resultants (Section 8.1)
$\mathbf{N}_2$ = matrix of in-plane stress resultants (see (8.29)) (Section 8.1)
$\mathbf{p}$ = nodal displacement; for Section 8.1, $\mathbf{p}^T = (\bar{\mathbf{u}}^T, \bar{\mathbf{v}}^T, \mathbf{w}^T, \boldsymbol{\theta}_x^T, \boldsymbol{\theta}_y^T) = (\mathbf{p}_m^T, \mathbf{w}^T, \mathbf{p}_\theta^T)$
$\mathbf{P}_k$ = matrix relating $\delta\mathbf{G}_k$ to $\delta\mathbf{p}$ (see (8.54)) (Section 8.2)
$\mathbf{Q}$ = transverse shear forces (Section 8.1)
$\mathbf{r}$ = position vector (Section 8.2)
$\bar{\mathbf{r}}$ = position vector of reference surface (Section 8.2)
$\mathbf{s}$ = vector containing slopes, $\partial w/\partial x$ and $\partial w/\partial y$ (Section 8.1)
$\mathbf{S}$ = Second Piola–Kirchhoff stresses corresponding to $\bar{\mathbf{E}}$ (Section 8.1)
$\mathbf{S}$ = Second Piola–Kirchhoff stresses (Section 8.2)
$\mathbf{T}$ = matrix of slopes (see (8.5)) (Section 8.1)
$\bar{u}$ = $u$ ($x$-direction) displacement at reference plane (Section 8.1)
$\bar{\mathbf{u}}$ = nodal values of $\bar{u}$ (Section 8.1)
$\bar{v}$ = $v$ ($y$-direction) displacement at reference plane (Section 8.1)
$\bar{\mathbf{v}}$ = nodal values of $\bar{v}$ (Section 8.1)
$\mathbf{v}_i$ = nodal 'unit director vectors' (Section 8.2)
$\mathbf{v}_{io}$ = nodal 'unit director vectors' in initial configuration (Section 8.2)
$\mathbf{w}$ = nodal values of $w$ (Section 8.1)
$\bar{\mathbf{x}}, \bar{\mathbf{y}}, \bar{\mathbf{z}}$ = nodal values of $x$-, $y$- and $z$ coordinates of reference surface (Section 8.2)
$\mathbf{z}_k$ = vector used to define $\mathbf{G}$ (see (8.48)) (Section 8.2)
$\boldsymbol{\gamma}$ = vector of shear strains (Section 8.1)
$\theta_x, \theta_y$ = rotations of the normal (Section 8.1)
$\boldsymbol{\theta}$ = vector containing $\theta_x$ and $\theta_y$ (Section 8.1)

$\boldsymbol{\theta}$ = vector containing displacement derivatives (see (8.44)) (Section 8.2)
$\boldsymbol{\theta}_x$ = nodal values of $\theta_x$ (Section 8.1)
$\boldsymbol{\theta}_y$ = nodal values of $\theta_y$ (Section 8.1)
$\psi_i, \omega_i$ = nodal parameter defining orientation of $\mathbf{v}_i$ (Section 8.2)
$\psi_{io}, \omega_{io}$ = nodal parameter defining orientation of $\mathbf{v}_{io}$ (Section 8.2)
$\boldsymbol{\chi}$ = curvatures (Section 8.1)
$\delta\boldsymbol{\psi}$ = vector containing nodal values of $\delta\psi$ (Section 8.2)
$\delta\boldsymbol{\omega}$ = vector containing nodal values of $\delta\omega$ (Section 8.2)

## 8.4 REFERENCES


[A1] Ahmad, S., Irons, B. M. & Zienkiewicz, O. C., Analysis of thick and thin shell structures by curved elements, *Int. J. Num. Meth. Engng.*, **2**, 419–451 (1970).

[A2] Allman, D. J., Improved finite element models for the large displacement bending and post-buckling analysis of thin plates, *Int. J. Solids & Structs.*, **18**, 737–762 (1982).

[A3] Allman, D. J., The constant strain triangle with drilling rotations: a simple prospect for shell analysis, *The Mathematics of Finite Elements and Applications, Mafelap 1987*, Vol. VI, ed. J. R. Whiteman, Academic Press, pp. 233–240 (1987).

[A4] Argyris, J., An excursion into large rotations, *Comp. Meth. Appl. Mech. & Engng.*, **32**, pp. 85–155 (1982).

[B1] Backlund, J., Finite element analysis of nonlinear structures, Ph.D. Thesis. Chalmers Univ., Gottenburg, Sweden (1973).

[B2] Bathe, K-J. & Bolourchi, S., A geometric and material nonlinear plate and shell element, Computers & Structs., **11**, 23–48 (1980).

[B3] Belytschko, T., Wong, B.L. & Chiang, H-Y., Improvements in low-order shell elements for explicit transient analysis, *Proc. Symp. on Anal. and Comp. Models for Shells*, ed. A. K. Noor, ASME, New York (1989).

[B4] Belytschko, T. & Lin, J. I., Explicit algorithms for the nonlinear dynamics of shells, *Computer Meth. Applied Mech. Engng*, **42**, 225–251 (1984).

[B5] Belytschko, T., Wong, B. L. & Stolarski, H., Assumed strain stabilization procedure for the 9-node Lagrange shell element, *Int. J. Num. Meth. Engng.*, **28**, 385–414 (1989).

[B6] Belytschko, T., Lin, J. & Tsay, C-S., Explicit algorithms for the nonlinear dynamics of shells, *Comp. Meth. Appl. Mech. & Engng.*, **42**, 225–251 (1984).

[B7] Belytschko, T., Stolarski, H., Liu, W. K., Carpenter, N. & Ong, J.S-J., Stress projection for membrane and shear locking in shell finite elements, *Comp. Meth. Appl. Mech. & Engng.*, **51**, 221–258 (1985).

[B8] Bergan, P. G. & Felippa, C. A., Efficient implementation of a triangular membrane element with drilling freedoms, *Finite Element Methods for Plate and Shell Structures*, ed. T.J.R. Hughes *et al.*, Pineridge, Swansea, pp. 128–152 (1986).

[C1] Calladine, C. R., The theory of thin shell structures, 1888–1988, *Proc. Instn. Mech. Engrs.*, **202**, 42 (1988).

[C2] Carpenter, N., Stolarski, H. & Belytschko, T., A flat triangular shell element with improved membrane interpolation. *Comm. Appl. Num. Meth.*, **1**, 161–168 (1985).

[C3] Clough, R. W. & Tocher, J. L., Analysis of thin arch dams by the finite element method, *Proc. Symp. Theory of Arch Dams*, Southampton University (1964).

[C4] Crisfield, M. A., Linear and non-linear finite element analysis of cylindrical shells, Transport & Road Res. Lab. Report LR 987, Crowthorne, Berks., England (1981).

[C5] Crisfield, M. A., A four-noded thin-plate bending element using shear constraints—a modified version of Lyons' element, *Comp. Meth. Appl. Mech. & Engng.*, **38**, 93–120 (1983).

[C6] Crisfield, M. A., A quadratic Mindlin element using shear-constraints, *Comp. & Struct.*, **18**, 833–852 (1984).

[C7] Crisfield, M. A., Full-range analysis of steel plates and stiffened plating under uniaxial compression, *Proc. Instn. Civ. Engrs.*, **59**, Part 2, 595–624 (1975).

[C8] Crisfield, M. A. & Wills, J., Numerical analysis of a half-scale beam and slab bridge deck, *Proc. 2nd Int. Conf. on Computer Aided Analysis and Design of Concrete Structures*, ed. N. Bicanic, Pineridge, Swansea, Vol. 1, pp. 365–378 (1990)

[C9] Crisfield, M. A., Explicit integration and the isoparametric arch and shell elements, *Comm. Appl. Num. Meth.*, **2**, 181–187 (1986).

[C10] Crisfield, M. A., Large-deflection elasto-plastic buckling analysis of plates using finite elements, Transport & Road Res. Lab. Report LR 593 (1973).

[C11] Crisfield, M. A., Numerical methods for the non-linear analysis of bridges, Comp. & Structs., **30**, 637–644 (1988).

[C12] Crisfield, M. A., Shear-constraints and folded-plated structures, *Engng. Comp.*, **2**, 237–246 (1985).

[D1] Dvorkin, E. N. & Bathe, K. J., A continuum mechanics based four-node shell element for general non-linear analysis, *Engng. Comp.* **1**, 77–88 (1984).

[F1] Frey, F. & Cescotto, S., Some new aspects of the incremental total Lagrangian description in nonlinear analysis, *Int. Conf. 'Finite Elements in Nonlinear Mechanics'* Vol. 1, Geilo, Norway, 5.1–5.20 (1977).

[H1] Horrigmoe, G. & Bergan, P. G., nonlinear analysis of free-form shells by flat finite elements, *Comp. Meth. Appl. Mech. & Engng.*, **16**, 11–35 (1978).

[H2] Huang, H. C. & Hinton, E., Lagrangian and serendipity plate and shell elements through thick and thin, *Finite Element Methods for Plate and Shell Structures*, ed. T.J.R. Hughes *et al.*, Pineridge Swansea, pp. 46–61 (1986).

[H3] Hughes, T. J. R. & Liu, W. K., Nonlinear finite element analysis of shells: Part 1, three-dimensional shells, *Comp. Meth. Appl. Mech. & Engng.*, **26**, 331–362 (1981).

[H4] Hughes, T. J. R. & Hinton, E. (eds.), *Finite Element Methods for Plate and Shell Structures*, Vol. 1: *Element Technology*; Vol. 2: *Formulation and Algorithms*, Pineridge, Swansea (1986).

[H5] Hughes, T. J. R., *The Finite Element Method: Linear Static and Dynamic Finite Element Analysis*, Prentice-Hall, Englewood Cliffs (1987).

[I1] Irons, B. M. & Ahmad, S., *Techniques of Finite Elements*, Ellis Horwood, Chichester, U.K. (1980).

[J1] Jang, J. & Pinsky, P. M., An assumed co-variant strain based 9-noded shell element *Int. J. Num. Meth. Engng.*, **24**, 2389–2411 (1987).

[J2] Jetteur, P. & Frey, F., A four node Marguerre element for non-linear shell analysis *Engng. Comput.*, **3**, 276–282 (1986).

[J3] Jetteur, P., Improvement of the Quadrangular 'Jet' Shell Element for a Particular Class of Shell Problems, IREM Internal Report 87/1, Feb., 1987.

[L1] Liu, W. K., Law, E. S., Lam, D. & Belytschko, T. Resultant-Stress degenerate-shell element, *Comp. Meth. Appl. Mech. & Engng.*, **55**, 259–300 (1986).

[M1] Macneal, R. H., The evolution of lower order plate and shell elements in MSC/NASTRAN, *Finite Element Methods for Plate and Shell Structures*, Vol. 1: *Element Technology*, Pineridge, Swansea, pp. 85–127 (1986).

[M2] Milford, R. V. & Schnobrich, W. C., Degenerated isoparametric finite elements using explicit integration, *Int. J. Num. Meth. Engng.*, **23**, 133–154 (1986).

[M3] Mindlin, R. D., Influence of rotary inertia and shear on flexural motions of isotropic elastic plates, *J. Appl. Mech.*, **18**, 31–38 (1951),

[M4] Morley, The constant-moment plate-bending element, *J. Strain Analysis*, **6**, 20–24 (1971).

[M5] Morley, L. S. D. Geometrically Nonlinear Constant Moment Triangular Which Passes the von Karman Patch Test, Tech. Report, BICOM 89/4, Brunnel University (July 1989).

[N1] Nygard, M. K., The free formulation for non-linear finite elements with applications to shells, Report 86-2, Div. of Struct. Mechs., The University of Trondheim Norway (1986).

[P1] Parisch, H., An Investigation of a Finite Rotation Four Node Shell Element, *Int. J. Num. Meth. Engng.* (to be published).

[P2] Parisch, H., Large displacements of shells including material nonlinearities, *Comp. Meth. Appl. Mech. & Engng.*, **27**, 183–214 (1981).

[P3] Pawsey, S. E. & Clough, R. W., Improved numerical integration of thick shell finite elements, *Int. Num. Meth. Engng.*, **3**, 545–586 (1971).

[P4] Providos, E., On the Geometrically Nonlinear Constant Moment Triangle (with a note on drilling rotations), Ph.D Thesis, Brunel University (1990).

[R1] Ramm, E., A plate/shell element for large deflections and rotations, *US–Germany Symp. on 'Formulations and Computational Algorithms in Finite Element Analysis*, MIT-Press (1976).

[R2] Ramm, E. & Matzenmiller, A., Large deformation shell analysis based on the degenerate concept, *Finite Element Methods for Plate and Shell Structures*, ed. T. J. R. Hughes *et al.*, Pineridge, Swansea, pp. 365–393 (1986).

[R3] Reissner, E., The effect of transverse shear deformation on the bending of elastic plates, *J. Appl. Mech.*, **12**, 1.69–1.77 (1945).

[S1] Simo, J. C. & Fox D. D. On a stress resultant geometrically exact shell model. Part I: Formulation and optimal parametrization, *Computer Methods in Applied Mechanics and Engineering*, **72**, 267–304 (1989).

[S2] Simo, J. C., Fox, D. D. & Rafai, M. S. On a stress resultant geometrically exact shell model. Part II: The linear theory; computational aspects, *Computer Methods in Applied Mechanics and Engineering*, **73**, 53–92 (1989).

[S3] Simo, J. C., Fox, D. D. & Rafai, M. S., On a stress resultant geometrically exact shell model. Part III: Computational aspects of the nonlinear theory, *Computer Methods in Applied Mechanics and Engineering*, **79**, 21–70 (1990).

[S4] Stander, N., Matzenmiller, A. & Ramm, E. An assessment of assumed strain methods in finite element rotation shell analysis *Engineering Computations*, **6**, 58–66 (1989).

[S5] Stanley, G. M., Continuum-based shell elements, Stanford University, Ph.D. Thesis, also Lockheed Report LMSC-FO35839 (1985).

[S6] Stanley, G. M., Park, K. C. & Hughes, T. J. R., Continuum-based resultant shell elements, *Finite Element Methods for Plate and Shell Structures*, ed. T. J. R. Hughes *et al.*, Pineridge, Swansea, pp. 1–45 (1986).

[S7] Stolarski, H., Belytschko, T., Carpenter, N. & Kennedy, J. M. A simple triangular curved shell element for collapse analysis, *Engng. Comp.*, **1**, 210–218 (1984).

[S8] Surana, K. S., Geometrically nonlinear formulations for the curved shell elements, *Int. J. Num. Meth. Engng.*, **19**, 353–386 (1983).

[T1] Taylor, R. L. & Simo, J. C., Bending and membrane elements for analysis of thick and thin shells, *Proc. NUMETA '85*, Swansea, 587–591 (1985).

[W1] Wempner, G. Mechanics and finite elements of shells, *Appl. Mech.*, **42**, 129–142 (1989)

[W2] Wriggers, P. & Gruttmann, F. Thin shell with finite rotations theory and finite element formulation, *Analytical and Computational Models of Shells*, ed. Noor, A. K., Belytschko, T. and Simo, J. C., ASME, CED, Vol. 3, pp. 135–159 (1989).

[Z1] Zienkiewicz, O. C., Taylor, R. L. & Too, J. M., Reduced integration techniques in general analysis of plates and shells, *Int. J. Num. Meth. Engng.*, **3**, 275–290 (1971).

[Z2] Zienkiewicz, O. C. Parekh, C. J. & King, I. P., Arch dams analysed by a linear finite element shell solution program, *Proc. Symp. Arch Dams, Ints. Civ. Engrs.*, London (1968).

# 9 More advanced solution procedures

The solution procedure developed in Chapters 1–3 involved a combination of incremental load or displacement control coupled with full or modified Newton–Raphson iterations. Some history relating to the early introduction of these techniques was discussed in Section 1.1 with references (up to 1972) being given in Section 1.6. Further work following similar developments can be found in [S7, B7, C20]. Although these techniques still provide the basis for most non-linear finite element computer programs, additional sophistications are required to produce effective, robust solution algorithms. In this chapter, we will describe a number of these procedures with emphasis on those techniques that I have found to be the most advantageous in practical non-linear calculations. References to other work will be given throughout but in the final Section 9.10, a summary is given of solution procedures that have not been covered earlier. Some of these will be considered in depth in Volume 2.

The first topic to be discussed is that of 'line searches'. This technique is widely used within 'mathematical programming'. There are a whole range of procedures that have been developed within this field which are extremely relevant to non-linear finite element analysis (see also the quasi-Newton and acceleration techniques of Sections 9.7 and 9.8). Good books on this topic are due to Fletcher [F7], Luenberger [L2] and Wolfe [W6].

A weakness of the full Newton–Raphson method is the high cost of each iteration. As discussed in Chapter 1, this can be reduced by adopting the 'modified Newton–Raphson method'. However, the convergence rate is then rather poor. A compromise involves the 'quasi-Newton techniques' [F7, L3, W6, D8] which are discussed in this chapter along with a set of 'acceleration methods' which are closely related.

Apart from these techniques which stem from mathematical programming, the other procedures that will be discussed in this chapter are related to the mathematical field of 'continuation techniques' [R3, R4, R6, R7, U1, C17] although, in relation to finite elements, they were often developed quite independently by engineers. It has already been indicated in Chapter 1 that 'continuation techniques' are aimed at tracing a complete path—for structures, an equilibrium path. It has also been shown in Chapter 1 that, when applying such techniques, severe difficulties can be encountered with 'limit points' where the load/deflection response becomes horizontal (or vertical for displacement control). In the present chapter, emphasis will be placed on 'arc-length' techniques for solving these problems. These techniques were originally introduced (in relation to mechanics) by Riks [R8, R9] and Wempner [W5] and later modified by a number of authors.

Other techniques required for a robust continuation method largely relate to the automatic selection of a suitable increment size as well as the automatic reduction of this step when trouble is encountered. In addition, re-start facilities should be introduced because, despite one's best attempts at automation, user intervention is often required.

Throughout the chapter, the theoretical concepts are related directly to computer applications by the amendment of the simple non-linear finite element computer program originally developed in Chapters 2–3. The reader is given the same advice as in these chapters; if he or she does not wish to get involved in the full detail of the computer implementation, it should suffice to study the flowcharts rather than the Fortran listings.

In Section 9.9, the enhanced computer program is used to re-analyse the 'NAFEMS problems' [D1.2, C1.2] originally introduced in Chapter 3. These examples are used to illustrate the application of all of the new techniques from 'line searches' to 'automatic increment reduction' and the use of 're-starts'.

## 9.1 THE TOTAL POTENTIAL ENERGY

An energy basis for non-linear structural analysis was briefly discussed in Section 1.3.3. One important advantage of adopting such a viewpoint is that it allows the introduction of various solution algorithms developed in the field of mathematical programming or unconstrained optimisation [F1, L1, W1]. Strictly, such techniques are not applicable to many structural problems such as those involving plasticity in which there is no elastic potential and the solution is path-dependent. Nonetheless, we can still beneficially adopt solution algorithms that stem from an energy approach even when the latter is not strictly applicable. With this in mind, we will now re-state in general terms the energy concepts of Section 1.3.3.

The problem of (elastic) non-linear analysis can be viewed as that of minimising the total potential energy $\phi$ which is a function of the total displacements, $\mathbf{p}$. A truncated Taylor series then leads to

$$\phi_{\mathrm{n}}(\mathbf{p}+\delta\mathbf{p}) = \phi_{\mathrm{o}}(\mathbf{p}) + \frac{\partial\phi}{\partial\mathbf{p}}\delta\mathbf{p} + \frac{1}{2}\delta\mathbf{p}^{\mathrm{T}}\frac{\partial^2\phi}{\partial\mathbf{p}^2}\delta\mathbf{p} + \cdots \tag{9.1}$$

where the subscript n means new while o means old. As indicated in Section 1.3.3, $(\partial\phi/\partial\mathbf{p})^{\mathrm{T}}$ can be identified as the out-of-balance forces or gradient, $\mathbf{g}$, of the total potential energy, while $\partial^2\phi/\partial\mathbf{p}^2$ is the tangent stiffness matrix. It follows that (9.1) can be re-expressed as:

$$\phi_{\mathrm{n}}(\mathbf{p}+\delta\mathbf{p}) = \phi_{\mathrm{o}}(\mathbf{p}) + \mathbf{g}(\mathbf{p})^{\mathrm{T}}\delta\mathbf{p} + \tfrac{1}{2}\delta\mathbf{p}^{\mathrm{T}}\mathbf{K}_{\mathrm{t}}(\mathbf{p})\delta\mathbf{p} + \cdots \tag{9.2}$$

where the principle of stationary total potential energy requires that, for equilibrium at $\mathbf{p}$,

$$\mathbf{g}(\mathbf{p}) = \frac{\partial \phi^{T}}{\partial \mathbf{p}} = \mathbf{0} \tag{9.3}$$

while for the energy to be a minimum and for the equilibrium state to be stable,

$$\delta\mathbf{p}^{T}\mathbf{K}_{t}(\mathbf{p})\delta\mathbf{p} = \delta\mathbf{p}^{T}\frac{\partial \mathbf{g}}{\partial \mathbf{p}}\delta\mathbf{p} > 0 \tag{9.4}$$

at the equilibrium point. Hence $\mathbf{K}_t$ should, for stable equilibrium, be positive define at the equilibrium point.

## 9.2 LINE SEARCHES

### 9.2.1 Theory

The line-search technique is an important feature of most numerical techniques for unconstrained optimisation and can be used with a wide range of iterative solution procedures. Detailed discussion is given in [F7, L3, W6, G4, C16]. Using such a technique, one would obtain a direction from an iterative procedure such as (in the present context) the modified Newton–Raphson iteration, i.e.

$$\delta\bar{\mathbf{p}} = -\bar{\mathbf{K}}_{t}^{-1}\mathbf{g} \tag{9.5}$$

where $\bar{\mathbf{K}}_t$ would be the tangent stiffness matrix at the end of the previous increment. The displacements would then be updated according to

$$\mathbf{p}_{n} = \mathbf{p}_{o} + \eta\delta\bar{\mathbf{p}} \tag{9.6}$$

where $\mathbf{p}_o$ would be the fixed displacements at the end of the previous iteration and $\delta\bar{\mathbf{p}}$ the fixed direction obtained from (9.5). For the simple iterative procedures of Chapters 1–3, the scalar $\eta$ in (9.6) would be set to unity. With the introduction of line searches, the scalar, $\eta$, becomes the iterative 'step length' which, for the line search, is the only variable. To derive the necessary conditions for $\phi$ to be a minimum at a particular value of $\eta$, we replace (9.1) with an equivalent Taylor expansion about the solution at $\eta$, i.e.

$$\phi_{n}(\eta + \delta\eta) = \phi_{o}(\eta) + \frac{\partial \phi}{\partial \eta}\delta\eta + \cdots = \phi_{o} + \frac{\partial \phi}{\partial \mathbf{p}}\frac{\partial \mathbf{p}}{\partial \eta}\delta\eta + \cdots = \phi_{o} + (\mathbf{g}(\eta)^{T}\delta\bar{\mathbf{p}})\delta\eta + \cdots \tag{9.7}$$

where use has been made of (9.3) and (9.6). For the solution at $\eta$ to be stationary, we require that

$$s(\eta) = \frac{\partial \phi}{\partial \eta} = \delta\bar{\mathbf{p}}^{T}\mathbf{g}(\eta) = 0. \tag{9.8}$$

In both (9.7) and (9.8), $\mathbf{g}$ has been written as a function of $\eta$ because (see (9.6)), $\mathbf{p}_0$ and $\delta\bar{\mathbf{p}}$ are fixed. A search to satisfy equation (9.8) should find the step length, $\eta$, at which the angle $\alpha$ in Figure 9.1, is zero. From (9.8) and Figure 9.1, the slope, $\tan\alpha_0$

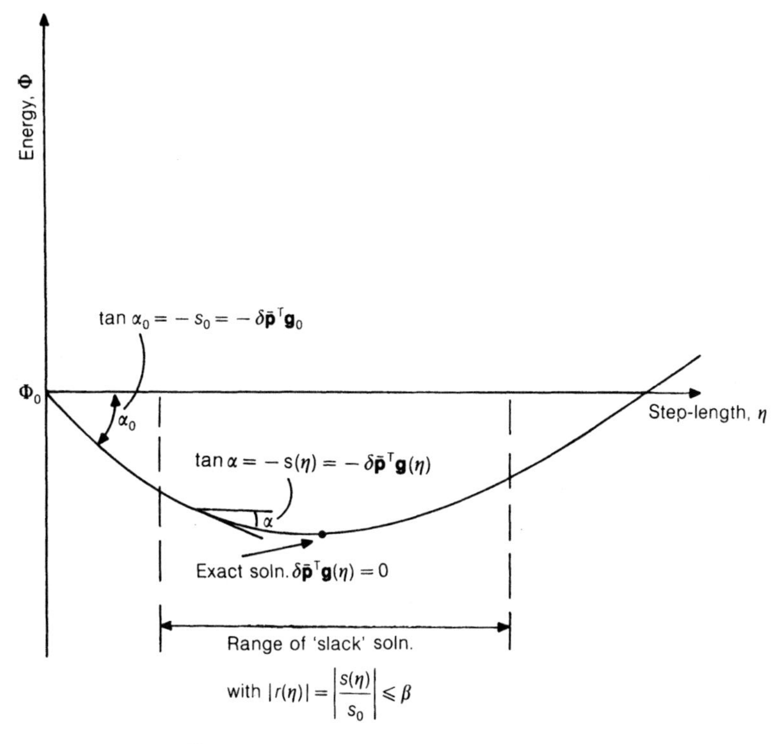


**Figure 9.1** Line searches and the energy $\phi$.

at $\eta = 0$ is $-s_0$, where

$$s_0 = s(\eta = 0) = \delta\bar{\mathbf{p}}^T\mathbf{g}(\eta = 0) = \delta\bar{\mathbf{p}}^T\mathbf{g}_0 \tag{9.9}$$

where $\mathbf{g}_0$ is the out-of-balance force vector at the end of the previous iteration. *If* we are adopting the modified Newton–Raphson method as in (9.5), it follows that

$$s_0 = -\mathbf{g}_0^T\bar{\mathbf{K}}_t^{-1}\mathbf{g}_0 = -\delta\bar{\mathbf{p}}^T\bar{\mathbf{K}}_t\delta\bar{\mathbf{p}} \tag{9.10}$$

where $\bar{\mathbf{K}}_t$ is the tangent stiffness matrix at the last converged equilibrium state. Assuming that the latter is a stable state, $\bar{\mathbf{K}}_t$ will be positive-definite and hence $s_0$ in (9.10) will be negative and the energy direction will, as illustrated in Figure 9.1, be 'downhill'.

If the full Newton–Raphson method were applied, (9.5) and (9.10) would still apply but, in place of $\bar{\mathbf{K}}_t$, from the last converged equilibrium state, we should use $\mathbf{K}_{t0}$ as computed from the last iterative solution, $\mathbf{p}_0$. Because the displacements, $\mathbf{p}_0$ relate to the last iteration and not an equilibrium state, one cannot use stability of the equilibrium state to infer that $\mathbf{K}_{t0}$ is positive-definite and there is no guarantee that, with the full N–R method, $s_0$ will be negative and the current iterative direction will be in a downhill energy direction. Various techniques have been devised for directly modifying $\mathbf{K}_t$ or otherwise changing the iterative procedure to ensure that it is positive definite [F7, L3, A2, B1]. These techniques are usually aimed to converge on another state which is stable (with positive definite $\mathbf{K}_t$). However, often in structural analysis, the situation is more complex because (see Section 9.3.1) we do not always aim at a stable equilibrium state.

For the present, we will simply assume that $s_0$ (see (9.9)) is negative and that we can apply a standard downhill line search. If this line search were to be exact, we would be looking for the smallest positive $\eta$-value to make $s(\eta)$ (see (9.8)) zero. In practice, it is inefficient to apply an 'exact' line search and, instead, we apply a 'slack' line search with the aim of making the modulus of $s(\eta)$ small in comparison with the modulus of $s_0$, i.e.

$$|\mathbf{r}(\eta)| = \left|\frac{s(\eta)}{s_0}\right| < \beta_{ls} \tag{9.11}$$

where $\beta_{ls}$ is the 'line-search tolerance'. From the author's experience, a suitable value for $\beta_{ls}$ is of the order of 0.8. The situation depicted in (9.11) is illustrated in Figure 9.1. For mathematical programming, (9.11) would be supplemented by a condition ensuring a sufficient reduction in $\phi$ which would involve the energy itself and not just the slopes, $s$ [F7, L3, G4].

In most finite element systems, we do not have the energy and hence we cannot apply this last condition. Also with plasticity, this quantity is questionable. Nonetheless, it would probably be very valuable for finite element computer programs to compute an estimate of $\phi$. This can be very economically produced at the Gauss-point level (as part of the stress updating) via

$$\Delta\phi_n \simeq \Delta\phi_o + \tfrac{1}{2}w(\boldsymbol{\sigma}_n + \boldsymbol{\sigma}_0)^T \Delta\boldsymbol{\varepsilon} \tag{9.12}$$

where $w$ includes the weighting and area contribution, $\boldsymbol{\sigma}_n$ are the current stresses, $\boldsymbol{\sigma}_o$ the stresses at the last converged equilibrium state and $\Delta\boldsymbol{\varepsilon}$ the strain increment (from the last converged equilibrium state). The author introduced such a procedure into a finite element system and used both the energy terms and slopes, $s$, with a cubic interpolation procedure in order to estimate the required step length, $\eta$. However, preliminary studies indicated that the resulting technique was very susceptible to round-off error [C9]. Hence, for the present, we will avoid the use of $\phi$ and assume only the existence of the slopes, $s$, which can be easily computed (see (9.8) and (9.9)) from the inner product of the out-of-balance forces $\mathbf{g}$ and iterative displacement direction, $\delta\bar{\mathbf{p}}$. We will also assume that we are including non-smooth non-linearity resulting from, say, plasticity or, more extremely, concrete cracking.

In these circumstances, it is simplest to apply a simple bracketed interpolation procedure as illustrated in Figure 9.2, which relates to real computations on reinforced concrete beams and slabs that were performed by the author (C16]. To produce the results in this figure, the line searches have been applied with a very tight tolerance (small $\beta_{1s}$ in (9.11)). In practice such a tight tolerance would not be used.

The main features of the procedure are illustrated in Figure 9.2(a). Having computed $\mathbf{g}$ in the standard way with $\eta_1 = \eta = 1$, the inner product, $s_1 = s(\eta = 1)$, of (9.8) was computed and related to the inner product of (9.9), $s_0$, (with $\eta = 0$) in order to obtain the ratio $r(\eta)$ of (9.11). This ratio was about $-1.4$ and is plotted as point 1 in Figure 9.2(a). Linear interpolation between point 0 and 1 then involved

$$\eta_2 = \frac{-s_0}{s_1 - s_0} = \frac{-1}{r(\eta = 1) - 1} \tag{9.13}$$

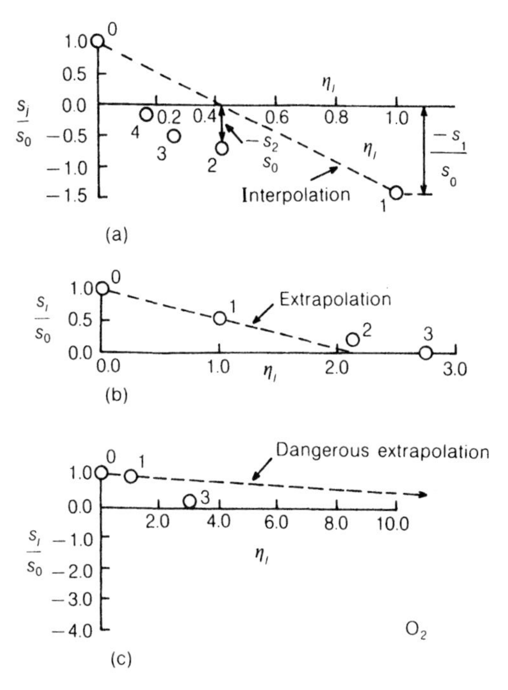


**Figure 9.2** Examples showing line-searches: (a) interpolated; (b) extrapolated; (c) limited extrapolation.

indicating a new estimate for the step length ($\eta_2$) of about 0.42. The out-of-balance force and inner product in (9.8) was then recomputed to give a ratio $r(\eta)$ (see (9.11)) of about $-0.7$. This solution is plotted as point 2 in Figure 9.2(a). In practice, with $\beta_{ls}$ in (9.11) being set to 0.8, this point would be deemed acceptable and the line-search procedure would be completed and the next iteration begun. Purely for illustration purposes, the procedure was continued with interpolation between point 0 and 2 in Figure 9.2(a), which leads to point 3 and, following interpolation between points 0 and 3, to point 4.

A generalisation of (9.13), relevant to the procedure in Figure 9.2(a), would be

$$\eta_{j+1} = \eta_j \left( \frac{-s_0}{s_j - s_0} \right). \tag{9.14}$$

This process involves an interpolation between the current 'slope' and the 'slope' at $\eta = 0$. Such a procedure will not always be appropriate.

Figure 9.2(b) shows that extrapolation can be used instead of interpolation while Figure 9.2(c) shows that this extrapolation should not be taken too far and that a maximum amplification (10 in Figure 9.2(c)) should be allowed. In a similar fashion, it is wise to introduce a minimum step-length, $\eta_{\min}$. In Figure 9.2(c), the final interpolation to obtain point 3 is between point 1 and point 2 and therefore does not fit in with the procedure of (9.14). Instead the interpolation is performed between the point (2) with a negative ratio $r = s(\eta)/s_0$ and the nearest point (1) with a positive ratio. This approach can be extended so as to involve the smallest $\eta$ value with a negative ratio.

Before detailing an algorithm to implement a line-search procedure, it should be emphasised that, for many iterations, the first 'trial' step length of $\eta = 1$ would immediately satisfy the line-search tolerance. For example, in Figure 9.2(b), the ratio $r$ of (9.11) corresponding to point 1 (with $\eta = 1$) was about 0.6. With a tolerance factor, $\beta_{ls}$ in (9.11) of 0.8, this solution would have been deemed satisfactory. Hence, no extra residual (or out-of-balance force vector) calculations would be required on account of the line-search procedure. The only extra work, in comparison with a formulation without line searches, would, in these circumstances, be the calculation of the inner product of (9.8) (with $\eta = 1$). In comparison with the other work, this work is almost negligible.

### 9.2.2 Flowchart and Fortran subroutine to find the new step length

The previous ideas are incorporated in the flowchart of Figure 9.3 and the following Fortran subroutine. The aim is to use linear interpolation as in Figure 9.2(a) and 9.2(c). However, this can only be applied once one has obtained in step length, $\eta$, relating to a negative ratio, $r$ (see (9.11)). In the absence of such a step length, the

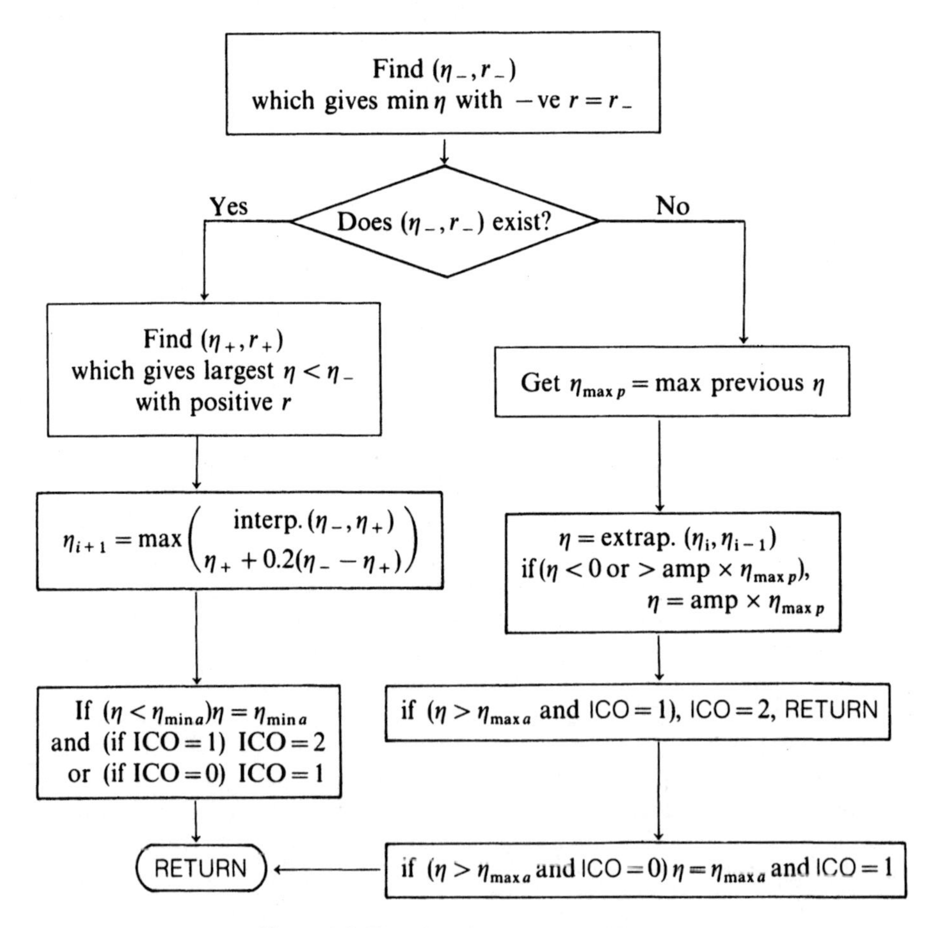


**Figure 9.3** Flowchart for subroutine SEARCH.

algorithm adopts extrapolation as in Figure 9.2(b) or, if the extrapolation goes too far (Figure 9.2(c)), simple step-length amplification using an input maximum amplification factor, 'amp'. The latter is illustrated in Figure 9.2(c).

The algorithm is aimed at computing $\eta_{i+1}$ after entering with a new step length, $\eta_i$, and equivalent ratio, $r_i$. Additional input is 'amp', the maximum amplification factor, $\eta_{\max a}$, the maximum allowed step length and, $\eta_{\min a}$, the minimum allowed step length as well as ICO, a counter that is normally zero but is set to unity once the maximum or minimum allowable step length is reached. The subroutine sets ICO to two once the maximum allowable step length has been reached twice. In these circumstances, the computer program resorts to increment reduction (see Section 9.5.1).

#### 9.2.2.1 *Fortran subroutine* SEARCH

The following FORTRAN subroutine implements the algorithm illustrated in the previous flowchart. The step lengths, $\eta$, are stored in ETA with corresponding ratios, $r$ (see (9.11)), in PRODR. These vectors are initially set up to contain

| | 1 | 2 | 3 |
|---|---|---|---|
| ETA ($\eta$) | 0 | 1 | |
| PRODR (r) | 1 | $\dfrac{s(\eta=1)}{s_0}$ | |

The routine is not optimised for efficiency so that, for example, searches are made through previous values to obtain a step length with a negative ratio. However, for practical sized problems, the time spent in computing the step length is almost negligible in comparison with the time required for the computation of a new out-of-balance force vector, $\mathbf{g}(\eta)$.

```
      SUBROUTINE SEARCH (ILS,PRODR,ETA,AMP,ETMXA,ETMNA,IWRIT,
     1                   IWR,ICO,NLSMXP)
C
C     PERFORMS LINE LOCAL LINE-SEARCH TO GET STEP LENGTH
C     IN ETA (ILS+2)
C     ETA 1-ILS HAS PREVIOUS STEP LENGTHS (ETA(1)=0.,ETA(2)=1.)
C     WITH EQUIVALENT INNER-PRODUCT RATIOS IN PRODR, (PRODR(1)=1.)
C     AMP HAS MAX AMP. FACTOR FOR STEP LENGTH,
C     ETMXA AND ETMNA HAVE MAX AND MIN ALLOWED STEP LENGTHS
C     ICO ENTERS=1 IF MAX OR MIN STEP LENGTH USED ON PREVIOUS
C     SEARCH
C     EXITS SET TO 1 IF USED ON PRESENT SEARCH
C              OR 2 IF ALSO USED ON LAST SEARCH
C
      IMPLICIT DOUBLE PRECISION (A-H,O-Z)
      DIMENSION PRODR(NLSMXP),ETA(NLSMXP)
C
C     OBTAIN INEG=NO OF PREVIOUS S-L WITH NEG. RATIO NEAREST
C     TO ORIGIN
C     AS WELL AS MAX PREVIOUS STEP LENGTH, ETMAXP
C     IF NO NEGATIVE PRODUCTS, INEG ENDS AS 999
```

```
C
      INEG=999
      ETANEG=1.D5
      ETMAXP=0.D0
      D0 10 I=1,ILS+1
      IF (ETA(I).GT.ETMAXP) ETMAXP=ETA(I)
      IF (PRODR(I).GE.0.D0) GO TO 10
      IF (ETA(I).GT.ETANEG) GO TO 10
      ETANEG=ETA(I)
      INEG=I
   10 CONTINUE
C
C     BELOW NOW ALLOWS INTERPOLATION
      IF (INEG.NE.999) THEN
C     FIND IPOS=NO OF PREVIOUS S-L WITH POS RATIO THAT IS
C     CLOSEST TO INEG (BUT WITH SMALLER S-L)
      IPOS=1
      DO 20 I=1,ILS+1
      IF (PRODR(I).LT.0.D0) GO TO 20
      IF (ETA(I).GT.ETA(INEG) ) GO TO 20
      IF (ETA(I).LT.ETA(IPOS) ) GO TO 20
      IPOS=I
   20 CONTINUE
C
C     INTERPOLATE TO GET S-L ETAINT
      ETAINT=PRODR(INEG)*ETA(IPOS)-PRODR(IPOS)*ETA(INEG)
      ETAINT=ETAINT/(PRODR(INEG)-PRODR(IPOS) )
C     ALTERNATIVELY GET ETAALT ENSURING A REASONABLE CHANGE
      ETAALT=ETA(IPOS)+0.2*(ETA(INEG)-ETA(IPOS) )
C     TAKE MAX
      IF (ETAINT.LT.ETAALT) ETAINT=ETAALT
C     OR MIN STEP LENGTH
      IF (ETAINT.LT.ETMNA) THEN
      ETAINT=ETMNA
      IF (ICO.EQ.1) THEN
      ICO=2
      WRITE (IWR,1010)
 1010 FORMAT(/,1X,'MIN STEP-LENGTH REACHED TWICE')
      ELSEIF (ICO.EQ.0) THEN
      ICO=1
      ENDIF
      ENDIF
C
      ETA(ILS+2)=ETAINT
      IF (IWRIT.EQ.1) THEN
      WRITE (IWR,1001) (ETA(I),I=1,ILS+2)
 1001 FORMAT(/,1X,'L-S PARAMETERS',/,1X,'ETAS ',(6G11.3) )
      WRITE(IWR,1002) (PRODR(I),I=1,ILS+1)
 1002 FORMAT(/,1X,'RATIOS',(6G11.3) )
      ENDIF
      RETURN
```

```
C
C
C        BELOW WITH EXTRAPOLATION
         ELSE IF (INEG.EQ.999) THEN
C        SET MAX TEMP STEP LENGTH
         ETMXT=AMP*ETMAXP
         IF (ETMXT.GT.ETMXA) ETMXT=ETMXA
C        EXTRAP. BETWEEN CURRENT AND PREVIOUS
         ETAEXT=PRODR(ILS+1)*ETA(ILS)-PRODR(ILS)*ETA(ILS+1)
         ETAEXT=ETAEXT/(PRODR(ILS+1)-PRODR(ILS))
         ETA(ILS+2)=ETAEXT
C        ACCEPT IF ETAEXT WITHIN LIMITS
         IF (ETAEXT.LE.0.D0.OR.ETAEXT.GT.ETMXT) ETA(ILS+2)=ETMXT
         IF (ETA(ILS+2).EQ.ETMXA.AND.ICO.EQ.1) THEN
         WRITE (IWR,1003)
    1003 FORMAT(/,1X,' MAX STEP-LENGTH AGAIN')
C        STOP 'SEARCH 1003'
         ICO=2
         RETURN
         ENDIF
         IF (ETA(ILS+2).EQ.ETMAXA) ICO=1
         IF (IWRIT.EQ.1) THEN
         WRITE (IWR,1001) (ETA(I),I=1,ILS+2)
         WRITE (IWR,1002) (PRODR(I),I=1,ILS+1)
         ENDIF
         ENDIF
         RETURN
         END
```

### 9.2.3 Implementation within a finite element computer program

We will now outline a procedure whereby the computer programs of Chapters 2 and 3 could be modified to include line searches using the previous subroutine SEARCH. In Sections 9.4.2 and 9.6, we will give a modified computer program that includes not only line searches but also the arc-length method, automatic increment sizes, accelerations and automatic increment reduction. In order to introduce all these options, a number of changes have had to be made to the structure of the original programs. At the present stage, we will merely outline the way in which the programs of Chapters 2 and 3 could be most simply modified to introduce line searches. The reader might like to make these changes himself (or herself) or might, at this stage, prefer to simply follow the ideas and wait until Sections 9.4.2 and 9.6 before considering detailed implementation.

In order to incorporate the previous subroutine within the finite element computer program of Chapters 2 and 3, we must firstly input the line-search parameters.

#### *9.2.3.1 Input*

In order to achieve this with the minimum disruption to the previous programs, we can introduce a COMMON block, DATLS into both the main program NONLTC of

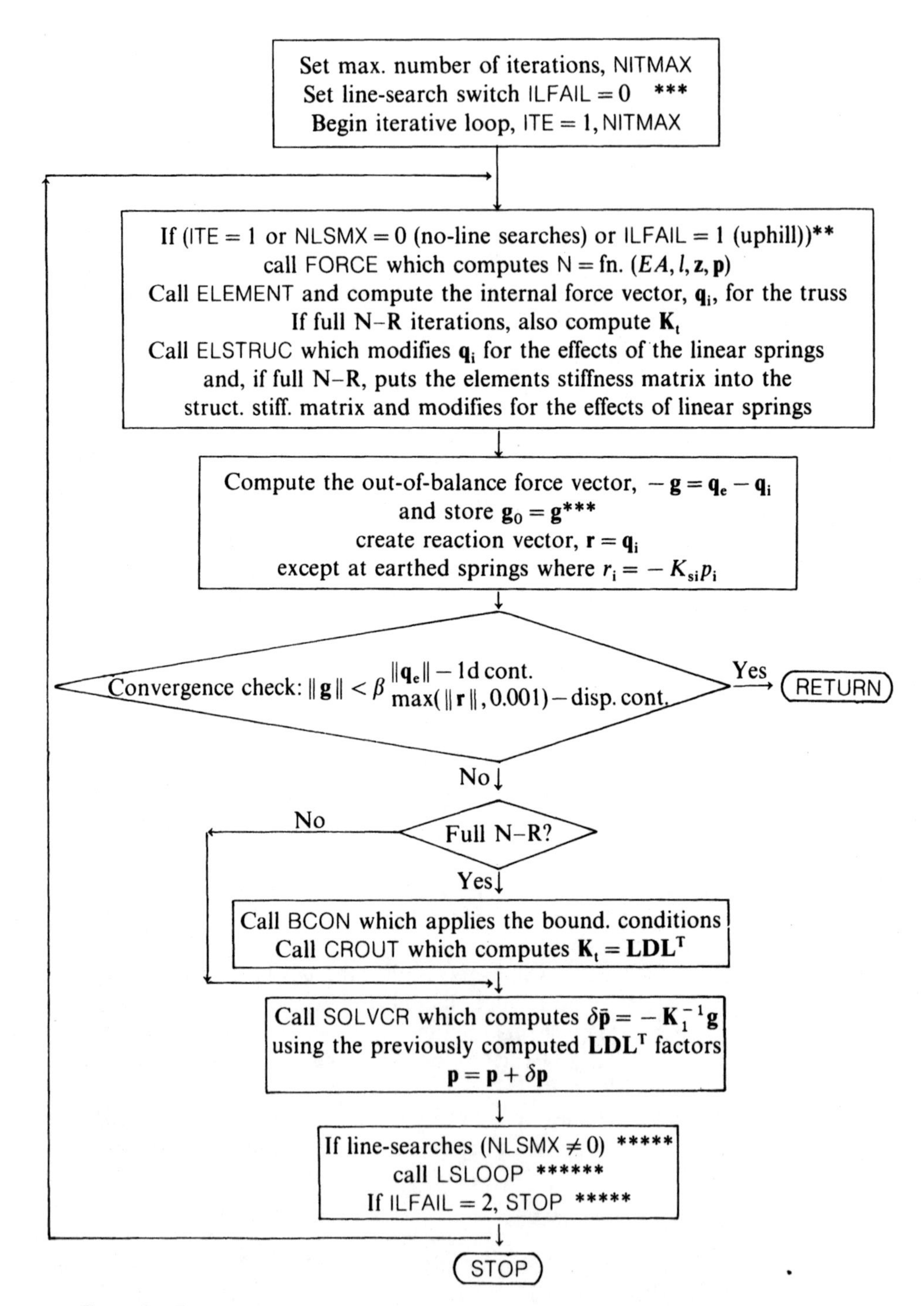


**Figure 9.4** Flowchart for subroutine ITER when modified to include line searches.

Section 2.5.1 and into the main iterative routine ITER of Section 2.4.2. This block is

```
COMMON/DATLS/NLSMX,PERMLS, AMPMX, ETMXA,ETMNA,GO(5)
```

where the constants will become apparent from the following and G0(5) will contain the initial (for any iteration) out-of-balance force vector, $\mathbf{g}_0$ (at $\eta = 0$). In addition the input line:

```
READ (IRE,*) BETOK,ITERTY
```

and accompanying output line of program NONLTC (see Section 2.5.1) can be amended to the following

```
      READ (IRE,*) BETOK,ITERTY,NLSMX
      WRITE (IWR,1003) BETOK,ITERTY,NLSMX
 1003 FORMAT(/,1X,'CONV. TOL FACTOR, BETOK=  ',G13.5,/,
     1          1X,'ITERATIVE SOLN. TYPE, ITERTY=  ',I5,/,
     2          5X,'= 1, FULL N-R;       =2, MOD. N-R',/,
     3          1X,'MAX. NO. OF L-SEARCHES=  ',I5)
C     BELOW SPECIFIC TO LINE SEARCHES
      IF (NLSMX.NE.0) THEN
      READ (IRE,*) PERMLS, AMPMX, ETMXA, ETMNA
      WRITE (IWR, 1009) PERMLS, AMPMX, ETMXA, ETMNA
 1009 FORMAT(/, 1X, 'LINE SEARCH PARAMS ARE',/,
     1          3X, 'TOLERANCE ON RATIO, PERMLS = ',G13.4/,
     2          3X, 'MAX. AMP. AT ANY STEP, AMPMX = ',G13.4,/,
     3          3X, 'MAX. TOTAL STEP-LENGTH, ETMXA = 'G13.4,/,
     4          3X, 'MIN. TOTAL STEP-LENGTH, EXTMNA = ',G13.4)
```

### 9.2.3.2 *Changes to the iterative subroutine* ITER

Figure 2.4 contains the flowchart for subroutine ITER which iterates, at the structural level, to equilibrium. In order to introduce line searches, this flowchart can be altered as indicated in Figure 9.4 where the asterisked sections relate to the changes.

Apart from the introduction of the COMMON block DATLS (see Section 9.2.3.1), the changes to the subroutine ITER of Section 2.4.2 could involve:

(1) At the very start, setting ILFAIL = 0
(2) Inserting before CALL FORCE,

```
IF (ITE.EQ.1.OR.NLSMX.EQ.0.OR.ILFAIL.EQ.1)
```

and after CALL ELSTRUC,

```
ENDIF
```

This avoids the recomputation of the internal force vectors immediately after a call to the line search loop (via CALL ISLOOP).

(3) Replacing the DO 10 loop to form the out-of-balance force vector with:

```
C     BELOW FORMS GM=OUT-OF-BALANCE FORCE VECTOR
C     AND REACTION VECTOR
C     IN ADDITION NOW SAVES GO FOR LINE SEARCHES
      DO 10 I=1,NV
      GM(I) = 0.D0
```

```
      REAC(I) = FI(I)
      IF(IBC(I).EQ.0) THEN
      GM(I)=QEX(I)-FI(I)
      ENDIF
      GO(I)= -GM(I)
   10 CONTINUE
```

where the amendments have been underlined.

(4) Just below the DO 30 loop to have:

```
C     ABOVE UPDATES DISPS.
C
      IF (NLSMX.NE.0) CALL LSLOOP(PT,GM,IBC,IWRIT,IWR,ITERTY,NV,
     1       FI,QEX,AKTS,AKTE,D)
      IF (ILFAIL.EQ.2) STOP C
C     IF (IWRIT.EQ.1) WRITE (IWR,1004) (PT(I),I=1,NV)
```

where, again, the amendments have been underlined.

#### 9.2.3.3 *Flowchart for line-search loop at the structural level*

In the above, we have called a subroutine =LSLOOP which calls subroutine SEARCH (see Section 9.2.2.1) and performs the line-search loop at the structural level. In the flowchart which is given in Figure 9.5, $\boldsymbol{\eta}$ is the vector ETA of Section 9.2.2 and $\mathbf{r}$ is the vector =PRODR containing the ratios $\mathbf{r}$ of (9.11). These arrays must be defined in subroutine =LSLOOP which, via the common /DATLS/ of Section 9.2.2.1 also has access to the array =GO = $\mathbf{g}_0$ formed in subroutine =ITER (see Section 9.2.3.2).

This routine returns to ITER with ILFAIL = 0 if the line-search procedure has been satisfactory. If the iterative direction is 'uphill' (see Section 9.2.1), it returns with ILFAIL = 1 and ITER continues using the default step length of 1 (having abandoned the line search). If the line-search procedure fails, ILFAIL is set to 2 and on the return to ITER, the safest option would be to STOP (see the flowchart in Figure 9.4). Later (in Section 9.5.1 and 9.6.5), we will instead adopt automatic increment reduction.

A Fortran subroutine relating to a slightly modified form of the above will be given in Section 9.4.2.1. It has already been pointed out in Section 9.2.1 that, for many iterations, the introduction of the line-search algorithm (with a 'slack tolerance') will introduce very little extra work. This point can be checked by studying the flowcharts of Figures 9.4 and 9.5. They show that if, on the first application of the satisfaction check in Figure 9.5 (with $l = 1$), the line-search tolerance is satisfied, the extra computational work in comparison with a standard N–R or mN–R procedure only involves the computation of $s_0(\eta = 0)$ and $s(\eta = 1)$. The advantage of introducing the line-search technique is that the ratio $r$ (9.11) gives a very effective indication of whether or not it is safe to proceed directly to the next iteration. In very many instances, with 'slack line searches' and $\beta_{ls}$ (see (9.11)) $\simeq 0.8$, the tolerance check of (9.11) will be satisfied within the line-search loop of Figure 9.5 with $l = 1$ so that no extra residual (or out-of-balance force) calculations are required. However, when $|r| > \beta_{ls}$, without the line-search facility, the iterative procedure to enforce equilibrium will often diverge.

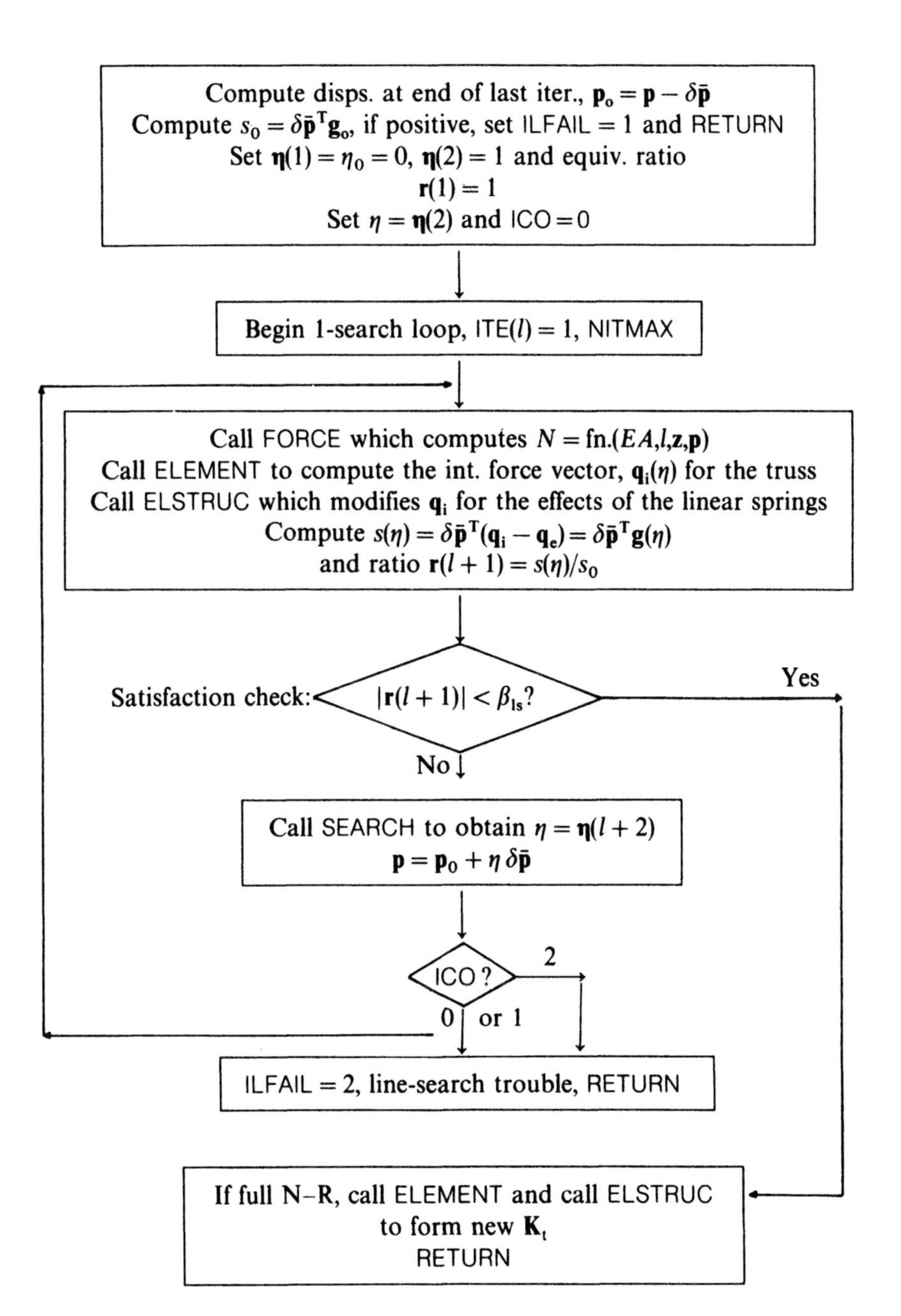


**Figure 9.5** Flowchart for subroutine LSLOOP.

## 9.3 THE ARC-LENGTH AND RELATED METHODS

The arc-length methods are intended to enable solution algorithms to pass limit points (maximum and minimum loads—see Chapter 1). Prior to their introduction, analysts either used artificial springs [S3.1, W9], switched from load to displacement control [S1.1] or abandoning equilibrium iterations in the close vicinity of the limit point [B7, B8]. In relation to structural analysis, the arc-length method was originally introduced by Riks [R8, R9] and Wempner [W5] with later modifications being made by the author [C11, C19] and other [R1, R2, S2, F9, G3, B5, S6]. Details will be discussed later. Closely related work can be found in the mathematical literature [K1, K2, R6] with the first such paper appearing to be due to Haslegrove [H2].

### 9.3.1 The need for arc-length or similar techniques and examples of their use

Figures 9.6(a)–(c) show three possible load/deflection curves involving limit points with both 'snap-throughs' (Figure 9.6(a)) and 'snap-backs' (Figure 9.6(b)). Simple examples involving both of these phenomena have already been discussed in Chapters 1–3. More complex examples include as wide-ranging phenomena as the 'buckling' of shells (Figure 9.7 and [C11]), stiffened-plated structures (Figure 9.8 and [C11]) and the cracking of reinforced concrete (Figure 9.9 and [C14]).

It has already been indicated in Chapters 1–3 that the true response in Figures 9.6(a) and (b) would involve both inertia effects and dynamics. Under load control the dynamic response in Figure 9.6(a) would follow the dashed line (possibly followed by a small damped oscillation around point C). In contrast, the solid, static line from A to C would maintain equilibrium but be unstable under load control but stable under displacement control. Under displacement control, the dynamic response in Figure 9.6(b) would follow the dashed line between A and C with the equivalent solid line being static but, again, unstable.

The reader might ask 'Why do we attempt to follow such unstable paths?' Why not (a) jump straight from point A to point C using statics or (b) follow the dashed line using dynamics? The latter should indeed be possible with a non-linear dynamic finite element program. However, this is not a simple solution and we are here concerned with static computer programs. As already discussed in Chapters 2 and 3, one can, in some circumstances, move directly from A to C but, in many other cases, the large jump is too much for the iterative solution procedure to handle. Also, the required response may be that shown in Figure 9.6(c) where there is no point C and the structure collapses at A. Again one might ask 'Why bother to proceed beyond point A when the structure has collapsed?' There are a number of answers:

(i) 'A' may only be the local maximum (see Figures 9.6(a) and (b)).
(ii) The 'structure' being analysed may be only a component. It may later be desirable to incorporate the load/deflection response of this component within a further analysis of a complete structure.
(iii) In the above and other situations, it may be important to know not just the collapse load but whether or not this collapse is of a 'ductile' (Figure 9.6(d)) or 'brittle' (Figure 9.6(c)) form.

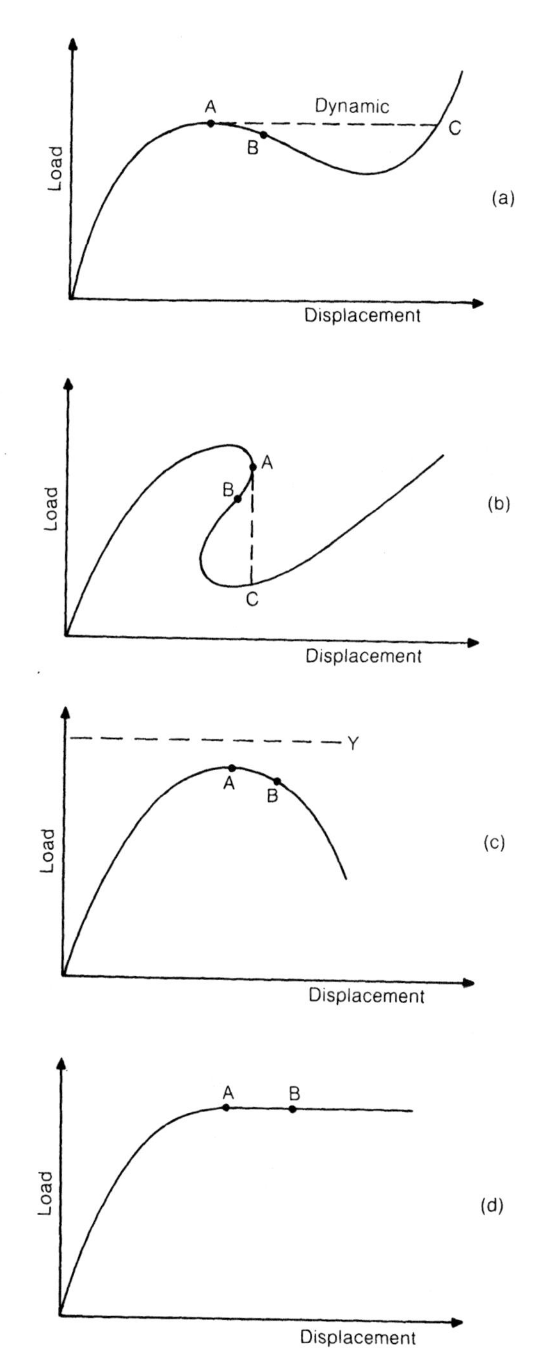


**Figure 9.6** Various load/deflection curves: (a) snap-through; (b) snap-back; (c) 'brittle' collapse; (d) 'ductile' collapse.

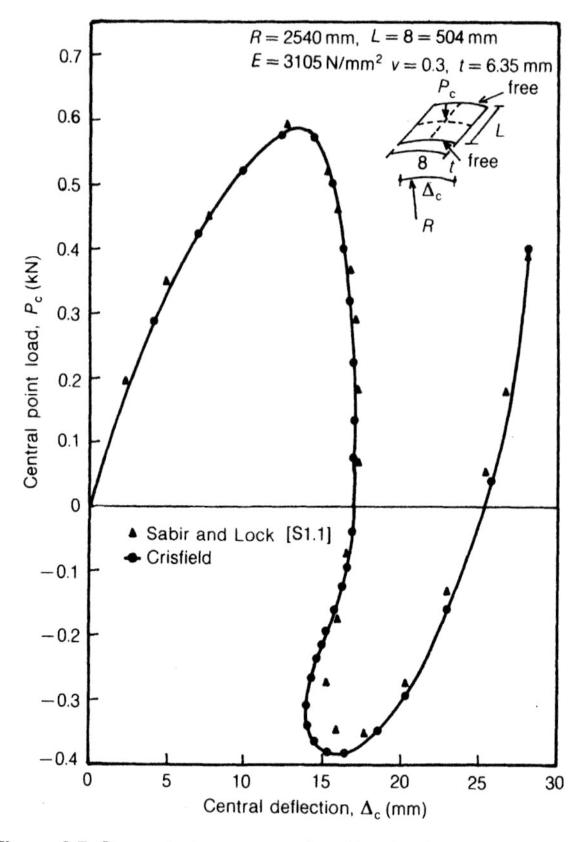


**Figure 9.7** Computed responses for thin simply supported shell.

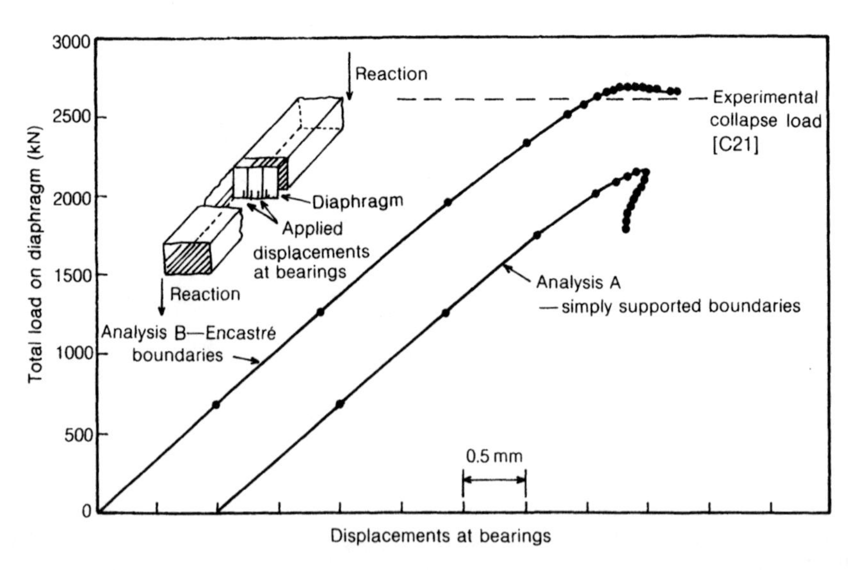


**Figure 9.8** Idealisation and computed responses for steel box-girder diaphragm (with experimental collapse load).

(iv) Even if A is the only maximum, it may be necessary to move to point B (just beyond the maximum) because
  (a) This confirms that we have indeed just passed a limit-point. Many analysts simply apply further load increments until the solution procedure fails to converge (say applying the increment to level Y in Figure 9.6(c)) and then assume that this iterative failure reflects a structural failure. Sadly, iterative failure can occur for other reasons and this approach is not recommended. Even to establish the plateau in Figure 9.6(d), load control is inappropriate.
  (b) Having converged on a point such as B, it is then possible to investigate (preferably with the aid of graphics) the structural state (stresses, strains, deflections, plastic zones, etc.) at B in order to gain insight into the mechanism or cause of the structural failure.

(v) Figure 9.6(d) illustrates the type of load–deflection response, stemming from an elastic/perfectly plastic, geometrically linear analysis. For this type of problem, the load corresponding to the plateau could also be obtained via a 'plastic mechanism' or 'yield-line method'. With standard load control, it would be very difficult to reproduce this 'limit load' and, without converging on a point such as B (Figure 9.6(d)), it would be impossible to fully establish the 'mechanism'. With the aid of the arc-length method, the author has used the finite element method to obtain plastic mechanism solutions for reinforced concrete structures [C18].

Before discussing the detail of the arc-length and other related methods, one should mention standard 'displacement control' for which a solution algorithm has already been given in Section 2.2.5. For many problems, it is possible to use such a technique to obtain, for example, the complete solid line in Figure 9.6(a). By and large, this technique can be applied when an equivalent displacement-control could be used in an experiment. There are, however, occasions when this is difficult or impossible. For example, one may wish to obtain the scalar multiple of an 'abnormal vehicle loading' on a bridge in which the 'abnormal loading' involved a set of, say, sixteen wheel loads of equal magnitude [C18] (or some other fixed loading pattern—Figure 9.9).

Other examples include structures for which the response involves a 'snap-back behaviour' as illustrated in Figure 9.6(b). Such a response is typical of shell structures. Figure 9.7 shows the load–deflection response that was computed by the author for the response of a thin cylindrical shell subject to a central point load [C11]. The solution obtained by the author involved the arc-length method coupled with automatic increments (see Section 9.5). The solution by Sabir and Lock [S1.1] was obtained by switching from load to displacement control.

Another example involves the large-deflection elasto-plastic analysis of the stiffened diaphragm from a box-girder bridge which was analysed by the author [C11, C21]. Because of the limited computer power, an elastic substructuring technique was combined with some structural idealisations [C11, C20] so that only the diaphragm was analysed in a non-linear manner (Figure 9.8). In order to bracket the experimental behaviour, two extreme boundary conditions were applied to the out-of-plane deflections at the edges of the diaphragm. For analysis A (curve A, Figure 9.8), the boundaries were assumed to be simply supported, while for analysis B (curve B, Figure 9.8), they were assumed to be encastré. (The solution for analysis A has been

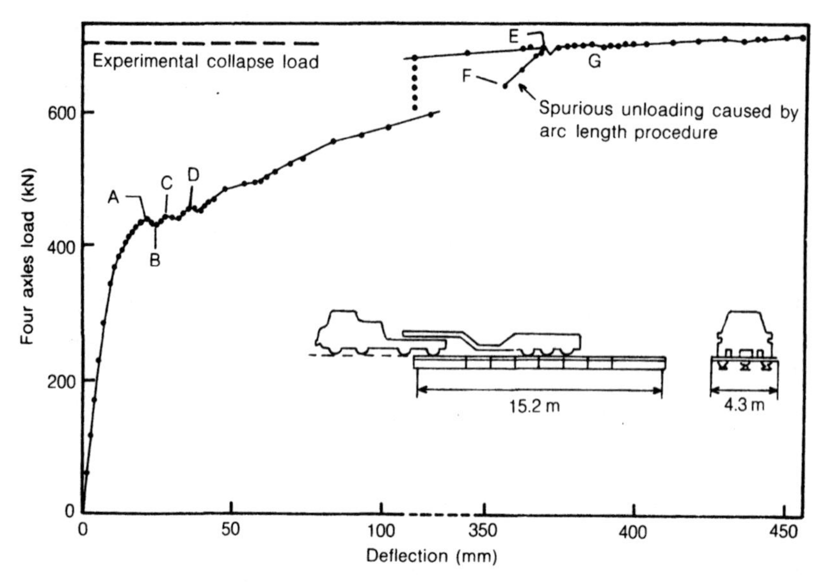


**Figure 9.9** Computed response for prestressed concrete bridge (with experimental collapse load).

offset from the origin in Figure 9.8 in order to avoid interference with the solution curve for analysis B.) For both analyses, the structure was 'loaded' by incrementing prescribed displacements across bearings and combining this displacement control with the arc-length method. In addition the length increments (see Section 9.3.2) were automatically computed (see Section 9.5). As illustrated in Figure 9.8, the resulting solutions bracketed the experimental collapse load. Solution A resulted in a snap-back form of response which could not be obtained without the addition of the arc-length constraint (Section 9.3.2). With ordinary displacement control, a point was reached at which numerical convergence could not be obtained. At this stage, because of the suddenness of the 'softening' (Figure 9.8), a maximum load was not properly defined. It was this example that led to the author's work on the arc-length procedure.

On switching the emphasis of my work from steel to concrete structures [C14, C15], I was surprised to find that snap-through and snap-back responses were equally relevant to concrete structures. Figure 9.9 shows an example involving the analysis of a prestressed beam-and-slab bridge [C14] for which the numerical solution clearly exhibits snap-throughs and local maxima. (The strange 'blip' in the solution was, at the time, attributed to a defect in the arc-length method. It now seems likely that the phenomenon was caused by the solution procedure assuming that a negative pivot in the factorised $\mathbf{K}_t$ was associated with a 'limit point' when it was probably caused by a bifurcation associated with material instability [C12].) Other work with concrete 'softening' has encountered snap-backs [C14, D5].

Procedures for directly computing the critical points have been discussed in [M5, R7, E2, F5]. From an engineering viewpoint, the precise computation of limit points does not seem to be of major importance—a continuation solution passing over the point will usually locate the point to sufficient engineering accuracy (plasticity will often limit the increment size). On the other hand, it may be important to locate bifurcation points with reasonable accuracy, in order to be able to switch to the post-buckling path (see Section 9.10).

### 9.3.2 Various forms of generalised displacement control

As a starting point to various 'continuation methods', we can write the equilibrium equations as

$$\mathbf{g}(\mathbf{p}, \lambda) = \mathbf{q}_i(\mathbf{p}) - \lambda \mathbf{q}_{ef} = \mathbf{0} \tag{9.15}$$

where $\mathbf{q}_i$ are the internal forces which are functions of the displacements, $\mathbf{p}$, the vector $\mathbf{q}_{ef}$ is a 'fixed external loading vector' and the scalar $\lambda$ is a 'load-level parameter' that multiplies $\mathbf{q}_{ef}$. Equation (9.15) defines a state of 'proportional loading' in which the loading pattern is kept fixed. Non-proportional loading will be discussed in Section 9.5.3.

In Section 9.5, we will introduce a simple method whereby the scalar loading parameter, $\lambda$, may be automatically incremented thus producing a 'load-controlled continuation method'. However, as already discussed, the major limitation of load control is that, near a limit point, there may be no intersection between the equilibrium path of (9.15) and the plane $\lambda$ = constant which represents the next 'load level'. Various forms of 'arc-length methods' have stemmed from the original work of Riks [R8, R9] and Wempner [W5] who aimed to find the intersection of (9.15) with $s$ = constant, where $s$ is the arc length, defined by:

$$s = \int \mathrm{d}s \tag{9.16}$$

and

$$\mathrm{d}s = \sqrt{(\mathrm{d}\mathbf{p}^T \mathrm{d}\mathbf{p} + \mathrm{d}\lambda^2 \psi^2 \mathbf{q}_{ef}^T \mathbf{q}_{ef})}. \tag{9.17}$$

The scaling parameter $\psi$ is required in (9.17) because the load contribution depends on the adopted scaling between the load and displacement terms. Having introduced the arc-length, $s$, one may attempt to directly solve

$$\mathbf{g}(s) = \mathbf{q}_i(\mathbf{p}(s)) - \lambda(s)\mathbf{q}_{ef} = \mathbf{0} \tag{9.18}$$

using a higher-order ODE method [W3]. However, with this approach it is often very difficult to successfully limit the 'drift from equilibrium' and hence 'predictor–corrector' methods are usually used. For load control, these would involve the techniques of Chapters 1–3 with an incremental, tangential, predictor being followed by Newton–Raphson or modified Newton–Raphson iterations which act as 'correctors'. For the arc-length methods, one would effectively replace the differential form of (9.17) with an incremental form:

$$a = (\Delta\mathbf{p}^T \Delta\mathbf{p} + \Delta\lambda^2 \psi^2 \mathbf{q}_{ef}^T \mathbf{q}_{ef}) - \Delta l^2 = 0 \tag{9.19}$$

where $\Delta l$ is the fixed 'radius' of the desired intersection (see Figure 9.10 where for brevity we have written $\mathbf{q}_{ef}$ as $\mathbf{q}$) which is an approximation to the incremental arc length. The vector $\Delta\mathbf{p}$ and scalar $\Delta\lambda$ are incremental (not iterative, for which we will us $\delta$s) and relate back to the last converged equilibrium state (see Figure 9.10).

The main essence of the arc-length methods is that the load parameter, $\lambda$, becomes

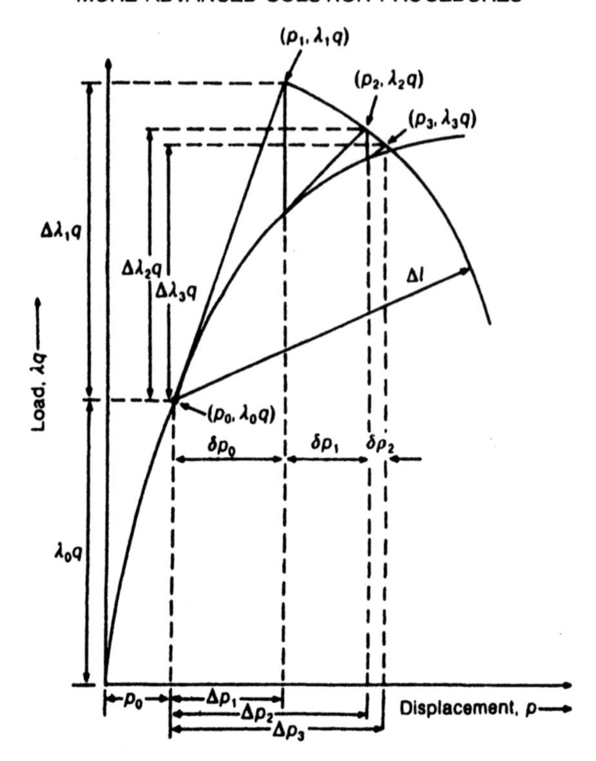


**Figure 9.10** Spherical arc-length procedure and notation for one degree of freedom system (with $\psi = 1$).

a variable. Hence, together with the $n$ displacement variables, we have a total of $n + 1$ variables. To solve for these, we have the $n$ equilibrium equations of (9.15) and the one constraint equation of (9.19). Following Riks [R8, R9] and Wempner [W5], we can solve for these $n + 1$ variables by directly applying the Newton–Raphson method to (9.15) and (9.19). From our previous developments, the Newton–Raphson method is best introduced via a truncated Taylor series with the subscript n meaning new and o meaning old. From (9.15) and (9.19) this leads to

$$\mathbf{g}_\mathrm{n} = \mathbf{g}_\mathrm{o} + \frac{\partial \mathbf{g}}{\partial \mathbf{p}}\delta \mathbf{p} + \frac{\partial \mathbf{g}}{\partial \lambda}\delta\lambda = \mathbf{g}_\mathrm{o} + \mathbf{K}_\mathrm{t}\,\delta\mathbf{p} - \mathbf{q}_\mathrm{ef}\,\delta\lambda = 0 \tag{9.20a}$$

$$a_\mathrm{n} = a_\mathrm{o} + 2\Delta\mathbf{p}^\mathrm{T}\delta\mathbf{p} + 2\Delta\lambda\,\delta\lambda\,\psi^2\mathbf{q}_\mathrm{ef}^\mathrm{T}\mathbf{q}_\mathrm{ef} = 0. \tag{9.20b}$$

Equations (9.20a) and (9.20b) can be combined and, after setting $\mathbf{g}_\mathrm{n}$ and $a_\mathrm{n}$ to zero, solved for $\delta\mathbf{p}$ and $\delta\lambda$, giving

$$\begin{pmatrix} \delta\mathbf{p} \\ \delta\lambda \end{pmatrix} = -\begin{bmatrix} \mathbf{K}_\mathrm{t} & -\mathbf{q}_\mathrm{ef} \\ 2\Delta\mathbf{p}^\mathrm{T} & 2\Delta\lambda\psi^2\mathbf{q}_\mathrm{ef}^\mathrm{T}\mathbf{q}_\mathrm{ef} \end{bmatrix}^{-1} \begin{pmatrix} \mathbf{g}_\mathrm{o} \\ a_\mathrm{o} \end{pmatrix}. \tag{9.21}$$

The augmented 'Jacobian' or 'stiffness matrix' within the square brackets in (9.21) remains non-singular even when $\mathbf{K}_\mathrm{t} = \partial\mathbf{g}/\partial\mathbf{p}$ is singular. (Equations of the form of (9.21) are also known as 'bordered' equations [R14]. Other interesting structural uses of bordered equations have been given by Kroplin *et al.* [K7, K9, K10]). Equations (9.21) can be used directly to find the changes $\delta\mathbf{p}$ and $\delta\lambda$. However, in contrast to $\mathbf{K}_t$, the augmented stiffness matrix in (9.21) is neither symmetric nor banded.

#### 9.3.2.1 *The 'spherical arc-length' method*

Instead of solving (9.21), one may directly introduce the constraint of (9.20b) by following Batoz and Dhatt [B4] for displacement control at a single point (see Section 9.3.2.3). To this end, the iterative displacement, $\delta\mathbf{p}$, is split into two parts. Hence the Newton change at the new unknown load level, $\lambda_n = \lambda_o + \delta\lambda$, becomes

$$\delta\mathbf{p} = -\mathbf{K}_t^{-1}\mathbf{g}(\mathbf{p}_o, \lambda) = -\mathbf{K}_t^{-1}(\mathbf{q}_i(\mathbf{p}_o) - \lambda_n\mathbf{q}_{ef}) = -\mathbf{K}_t^{-1}(\mathbf{g}(\mathbf{p}_o, \lambda_o) - \delta\lambda\mathbf{q}_{ef}). \tag{9.22}$$

One can work with either of the forms on the far right-hand side of (9.22). The penultimate form involves a complete split into internal forces, $\mathbf{q}_i$, and external forces, $\lambda\mathbf{q}_{ef}$, while the final form can be expressed as

$$\delta\mathbf{p} = -\mathbf{K}_t^{-1}\mathbf{g}_o + \delta\lambda\mathbf{K}_t^{-1}\mathbf{q}_{ef} = \delta\bar{\mathbf{p}} + \delta\lambda\delta\mathbf{p}_t \tag{9.23}$$

with $\delta\mathbf{p}_t = \mathbf{K}_t^{-1}\mathbf{q}_{ef}$. Using this form, $\delta\bar{\mathbf{p}}$ is the iterative change that would stem from the standard load-controlled Newton–Raphson method (at a fixed load level, $\lambda_o$), while $\delta\mathbf{p}_t$ is the displacement vector corresponding to the fixed load vector, $\mathbf{q}_{ef}$. If the modified Newton–Raphson method is adopted, $\delta\mathbf{p}_t$ must be computed for the initial 'predictor' (Section 9.4.3) step but (because $\mathbf{K}_t$ is fixed) does not change during the iterations.

Having obtained $\delta\mathbf{p}$ from (9.23) (with $\delta\lambda$ still unknown), the new incremental displacements are

$$\Delta\mathbf{p}_n = \Delta\mathbf{p}_o + \delta\mathbf{p} = \Delta\mathbf{p}_o + \delta\bar{\mathbf{p}} + \delta\lambda\,\delta\mathbf{p}_t \tag{9.24}$$

where $\delta\lambda$ is the only unknown. It can be found from the constraint of (9.19), which can be re-expressed as

$$(\Delta\mathbf{p}_o^T\Delta\mathbf{p}_o + \Delta\lambda_o^2\psi^2\mathbf{q}_{ef}^T\mathbf{q}_{ef}) = (\Delta\mathbf{p}_n^T\Delta\mathbf{p}_n + \Delta\lambda_n^2\psi^2\mathbf{q}_{ef}^T\mathbf{q}_{ef}) = \Delta l^2. \tag{9.25}$$

Substitution from (9.24) for $\Delta\mathbf{p}_n$ into (9.25) leads to the scalar quadratic equation:

$$a_1\delta\lambda^2 + a_2\delta\lambda + a_3 = 0 \tag{9.26}$$

where

$$a_1 = \delta\mathbf{p}_t^T\delta\mathbf{p}_t + \psi^2\mathbf{q}_{ef}^T\mathbf{q}_{ef} \tag{9.27a}$$

$$a_2 = 2\delta\mathbf{p}_t(\Delta\mathbf{p}_o + \delta\bar{\mathbf{p}}) + 2\Delta\lambda_o\psi^2\mathbf{q}_{ef}^T\mathbf{q}_{ef} \tag{9.27b}$$

$$a_3 = (\Delta\mathbf{p}_o + \delta\bar{\mathbf{p}})^T(\Delta\mathbf{p}_o + \delta\bar{\mathbf{p}}) - \Delta l^2 + \Delta\lambda_o^2\psi^2\mathbf{q}_{ef}^T\mathbf{q}_{ef} \tag{9.27c}$$

which can be solved for $\delta\lambda$ so that, from (9.24), the complete change is defined. (The issue of the choice of root will be discussed in Section 9.4.1.) In contrast to the use of (9.2.1), this technique only requires the inversion (or, in practice, factorisation) of the banded symmetric tangent stiffness matrix, $\mathbf{K}_t$. In theory, the method suffers from the limitation that, precisely at the limit point, $\mathbf{K}_t$ will be singular and the equations cannot be solved. In reality, the author has not found this to be a significant problem, because one appears never to arrive precisely at the limit point (see also [B4]).

Nonetheless, a number of authors have addressed the issue of 'stabilising' the stiffness matrix near to limit points. Riks and Rankin [R14] have proposed two techniques, one of which requires knowledge of the lowest eigenmode of $\mathbf{K}_t$. Felippa [F1, F2] has combined a partitioning device of Rheinboldt [R6] with the original 'fictitious spring approach' of Sharifi and Popov [S3.1]. As already discussed, in the mathematical programming literature there are approaches to modify $\mathbf{K}_t$ in the presence of 'uphill directions' [F7, L3]. These methods must also be relevant.

We have not yet discussed the 'scaling parameter', $\psi$, in the constraints of (9.19) and (9.25). Both the author [C11] and Ramm [R1, R2] independently concluded that, for practical problems involving a realistic number of variables, the 'loading terms' (those involving $\psi$) had little effect and hence advocated setting $\psi$ to zero. As a result, the constraint should be considered as 'cylindrical' rather than 'spherical'. This cylindrical constraint will be adopted in the detailed treatment of Section 9.4. Padovan and Arechaga [P1] and Park [P7] proposed adopting a variable $\psi$ which is large in the initial stages (so that the technique then tends towards load control) and small when the limit points is approached. In Section 9.5.2, we will discuss a procedure that produces a similar effect by switching to 'arc-length control' as the limit point is reached. Felippa [F1, F2] and Simo *et al.* [S6] have taken the scaling or weighting further by replacing the $\Delta\mathbf{p}^T\Delta\mathbf{p}$ term in (9.19) $\Delta\mathbf{p}^T\mathbf{S}\,\Delta\mathbf{p}$, where one suggestion for the scaling matrix $\mathbf{S}$ in Diag ($\mathbf{K}_t$). Weighting schemes have also been advocated by de Borst [D5] and Gierlinski and Graves-Smith [G3].

At this stage, it would be a good idea to describe the progress of the arc-length method in relation to Figure 9.10 (where, for brevity, we have written $\mathbf{q}$ instead of $\mathbf{q}_{ef}$). Having converged on the equilibrium point ($\mathbf{p}_o$, $\lambda_o\mathbf{q}_{ef}$), an incremental, tangential predictor ($\Delta\mathbf{p}_1, \Delta\lambda_1$) would be computed (see Section 9.4.3 for further details on the predictor) leading to the point ($\mathbf{p}_1, \lambda_1\mathbf{q}_{ef}$). The first iteration would then use (9.26) and (9.27) with the 'old' $\Delta\mathbf{p}_o$ as $\Delta\mathbf{p}_1$ and the 'old' $\Delta\lambda_o$ as $\Delta\lambda_1$ to obtain $\delta\mathbf{p}_1$ and $\delta\lambda_1$, after which the updating procedure (see 9.24)) would lead to

$$\Delta\mathbf{p}_2 = \Delta\mathbf{p}_1 + \delta\mathbf{p}_1, \qquad \Delta\lambda_2 = \Delta\lambda_1 + \delta\lambda_1. \tag{9.28}$$

When added to the displacements, $\mathbf{p}_o$ and load level, $\lambda_o$, at the end of the previous increment this process would lead to the point ($\mathbf{p}_2, \Delta\lambda_2\mathbf{q}_{ef}$) in Figure 9.10.

The next iteration would then re-apply (9.26) and (9.27) with the old value $\Delta\mathbf{p}_o$ as $\Delta\mathbf{p}_2$ and the old $\Delta\lambda_o$ as $\Delta\lambda_2$ to obtain $\delta\mathbf{p}_2$ and $\delta\lambda_2$, after which the updating procedure would lead to $\Delta\mathbf{p}_3 = \Delta\mathbf{p}_2 + \delta\mathbf{p}_2$ and $\Delta\lambda_3 = \Delta\lambda_2 + \delta\lambda_2$. The iterations would cease once the convergence criterion (see Section 9.5.4) was satisfied.

#### *9.3.2.2 Linearised arc-length methods*

A number of authors [R1, R2, R8, R9, W5, S2, F9, F12] have advocated linearised forms of arc-length method. From (9.20b), we can write

$$\Delta\mathbf{p}_o^T\delta\mathbf{p} + \delta\lambda(\Delta\lambda_o\psi^2\mathbf{q}_{ef}^T\mathbf{q}_{ef}) = -a_o/2 \tag{9.29}$$

where $a_o$ is the 'old' value of the arc-length mismatch (see (9.19)). If $a_o$ is taken as zero, we have Ramm's approach [R1, R2] which ensures that the iterative change ($\delta\mathbf{p}$, $\delta\lambda\psi\mathbf{q}_{ef}$) is orthogonal to the 'secant change' ($\Delta\mathbf{p}_o$, $\Delta\lambda_o\psi\mathbf{q}_{ef}$) (see Figure 9.11(b)).

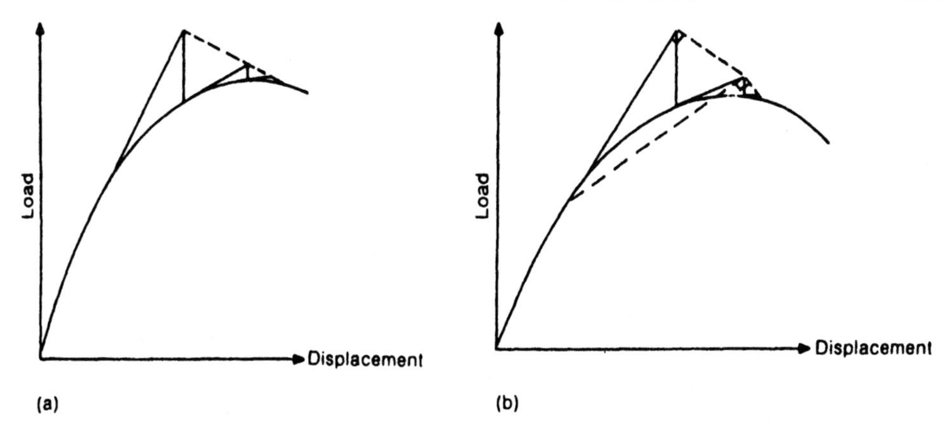


**Figure 9.11** Linearised arc-length methods: (a) the Riks–Wempner method; (b) Ramm's method.

Using (9.23), this leads to

$$\delta\lambda(\Delta\mathbf{p}_o, \Delta\lambda_o) = \frac{-(a_o/2) - \Delta\mathbf{p}_o^T\delta\bar{\mathbf{p}}}{(\Delta\mathbf{p}_o^T\delta\mathbf{p}_t + \Delta\lambda_o\psi^2\mathbf{q}_{ef}^T\mathbf{q}_{ef})}. \tag{9.30}$$

This technique is closely related to the original procedures due to Riks [R8, R9] and Wempner [W5] which, as illustrated in Figure 9.11(a), involve making the iterative change $(\delta\mathbf{p}, \delta\lambda\psi\mathbf{q}_{ef})$ orthogonal to the predictor solution $(\Delta\mathbf{p}_p, \Delta\lambda_p)$ (again with $a_o=0$). The relevant formulae are simply obtained by replacing the 'old incremental' $(\Delta p_o, \Delta\lambda_o)$ with the 'initial predictor', $(\Delta\mathbf{p}_n, \Delta\lambda_n)$ in (9.29) and (9.30). A further variation has been given by Fried [F12] which uses $(\delta\mathbf{p}_t, 1/(\psi^2\mathbf{q}_{ef}^T\mathbf{q}_{ef}))$ in place of $(\Delta\mathbf{p}_o, \Delta\lambda_o)$ with also $a_o = 0$. In this approach, which is related to a procedure by Haselgrove [H2], the solution process does not depend on the predictor solution. The idea of including the $a_o$ terms in (9.29) and (9.30) is due to Schwiezerhof and Wriggers [S2] and Riks [R11] with a further modification by Forde and Stiemer [F9].

The linearised versions are simpler than the spherical form of Section 9.3.2.1 because there is no issue of the choice of root in the solution to (9.26). However, the 'spherical form' has the advantage that throughout the iterations the solution is alway aimed at the same point (although some improvements can be made, in this respect, to the linear forms by including the $a_o$ terms in (9.30)). Hence, it is more stable and can converge when the linearised form misses the equilibrium path [W4].

#### 9.3.2.3 *Generalised displacement control at a specific variable*

All the methods described in this section may be considered as forms of generalised displacement control which can be applied although, physically, the problem does not involve displacements control. This has effectively been achieved with the spherical arc-length method (with $\psi = 0$) by constraining the Euclidean norm of the incremental displacement to a fixed quantity (equation (9.25)). Instead, following Batoz and Dhatt [B4], one may constrain the displacement increment at a particular variable to a specified quantity so that in place of (9.25) one would have

$$\Delta \mathbf{p}_n(k) = \Delta \mathbf{p}_o(k) + \delta \mathbf{p}(k) = \Delta \mathbf{p}_o(k) = \Delta p_k \tag{9.31}$$

where $\Delta \mathbf{p}(k)$ is the $k$th (scalar) component from the vector $\Delta \mathbf{p}$ and $\Delta \mathbf{p}_k$ is the prescribed magnitude of the $k$th incremental displacement variable. Using, as before, (9.23) for the iterative change $\delta \mathbf{p}$, the constraint (9.31) then leads to

$$\delta\lambda = \frac{\Delta p_k - \Delta \mathbf{p}_o(k) - \delta \bar{\mathbf{p}}(k)}{\delta \mathbf{p}_t(k)}. \tag{9.32}$$

The tangential, predictor solution would simply be achieved by finding

$$\delta \mathbf{p}_t = -\mathbf{K}_t \mathbf{q}_{ef}^{-1} \tag{9.33}$$

with $\mathbf{K}_t$ as the tangent stiffness matrix at the beginning of the increment and then $\Delta\lambda$ from

$$\Delta\lambda\, \delta \mathbf{p}_t(k) = \Delta p_k \tag{9.34}$$

where $\delta \mathbf{p}_t(k)$ is the $k$th (scalar) component from the vector $\delta \mathbf{p}_t$. In contrast to standard displacement control (Section 2.2.5), the variable $k$ would be one where, physically, there would be no real displacement control (and hence no reaction). Rheinoldt [R3, R6] has adopted this approach with the variable $k$ being changed for each increment so as to relate to the largest tangential (predictor) component.

Simons and Bergan [S5], expanding on the work of Powell and Simons [P11] and Bergan and Mollestad [B10], have advocated a 'hyperplane control method' which effectively involves an extension of Batoz and Dhatt's method with a weighted linear combination of 'individual specified displacements' with, say, half-a-dozen key displacement variables and weights being specified by the user. A procedure lying between the arc-length method and this 'hyperplane control' has been proposed by Gierlinski and Graves-Smith [G3].

## 9.4 DETAILED FORMULATION FOR THE 'CYLINDRICAL ARC-LENGTH' METHOD

In Section 9.3.2.1 we introduced the 'spherical arc-length method' and (with $\psi = 0$ in (9.25)) the 'cylindrical method'. In the present section, we will complete the detail (with $\psi = 0$) and provide flowcharts and a computer implementation.

### 9.4.1 Flowchart and Fortran subroutines for the application of the arc-length constraint

We will now apply equations (9.26) and (9.27) (with $\psi = 0$) which lie at the heart of the arc-length method. However, we must firstly address the issue of finding an appropriate root to (9.26). The idea is to compute both solutions ($\delta\lambda_1$ and $\delta\lambda_2$) and hence to have both

$$\Delta \mathbf{p}_{n1} = \Delta \mathbf{p}_o + \delta \bar{\mathbf{p}} + \delta\lambda_1 \delta \mathbf{p}_t \tag{9.35a}$$

and

$$\Delta \mathbf{p}_{n2} = \Delta \mathbf{p}_o + \delta \bar{\mathbf{p}} + \delta\lambda_2 \delta \mathbf{p}_t \tag{9.35b}$$

and then to find which of $\Delta \mathbf{p}_{n1}$ and $\Delta \mathbf{p}_{n2}$ lies closest to the old incremental direction $\Delta \mathbf{p}_o$. This should prevent the solution from 'doubling back on its tracks'. This procedure can be implemented by finding the solution with the minimum angle between $\Delta \mathbf{p}_o$ and $\Delta \mathbf{p}_n$ and hence the maximum cosine of the angle, using

$$\cos\theta = \frac{\Delta \mathbf{p}_o^T \Delta \mathbf{p}_n}{\Delta l^2} = \frac{\Delta \mathbf{p}_o^T (\Delta \mathbf{p}_o + \delta \bar{\mathbf{p}})}{\Delta l^2} + \delta\lambda \frac{\Delta \mathbf{p}_o^T \delta \mathbf{p}_t}{\Delta l^2} = \frac{a_4 + a_5 \delta\lambda}{\Delta l^2} \tag{9.36}$$

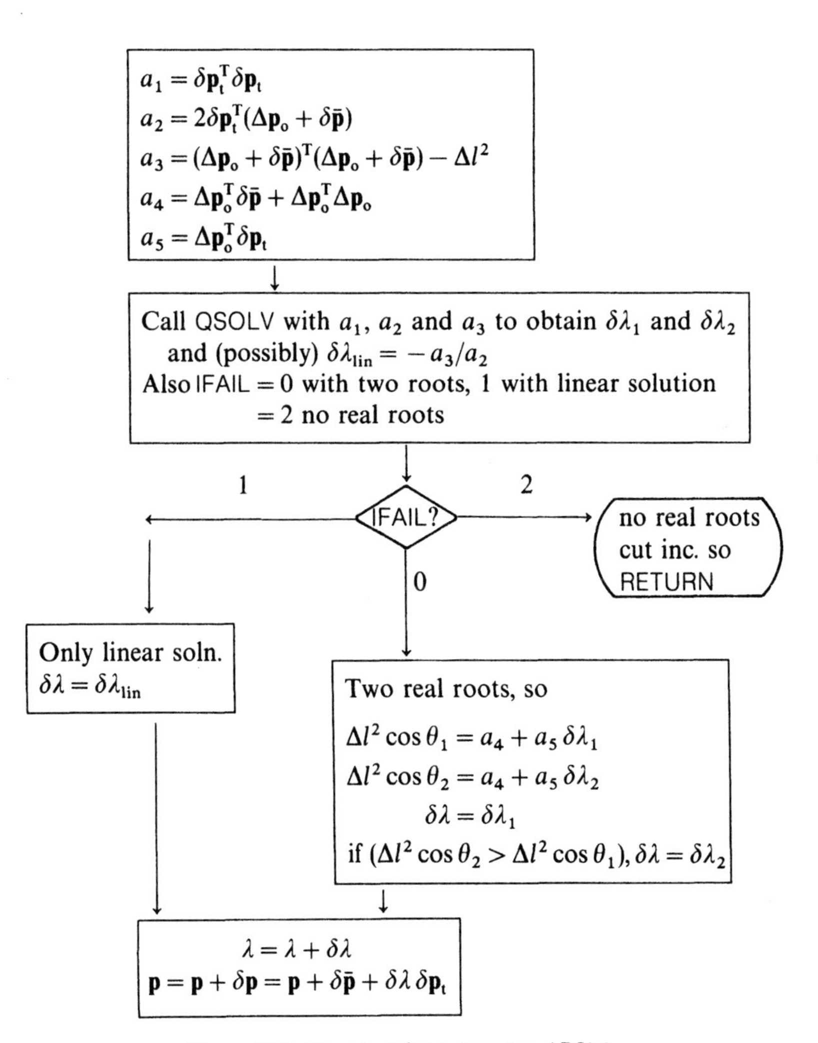


**Figure 9.12** Flowchart for subroutine ARCL1.

where we have used (9.24) for $\Delta\mathbf{p}_n$. Assuming that we have computed

$$\delta\bar{\mathbf{p}} = -\mathbf{K}_t^{-1}\mathbf{g}_o, \qquad \delta\mathbf{p}_t = \mathbf{K}_t^{-1}\mathbf{q}_{ef} \tag{9.37}$$

a flowchart to implement (9.26), (9.27) (with $\psi = 0$) and (9.36) is given as Figure 9.12.

In the accompanying Fortran subroutine ARCL1, instead of $\Delta\mathbf{p}_o$, we have PT which contains the current displacement, $\mathbf{p}$ and PTOL which contains the old total displacements at the end of the last increment. The difference between the two gives $\Delta\mathbf{p}_o$. Also, at the end of the routine we update the total displacement rather than updating the incremental displacement, $\Delta\mathbf{p}$.

The subroutine ARCL1 calls a routine QSOLV which solves the quadratic equation (9.26). The reader might prefer to introduce his own routine for this purpose. If, because $a_1$ is very small (see (9.26) and (9.27)), the constraint relationship is effectively linear, we adopt the linear solution. If we can find no real roots to the constraint equation, we exit from ARCL1 with the variable IFAIL set to 2 and (via subroutine NEXINC—see Sections 9.5.1 and 9.6.5), we cut the increment. Other remedies are possible [C6, C16].

#### 9.4.1.1 *Fortran subroutines* ARCL1 *and* QSOLV

```
      SUBROUTINE ARCL1 (DT, DELBAR, PT, PTOL, IBC, NV, DL2, FACT, IFAIL)
C     FOR ARC-LENGTH SOLN
C     DT=TANGENTIAL DISP WITH FIXED LOADS
C     DELBAR=TANGENTIAL DISP WITH O.B. FORCES
C     PT=TOTAL DISPS
C     PTOL=TOTAL DISPS AT END OF LAST INC
C     DL2=DESIRED INC LENGTH SQUARED
C     FACT=TOTAL LOAD FACTOR
C     IFAIL=OUTPUT AS 2 IF NEED TO CUT INC.
C
      IMPLICIT DOUBLE PRECISION (A-H, O-Z)
      DIMENSION DT(NV), DELBAR(NV), PT(NV), IBC(NV), PTOL(NV)
C
      A1=0.D0
      A2=0.D0
      A3=-DL2
      A4=0.D0
      A5=0.D0
      DO 10 I=1, NV
      IF (IBC(I).NE.1) THEN
      A1=A1+DT(I)*DT(I)
      DPBAR=PT(I)-PTOL(I)+DELBAR(I)
      A2=A2+2.D0*DT(I)*DPBAR
      A3=A3+DPBAR*DPBAR
      A4=A4+DPBAR*(PT(I)-PTOL(I))
      A5=A5+DT(I)*(PT(I)-PTOL(I))
      ENDIF
   10 CONTINUE
```

```
C
      CALL QSOLV(A1, A2, A3, R1, R2, RLIN, IFAIL)
C     BELOW NEEDS INC. CUTTING
      IF (IFAIL.EQ.2) RETURN
      IF (IFAIL.EQ.1) THEN
C     ONLY LINEAR SOLN. POSSIBLE
      SOL=R1
      ELSEIF(IFAIL.EQ.0) THEN
      COST1=A4+A5*R1
      COST2=A4+A5*R2
      SOL=R1
      IF (COST2.GT.COST1) SOL=R2
      ENDIF
      FACT=FACT+SOL
      DO 20 I=1, NV
   20 PT(I)=PT(I)+DELBAR(I)+SOL*DT(I)
      RETURN
      END


      SUBROUTINE QSOLV(A1, A2, A3, R1, R2, RLIN, IFAIL)
      IMPLICIT DOUBLE PRECISION (A-H, O-Z)
C     SOLVES QUADRATIC A1 X**2+A2 X+A3
C     IF A2.NE.O LINEAR SOLN, -A3/A2 in RLIN
C     IF IFAIL OUT AS ZERO TWO REAL ROOTS IN R1, R2
C     IF IFAIL OUT AS UNITY A1 TENDS TO ZERO AND R1 OUT AS RLIN
C     IF IFAIL OUT AS 2, NO REAL ROOTS
C
      SMALL=1.D-10
      IFAIL=0
      IF (A2.NE.O) RLIN= -A3/A2
      FAC=A2*A2-4.DO**A1*A3
      IF (FAC.LT.0.D0) THEN
C        NO REAL ROOTS
         IFAIL=2
         RETURN
      ELSE
C        REAL ROOTS
         FAC=DSQRT(FAC)
         IF (A1.EQ.0.DO) THEN
            IF (A2.NE.0) THEN
               R1=RLIN
               IFAIL=1
               RETURN
            ELSE
                 STOP 'QSOLV 1'
C                A1=0 AND A2=0
            ENDIF
         ELSE
```

```
C       REAL ROOTS AND AA.NE.0
          R1=-0.5D*(FAC+BB)/AA
          R2=0.5D0*(FAC-BB)/AA
        ENDIF
      ENDIF
      RETURN
      END
```

### 9.4.2 Flowchart and Fortran subroutine for the main structural iterative loop (ITER)

The routine ARCL1 of Section 9.4.1 will be called from the main iterative subroutine ITER that was initially introduced in Section 2.4.2 and, in outline, modified for line searches in Section 9.2.3.2. We will now further adapt this routine so that it can be used with either line searches or the arc-length method (the combination of the two is discussed in [C19] and will be considered in Volume 2). Apart from the direct call to subroutine ARCL1, the main modification to the routine involves the use of $\mathbf{g} = \mathbf{q}_i - \lambda \mathbf{q}_{ef}$ rather than, as before (Chapter 2), $\mathbf{g} = \mathbf{q}_i - \mathbf{q}_e$ where $\mathbf{q}_e$ was the fixed external load vector relating to load level, $\lambda$. The vector $\mathbf{q}_{ef}$ is the fixed loading vector (see Chapter 2) that was before (see Section 2.3.1) and is now called QFI. The loading parameter, $\lambda$, is called FACT (total factor). Having introduced these changes, the final modified flowchart is shown in Figure 9.13.

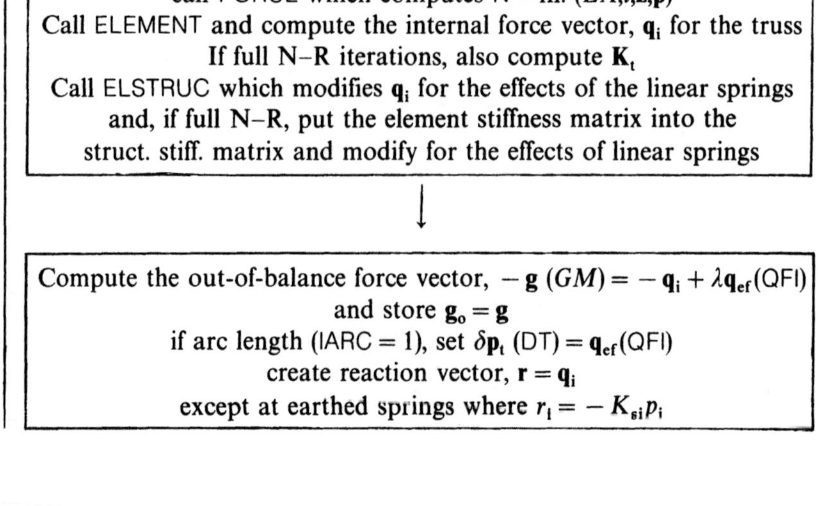

ILFAIL $= 0$
Begin iterative loop, ITE $=$ 1,NITMAX

↓

If (ITE $= 1$ or NLSMX $= 0$ (no line searches) or ILFAIL $= 1$ (l.s. uphill)
call FORCE which computes N $=$ fn. $(EA,l,\mathbf{z},\mathbf{p})$
Call ELEMENT and compute the internal force vector, $\mathbf{q}_i$ for the truss
If full N–R iterations, also compute $\mathbf{K}_t$
Call ELSTRUC which modifies $\mathbf{q}_i$ for the effects of the linear springs
and, if full N–R, put the element stiffness matrix into the
struct. stiff. matrix and modify for the effects of linear springs

↓

Compute the out-of-balance force vector, $-\mathbf{g}$ $(GM) = -\mathbf{q}_i + \lambda \mathbf{q}_{ef}$(QFI)
and store $\mathbf{g}_o = \mathbf{g}$
if arc length (IARC $= 1$), set $\delta \mathbf{p}_t$ (DT) $= \mathbf{q}_{ef}$(QFI)
create reaction vector, $\mathbf{r} = \mathbf{q}_i$
except at earthed springs where $r_i = -K_{si} p_i$

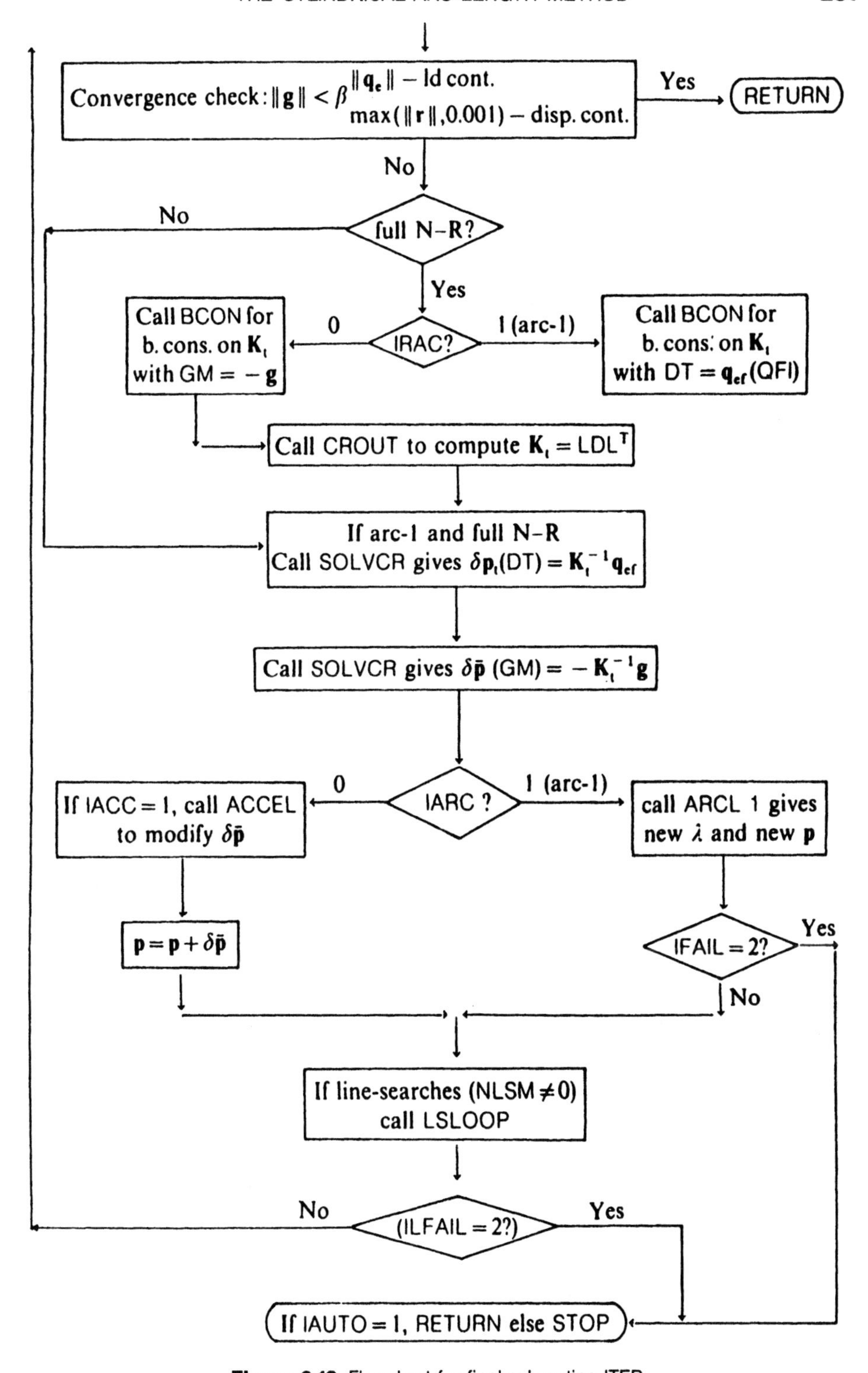


**Figure 9.13** Flowchart for final subroutine ITER.

The last box in the flowchart implements (with IAUTO = 1) automatic increment reduction if convergence is not achieved, if no real roots are found for the arc-length constraint or if difficulties are encountered with the line-search procedure. These issues are discussed further in Section 9.5. The flowchart also refers to a subroutine ACCEL (following the computations of $\delta\bar{\mathbf{p}}$ in subroutine SOLVCR). This subroutine applies an 'acceleration' to the modified Newton–Raphson method and is discussed in Section 9.8.

The arc-length procedure will work with either applied load-control or displacement control. However, as implemented here, the latter can only be used with the full N–R method and not the modified N–R method. In the latter case, the new $\mathbf{K}_t$ matrix is not formed and nor is $\delta\mathbf{p}_t = \mathbf{K}_t^{-1}\mathbf{q}_{ef}$. However, with displacement control, we need the new $\mathbf{K}_t$ matrix (or part of it) in order to produce the 'effective load vector' in subroutine BCON (see Section 2.2.5).

#### *9.4.2.1 Fortran subroutine* ITER

```
      SUBROUTINE ITER(PT,AN,BETOK,QFI,IBC,IWRIT,IWR,AKTS,D,ITERTY,NV,
     1                GM,FI,REAC,PTOL,DT,FACT,DL,IARC)
C
C     THIS FINAL VERSION HAS EITHER LOAD,DISP OR ARC-L CONTROL
C     ALSO INCLUDES LINE-SEARCHES (BUT NIT FOR ARC-L) VIA
C       A) COMMON /DATLS/
C       B) POSSIBLE CALL TO LSLOOP
C     INPUTS PREDICTOR DISPS. PT(NV) AND TOTLA. FIXED FORCE VEC QFI(NV)
C     ALSO BETOK=CONV. TOL, IBC=B. CON COUNTER
C     ITERATES TO EQUILIBRIUM: OUTPUTS NEW PT AND FORCE IN BAR,AN
C     IF ITERTY (INPUT)=1 USES FULL N-R, =2 USES MODN-R
C     IN LATTER CASE, AKTS AND D INPUT AS CROUT FACTORS (D=PIVOTS)
C     LOCAL ARRAY IS AKTE=EL. STIFF. MATRIX
C     GM USED FOR O.B. FORCES, FI FOR INT FORCES,
C     PTOL HAS OLD DISPS. AT END OF LAST INC.,
C     DT FOR TANGENTIAL (DUE TO TOTAL LD) DISPS (FIXED IF MOD N-R)
C     ARGUMENTS IN COMMON/DAT2/ AND ARRAY X NOT USED FOR SHALLOW TRUSS
C     ARGUMENTS IN COMMON/ACEL/ ONLY FOR ACCELERATION
C
      IMPLICIT DOUBLE PRECISION (A-H,O-Z)
      COMMON /DAT/ X(2),Z(2),E,ARA,AL,ID14S(4),AK14S(4),NDSP,ANIT,AK15
      COMMON /DAT2/ ARN,POISS,ALN,ITYEL
      COMMON /DATLS/ NLSMX,PERMLS,AMPMX,ETMXA,ETMNA,GO(5)
      COMMON /AUTOINC/ IAUTO,ITE,NITMAX,BET
      COMMON /ACEL/DELO(5),GOO(5),IACC,R1C,R2C
      DIMENSION PT(NV),QFI(NV),IBC(NV),REAC(NV)
      DIMENSION FI(NV),GM(NV),AKTS(NV,NV),D(NV),AKTE(4,4)
      DIMENSION DT(NV),PTOL(NV)
C
      ILFAIL=0
      SMALL=0.1D-2
C     BELOW OLD STEP-LENGTH FOR ACCN.
```

```
      SLOL=1.DO
      DL2=DL*DL
      IMOD=1
      IF (ITERTY.EQ.1) IMOD=3
C
      DO 100 ITE=1,NITMAX
C
      IF (IWRIT.EQ.1) WRITE (IWR,1005) ITE
 1005 FORMAT(/,1X,'ITERATIVE LOOP WITH ITE=  ',I5)
      IF (ITE.EQ.1.OR.NLSMX.EQ.0.OR.ILFAIL.EQ.1) THEN
C     ILFAIL=1 IF LS LOOP GAVE UPHILL
C     BELOW CALCS FORCE IN BAR (AN)
      CALL FORCE(AN,ANIT,E,ARA,AL,X,Z,PT,IWRIT,IWR,
     1      ITYEL,ARN,ALN,POISS)
C     ABOVE ARGUMENTS NOT USED FOR SHALLOW TRUSS
C
C     ABOVE CALCS FORCE IN BAR, AN: BELOW TAN STIFF AKT
C     (IF NR) AND INT. FORCE VECT. FI
      CALL ELEMENT(FI,AKTE,AN,X,Z,PT,E,ARA,AL,IWRIT,IWR,IMOD,
     1      ITYEL,ALN,ARN)
C     ABOVE ARGUMENTS NOT USED FOR SHALLOW TRUSS
C
C     BELOW PUTS EL. STIFF. MAT., AKTE, IN STR. STIFF., AKTS AND
C     ADDS IN EFFECTS OF VARIOUS LINEAR SPRINGS (IN NR)
C     ALSO MODIFIES INT. FORCE VECT. FI FOR SPRING EFFECTS
      CALL ELSTRUC(AKTE,AKTS,NV,AK15,ID14S,AK14S,NDSP,FI,PT,
     1                 IMOD,IWRIT,IWR)
      ENDIF
C
C     BELOW FORMS GM=OUT-OF-BALANCE FORCE VECTOR
C     AND REACTION VECTOR
C     IN ADDITION, NOW SAVES GO FOR LINE SEARCHES
      DO 10 I=1,NV
      GM(I)=0.DO
      REAC(I)=FI(I)
      IF (IARC.EQ.1.AND.ITERTY.EQ.1) DT(I)=QFI(I)
      IF (IBC(I).EQ.0) THEN
      GM(I)=FACT*QFI(I)-FI(I)
      ENDIF
      GO(I)=-GM(I)
   10 CONTINUE
   67 FORMAT(6G13.5)
   47 FORMAT(5I5)
C
C     OVERWRITE SPRING REACTION TERMS
      IF (NDSP.NE.O) THEN
      DO 50 I=1,NDSP
   50 REAC(ID14S(I))=-AK14S(I)*PT(ID14S(I))
      ENDIF
C
C     BELOW CHECKS CONVERGENCE
```

```
      FNORM=0.DO
      GNORM=0.DO
      RNORM=0.DO
      INDSP=0
      DO 20 I=1,NV
      IF (IBC(I).EQ.0) FNORM=FNORM+QFI(I)*QFI(I)
      IF (IBC(I).EQ.-1) IDSP=1
      RNORM=RNORM+REAC(I)*REAC(I)
   20 GNORM=GNORM+GM(I)*GM(I)
      FNROM=FACT*DSQRT(FNORM)
      GNORM=DSQRT(GNORM)
      RNORM=DSQRT(RNORM)
      BAS=MAX(FNORM,SMALL)
C     BELOW DISP. CONTROL
      IF (IDSP.EQ.1) BAS=MAX(RNORM,SMALL)
      BET=GNORM/BAS
      ITEM=ITE-1
      WRITE (IWR,1001) ITEM,BET
 1001 FORMAT(/,1X,'ITERN. NO.=',I5,'CONV.FAC.=',G13.5)
      IF (IWRIT.EQ.1) WRITE (IWR,1003) (GM(I),I=1,NV)
 1003 FORMAT(/,1X,'OUT-OF-BAL.FORCE VECTOR=',/,1X,6G13.5)
      IF (BET.LE.BETOK) GO TO 200
C
C     BELOW FOR FULL N-R
      IF (ITERTY.EQ.1) THEN
C     BELOW FOR ARC-L ON LOADING TERM
      IF (IARC.EQ.1) THEN
      CALL BCON(AKTS,IBC,NV,DT,IWRIT,IWR)
      ELSEIF (IARC.EQ.0) THEN
C     BELOW NON-ARC L
      CALL BCON(AKTS,IBC,NV,GM,IWRIT,IWR)
C     ABOVE APPLIES B. CONDITIONS
      ENDIF
      CALL CROUT(AKTS,D,NV,IWRIT,IWR)
C     ABOVE FORMS LDL(TRAN) FACTORISATION INTO AKTS AND D
      ENDIF
C
      IF (IARC.EQ.1.AND.ITERTY.EQ.1)
     1     CALL SOLVCR(AKTS,D,DT,NV,IWRIT,IWR)
C     ABOVE GIVES TANGENTIAL CHANGE DT DUE TO LOADING
      CALL SOLVCR(AKTS,D,GM,NV,IWRIT,IWR)
C     ABOVE GETS ITER. DISP. CHANGE IN GM DUE TO O.B. FORCES
C
      IF (IARC.EQ.1) THEN
      CALL ARCL1(DT,GM,PT,PTOL,IBC,NV,DL2,FACT,IFAIL)
      IF (IFAIL.EQ.2) STOP 'ITER 30'
C     NO ROOTS TO ARC-L CONSTRAINT
C     BELOW NON ARC-L
      ELSEIF (IARC.EQ.0) THEN
C     IF ACCEL., MODIFIES ITER. DISP. VECT., GM
      IF (IACC.EQ.1)
```

```
     1      CALL ACCEL(GM,GO,GOO,IBC,NV,DELO,SLOL,R1C,R2C,ITE,IWRIT,
     2                        IWR)
        DO 30 I=1,NV
        IF (IBC(I).EQ.0) THEN
        PT(I)=PT(I)+GM(I)
        ELSE
        PT(I)=FACT*QFI(I)
        ENDIF
   30   CONTINUE
C       ABOVE UPDATES DISPS.
C
        IF (NLSMX.NE.O) CALL LSLOOP(PT,GM,IBC,IWRIT,IWR,ITERTY,NV
     1                  ,FI,QFI,AKTS,AKTE,D,FACT,AN,SLOL,ILFAIL)
        IF (ILFAIL.EQ.2) GO TO 110
        ENDIF
C
        IF (IWRIT.EQ.1) WRITE (IWR,1004) (PT(I),I=1,NV)
 1004   FORMAT(/,1X,'TOTAL DISPS ARE',/,1X,6G13.5)
C
  100   CONTINUE
C
  110   WRITE (IWR,1002)
        ITE=NITMAX
 1002   FORMAT(/,1X,'FAILED TO CONVERGE OR L.S. TROUBLE****')
        IF (IAUTO.EQ.0) THEN
        STOP 'ITER 100'
        ELSE
        RETURN
        ENDIF
C
  200   CONTINUE
        RETURN
        END
```

In the above subroutine, both the COMMON /ACEL/ and the call to subroutine ACCEL relate to an accelerated modified Newton–Raphson method that will be discussed in Section 9.8.

### 9.4.3 The predictor solution

We have so far described the implementation of the 'spherical arc-length method' within the overall iterative loop for equilibrium but have not discussed in any detail the implementation of the 'predictor' solution. Assuming, as before, the adoption of a forward-Euler tangential predictor, the latter is given by

$$\Delta\mathbf{p}_p = \mathbf{K}_t^{-1}\Delta\mathbf{q}_e = \Delta\lambda_p\mathbf{K}_t^{-1}\mathbf{q}_{ef} = \Delta\lambda_p\delta\mathbf{p}_t \tag{9.38}$$

where $\mathbf{K}_t$ is the tangent stiffness matrix at the beginning of the increment. Substituting

(9.38) into the constraint of (9.25) (with $\psi = 0$) gives

$$\Delta\lambda_{\mathrm{p}} = \pm \frac{\Delta l}{\sqrt{(\delta \mathbf{p}_t^{\mathrm{T}} \delta \mathbf{p}_t)}} = s \frac{\Delta l}{\sqrt{(\delta \mathbf{p}_t^{\mathrm{T}} \delta \mathbf{p}_t)}} \tag{9.39}$$

where $\Delta l$ is the given incremental length. Because of the plus or minus sign in (9.39) we have two possible predictors. Following [C11], in the present work we will let $s$ be $+1$ when $\mathbf{K}_t$ (at the beginning of the increment) is positive definite. In relation to the adopted solution procedure the latter will occur when all the terms in $\mathbf{D}$, the diagonal matrix of the $\mathbf{LDL}^{\mathrm{T}}$ factorisation of $\mathbf{K}_t$ are positive. When one of these terms is negative, we have one 'negative pivot', which implies one negative eigenvalue for $\mathbf{K}_t$. This will occur when we have overcome a limit point (see Figure 9.6(a)) and we then set $s$ to $-1$. Unfortunately, a negative pivot will also be found (see Sections 2.6.3, 3.10.4 and 9.9.4) when we have passed a bifurcation point rather than a limit point. Hence, as will be shown in Section 9.9.4, in the presence of a bifurcation, the present algorithm will lead to a solution that oscillates about this bifurcation point. If one simply wishes to continue following the unstable post-bifurcation path, one may, instead of switching with a negative pivot in $\mathbf{K}_t$, switch when the predictor 'work increment', $\Delta \mathbf{q}^{\mathrm{T}} \Delta \mathbf{p}$ becomes positive [C22, M2–M4]. The latter is essentially the 'current stiffness parameter' which will be discussed in Section 9.5.2. As shown in Section 9.9.4, this parameter does not respond to 'bifurcations'.

Ideally, however, one would prefer to automatically switch to the stable (or more stable) post-buckling path. These issues will be briefly discussed in Section 9.10 and, in more detail, in Volume 2. This volume will also consider the problems in which one encounters more than one negative pivot [C12, C15, M2–M4]. For the present, the solution algorithm will automatically stop if more than one negative pivot is encountered on the factorisation of the tangent stiffness matrix at the beginning of an increment. (The author recommends that, until more advanced path-following techniques are available within commercial finite element codes, a similar approach should be adopted therein. It should, of course, be possible to override this requirement and, with the aid of restarts (Section 9.5.5) and manual intervention, the problems may be overcome but the user should, at the very least, be made aware that he is treading in dangerous waters.)

More sophisticated predictors can be used instead of the 'forward-Euler' predictor of (9.38) [R13]. However, den Heijer and Rheinboldt [D7] have argued that 'higher order predictions are very rarely effective.' In order to avoid duplication, we will not give the precise details of the predictor solution for the arc-length method at this stage but will wait until Section 9.6, when we have discussed the provision of automatic increments and automatic increment reduction. For the non-linear analysis of skeletal space structures, Kondoh and Atluri [K5] have adjusted the 'incremental length', $\Delta l$, to be no greater than the step required to initiate 'local buckling'.

## 9.5 AUTOMATIC INCREMENTS, NON-PROPORTIONAL LOADING AND CONVERGENCE CRITERIA

We will firstly discuss automatic increments in relation to standard load or displacement control, for which we wish to find a way of choosing a suitable 'load increment factor', $\Delta\lambda$. (The solution algorithms of Chapters 2 and 3 were all based on the use of fixed, equal, increments (as input by FACI $= \Delta\lambda$).) A number of procedures have been advocated for calculating a changing increment size [D7, S1, B7, B9, C11, R1, R2]. Den Heijer and Rheinboldt [D7] relate the increment size to the curvature of the non-linear path, with the latter requiring both the tangential predictor and the difference between the displacement vectors at the current and previous load levels. This procedure has much in common with a technique advocated by Bergan and Soreide [B9]. Later, Bergan and co-workers [B6, B7, D8] advocated an approach based on the 'current stiffness parameter', which will be discussed in Section 9.5.2. Following numerical experiments, they observed [B8] that both techniques led to nearly the same number of iterations being required to restore equilibrium. The author (C11) advocated a procedure whereby this was aimed at directly. To this end, the new increment factor, $\Delta\lambda_n$ was set to

$$\Delta\lambda_n = \Delta\lambda_o \left(\frac{I_d}{I_o}\right)^n \tag{9.40}$$

where $\Delta\lambda_o$ is the old increment factor for which $I_o$ iterations were required and $I_d$ is the input, desired, number of iterations ($I_d \simeq 3$). The parameter $n$ was set to unity. Ramm [R1, R2] suggested that $n$ should be set to $\frac{1}{2}$ and this approach has since been adopted by the author and will be used here. This technique leads to the provision of small increments when the response is most non-linear and large increments when the response is most linear.

This simple technique can very easily be extended to the arc-length method so that instead of (9.40) with $n = \frac{1}{2}$, we would have

$$\Delta l_n = \Delta l_o \left(\frac{I_d}{I_o}\right)^{1/2} \tag{9.41}$$

where $\Delta l_n$ and $\Delta l_o$ are 'incremental lengths' (Section 9.3.2). As well as inputting the desired number of increments, the user should provide a maximum and, possibly, a minimum increment size. (Note, however, that the provision of too high a minimum increment size can interfere with the increment-cutting procedure of Section 9.5.1.)

The user of a non-linear finite element computer program will (or should) usually have some idea of a suitable starting load increment so that once this is specified the technique of (9.40) (with the addition of maximum and minimum step-sizes) can lead to a fully automatic solution. However, the user will have little idea of an appropriate magnitude for a starting length increment, $\Delta l = \sqrt{(\Delta \mathbf{p}^T \Delta \mathbf{p})}$. There are (at least) three possible solutions. The first is to apply a preliminary load-controlled step and from the output $\Delta l$ (this should be output even under load control) a suitable starting value can be estimated. Alternatively, the user may start by specifying a load increment $\Delta\lambda$. The incremental displacement vector, $\Delta \mathbf{p}$, can then be computed from (9.38) and, via (9.39), a starting length increment, $\Delta l$, can be obtained. This is one of the procedures adopted in the current computer program.

A third, very useful, tactic is to apply standard load (or displacement) control for the early increments and only switch to arc-length control once a limit point is approached. A procedure for automatically introducing such a switch is given in Section 9.5.2.

### 9.5.1 Automatic increment cutting

If convergence of the structural equilibrium iterations is not achieved within the specified number of iterations, a simple strategy is to cut the increment size. In the present computer program, we have introduced the simple algorithm

$$\frac{\Delta\lambda_n}{\Delta\lambda_o} \quad \text{or} \quad \frac{\Delta l_n}{\Delta l_o} = \frac{\beta_d}{\beta} \quad \text{but} \quad \geqslant 0.1 \quad \text{and} \quad \leqslant 0.5 \tag{9.42}$$

where $\beta$ is the convergence factor of (2.30) and $\beta_d$ is the input, desired convergence factor (BETOK in the Fortran). Automatic increment cutting can be adopted in other situations [C6] such as the failure of the 'spherical' or 'cylindrical arc-length method to find real roots to the constraint equation [C11, M2–M4].

### 9.5.2 The current stiffness parameter and automatic switching to the arc-length method

The current stiffness parameter [B6, B7, B8, B10] is a very useful index to give some scalar measure of the degree of non-linearity. In its unscaled form, it effectively measures the 'stiffness' of the system as related to the tangential predictor, i.e. '$k$' $= \Delta\mathbf{q}/\Delta\mathbf{p}$, where $\Delta\mathbf{q}$ is the incremental applied load vector and $\Delta\mathbf{p}$ the resulting tangential displacements. However, because $\Delta\mathbf{q}$ and $\Delta\mathbf{p}$ are vectors, we must multiply the top and bottom by $\Delta\mathbf{p}$ so that

$$`k' = \frac{\Delta\mathbf{q}^T\Delta\mathbf{p}}{\Delta\mathbf{p}^T\Delta\mathbf{p}} = \frac{\mathbf{q}_{ef}^T\delta\mathbf{p}_t}{\delta\mathbf{p}_t^T\delta\mathbf{p}_t} \tag{9.43}$$

where we have used $\Delta\mathbf{q} = \Delta\lambda\mathbf{q}_{ef}$ and (9.38). For displacement control, instead of using the 'fixed load vector', $\mathbf{q}_{ef}$ in (9.43), we must use the 'effective fixed load vector' as produced by the process of Section 2.2.5.

To obtain the current stiffness parameter, $C_s$, we simply scale the current '$k$'-value by the initial '$k$'-value, so that

$$C_s = \frac{`k'}{`k_o'}. \tag{9.44}$$

Bergan and co-workers [B6–B8] advocated a technique for automatic incrementation whereby, instead of (9.40), they would place on the right hand-side of that equation, $\Delta C_{sd}/\Delta C_{so}$ with $\Delta C_{sd}$ as the desired change in 'current stiffness' and $\Delta C_{so}$ would be the previously achieved change.

Many structures exhibit a response in which the structure softens as the load is applied (i.e. Figure 9.6(c)). In such situations, it is very useful to force the solution procedure to automatically switch from load (or displacement) control to arc-length control as the limit point is reached ($C_s$ will be zero at the limit point). This can be achieved by introducing a value for the current stiffness parameter (say $\bar{C}_s$) below which this switch is automatically introduced. Such a feature is included in the computer program to be described in Section 9.6.

A range of alternative or supplementary 'path-measuring parameters' has been advocated by Eriksson [E3].

### 9.5.3 Non-proportional loading

Most of the solution procedures in this chapter have been based on the equilibrium relationship of (9.15) which implies a single loading (or displacing) vector, $\mathbf{q}_{ef}$, is proportionally scaled via $\lambda$. For many practical structural problems, this loading regime is too restrictive. For example, we often wish to apply the 'dead load' or 'self-weight' and then monotonically increase the live load. In other instances, a whole range of loading stages may be required [C18]. Fortunately, many such loading regimes can be applied by means of a series of loading sequences involving two loading vectors, one that will be scaled (the previous $\mathbf{q}_{ef}$) and one that will be fixed ($\bar{\mathbf{q}}_{ef}$). The external loading can then be represented by

$$\mathbf{q}_e = \bar{\mathbf{q}}_{ef} + \lambda \mathbf{q}_{ef} \tag{9.45}$$

so that the out-of-balance force vector becomes

$$\mathbf{g} = \mathbf{q}_i - \bar{\mathbf{q}}_{ef} - \lambda \mathbf{q}_{ef}. \tag{9.46}$$

An equation such as (9.22) then becomes

$$\delta\mathbf{p} = -\mathbf{K}_t^{-1}\mathbf{g} = -\mathbf{K}_t^{-1}(\mathbf{q}_i(\mathbf{p}_o) - \bar{\mathbf{q}}_{ef} - \lambda \mathbf{q}_{ef}) \tag{9.47a}$$

or

$$\delta\mathbf{p} = -\mathbf{K}_t^{-1}(\mathbf{q}_i(\mathbf{p}_o) - \bar{\mathbf{q}}_{ef} - \lambda_o \mathbf{q}_{ef}) - \delta\lambda \mathbf{K}_t^{-1}\mathbf{q}_{ef} = \delta\bar{\mathbf{p}} + \delta\lambda\delta\mathbf{p}_t \tag{9.47b}$$

so that with these new definitions, the basic structure of the previous algorithms can be maintained.

These modifications are not difficult to implement in a general-purpose finite element system but, to avoid clouding the other issues, will be omitted from the present computer program.

### 9.5.4 Convergence criteria

In Section 2.4, we introduced a convergence criterion for the overall structural iterations that was effectively based on the magnitude of the Euclidean norm of out-of-balance force vector, $\mathbf{g}$. For the computer program to be given in Section 9.6, we will stick to this criterion. In the present section, we will briefly discuss some alternative criteria.

Obvious alternatives involve the use of different norms such as 'the maximum norm'. Other alternatives involve some scaling so that, for example, in place of (2.30a) for load control, we could have

$$\sqrt{(\mathbf{g}^T\mathbf{S}\mathbf{g})} < \beta\sqrt{(\mathbf{q}_e^T\mathbf{S}\mathbf{q}_e)} \tag{9.48}$$

where $\mathbf{S}$ is a scaling matrix that could, for example, be used to ensure that, for a problem involving rotational variables, all parameters had the same dimensions. In [C15], the author used $\mathbf{S} = \mathbf{C}^{-1}$, where $\mathbf{C}$ was the diagonal matrix containing the leading diagonal terms from the tangent stiffness matrix at the beginning of the increment.

Instead of, or in addition to, force-based convergence criteria, displacement-based criteria can be adopted so that, for example, we could have

$$\|\delta \mathbf{p}\| < \beta \|\mathbf{p}\| \tag{9.49}$$

where $\delta\mathbf{p}$ are the iterative displacement changes and $\mathbf{p}$ the total displacements. As shown in [C6], the norm of the iterative displacement change can be very small while the out-of-balance force norm is very large. Hence it is unwise to adopt a displacement-based criterion such as (9.49) on its own without supplementing it with some force-based criterion.

. An apparently attractive alternative to force or displacement-based convergence criteria is to use an energy-based criterion of the form

$$|\delta \mathbf{p}^{\mathrm{T}} \mathbf{g}| < \beta |\mathbf{p}^{\mathrm{T}} \mathbf{q}_{\mathrm{e}}|. \tag{9.50}$$

There are various ways in which such a criterion could be introduced but the author believes they should be used with great caution. For example, suppose that, more specifically, we had

$$|\delta \mathbf{p}^{\mathrm{T}} \mathbf{g}_{\mathrm{o}}| = |-\delta \mathbf{p}^{\mathrm{T}} \mathbf{K}_{\mathrm{t}}^{-1} \delta \mathbf{p}| < \beta |\mathbf{p}^{\mathrm{T}} \mathbf{q}_{\mathrm{e}}| \tag{9.51}$$

where the iterative change was $\delta\mathbf{p} = -\mathbf{K}_{\mathrm{t}}^{-1}\mathbf{g}_{\mathrm{o}}$. Equation (9.51) merely gives some measure of the 'stiffness' of $\mathbf{K}_{\mathrm{t}}$. Clearly as a limit point is approached, this can be small and yet the solution procedure may not have converged at all. (With a full Newton–Raphson iteration, away from equilibrium, $\mathbf{K}_{\mathrm{t}}$ may have no structural significance at all.) Alternatively, one could have

$$|\delta \mathbf{p}^{\mathrm{T}} \mathbf{g}_{\mathrm{n}}| = |\delta \mathbf{p}^{\mathrm{T}} \mathbf{g}(\mathbf{p}_{\mathrm{o}} + \eta\, \delta \mathbf{p})| < \beta |\mathbf{p}^{\mathrm{T}} \mathbf{q}_{\mathrm{e}}| \tag{9.52}$$

where we have introduced some of the notation of Section 9.1 on line searches. The criteria for $\delta\mathbf{p}^{\mathrm{T}}\mathbf{g}_{\mathrm{n}}$ to be zero is merely the criterion for an 'exact line search'. This can be achieved even by chance with a step length $\eta = 1$ and merely implies that a stationary energy position has been reached *in the current iterative direction*, $\delta\mathbf{p}$. This can, and frequently does, occur when the solutions is still a very long way from equilibrium. (A tentative thought; it appears that such energy-based criteria are often used in dynamics—surely the same limitations apply.)

### 9.5.5 Restart facilities and the computation of the lowest eigenmode of $\mathbf{K}_{\mathrm{t}}$

It almost goes without saying that a non-linear finite element computer program should have restart facilities. Despite all efforts to introduce a fully robust system, there will be many occasions when it is necessary to restart using, say, a different iterative procedure. On some occasions, the author has found it necessary to retrack quite a few increments and restart with a tighter convergence tolerance in order to avoid eventual divergence. When even such measures fail, it is often extremely useful at an equilibrium point beyond which no progress can be made, to compute and plot the lowest eigenmode of the system. The author has, in this way, discovered such diverse phenomena as (a) spurious mechanisms (b) errors in the computer program (c) errors in the input data [C6].

For the present computer program, we have included a very basic restart facility whereby (when using automatic increments), the solution can be restarted from the solution obtained at the end of the previous run. If one wishes to restart from an earlier position, one must firstly rerun the original problem with a reduced number of increments.

## 9.6 THE UPDATED COMPUTER PROGRAM

In this section, we will give both flowcharts and Fortran coding for the final complete computer program (apart from subroutine ACCEL—see Section 9.8.2) which includes:

(1) line searches
(2) the 'spherical' arc-length method
(3) automatic increments
(4) automatic increment cutting
(5) the current stiffness parameter
(6) automatic switching to the arc-length method
(7) acclerations to the mN–R method (Section 9.8)
(8) restarts.

Many of the subroutines have already been given. In particular, from Chapter 2, we require

ELSTRUC (2.2.4)
BCON (2.2.5)
CROUT (2.2.6)
SOLVCR (2.2.7)

Assuming that we are to use the general truss elements of Chapter 3 (rather than the shallow elements of Chapter 2, although these could be used), we also require from Chapter 3,

ELEMENT (3.9.1)
INPUT (3.9.2)
FORCE (3.9.3)

From the present chapter we require

SEARCH (9.2.2.1)
ARCL1 (9.4.1.1)
QSOLV (9.4.1.1)
ITER (9.4.2.1)
LSLOOP (9.6.1)
INPUT2 (9.6.2.1)
NONLTD (9.6.3.1)

SCALUP (9.6.4.1)
NEXINC (9.6.5.1)
ACCEL (9.8.2)

where the last six routines have yet to be given.

In Section 9.2.3.3, we gave a flowchart for the line-search loop at the structural level. We did not then give the Fortran because this had to be related to the modified subroutine ITER which was altered in Section 9.4.2.1 to allow the introduction of the arc-length method. We are now in a position to give the Fortran for LSLOOP. In relation to the flowchart of Section 9.2.3.3, the only significant change is the introduction of the out-of-balance forces in the form $\mathbf{g} = \mathbf{q}_i - \lambda \mathbf{q}_{ef}$ (as in (9.15)) so that both $\mathbf{q}_{ef}$ (QFI) and $\lambda$ (FACT) are required.

### 9.6.1 Fortran subroutine LSLOOP

```
      SUBROUTINE LSLOOP(PT,PBAR,IBC,IWRIT,IWR,ITERTY,NV,FI,QFI,
     1   AKTS,AKTE,D,FACT,AN,SLOL,ILFAIL)
C
C     PERFORMS LINE SEARCH LOOP
C     INPUTS TOTAL DISPS IN PT(NV), ITERATIVE CHANGE IN PBAR(NV)
C     QFI=TOTAL EXTERNAL LOADING (UNFACTORED)
C     GO=OLD O.B. FORCE VECTOR
C     IBC=B.CON COUNTER, FI FOR INTERNAL FORCE VECT.,
C     IF ITERTY (INPUT)=1 USES FULLY N-R,   =2 USES MOD N-R
C     FACT=TOTAL LOAD FACTOR LEVEL
C     ARGUMENTS IN COMMON/DATA2/ AND ARRAY X NOT USED FOR SHALLOW TRUSS
C     ILFAIL EXITS WITH ZERO IF 0.K, 1 IF UPHILL (ABANDON L.S.)
C     WITH 2 IF L.S. PROBLEMS (CUT INC IF IAUTO=1)
C
      IMPLICIT DOUBLE PRECISION (A-H,O-Z)
      COMMON /DAT/ X(2),Z(2),E,ARA,AL,ID14S(4),AK14S(4),NDSP,ANIT,AK15
      COMMON /DAT2/ ARN,POISS,ALN,ITYEL
      COMMON /DATLS/ NLSMX,PERMLS,AMPMX,ETMXA,ETMNA,GO(5)
      COMMON /AUTOINC/IAUTO,ITE,NITMAX,BET
      DIMENSION PT(NV),QFI(NV),IBC(NV)
      DIMENSION FI(NV),PBAR(NV),AKTS(NV,NV),AKTE(4,4),D(NV)
      DIMENSION PRODR(10),ETA(10),PTO(5)
C
C     CHECK ON SIZE OF PRODR AND ETA
      IF (NLSMX.GT.8) STOP 'LSLOOP 1'
C     PRODR AND ETA MUST BE OF DIM. GE. NLSMX+2
C     COMPUTE INNER PRODUCT AT START AND STOP IF POS. (UPHILL)
      SO=0.DO
      ILFAIL=0
      DO 10 I=1,NV
      IF (IBC(I).EQ.0) SO=SO+PBAR(I)*GO(I)
   10 CONTINUE
      IF (SO.GE.O.DO) THEN
```

```
      ILFAIL=1
      WRITE (IWR,1001) SO
 1001 FORMAT(/,1X,'RETURNS BECAUSE START INNER-PRODUCT UPHILL=',G13.5)
      RETURN
      ENDIF
C
C     PREPARE STARTING PRODUCT RATIOS, PRODR AND STEP LENGTHS, ETA
      PRODT(1)=1.DO
      ETA(1)=0.DO
      ETA(2)=1.DO
      ICO=0
C     ABOVE COUNTER WILL BECOME 1 WHEN MAX OR MIN S-LENGTH IS REACHED
C     OR 2 WHEN REACHED TWICE RUNNING
C
C     GET FIXED TOTAL DISPS AT END OF LAST ITER IN PTO
C     NOTE, FIXED ITER. DISP CHANGE IN PBAR
      DO 20 I=1,NV
      PTO(I)=0.DO
      IF (IBC(I).EQ.0) THEN
      PTO(I)=PT(I)-PBAR(I)
      ENDIF
   20 CONTINUE
C
C     BEGIN LINE-SEARCH LOOP
      DO 100 ILS=1,NLSMX
C
      IF (IWRIT.EQ.1) WRITE (IWR,1005) ILS
 1005 FORMAT(/,1X,'LINE-SEARCH LOOP WITH ILS=',I5)
C     BELOW CALCS FORCE IN BAR (AN)
      CALL FORCE(AN,ANIT,E,ARA,AL,X,Z,PT,IWRIT,IWR,
     1  ITYEL,ARN,ALN,POISS)
C     ABOVE ARGUMENTS NOT USED FOR SHALLOW TRUSS
C
C     ABOVE CALCS FORCE IN BAR, AN: BELOW INT. FORCE VECT, FI
      CALL ELEMENT(FI,AKTE,AN,X,Z,PT,E,ARA,AL,IWRIT,IWR,1,
     1  ITYEL,ALN,ARN)
C     ABOVE ARGUMENTS NOT USED FOR SHALLOW TRUSS
C
C     BELOW MODIFIES INT. FORCE VECT. FI FOR SPRING EFFECTS
      CALL ELSTRUC(AKTE,AKTS,NV,AK15,ID14S,AK14S,NDSP,FI,PT,
     1     1,IWRIT,IWR)
C
C     BELOW FORMS CURRENT INNER-PROD RATIO
      SETA=0.DO
      DO 30 I=1,NV
      IF (IBC(I).EQ.0) SETA=SETA+PBAR(I)*(FI(I)-FACT*QFI(I))
   30 CONTINUE
      SETA=SETA/SO
      PRODR(ILS+1)=SETA
   67 FORMAT(6G13.5)
```

```
   47 FORMAT(5I5)
C
C
C     BELOW CHECKS FOR SATISFACTION OF L-S TOLERANCE
      IF (ABS(SETA).LT.PERMLS) GO TO 300
C
C     CALL L-S ROUTINE TO GET NEW ESTIMATE ETA IN ETA(ILS+2)
      CALL SEARCH(ILS,PRODR,ETA,AMPMX,ETMXA,ETMNA,IWRIT,IWR,ICO,NLSMS+2)
      IF (ICO.EQ.2) GO TO 110
C     GET CURRENT DISPS.
      DO 40 I=1,NV
      IF (IBC(I).EQ.0) THEN
      PT(I)=PTO(I)+ETA(ILS+2)*PBAR(I)
      ENDIF
   40 CONTINUE
C
  100 CONTINUE
C
      WRITE (IWR,1002)
 1002 FORMAT(/,1X,'MAX NO OF L-SEARCHES EXCEEDED')
  110 CONTINUE
      IF (IAUTO.EQ.0) THEN
      STOP 'LSLOOP 1002'
      ELSEIF (IAUTO.EQ.1) THEN
      ILFAIL=2
      RETURN
      ENDIF
C
  300 CONTINUE
      IF (ILS,GT.1) WRITE (IWR,1003) ILS-1,ETA(ILS+1)
 1003 FORMAT(/,1X,'L-S, EXTRA RES. CALCS=',I5,' S-L=',G13.4)
      SLOL=ETA(ILS+1)
C     BEFORE RETURNING TO ITER MUST COMPUTE TANGET STIFF IF
C     USING FULL N-R ITERATION
      IF (ITERTY.EQ.1) THEN
      CALL ELEMENT (FI,AKTE,AN,X,Z,PT,E,ARA,AL,IWRIT,IWR,2,
     1     ITYEL,ALN,ARN)
      CALL ELSTRUC(AKTE,AKTS,NV,AK15,ID14S,AK14S,NDSP,FI,PT,
     1     2,IWRIT,IWR)
      ENDIF
      RETURN
      END
```

The final step-length, $\eta$, is output as =SLOL so that it can be used with the accelerated modified Newton–Raphson method (Section 9.8).

### 9.6.2 Input for incremental/iterative control

In the computer programs of Chapters 2 and 3, the incremental/iterative control parameters were input into the main program module, NONLTC (Section 2.5.1).

Because the present non-linear control is more complex, we now introduce a separate subroutine. This subroutine, INPUT2, which is given in Section 9.6.2.1, is almost self-explanatory (particularly when read in conjunction with the examples of Section 9.9). Nonetheless, we will here give some brief notes relating to the various input 'cards' in that routine.

*Card 1*
In the first card, the factor,

FACI is the initial value of $\Delta\lambda$. If the arc-length method is to be adopted (IARC = 1) and the 'desired length', $\Delta l_d$, for the first increment on Card 6 is input as zero, $\Delta\lambda$ = FACI is automatically converted to a 'length'.
NINC is the desired number of increments (if the increment is cut, it will count as two increments).
IWRIT Write control for extra information, 0 is off, 1 is on.
IAUTO If the parameter IAUTO is set to unity, automatic increment sizes are computed. Such automatic increments must be used with the arc-length method. If IAUTO is set to zero, equal increments of magnitude $\Delta\lambda$ = FACI are adopted as in Chapters 2 and 3.
IARC is set to zero for load (or displacement control) and 1 for arc-length control. To start with load control and later switch (Section 9.5.2) IARC is input as 0.
IACC is usually set to zero but to 1 if the acceleration method of Section 9.8 is to be used with the mN–R method (not for arc-length method).
IRES is set to zero for a standard solution and to 1 if the present solution is to be re-started from an earlier solution (see Section 9.5.5). Whenever IAUTO is set to 1, the program will, following the last increment, output an unformatted restart file 'RESOUT'. In order to restart using this information, the file 'RESOUT' must be copied to a file 'RESIN', which will be input if IRES is set to 1. Whenever IAUTO = 1, the main (formatted) output will contain, at the end, the parameters that the automatic incrementing routine, NEXINC (see Section 9.6.5) would use for the next increment. In particular, these include some of the parameters required on Cards 3 and 4 below.

*Card 2*
For this card:

BETOK = convergence tolerance factor (see Section 2.4)
ITERTY = 1 for full N–R; = 2 for modified N–R
NITMAX = maximum number of iterations
NLSMX = maximum number of extra residual calculations during line-searches (set to zero if no line-searches are required).

*Card 3*
This card relates to the line searches and inputs the parameters discussed in Section 9.2.3.1. Note that, at present, line searches cannot be applied with the arc-length method. (This facility will be introduced in Volumes 2.)

*Card 4*
This card relates to automatic increments (Section 9.5) and requires

$$I_d, \Delta\lambda_{max}, \Delta\lambda_{min}, \text{ISWCH}.$$

(Note that the provision of too high a $\Delta\lambda_{min}$ can interfere with the automatic increment cutting procedure.) With ISWCH = 1, the algorithm will automatically switch to the arc-length method using the procedure of Section 9.5.2, which requires Card 5.

*Card 5*
The desired current stiffness parameter for switching ($\bar{C}_s$).

*Card 6*
If the arc-length method is to be applied from the very start (IARC = 1), this card requires

$$\Delta l_d, \Delta l_{max}, \Delta l_{min}$$

where $\Delta l_d$ is the desired length for the first increment and the other two factors are self-explanatory. (Note that the provision of too high a $\Delta l_{min}$ can interfere with the automatic increment cutting procedure.) If the parameters on this card are set to zero, the program will use the load parameters of Cards 1 and 4 to compute appropriate $\Delta l$ quantities.

*Card 7*
This card relates to the cut-outs required for the accelerated modified Newton–Raphson method (Section 9.8.1) which is activated if IACC (card 1) is set to unity.

### 9.6.2.1 *Subroutine* INPUT2

```
      SUBROUTINE INPUT2(FACI,NINC,IARC,BETOK,ITERTY,IDES,FACMX,FACMN,
     1   DLDES,DLDMX,DLDMN,ISWCH,CSTIFS,IBC,NV,IRE,IWR,IWRIT,IRES)
C
C     INPUTS INCREMENTAL/ITERATIVE CONTROL
C     ARGUMENTS IN COMMON/DATLS/ FOR L-SEARCHES
C     ARGUMENTS IN COMMON/AUTOINC/ FOR AUTO-INCS.
C     ARGUMENTS IN COMMON/ACEL/ FOR ACCEL. MOD. N-R
C
      IMPLICIT DOUBLE PRECISION (A-H,O-Z)
      COMMON /DATLS/ NLSMX,PERMLS,AMPMX,ETMXA,ETMNA,GO(5)
      COMMON /AUTOINC/ IAUTO,ITE,NITMAX,BET
      COMMON /ACEL/ DELO(5),GOO(5),IACC,R1C,R2C
      DIMENSION IBC(NV)
C
C
      READ (IRE,*) FACI,NINC,IWRIT,IAUTO,IARC,IACC,IRES
      WRITE (IWR,1000) FACI,NINC,IWRIT,IAUTO,IARC,IACC,IRES
 1000 FORMAT(/,1X,'INCREMENTAL LOAD FACTOR=',G13.5,/,1X,
     1 'NO. OF INCS. (NINC)=',I5,/,1X
     2 'WRITE CONTROL (IWRIT)=',I5,/,3X,
     3 '0=LIMITED ; 1=FULL',/,1X,
     4 'IAUTO=',I5,/,3X,
     5 '0=FIXED INCS., 1=AUTOMATIC',/,1X,
     6 'IARC=',I5,/,3X,
```

```
         7  '0=LOAD INCS., 1=ARC-LENGTH',/,1X,
         8  'IACC=',I5,/3X,
         9  '0=NO ACCEL, 1=ACCEL WITH MOD. N-R',1,1X,
          A  'IRES=',I5,/,3X
          B  '0=NOT A RE-START, 1=IS A RE-START')
C
      READ (IRE,*) BETOK,ITERTY,NITMAX,NLSMX
      WRITE (IWR, 1003) BETOK, ITERTY, NITMAX, NLSMX
 1003 FORMAT(/,1X,'CONV. TOL FACTOR, BETOK=',G13.5,/,
     1  1X,'ITERATIVE SOLN. TYPE, ITERTY=',I5,/,
     2  5X,'=1, FULL N-R;   =2, MOD. N-R',/,
     3  1X,'MAX NO OF ITERATIONS=',I5,/,
     4  IX,'MAX NO. OF L-SEARCHES=',I5)
C
      IF (ITERTY. EQ.1).OR.(IARC.EQ.1)) THEN
      IF(IACC,EQ.1) STOP 'INPUT2 1009'
C     NO ACCN. WITH FULL N-R OR WITH ARC-L
      ENDIF
C
C     BELOW SPECIFIC TO LINE SEARCHES
      IF (NLSMX.NE.0) THEN
      READ (IRE,*) PERMLS,AMPMX,ETMXA,ETMNA
      WRITE (IWR,1009) PERMLS,AMPMX,ETMXA,ETMNA
 1009 FORMAT(/,1X,'LINE SEARCH PARMS ARE',/,
     1  3X,'TOLERANCE ON RATIO, PERMLS=',G13.4,/,
     2  3X,'MAX. AMP. AT ANY STEP, AMPMX=',G13.4,/,
     3  3X,'MAX. TOTAL STEP-LENGTH, ETMXA=',G13.4,/,
     4  3X,'MIN. TOTAL STEP-LENGTH, ETMNA=',G13.4)
      ENDIF
C
C     BELOW SPECIFIC TO AUTOMATIC INCS.
      IF (IAUTO.EQ.1) THEN
      READ (IRE,*) IDES,FACMX,FACMN,ISWCH
      WRITE (IWR,1008) IDES,FACMX,FACMN,ISWCH
 1008 FORMAT(/,1X,'DATA FOR AUTOMATIC INCREMENTS',/,
     1  1X,'DESIRED NO. OF ITERATIONS=',I5,/,
     2  1X,'MAX. LOAD INC.=    'G13.4,/,
     3  1X,'MIN. LOAD INC.=    ',G13.4,/,
     4  1X,'PARAM FOR ARC-L, ISWCH=    ',I5,/,
     5  3X,'=0 NO SWITCH,=1 SWITCH')
      IF (ISWCH.EQ.1) THEN
      READ (IRE,*) CSTIFS
      WRITE (IWR,1004) CSTIFS
 1004 FORMAT(/,1X,'SWITCHES TO ARC-L WHEN CSTIF.LE.CSTIFS=',G13.4)
      ENDIF
      ENDIF
C
C     BELOW SPECIFIC TO ARC-L METHOD
      IF (IARC EQ.1) THEN
      READ (IRE,*) DLDES,DLDMX,DLDMN
      WRITE (IWR,1005) DLDES,DLDMX,DLDMN
```

```
 1005 FORMAT(/,1X,'FOR ARC-LENGTH CONTROL',/,
     1  3X,'DESIRED LENGTH INC., DLDES=',G13.4,/,
     2  3X,'MAX. LENGTH INC.,    DLMX=',G13.4,/,
     3  3X,'MIN. LENGTH INC.,   DLMN=',G13.4,/,5X,
     4  'NOTE**, IF DLDES=DLMX=DLMN=0., USES LOAD INC FACT FOR 1ST INC')
C       CHECKS NOT USING MOD N-R WITH DISP. CONTROL
        IDSP=0
        DO 10 I=1,NV
        IF (IBC(I).LT.0) IDSP=1
   10   CONTINUE
        IF (IDSP.EQ.1.AND.ITERTY.EQ.2) THEN
        WRITE (IWR,1100)
 1100   FORMAT(/,1X,'**STOPS,CANNOT HAVE PRESC DISPS+ARC-L+MOD. N.R')
        STOP 'INPUT2 1100'
        ENDIF
        ENDIF
C
C       BELOW SPECIFIC TO ACCELERATION
        IF (IACC.EQ.1) THEN
        READ (IRE,*) R1C,R2C
        WRITE (IWR,1110) R1C,R2C
 1110   FORMAT(/,1X,'CUT-OFF PARAMS FOR ACCN. ARE',/,
     1  'R1C=',G13.4,' R2C=',G13.4)
        ENDIF
C
        RETURN
        END
```

### 9.6.3 Flowchart and Fortran subroutine for the main program Module NONLTD

The main program module NONLTD is a modification of the module NONLTC given in Section 2.5. A flowchart is given in Figure 9.14, while the Fortran is given in Section 9.6.3.1.

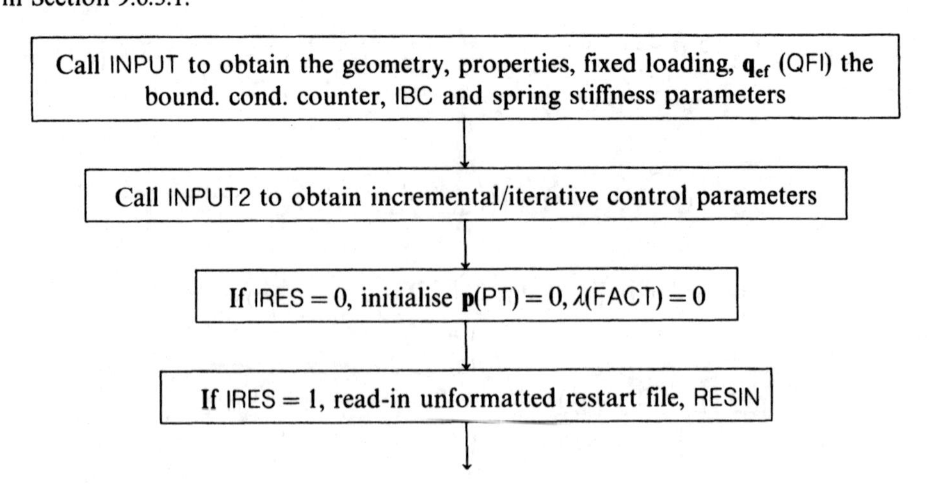

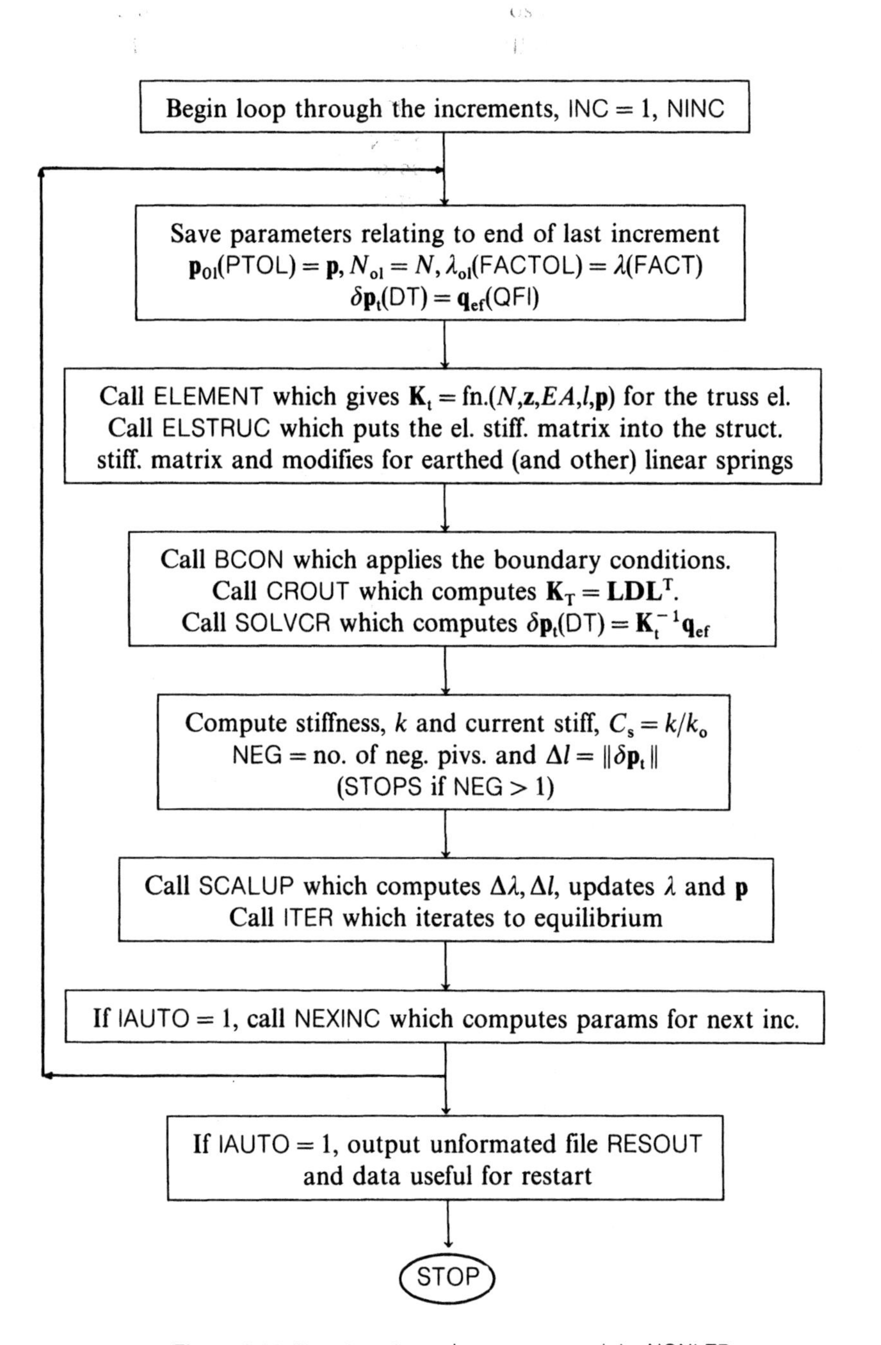


**Figure 9.14** Flowchart for main program module, NONLTD.

*9.6.3.1 Fortran for main program module* NONLTD.

```
      PROGRAM NONLTD
C
C     THIS VERSION INCLUDES FULL AUTO-INCS, LINE-SEARCHES, ARC-L
C     AND AUTOMATIC INCREMENTS (WITH INC. CUTTING)
C     PERFORMS NON-LIN. INCREMENTAL/ITERATIVE SOLN. FOR TRUSS
C     NV=NO. OF VARIABLES (4 OR 5)
C     QFI=FIXED LOAD VECTOR
C     IBC=B. COND. COUNTER (0=FREE, 1=FIXED)
C     Z=Z COORDS OF NODES
C     PT=TOTAL DISP. VECTOR
C     PTOL=TOTAL DISP VECTOR AT END OF LAST INC
C     DT=TANGENT DISP VECTOR FROM TOTAL LOAD (QFI)
C     GO=OLD O.B. FORCE VECTOR (FOR L-SEARCHES)
C     AKTS=STR. TAN. STIFF. MATRIX
C     AKTE=ELE. TAN. STIFF. MATRIX
C     FI=INTERNAL FORCES
C     D=DIAGONAL PIVOTS FROM LDL(TRAN) FACTORISATION
C     ID14S=VAR. NOS. (1-4) AT WHICH LINEAR EARTHED SPRINGS
C     AK14S=EQUIV. LINEAR SPRING STIFFNESS
C     AK15=LIN SPRING STIFF. BETWEEN VARBS. 1 AND 5 (IF NV=5)
C     GM=WILL BE OUT-OF-BALANCE FORCES (IN S/R ITER)
C     REAC=REACTIONS
C     X=X COORDS
C     FACT=TOTAL LOAD FACTOR, FACI=INC. LOAD FACTOR
C     DL=INCREMENTAL 'LENGTH' (DLDES=DESIRED VALUE)
C     ARGUMENTS IN COMMON/DAT2/ AND ARRAY X NOT USED FOR
C     SHALLOW TRUSS
C     ARGUMENTS IN COMMON/DATLS/ FOR L-SEARCHES
C     ARGUMENTS IN COMMON/AUTOINC/ FOR AUTO-INCS.
C     ARGUMENTS IN COMMON/ACEL/ FOR ACCELERATED MOD N-R.
C
      IMPLICIT DOUBLE PRECISION (A-H,O-Z)
      COMMON /DAT/ X(2),Z(2),E,ARA,AL,ID14S(4),AK14S(4),NDSP,ANIT,AK15
      COMMON /DAT2/ ARN,POISS,ALN,ITYEL
      COMMON /DATLS/ NLSMX,PERMLS,AMPMX,ETMXA,ETMNA,GO(5)
      COMMON /AUTOINC/ IAUTO,ITE,NITMAX,BET
      COMMON /ACEL/ DELO(5),GOO(5),IACC,R1C,R2C
      DIMENSION QFI(5),IBC(5),DT(5),PT(5),AKTE(4,4)
      DIMENSION FI(5),D(5),QEX(5),GM(5),AKTS(25),REAC(5)
      DIMENSION PTOL(5)
C
      IRE=5
      IWR=6
      OPEN (UNIT=5,FILE=' ')
      OPEN (UNIT=6,FILE=' ')
      OPEN (UNIT=17,FILE='RESIN',FORM='UNFORMATED')
      OPEN (UNIT=18,FILE='RESOUT',FORM='UNFORMATED')
```

```
C
      CALL INPUT(E,ARA,AL,QFI,X,Z,ANIT,IBC,IRE,IWR,AK14S,ID14S,NDSP,
     1           NV,AK15,
     2           POISS,ITYEL)
C     ARGUMENTS IN LINE ABOVE NOT USED FOR SHALLOW TRUSS
C     BELOW RELEVANT TO DEEP TRUSS BUT LEAVE FOR SHALLOW TRUSS
      AKN=AL
      ARN=ARA
C
C     READS IN DATA FOR INC/ITERATIVE CONTROL
      CALL INPUT2(FACI,NINC,IARC,BETOK,ITERTY,IDES,FACMX,FACMN,
     1  DLDES,DLDMX,DLDMN,ISWCH,CSTIFS,IBC,NV,IRE,IWR,IWRIT,IRES)
C
      IF (IRES.EQ.0) THEN
      AN=ANIT
      FACT=O.DO
      DO 5 I=1,NV
    5 PT(I)=O.DO
      ELSEIF (IRES.EQ.1) THEN
C     BELOW FOR RESTART FROM PREVIOUS RUN
      READ (17) AN,ALN,ARN,STIFI,FACT,PT
      CLOSE (UNIT=17)
      ENDIF
C
C
      DO 100 INC=1,NINC
      WRITE (IWR,1001) INC
 1001 FORMAT(//,1X,'INCREMENT NO.=  ',I5)
      DO 10 I=1,NV
C     SAVE DISPS. PT IN PTOL
      PTOL(I)=PT(I)
C     DT WILL BE TANGENTIAL DISPS DUE TO TOTAL LOAD QFI
      DT(I)=QFI(I)
   10 CONTINUE
C     SAVE OLD FORCE FOR CUT INC AND OLD TOTAL LOAD LEVEL
      FACTOL=FACT
      ANOL=AN
C
C     BELOW FORMS EL. TAN. STIFF MATRIX AKTE
      CALL ELEMENT (FI,AKTE,AN,X,Z,PT,E,ARA,AL,IWRIT,IWR,2,
     1   ITYEL,ALN,ARN)
C     ARGUMENTS IN LINE ABOVE NOT USED FOR SHALLOW TRUSS
C     BELOW PUTS EL. STIFF. AKTE IN STRUCT. STIFF. AKTS
C     AND ADDS EFFECT OF VARIOUS LINEAR SPRINGS
      CALL ELSTRUCT(AKTE,AKTS,NV,AK15,ID14S,AK14S,NDSP,FI,PT,
     1              2,IWRIT,IWR)
C
C
      CALL BCON(AKTS,IBC,NV,DT,IWRIT,IWR)
C     ABOVE APPLIES B. CONDITIONS
```

```
C        BELOW PUTS EFFECTIVE TANGENT LOAD VECTOR, DT IN GM (FOR CUR
C        STIFF)
         DO 15 I=1,NV
   15    GM(I)=DT(I)
         CALL CROUT(AKTS,D,NV,IWRIT,IWR)
C        ABOVE FORMS LDL(TRAN) FACTORISATION INTO AKTS AND D
         CALL SOLVCR(AKTS,D,DT,NV,IWRIT,IWR)
C        ABOVE SOLVES EQNS. AND GETS TAN. DISPS IN DT (FOR UNSCALED
C        LD)
C
C        BELOW COMPUTES NO OF NEG PIVOTS AND PARAMS FOR CURRENT
C        STIFF PARAM AND LENGTH INCS.
         STIFT=0.D0
         NEG=0
         DL=0.D0
         DO 20 I=1,NV
         IF (IBC(I).EQ.0) THEN
         STIFT=STIFT+DT(I)*GM(I)
         DL = DL + DT(I)*DT(I)
         IF (D(I).LT.0) NEG=NEG1
         ENDIF
   20    CONTINUE
C
C        BELOW COMPUTES 'STIFFNESS', STIF AND CSTIF=RATIO
C        OF VERY FIRST 'STIFFNESS'
         STIF=STIFT/DL
         IF (INC.EQ.1.AND.IRES.EQ.0) STIFI=STIF
         CSTIF = STIF/STIFI
         DL=DSQRT(DL)
C
         WRITE (IWR,1010) CSTIF,NEG
 1010    FORMAT(/,1X,'CURRENT STIFF. FACTOR=  ',G13.4,1X,
    1              'NO. OF NEG. PIVS.=  ',I5)
         IF (NEG.GT.1) THEN
         WRITE (IWR,1002)
 1002    FORMAT(/,'STOP BECAUSE NO OF NEG PIVS. GT. 1')
         STOP 'NONLTC 1002'
         ENDIF
C
         CALL SCALUP(IAUTO,IARC,NEG,FACt,FACT,DL,DLDES,PT,PTOL,DT,
    1       ISWCH,CSTIF,CSTIFS,NV,IWR,DLDMX,DLDMN)
C
C        BELOW ITERATES TO EQUILIBRIUM
         CALL ITER(PT,AN,BETOK,QFI,IBC,IWRIT,IWR,AKTS,D,ITERTY,NV,
    1            GM,FI,REAC,PTOL,DT,FACT,DL,IARC)
C
         IF (IARC.EQ.1) WRITE (IWR,1003) FACT
 1003    FORMAT(/,1X, 'TOTAL LOAD FACTOR AFTER ARC-L ADJUST=  ',G13.4)
         WRITE (IWR,1004) (PT(I),I=1,NV)
 1004    FORMAT(/,1X, 'FINAL TOTAL DISPLACEMENTS ARE',/,1X,6G12.5)
         WRITE (IWR,1006) (REAC(I),I=1,NV)
```

```
 1006 FORMAT(/,1X,'FINAL REACTIONS ARE',/,1X,5G12.5)
      WRITE (IWR,1005) AN
 1005 FORMAT(/,1X,'AXIAL FORCE IN BAR IS',G12.5)
C
C     BELOW COMPUTES INC. FACTORS FOR NEXT INCREMENT
      IF (IAUTO.EQ.1) THEN
      CALL NEXINC(IARC,ITE,NITMAX,FACI,FACMX,FACMN,DL,DLDES,
     1DLDMX,DLDMN,BETOK,BET,PT,PTOL,AN,ANOL,NV,IDES,FACT,FACTOL)
      ENDIF
C
  100 CONTINUE
C
      IF (IAUTO.EQ.1) THEN
      WRITE (IWR,1007) FACI,FACMX,FACMN,DLDES,DLDMX,DLDMN
 1007 FORMAT(/,1X,'AT END OF RUN',/,
     1   1X,'FACI=  ',G13.4,'FACMX=  ',G13.4,'FACMN=  ',G13.4,/,
     2   1X,'DLDES=  ',G13.4,'DLDMX=  ',G13.4,'DLDMN=  ',G13.4)
C     WRITE RESTART TAPE
      WRITE (18) AN,ALN,ARN,STIFI,FACT,PT
      CLOSE (UNIT=18)
      ENDIF
C
      STOP 'NONLTC'
      END
```

### 9.6.4 Flowchart and Fortran subroutine for routine SCALUP

This routine is called by program NONLTD (see last Section) following the computation of $\delta \mathbf{p}_t = \mathbf{K}_t^{-1}\mathbf{q}_{ef}$ in the array DT. Its main function is to compute $\Delta\lambda$ and $\Delta l$, and update $\lambda$ and the displacements, $\mathbf{p}$, following the predictor solution. The flow chart is given in Figure 9.15 and the Fortran in Section 9.6.4.1.

#### 9.6.4.1 *Fortran for routine* SCALUP

```
      SUBROUTINE SCALUP(IAUTO,IARC,NEG,FACI,FACT,DL,DLDES,PT,PTOL,
     1   DT,ISWCH,CSTIF,CSTIFS,NV,IWR,DLDMX,DLDMN)
C
C     UPDATES TOTAL LOAD FACTOR, FACT VIA INC. FACTOR FACI
C     AND OBTAINS INC LENGTH (POSSIBLY USING DESIRED VALUE)
C     UPDATES TOTAL DISPS. IN PT
      IMPLICIT DOUBLE PRECISION (A-H,O-Z)
      DIMENSION PT(NV),PTOL(NV),DT(NV)
C
      IF (IAUTO.EQ.0) GO TO 100
C     BELOW AUTOMATIC INCREMENTS
C
C     CHECKS FOR SWITCH TO ARC-L
      IF (ISWCH.EQ.1) THEN
```

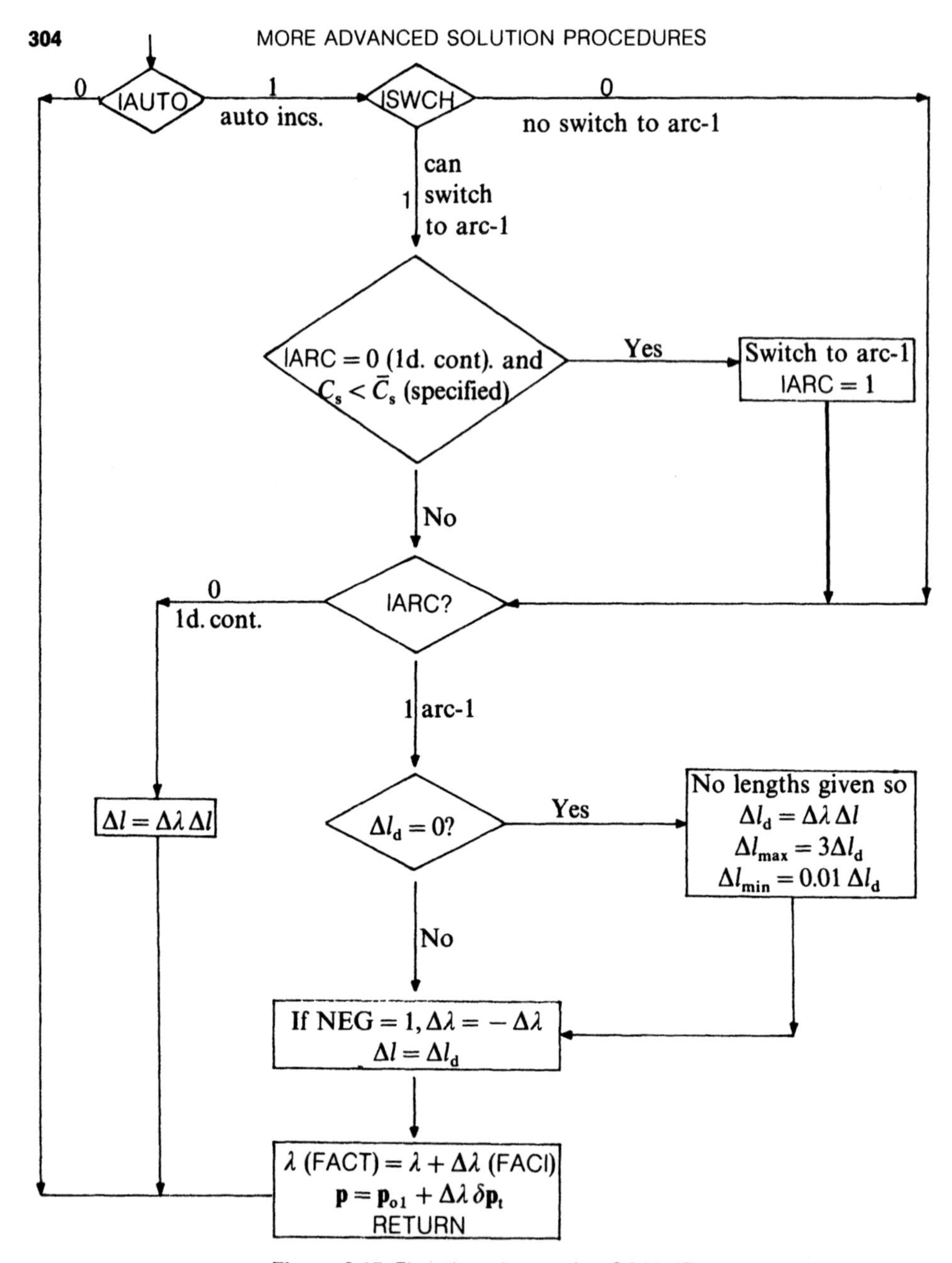


**Figure 9.15** Flowchart for routine SCALUP.

```
      IF ((IARC.EQ.0). AND. (CSTIF.LT.CSTIFS)) THEN
      IARC=1
      DLDES=DL
      DLDMX=5.D0*DL
      DLDMN=0.01D0*DL
      WRITE (IWR,1001) CSTIF,CSTIFS
1001  FORMAT(/,1X,'SWITCHED TO ARC-L BECAUSE CSTIF=  ',G13.4,/3X,
```

```
     1        ' LESS THAN CSTIFS=  ',G13.4)
      ISWCH=0
      ENDIF
      ENDIF
C
      IF (IARC.EQ.0) THEN
C     BELOW LOAD-CONTROL
      DL=FACI*DL
      ELSEIF (IARC.EQ.1) THEN
C     BELOW ARC-LENGTH CONTROL
C     FIRSTLY SET UP LENGTHS IF NONE WERE GIVEN
         IF (DLDES.EQ.0.D0) THEN
         DLDES=FACI*DL
         DLDMX=5.D0*DLDES
         DLDMN=0.01*DLDES
         ENDIF
C     COMPUTE INC LOAD FACTOR, FACI
      ASIGN=1.D0
      IF (NEG.EQ.1) ASIGN=-1.D0
      IF (NEG.EQ.1) WRITE (IWR,1003)
 1003 FORMAT(/,1X,'SWITCHING SIGN OF LOAD INCREMENT')
      FACI=ASIGN*(DLDES/DL)
      DL=DLDES
      ENDIF
C
  100 CONTINUE
      FACT=FACT+FACI
      WRITE (IWR,1002) FACT,FACI,DL
 1002 FORMAT(/,1X, 'TOTAL LD FACTOR=  ',G13.4,'INC FACTOR=  ',G13.4,
     1        'INC LENGTH=  ',G13.4)
      DO 10 I=1,NV
   10 PT(I)=PTOL(I)+FACI*DT(I)
      WRITE (IWR,1004) (PT(I),I=1,NV)
 1004 FORMAT(/1X,'TOTAL DISPS.AFTER TANG.SOLN ARE',/,1X,7G13.4)
      RETURN
      END
```

### 9.6.5 Flowchart and Fortran for subroutine NEXINC

Subroutine NEXINC is called by the main module NONLTD (see Section 9.6.3) when IAUTO is unity in order to compute the parameters for the next increment. The routine implements the techniques discussed in Sections 9.5 and 9.5.1 and, in particular, equations (9.40)–(9.42). The flowchart is given in Figure 9.16 and the Fortran in Section 9.6.5.1.

#### *9.6.5.1 Fortran for subroutine* NEXINC

```
      SUBROUTINE NEXINC(IARC,ITE,NITMAX,FACI,FACMX,FACMN,DL,DLDES,
     1 DLDMX,DLDMN,BETOK,BET,PT,PTOL,AN,ANOL,NV,IDES,FACT,FACTOL)
C     WITH AUTOMATIC INCREMENTS, COMPUTES FACTORS FOR NEXT INC
```

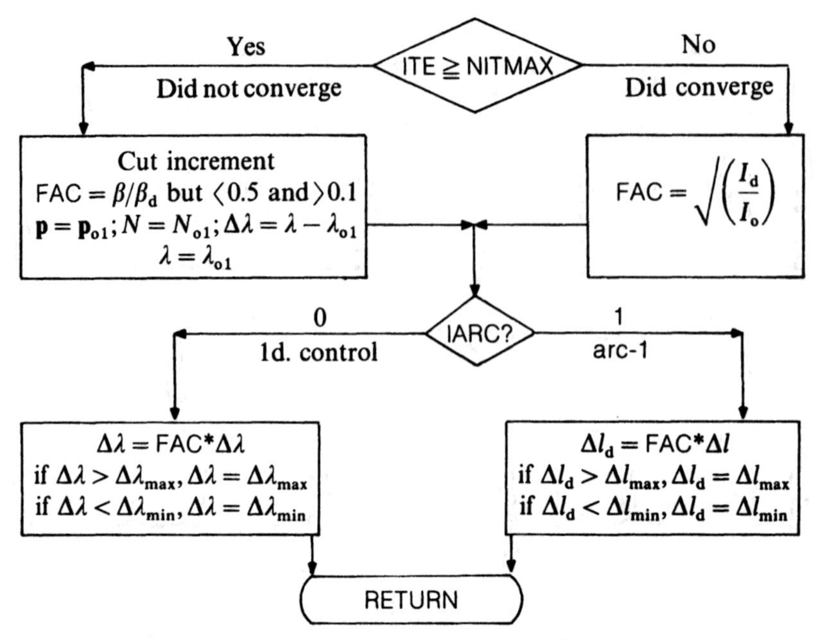


**Figure 9.16** Flowchart for subroutine NEXINC.

```
C
      IMPLICIT DOUBLE PRECISION (A-H,O-Z)
      DIMENSION PT(NV),PTOL(NV)
C
      IF (ITE.GE.NITMAX) THEN
C     DID NOT CONVERGE AT LAST INC, GET REDUCTION FACTOR, FAC
      FAC=BETOK/BET
      IF (FAC.GT.0.5D0) FAC=0.5D0
      IF (FAC.LT.0.1D0) FAC=0.1D0
C     RETURN DISPS AND FORCES TO VALUES BEFORE FAILED INC
      DO 30 I=1,NV
      PT(I)=PTOL(I)
   30 CONTINUE
      AN=ANOL
      FACI=FACT-FACTOL
      FACT=FACTOL
      ELSE
C     DID CONVERGE AT LAST INC., GET CHANGE FACTOR, FAC
C     MAKE LARGE IF NO REAL ITERS ON LAST INC.
      FAC=1000.D0
        IF (ITE.GT.1) THEN
        RITER=ITE-1
        RIDES=IDES
```

```
      FAC=RIDES/RITER
      FAC = DSQRT(FAC)
      ENDIF
    ENDIF
C
    IF (IARC.EQ.0) THEN
C   LOAD CONTROL
    FACI=FAC*FACI
    IF (FACI.GT.FACMX) FACI=FACMX
    IF (FACI.LT.FACMN) FACI=FACMN
    ELSEIF (IARC.EQ.1) THEN
C   ARC LENGTH
    DLDES=FAC*DL
    IF (DLDES.GT.DLDMX) DLDES=DLDMX
    IF (DLDES.LT.DLDMN) DLDES=DLDMN
    ENDIF
    RETURN
    END
```

## 9.7 QUASI-NEWTON METHODS

In the fields of methametical programming and unconstrained optimisation, much work has been devoted to the development of quasi-Newton solution procedures [F6, F7, L3, W6, B11, B12, B13, D2, D3, D8, S4, G5] (also known as the 'variable metric method' [D2]). A good review is given by Dennis and Moore [D8]. These quasi-Newton methods resemble the N–R technique but do not require the explicit re-formation of the tangent stiffness matrix. Instead, the stiffness matrix, or its inverse (or its Cholesky factors [G4]) are continuously updated as the iterations proceed. One of the earliest applications of the quasi-Newton method to finite elements involved linear analysis and was due to Fox and Stanton [F10].

In order to explain the method, it is best to replace the previous 'o' for 'old' and 'n' for 'new' with an iterative counter, $i$, so that in place of (9.6), the iterative update is

$$\mathbf{p}_{i+1} = \mathbf{p}_i + \eta_i \delta \bar{\mathbf{p}}_i \tag{9.53}$$

where, using a Newton-like algorithm (as in (9.5)),

$$\delta \bar{\mathbf{p}}_i = -\mathbf{K}_i^{-1} \mathbf{g}_i. \tag{9.54}$$

For a pure Newton–Raphson iteration $\mathbf{K}_i$ in (9.54) would be the true tangent stiffness matrix computed from $\mathbf{p}_i$. With the quasi-Newton methods, $\mathbf{K}_i$, is instead an approximation that satisfies the 'quasi-Newton equation':

$$\eta_{i-1} \delta \bar{\mathbf{p}}_{i-1} = \mathbf{K}_{i-1}^{-1} \boldsymbol{\gamma}_i \tag{9.55}$$

where

$$\boldsymbol{\gamma}_i = \mathbf{g}_i(\mathbf{p}_i) - \mathbf{g}_{i-1}(\mathbf{p}_{i-1}). \tag{9.56}$$

This equation is exactly satisfied for quadratic energy functions (linear structural analysis):

$$\phi = \tfrac{1}{2}\mathbf{p}^{\mathrm{T}}\mathbf{K}\mathbf{p} - \mathbf{q}_{\mathrm{e}}^{\mathrm{T}}\mathbf{p} \tag{9.57}$$

for which

$$\mathbf{g}_{i-1}(\mathbf{p}_{i-1}) = \left.\frac{\partial \phi}{\partial \mathbf{p}}\right|_{i-1}^{\mathrm{T}} = \mathbf{K}\mathbf{p}_{i-1} - \mathbf{q}_{\mathrm{e}} \tag{9.58}$$

so that, for the new displacements $\mathbf{p}_i = \mathbf{p}_{i-1} + \eta_{i-1}\delta\bar{\mathbf{p}}_{i-1}$ (see (9.53)),

$$\begin{aligned}\mathbf{g}_i(\mathbf{p}_{i-1} + \eta_{i-1}\delta\bar{\mathbf{p}}_{i-1}) &= \mathbf{g}_{i-1} + \boldsymbol{\gamma}_i = \mathbf{K}\mathbf{p}_i - \mathbf{q}_{\mathrm{e}} \\ &= \mathbf{K}*(\mathbf{p}_{i-1} + \eta_{i-1}\delta\bar{\mathbf{p}}_{i-1}) - \mathbf{q}_{\mathrm{e}} = \mathbf{g}_{i-1} + \eta_{i-1}\mathbf{K}\,\delta\bar{\mathbf{p}}_{i-1}. \end{aligned} \tag{9.59}$$

(The * multiplication sign has been used in the above to distinguish the following bracket from the 'function of' form.)

Many formulae have been derived that produce $\mathbf{K}_i$s satisfying (9.55). One of the most successful is the rank-2 BFGS update [F6, B12, S4] which can be written in the form

$$\mathbf{K}_i^{-1} = \mathbf{K}_{i-1}^{-1} - \frac{\delta\bar{\mathbf{p}}_{i-1}\boldsymbol{\gamma}_i^{\mathrm{T}}\mathbf{K}_{i-1}^{-1}}{\delta\bar{\mathbf{p}}_{i-1}^{\mathrm{T}}\boldsymbol{\gamma}_i} - \frac{\mathbf{K}_{i-1}^{-1}\boldsymbol{\gamma}_i\delta\bar{\mathbf{p}}_{i-1}^{\mathrm{T}}}{\delta\bar{\mathbf{p}}_{i-1}^{\mathrm{T}}\boldsymbol{\gamma}_i} + \left(1 + \frac{\boldsymbol{\gamma}_i^{\mathrm{T}}\mathbf{K}_{i-1}^{-1}\boldsymbol{\gamma}_i}{\eta_{i-1}\delta\bar{\mathbf{p}}_{i-1}^{\mathrm{T}}\boldsymbol{\gamma}_i}\right)\frac{\eta_{i-1}\delta\bar{\mathbf{p}}_{i-1}\delta\bar{\mathbf{p}}_{i-1}^{\mathrm{T}}}{\delta\bar{\mathbf{p}}_{i-1}^{\mathrm{T}}\boldsymbol{\gamma}_i}. \tag{9.60}$$

For the finite element method, the main difficulty with the quasi-Newton methods relates to the banded nature of the finite element stiffness matrix. Conventional applications will destroy this banded nature (away from equilibrium). Fortunately, as shown by Matthies and Strang [M1], an indirect form of solution can be adopted in which $\mathbf{K}_i^{-1}$ is not directly formed. (Mattheis and Strang's method is closely related to work in the mathematical programming literature by Buckley [B14], Nazareth [N1] and Nocedal [N2].) The method is used with an alternative, but equivalent, update to (9.60) whereby [M1, B11]:

$$\mathbf{K}_i^{-1} = (\mathbf{I} + \mathbf{w}_i\mathbf{v}_i^{\mathrm{T}})\mathbf{K}_{i-1}^{-1}(\mathbf{I} + \mathbf{v}_i\mathbf{w}_i^{\mathrm{T}}) = \mathbf{U}_i^{\mathrm{T}}\mathbf{K}_{i-1}^{-1}\mathbf{U}_i \tag{9.61}$$

where

$$\mathbf{w}_i = \frac{1}{\delta\bar{\mathbf{p}}_{i-1}^{\mathrm{T}}\boldsymbol{\gamma}_i}\delta\bar{\mathbf{p}}_{i-1} \tag{9.62}$$

$$\mathbf{v}_i = a_i^{1/2}\eta_{i-1}\mathbf{K}_{i-1}\delta\bar{\mathbf{p}}_i - \boldsymbol{\gamma}_i \tag{9.63}$$

and

$$a_i = \left(\frac{\delta\bar{\mathbf{p}}_{i-1}^{\mathrm{T}}\boldsymbol{\gamma}_i}{\eta_{i-1}\delta\bar{\mathbf{p}}_{i-1}^{\mathrm{T}}\mathbf{K}_{i-1}\delta\bar{\mathbf{p}}_{i-1}}\right). \tag{9.64}$$

It is not difficult to show that (9.61) satisfies (9.55). By substituting (9.55) into (9.64) one can show that $a_i$ is positive provided $\mathbf{K}_i$ and $\mathbf{K}_{i-1}$ are positive-definite. (Positive-definiteness is a basic assumption with the BFGS method [F7, L3]. For systems that are not necessarily positive definite (i.e. those beyond limit points as solved by the arc-length method), it might be more appropriate to use a rank 1 update such as that due to Davidon [D3] and Broyden [B13] or to Broyden as given by Dennis and Moore [D8]—the latter is unsymmetric and might therefore be appropriate for non-symmetric systems such as those arising from non-associative plasticity.) For positive-definite systems, there are, in concept, no problems in taking the square root of $a_i$ in (9.63).

Because

$$\delta\bar{\mathbf{p}}_{i-1} = -\mathbf{K}_{i-1}^{-1}\mathbf{g}_{i-1}. \tag{9.65}$$

(9.63) and (9.64) can be written in the more convenient form:

$$\mathbf{v}_i = -(1 + a_i^{1/2}\eta_{i-1})\mathbf{g}_{i-1} - \mathbf{g}_i \tag{9.66}$$

with

$$a_i = \frac{-\delta\bar{\mathbf{p}}_{i-1}^{\mathrm{T}}\boldsymbol{\gamma}_i}{\eta_{i-1}\delta\bar{\mathbf{p}}_{i-1}^{\mathrm{T}}\mathbf{g}_{i-1}}. \tag{9.67}$$

In order to 'indirected' apply the quasi-Newton technique, with $i = 1$, the iterative change of (9.54) and (9.61) is

$$\delta\bar{\mathbf{p}}_1 = -\mathbf{K}_1^{-1}\mathbf{g}_1 = -(\mathbf{I} + \mathbf{w}_1\mathbf{v}_1^{\mathrm{T}})\mathbf{K}_o^{-1}(\mathbf{I} + \mathbf{v}_1\mathbf{w}_1^{\mathrm{T}})\mathbf{g}_1 \tag{9.68}$$

which can be solved without directly computing $\mathbf{K}_1^{-1}$, via

$$\mathbf{b}_1 = \mathbf{g}_1 + (\mathbf{w}_1^{\mathrm{T}}\mathbf{g}_1)\mathbf{v}_1 \tag{9.69}$$

$$\mathbf{c}_1 = \mathbf{K}_0^{-1}\mathbf{b}_1 \tag{9.70}$$

and

$$\delta\bar{\mathbf{p}}_1 = -\mathbf{c}_1 - (\mathbf{v}_1^{\mathrm{T}}\mathbf{c}_1)\mathbf{w}_1. \tag{9.71}$$

Having obtained $\delta\bar{\mathbf{p}}_1$, a line-search would lead to $\eta_1$ (which, with a slack line search, may well be unity) after which the procedure would continue with

$$\delta\bar{\mathbf{p}}_2 = -(\mathbf{I} + \mathbf{w}_2\mathbf{v}_2^{\mathrm{T}})(\mathbf{I} + \mathbf{w}_1\mathbf{v}_1^{\mathrm{T}})\mathbf{K}_o^{-1}(\mathbf{I} + \mathbf{v}_1\mathbf{w}_1^{\mathrm{T}})(\mathbf{I} + \mathbf{v}_2\mathbf{w}_2^{\mathrm{T}})\mathbf{g}_2 \tag{9.72}$$

which can again be solved indirectly without ever forming $\mathbf{K}_2^{-1}$ or $\mathbf{K}_2^{-1}$.

The main disadvantage of the method is the continued accumulation of more and more vectors that have to be stored. However, depending on the storage availability, one can discard old vectors and merely accumulate new ones.

To avoid numerically dangerous updates, Matthies and Strang [M1] recommend that the updates of (9.61) are omitted if an estimate of the increase in condition number of $\mathbf{K}^{-1}$ exceeds some tolerance, tol ($\simeq 10^5$). From the work of Brodlie *et al.* [B11], this leads to

$$\text{est} = \text{cond}(\mathbf{U}_i) = \frac{[\|\mathbf{v}\|\cdot\|\mathbf{w}\| + \{\|\mathbf{v}\|^2\|\mathbf{w}\|^2 + 4(1 + \mathbf{v}^{\mathrm{T}}\mathbf{w})\}^{1/2}]^2}{4(1 + \mathbf{v}^{\mathrm{T}}\mathbf{w})} \geqslant \text{tol} \tag{9.73}$$

where $\|\mathbf{v}\|$ is the Euclidean norm of $\mathbf{v}$ and, for brevity, the subscript $i$ on the vectors $\mathbf{v}_i$ and $\mathbf{w}_i$ in (9.63) (or (9.66)) and (9.62) have been omitted.

Before leaving the quasi-Newton methods, and moving to related iterative techniques, it is useful to derive a weaker form of the quasi-Newton equation (9.55)

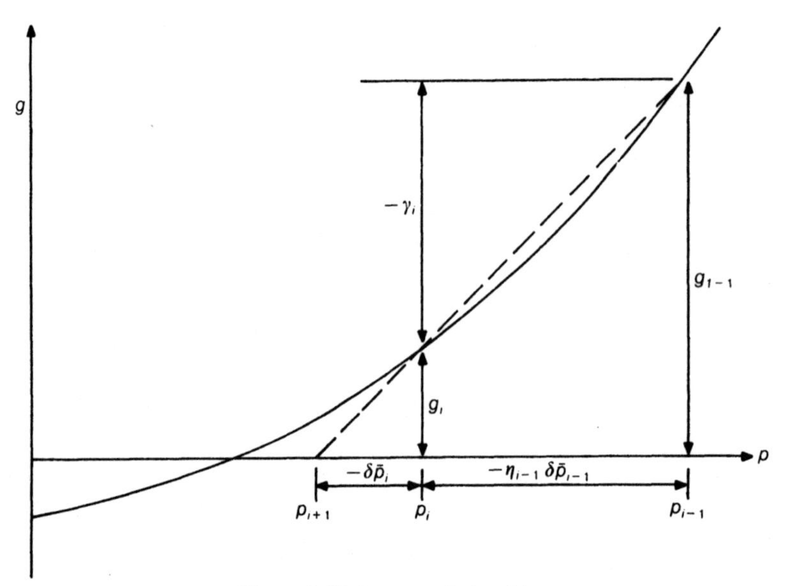


**Figure 9.17** Secant relationship.

by multiplying both sides by $\mathbf{g}_i$ and using (9.54), so that

$$\delta\bar{\mathbf{p}}_i^{\mathrm{T}}\boldsymbol{\gamma}_i = -\eta_{i-1}\delta\bar{\mathbf{p}}_{i-1}^{\mathrm{T}}\mathbf{g}_i. \tag{9.74}$$

In one-dimension, this scalar relationship becomes the 'secant formula' illustrated in Figure 9.17.

Quasi-Newton methods have been applied to non-linear finite elements by a number of workers [B3, P9, G2, H4].

## 9.8 SECANT-RELATED ACCELERATION TECHNIQUES

In [C7, C9, C10, C13, C17], the author developed a range of 'faster modified Newton–Raphson iterations' or 'secant-Newton techniques' which are closely related to the previous BFGS procedure. In particular, suppose we apply the technique of Section 9.7 but always discard all old updates except the current one. Hence, the iterative direction would from (9.54) and (9.61) take the form

$$\delta\bar{\mathbf{p}}_i = -[\mathbf{I} + \mathbf{w}_i\mathbf{v}_i^{\mathrm{T}}]\mathbf{K}_{\mathrm{o}}^{-1}[\mathbf{I} + \mathbf{v}_i\mathbf{w}_i^{\mathrm{T}}]\mathbf{g}_i \tag{9.75}$$

where $\mathbf{K}_{\mathrm{o}}$ is, say, the tangent stiffness matrix from the beginning of the increment so that if $\mathbf{w}_i$ and $\mathbf{v}_i$ were zero, we dould obtain the standard modified Newton–Raphson method. In these circumstances, it is easy to show that the iteration direction of $\delta\bar{\mathbf{p}}_i$ of (9.54) can be written as

$$\delta\bar{\mathbf{p}}_i = A\,\delta\overset{*}{\mathbf{p}}_i + B\eta_{i-1}\delta\bar{\mathbf{p}}_{i-1} + C\,\delta\overset{*}{\mathbf{p}}_{i-1} \tag{9.76}$$

where

$$\delta\overset{*}{\mathbf{p}}_i = -\mathbf{K}_{\mathrm{o}}^{-1}\mathbf{g}_i, \qquad \delta\overset{*}{\mathbf{p}}_{i-1} = -\mathbf{K}_{\mathrm{o}}^{-1}\mathbf{g}_{i-1}. \tag{9.77}$$

The scalars $A$, $B$ and $C$ are given by

$$C = \frac{\delta \bar{\mathbf{p}}_{i-1}^{\mathrm{T}} \mathbf{g}_i}{\delta \bar{\mathbf{p}}_{i-1}^{\mathrm{T}} \boldsymbol{\gamma}_i} \tag{9.78}$$

$$A = 1 - C \tag{9.79}$$

$$B = -C - \frac{(\delta \overset{*}{\mathbf{p}}_i - \delta \overset{*}{\mathbf{p}}_{i-1})^{\mathrm{T}} \mathbf{g}_i}{\eta_{i-1} \delta \bar{\mathbf{p}}_{i-1}^{\mathrm{T}} \boldsymbol{\gamma}_i} + C \frac{(\delta \overset{*}{\mathbf{p}}_i - \delta \overset{*}{\mathbf{p}}_{i-1})^{\mathrm{T}} \boldsymbol{\gamma}_i}{\eta_{i-1} \delta \bar{\mathbf{p}}_{i-1}^{\mathrm{T}} \boldsymbol{\gamma}_i}. \tag{9.80}$$

The three-vector iterative change in (9.76) can equally be derived from (9.54) and (9.60) with $\mathbf{K}_o$ instead of $\mathbf{K}_{i-1}$. It is therefore directly derived from the BFGS 'quasi-Newton formula' and is a 'memoryless single cycle' version of the Matthies and Strang BFGS procedure. It is also related to conjugate-gradient-like procedures by Shanno [S3] and others [B14, B15, N1, N2].

A two-vector update can be produced by making the approximation:

$$\eta_{i-1} \delta \mathbf{p}_{i-1} \simeq -\mathbf{K}_o^{-1} \mathbf{g}_{i-1}. \tag{9.81}$$

This leads to

$$\delta \bar{\mathbf{p}}_i = A\, \delta \overset{*}{\mathbf{p}}_i + \bar{B} \eta_{i-1} \delta \bar{\mathbf{p}}_{i-1} \tag{9.82}$$

where $A$ is again given by (9.79) (with $C$ from (9.78)) and $\bar{B}$ by

$$\bar{B} = -C - A \frac{\delta \bar{\mathbf{p}}_i^{\mathrm{T}} \boldsymbol{\gamma}_i}{\eta_{i-1} \delta \bar{\mathbf{p}}_{i-1}^{\mathrm{T}} \boldsymbol{\gamma}_i} \tag{9.83}$$

(with $A$ and $C$ from (9.78) and (9.79)). The update of (9.82) is not directly derivable from the BFGS method but does satisfy the 'secant relationship' of (9.74). Also, if $\mathbf{K}_o$ in (9.77) is replaced by the identity matrix, then for a quadratic energy function, with exact line searches both of the updates of (9.76) and (9.82) correspond with the conjugate-gradient method. With $\mathbf{K}_o$ as the tangent stiffness matrix at the beginning of the increment, the methods can be seen as forms of preconditioned conjugate gradient method [C5, C7, C8]. When seen in this light, they are closely related to the conjugate-Newton method of Irons and Elsawaf [I1]. It is work noting that numerical experiments [C13] showed a good performance using only a single-parameter accelerator via (9.82), with $A$ being taken from (9.79) and $\bar{B}$ being set to zero.

### 9.8.1 Cut-outs

It has already been indicated in Section 9.7 that there are iterations for which it is better to avoid the quasi-Newton update. These occasions also occur with the acceleration methods of Section 9.8 so that, in some circumstances, instead of using (9.76) or (9.82), one would use the standard modified Newton–Raphson direction of

$$\delta \bar{\mathbf{p}}_i = \delta \overset{*}{\mathbf{p}}_i \tag{9.84}$$

with $\delta \overset{*}{\mathbf{p}}_i$ from (9.77). Purely on the basis of numerical experiments [C8], the author devised the 'cut-out' criteria whereby the accelerated iterations ((9.76) or (9.82)) are only used if

$$R_1 > A > \frac{1}{R_1} \tag{9.85}$$

or, with (9.76),

$$R_2 > \frac{B + C}{A} > -\tfrac{1}{2}R_2 \tag{9.86a}$$

or with (9.82),

$$R_2 > \frac{\bar{B}}{A} > -\tfrac{1}{2}R_2. \tag{9.86b}$$

Otherwise, the mN–R change of (9.84) is used instead. In the work of [C8], it was recommended that $R_1$ lies between 2 and 3 and $R_2$ between 0.3 and 1. Subsequent work suggested a higher cut-off with $R_1$ and for the present work, $R_1$ has been taken as 3.5 and $R_2$ as 0.3.

### 9.8.2 Flowchart and Fortran for subroutine ACCEL

For the present computer program, we will implement the two-vector acceleration of (9.82). To this end, as already discussed in Section 9.4.2, subroutine ITER calls a routine ACCEL to change the mN–R direction $\delta\bar{\mathbf{p}}_i = \delta\overset{*}{\bar{\mathbf{p}}}_i$ to account for the acceleration (see flowchart in Figure 9.13). A flowchart for subroutine ACCEL is given in Figure 9.18, where we are using the subscript 'n' for 'new' and 'o' for 'old'.

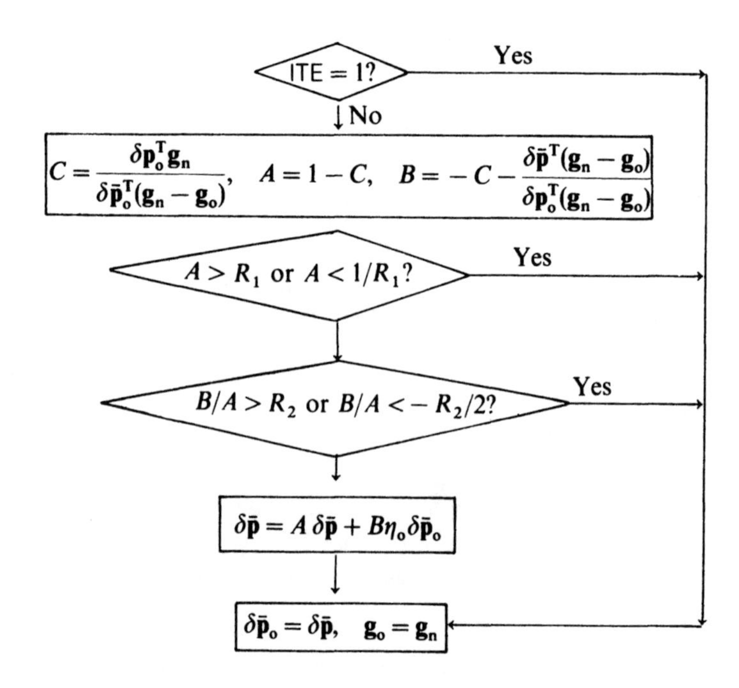


**Figure 9.18** Flowchart for subroutine ACCEL.

*9.8.2.1 Fortran for subroutine ACCEL*

```
      SUBROUTINE ACCEL(DPB,GO,GOO,IBC,NV,DELO,SLOL,R1C,R2C,ITE,IWRIT,
     1              IWR)
C
C     APPLIES SECANT ACCELERATION TO THE MOD N-R METHOD
C     DPB=ITER DISP TO BE MODIFIED ON ACCOUNT OF ACEL.
C     GO=CURRENT GRAD, GOD=OLD
C     DELO=OLD DISP CHANGE (WITHOUT S-L), SLOL=OLD STEP-LENGTH
C     R1C AND R2C ARE INPUT CUT-OUT PARAMS.
C
      IMPLICIT DOUBLE PRECISION (A-H,O-Z)
      DIMENSION DPB(NV),GO(NV),GOO(NV),IBC(NV),DELO(NV)
C
C     SET CUT-OUT VALUES
      R1I=1.D0/R1C
      R2E=-0.5D0*R2C
      IF (ITE.EQ.1) GO TO 100
C     BELOW REAL ACCN.
      BAS=0.D0
      TOP=0.D0
      TOP2=0.D0
      DO 10 I=1,NV
      IF (IBC(I).EQ.0) THEN
      BAS=BAS+DELO(I)*(GO(I)-GOO(I))
      TOP=TOP+DELO(I)*GO(I)
      TOP2=TOP2+DPB(I)*(GO(I)-GOO(I))
      ENDIF
   10 CONTINUE
      IF (ABS(BAS).LT.0.D-10) GO TO 100
      C=TOP/BAS
      A=1.D0-C
      B=-C-(A*TOP2/(SLOL*BAS))
      IF (A.LT.R1I) GO TO 100
      IF (A.GT.R1C) GO TO 100
      RAT=B/A
      IF (RAT.LT.R2E) GO TO 100
      IF (RAT.GT.R2C) GO To 100
      DO 20 I=1,NV
      IF (IBC(I).EQ.0) THEN
      DPB(I)=A*DPB(I)+B*SLOL*DELO(I)
      ENDIF
   20 CONTINUE
C     END OF ACCN. UPDATE
      IF (IWRIT.EQ.1) WRITE (IWR,1007) A,B
 1007 FORMAT(/,1X,'ACCN. WITH A=  ',G13.4,' B=  ',G13.4)
C
  100 CONTINUE
C     STORE OLD VALUES
      DO 30 I=1,NV
      IF (IBC(I).EQ.0) THEN
```

```
      DELO(I)=DPB(I)
      GOO(I)=GO(I)
      ENDIF
   30 CONTINUE
      RETURN
      END
```

## 9.9 PROBLEMS FOR ANALYSIS

In Section 3.10, we have already described a range of 'Benchmark tests for solution procedures for geometric non-linearity' that have been proposed by NAFEMS, the National Agency of Finite Elements [C1.2, D1.2]. We have also applied the basic Newton–Raphson (N–R) and modified Newton–Raphson (mN–R) methods to these problems. To this end, we used the computer program of Chapter 2 in conjunction with the deep-truss routines of Section 3.9. This computer program only allowed fixed increments. In the present section, we will apply some of the more advanced solution procedures of this chapter to these same problems.

In studying the results, it should be remembered that the solution algorithms have not been aimed specifically at these small problems but rather at more realistic large-scaled problems. Hence, one cannot necessarily draw useful conclusions regarding the relative efficiency of the different methods from these small problems. Nonetheless, they should illustrate the main features. In addition, they provide valuable examples for checking one's understanding of the various techniques. If the reader has implemented the computer programs, he (or she) may find it very useful to follow the program through some of the examples using, say, an interactive 'debugger'. He or she can, of course, experiment by trying different combinations of solution algorithm or, indeed, by solving different problems.

Unless stated otherwise, the convergence criterion will be that of (2.30) with $\beta = 0.001$ while, if line searches are used the tolerance, $\beta_{1s}$ of (9.11) will be set to 0.8. Also, the maximum number of iterations will be set to 21.

### 9.9.1 The problems

From Section 9.9.2 onwards, we will follow the convention whereby Section 9.9.x refers to the NAFEMS Example x. This procedure was followed in Section 3.10. It will be useful to return to the relevant subsection of Section 3.10 in order to recall the problems and their attributes. In order to avoid repetition, reference will be made to some of the figures (and tables) in that chapter. Figure 9.19 summarises the different tests and refers to relevant earlier work in the book.

### 9.9.2 Small-strain, limit-point example with one variable (Example 2.2)

This example is illustrated in Figure 9.19(a) and was previously discussed in Section 3.10.2.2. Table 9.1 supplements that of Table 3.1 by including the effects of line searches and accelerations.

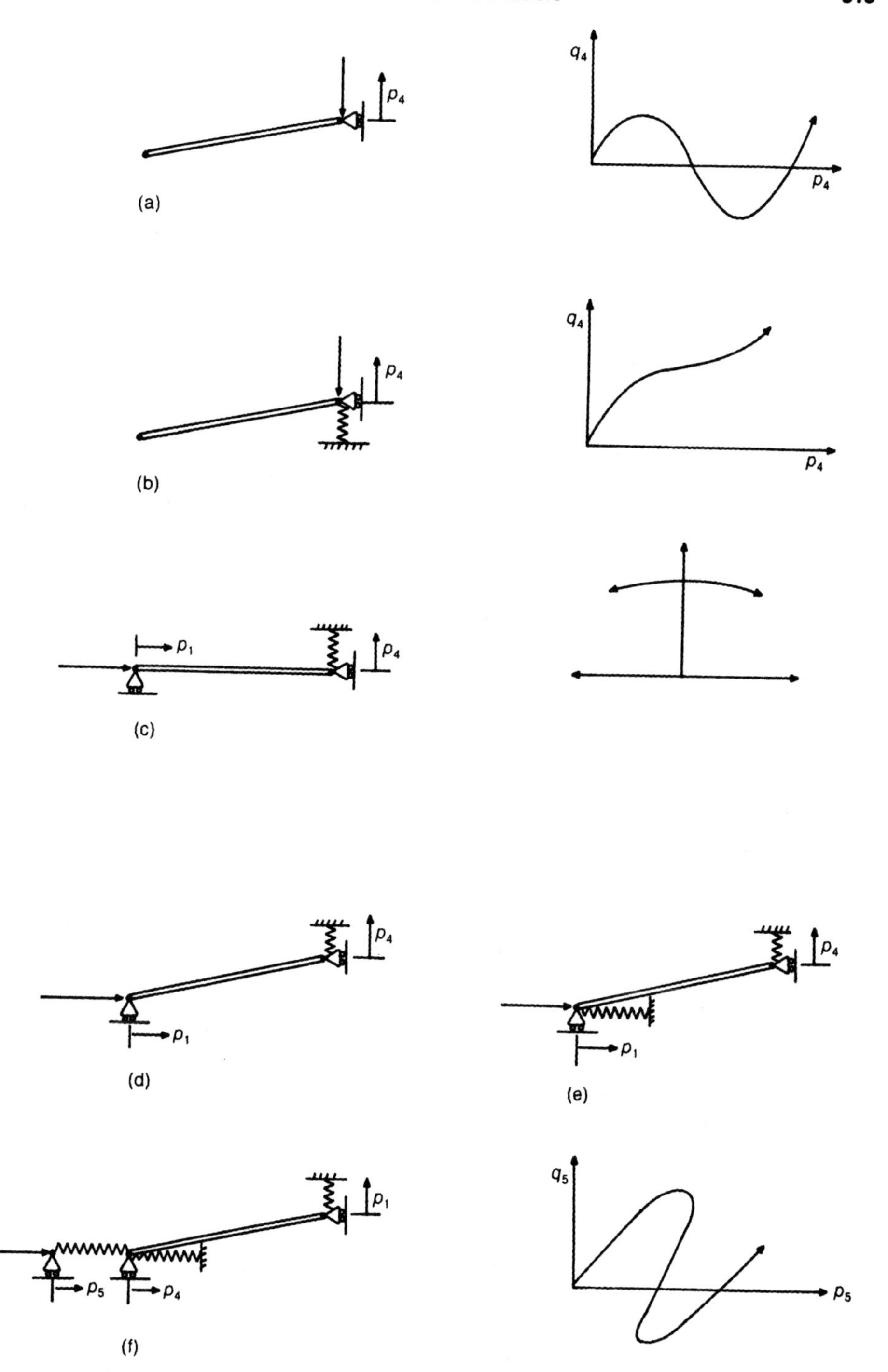


**Figure 9.19** The range of examples: (a) Example 2 (see Figures 3.12(a), 3.13); (b) Example 3 (see Figures 3.12(a), 3.14); (c) Example 4 (see Figure 3.15); (d) Example 5 (Figures 3.12(b) + 3.18); (e) Example 6 (Figures 3.12(b) + 3.19); (f) Example 7 (see Figures 3.12(c) and 3.20).

**Table 9.1.** Iterative performance for Example 2.2 (see Figures 3.12(a) and 3.13).

| Method | Iterations at load step 1 | 2 | 3 | 4 | 5 | 6 |
|---|---|---|---|---|---|---|
| mN–R | 2 | 2 | 3 | 3 | 12 | fail |
| mN–R + accn. | 2 | 2 | 2 | 2 | 4 | fail |
| mN–R + 1.s. | 2 | 2 | 3 | 3 | 12 | 3(3) |
| N–R | 1 | 1 | 1 | 2 | 3 | fail |
| N–R + 1.s | 1 | 1 | 1 | 2 | 3 | 17(7) |

In the above table and for the rest of the chapter when giving results for a solution procedure with line searches, the figure in brackets will be the number of extra calculations of the out-of-balance force vector, **g**, resulting from the use of the line searches. Hence in the above table, the addition of line searches to the mN–R method allowed point 6 (Figure 3.13) to be reached via three iterations and three extra 'residual' calculations.

The following data is for the mN–R method with line searches.

4 2 50000000. 1. 0. 0. ; NV, ITYE (rot eng.), ARA, POIS, ANIT
0. 2500. *x* coords.
0. 25. *z* coords.
0. 0. 0. −1.0 ; load of −1.0 at variable 4 (vertical at node 2)
1 1 1 0 ; only variable 4 is free
0 ; no earthed springs
1.9 6 0 0 0 0 0 ; FACI,NINC,IWRIT,IAUTO,IARC,IACC,IRES
0.001 2 21 6 ; BETOK,ITERTY(mN–R),NITMAX,NLSMX
0.8 5.0 25. 0.01 PERMLS,AMPMX,ETMXA,ETMNA

This problem could be successfully solved using either displacement control or the arc-length method (effectively the same for this trivial problem with a single variable).

### 9.9.3 Hardening problem with one variable (Example 3)

For this problem, a linear spring, has been added (Figure 9.19(b)) so that the response is continuously hardening although with a softer and then a stiffer region. On introducing the newer solution procedure, Table 3.2 becomes augmented to give Table 9.2.

Data for the solution using the accelerated mN–R method is given below:

4 2 50000000. 1 0. 0. ; NV, ITYE (rot. eng.), E, ARA, POIS, ANIT
0. 2500. *x* coords.
0. 25. *z* coords.
0. 0. 0. −1.0 ; load of −1 at variable 4 (vertical at node 2)
1 1 1 0 ; only variable 4 is free

**Table 9.2.** Iterative performance for Example 3.10.3 (see Figures 3.12(a) and 3.14).

| Method | Iterations at load step 1 | 2 | 3 | 4 | 5 | 6 | 7 |
|---|---|---|---|---|---|---|---|
| mN–R | 3 | 3 | 3 | 5 | 10 | fail | |
| mN–R + accn. | 2 | 2 | 2 | 3 | 8 | 9 | 3 |
| mN–R + 1.s. | 3 | 3 | 3 | 5 | 7(1) | 3(1) | 6 |
| N–R | 1 | 1 | 2 | 2 | 2 | 3 | 2 |

1 one earthed spring
4 at variable 4
1.125 ; of magnitude 1.125
6. 7 0 0 0 1 0 ; FACI,NINC,IWRIT,IAUTO,IARC,IACC,IRES
0.001 2 21 0 ; BETOK,ITERTY NITMAX,NLSMX
3.5 0.3 ; R1C, R2C

This example can also be used to illustrate the use of 'automatic increments' and 'automatic increment reduction'. The following data relates to such a solution with the basic mN–R method.

4 2 50000000. 1. 0. 0. ; NV, ITYE (rot. eng.), E,ARA, POIS, ANIT
0. 2500. *x* coords.
0. 25. *z* coords.
0. 0. 0. −1.0 ; load of −1 at variable 4 (vertical at node 2)
1 1 1 0 ; only variable 4 is free
1 one earthed spring
4 at variable 4
1.125 ; of magnitude 1.125
6. 30 0 1 0 0 0 ; FACI,NINC,IWRIT,IAUTO,IARC,IACC,IRES
0.001 2 10 0 ; BETOK,ITERTY NITMAX,NLSMX
3 8 0.5 0 ; IDES,FACMX,FACMN,ISWCH

The results are plotted in Figure 9.20. With the maximum number of iterations (NITMAX) being set to 10 (see above data), the increments were automatically cut on two occasions (Figure 9.20).

### 9.9.4 Bifurcation problem (Example 4)

The bifurcation problem (Figure 9.19(c)) has already been discussed in Sections 1.2, Section 2.6.3 and 3.10.4. In the present context, it is useful to apply the arc-length method to this problem and to imagine that we did not know of the bifurcation. As a precursor to the arc-length solution we might have applied a single load increment of 1000 which would give a length increment, $\Delta l$ of 0.05. Suppose we then wished to start from the beginning with the arc-length method with increments of $\Delta l = 0.05$. Appropriate data is given below.

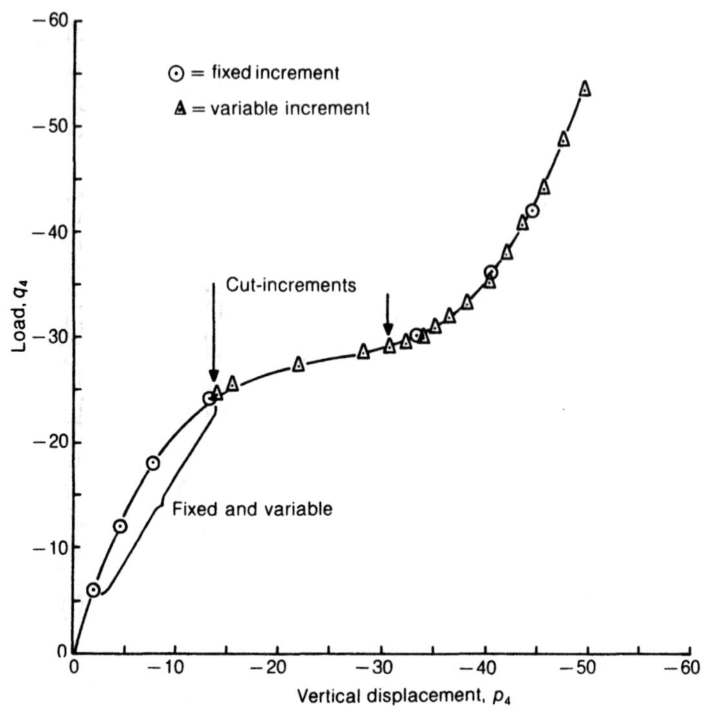


**Figure 9.20** Solution points for Example 3.

```
4 2 50000000. 1. 0.0 0.0 ; NV,ITYE(Engng.).E,ARA,POIS,ANIT
0. 2500.  ; x coords.
0. 0.0    ; z coords.
1000. 0. 0. 0. ; fixed LOAD vector
0  1  1  0 ; Bdry.condn. code
1  ; one earthed spring
4  ; at variable 4
1.5 ; of mag. 1.5
1.0 7 0 1 1 0 0 ; FACI,NINC,IWRIT,IAUTO,IARC,IACC,IRES
0.001 2 21 0  BETOK,ITERTY(mN-R),NITMAX,NLSMX
3  1. 1. 0   IDES,FACMX,FACMN,ISWCH
0.05 0.05 0.05  DLDES,DLDMX,DLDMN
```

Because DLDES $= \Delta l_{des}$ is specified as 0.05 on the last 'card', the load increment factors on the penultimate card are not used. The first increment would result in a $\lambda$-value of 1 or, with the given loading, a load of 1000. Because $\Delta l_{max}$ and $\Delta l_{min}$ are also set to 0.05, the next increment would be of the same length, leading to $\lambda = 2$ and then $\lambda = 3$ and on the fourth increment $\lambda = 4$. At this stage, the critical buckling load of 3750 would have been passed and on factorising $\mathbf{K}_t$ one negative pivot would be found. Hence the switching procedure in subroutine SCALUP (see Section 9.6.4) would reverse the sign for the next (fifth) increment leading to $\lambda = 4 - 1 = 3$ at which level $\mathbf{K}_t$ would be positive-definite so that the next increment would be positive. This would continue with the solution oscillating about the critical buckling load. The reason for this behaviour has already been discussed in Section 9.4.3 and relates to a bifurcation point being mistaken for a limit point. It is worth observing from the solution to this problem that the current stiffness parameter (Section 9.5.2) does not respond at all to the bifurcation and remains at unity throughout the analysis.

### 9.9.5 Limit point with two variables

Problem 5 is illustrated in Figure 9.19(d) and has been discussed before in Section 3.10.5. Because it involves a limit point, it would seem a good example to illustrate the automatic switch to the arc-length method. Appropriate data is given below:

```
4 2 50000000. 1. 0.0 0.0 ; NV,ITYE(Engng.).E,ARA,POIS,ANIT
0. 2500.  ; x coords.
0. 25.    ; z coords.
1000. 0. 0. 0. ; fixed LOAD vector
0  1  1  0 ; Bdry.condn. code
1  ; one earthed spring
4  ; at variable 4
1.5 ; of mag. 1.5
0.76 30 0 1 0 0 0 ; FACI,NINC,IWRIT,IAUTO,IARC,IACC,IRES
0.001 1 21 0   BETOK,ITERTY,NITMAX,NLSMX
3  1.0 0.1 1    IDES,FACMX,FACMN,ISWCH
0.3 ; CSTIFS
```

This data is designed to start under load control with a first increment of $\Delta\lambda = 0.76$ (leading to a load of 760), to apply automatic increments with a maximum factor of $\Delta\lambda = 1$ and a minimum of $\Delta\lambda = 0.1$. The solution should switch automatically (Section 9.5.2) to the arc-length method once the current stiffness parameter reduces below 0.3.

In order to record the solution, it is useful to plot an end-on view of the solution in Figure 3.17, i.e. in the $q_1$–$p_4$ plane (see Figure 3.12(b)). The results of the previous load-controlled solution (Section 3.10.5) are plotted as circles in Figure 9.21 and clearly indicate the branch switching. The triangles were obtained using the previous data and illustrate the success of the automatic switch to the arc-length procedure on increment 4. The solution then proceeded without difficulty under arc-length control until increment 26 (see Figure 9.21). At the following increment the solution switched to the secondary branch in the region Z on Figure 3.17 and then oscillated between points 26 and 27 (Figure 9.21). From the exact solutions [C1.2], we know of the second branch in this region and so, with this knowledge, the solution is at least believable although not desirable.

Before giving a reason for this behaviour, we should explain how we continued on from point 26 (Figure 9.21) to obtain the 'correct solutions' which are the squares in Figure 9.21. This was achieved by firstly rerunning the previous data but reducing the number of increments from 30 to 26. Hence the 'unformatted' restart file, RESOUT was available for a restart from point 26. The data used to obtain the squares in Figure 9.21 is given below:

```
4 2 50000000. 1. 0.0 0.0 ; NV,ITYE,E,ARA,POIS,ANIT
```

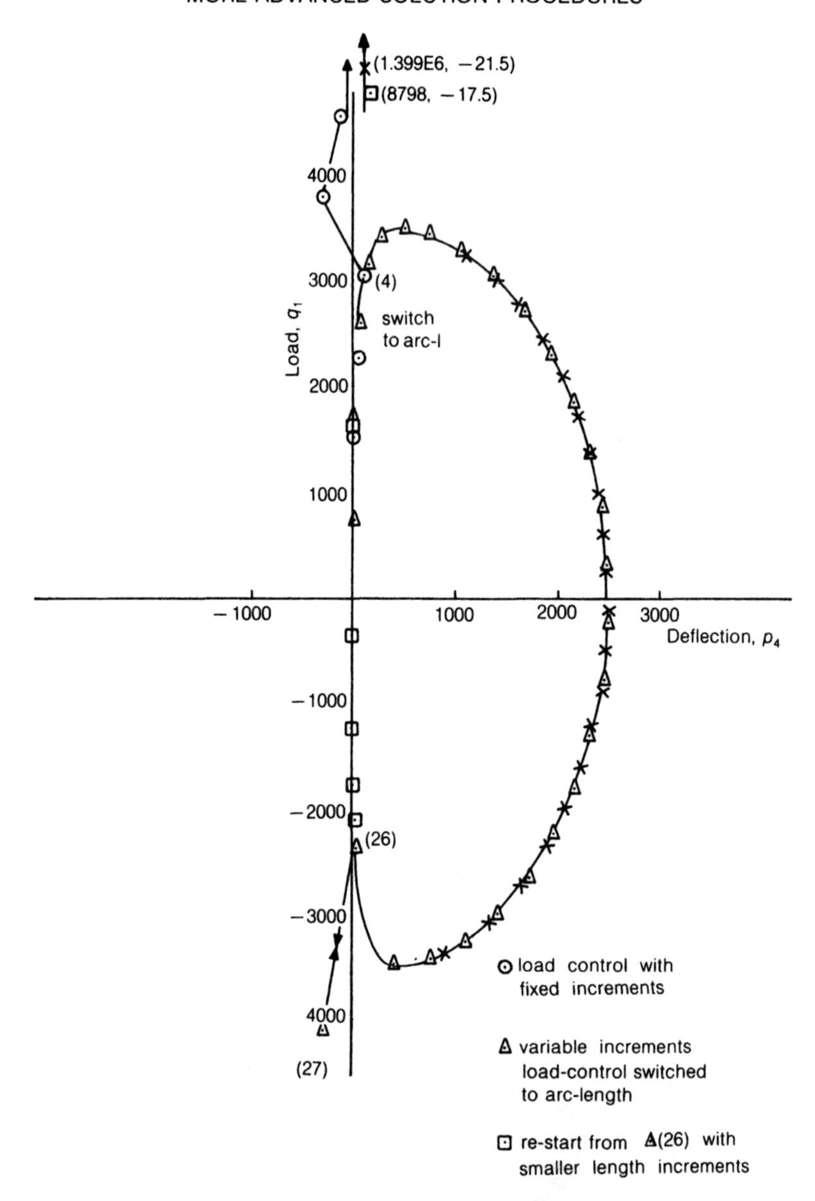


**Figure 9.21** Solution points for Example 5.

0. 2500. ; *x* coords.
0. 25. ; *z* coords.
1000. 0. 0. 0. ; fixed LOAD vector
0 1 1 0 ; Bdry.condn. code
1 ; one earthed spring
4 ; at variable 4

```
1.5 ; of mag. 1.5
0.08583 10 0 1 1 0 1 ; FACI,NINC,IWRIT,IAUTO,IARC,IACC,IRES
0.001 1 21 0  BETOK,ITERTY,NITMAX,NLSMX
3 1.0 0.1 0   IDES,FACMX,FACMN,ISWCH
10.0 10.0 0.5     DLDES,DLDMX,DLDMN
```

This data forces the solution to start under arc-length control with a length increment, $\Delta l$ of 10 and does not let the incremental length grow beyond this value. This increment compares with the value of $\Delta l = 370$ that was used in going between the triangles 26 and 27 in Figure 9.21. As a result of this dramatic reduction in increment size, the squared points in Figure 9.21 were obtained (although off the graph, the seventh increment again introduced a branch switch).

In explaining the strange behaviour associated with the previous branch switching, it is useful to return to the solutions in Figure 3.15 which relate to the 'perfect system' and to reconsider the explanation of these solutions that was originally given in Section 3.10.4 and is related to the system of Figure 3.12(b) or 9.19(d) (with $z = 0$). This explanation related the load-deflection response OAC (or C′) DEF (Figure 3.15) to the configurations in Figure 3.16. While the present analysis relates to an imperfect rather than a 'perfect' system, the two are closely related (Figure 3.17). In relation to the current analysis, the low point in Figure 9.21 relates to point D in Figure 3.15 and the rising part to the line DEF. At point D in Figure 3.15, the configuration of the bar is as in Figure 3.16(iii) with point 'a' now lying horizontally on the right of the pivot point b. The rising portion DEF in Figure 3.15 then involves further movement of the 'load point a' to the right, thus reducing the compression in the bar and eventually taking it into tension. Movement downwards from point D, Figure 3.15 towards point G involves an 'unstable' compression of the bar with the load point a now moving to the left. The differences between a small movement from D towards F and one towards G (Figure 3.15) are very minor in comparison with the total movements to reach that configuration. In particular, the movement $p_1$ (Figure 9.19c) to reach the configuration (iii) in Figure 3.16 is about 50 000 or twice the length of the bar. In these extreme circumstances, the two alternative solutions are, as far as the arc-length method is concerned, very close together so that the method may converge on one or the other. In relation to Figure 3.18, which relates to the present imperfect system, the stable solution is towards point 5, while the unstable solution induces further compression towards point x. The solution with the triangle 27 in Figure 9.21 lies on the latter unstable equilibrium curve. Because it is unstable, a negative pivot was detected and hence the oscillation (Figure 9.21) between points 26 and 27.

Clearly, the most sensible way to solve this problem is to use displacement control at the 'loading point'. The following data relates to such a solution:

```
4 2 50000000. 1. 0.0 0.0 ; NV,ITYE(Engng),E,ARA,POIS,ANIT
0. 2500.  ; x coords.
0. 25.    ; z coords.
1000. 0. 0. 0. ; fixed Displ. vector
-1  1  1  0 ; Bdry.condn. code
1  ; one earthed spring
4  ; at varbl. 4
```

```
1.5 ; of mag. 1.5
0.25 25 0 1 0 0 0 ; FACI,NINC,IWRIT,IAUTO,IARC,IACC,IRES
0.001 1 21 0  BETOK,ITERTY,NITMAX,NLSMX
3 0.25 0.05 0   IDES,FACMX,FACMN,ISWCH
```

This solution involves applying a first increment of $\Delta p_1 = 250$ and forcing subsequent displacement increments to be no larger. The results are plotted as the crosses in Figure 9.21.

### 9.9.6 Hardening solution with two variables (Example 6)

This problem has previously been discussed in Section 3.10.6. In relation to the results for fixed load-increments that were given in Table 3.4 and relate to Figure 3.19, the new solution procedures, such as line searches and accelerations, were not very successful. Successful solutions were obtained using the following data for a solution with automatic load increments and the mN–R method with line searches although the increments became very small so that progress was very slow.

```
4 2 50000000. 1. 0.0 0.0 ; NV,ITYE=Engng.,E,ARA,POIS,ANIT
0. 2500.  ; x coords.
0. 25.    ; z coords.
1000. 0. 0. 0. ; fixed LOAD Vector
0  1  1  0 ; Bdry.condn. code
2  ; Two earthed springs
1 4 ; At varbls. 1 and 4
2.0 1.5 ; of mag. 2.0 and 1.5 respectively
1.0 25 0 1 0 0 0 ; FACI,NINC,IWRIT,IAUTO,IARC,IACC,IRES
0.001 2 21 4  BETOK,ITERTY(mN-R),NITMAX,NLSMX
0.8 5.0 25. 0.01 PERMLS,AMPMX,ETMXA,ETMNA
3 2.0 0.02 0 IDES,FACMX,FACMN,ISWCH
```

A solution was also obtained using the automatic switch to the arc-length method via

```
4 2 50000000. 1. 0.0 0.0 ; NV,ITYE=Engng.,E,ARA,POIS,ANIT
0. 2500.  ; x coords.
0. 25.    ; z coords.
1000. 0. 0. 0. ; fixed LOAD Vector
0  1  1  0 ; Bdry.condn. code
2  ; Two earthed springs
1 4 ; At varbls. 1 and 4
2.0 1.5 ; of mag. 2.0 and 1.5 respectively
1.0 60 0 1 0 0 0 ; FACI,NINC,IWRIT,IAUTO,IARC,IACC,IRES
0.001 1 21 0  BETOK,ITERTY,NITMAX,NLSMX
3  2.0 0.02 1   IDES,FACMX,FACMN,ISWCH
0.3  ; CSTIFS
0.0 0.0 0.0  ; DLDES,DLDMX,DLDMN
```

However, as with the previous example (9.9.5), branch-switching and oscillation occurred in the very late stages when the bar was almost fully inverted. Again, as with the previous example, the difficulties could be overcome via displacement control.

### 9.9.7 Snap-back (Example 7)

This example has previously been discussed in Section 3.10.7. Because of the snap-back, this is an ideal example for the arc-length method. Figure 9.22 shows the relationship between end load, $q_5$ and end-shortening, $p_5$ that was obtained with the aid of the following data:

```
5 1 50000000. 1. 0.0 0.0 ; NV,ITYE=(Engng.),E,ARA,POIS,ANIT
0. 2500.  ; x coords.
0. 25.    ; z coords.
0. 0. 0. 0. 100. Fixed LOAD Vector, loading at variable 5
0  1  1  0  0 ; Bdry.condn. code
2  ; Two earthed springs
1 4 ; At varbls. 1 and 4
0.25 1.5 ; of mag 0.25 and 1.5 respectively
1.0    ; linear spring of stiff 1.0 between variables 1 and 5
```

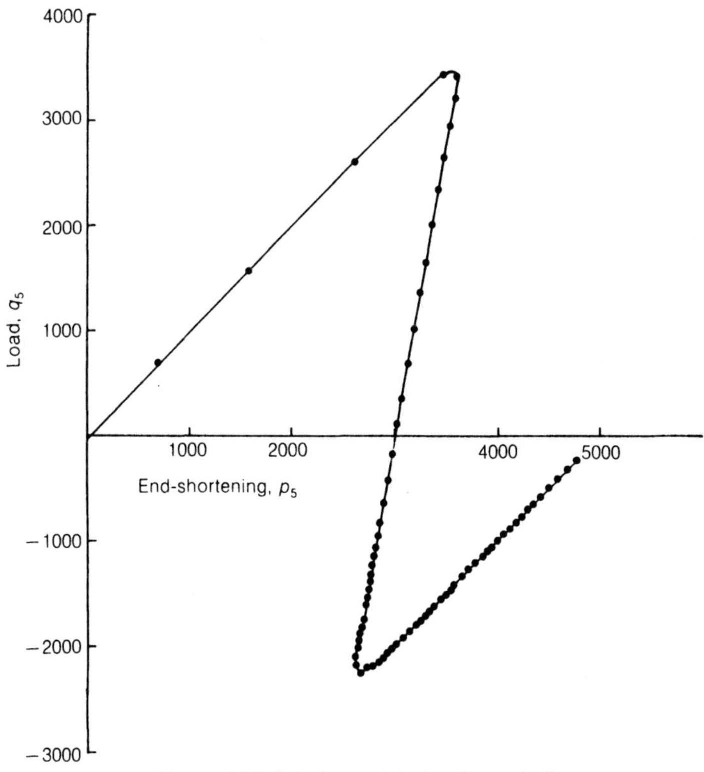


**Figure 9.22** Solution points for Example 7.

1.0 100 0 1 1 0 0 ; FACI,NINC,IWRIT,IAUTO,IARC,IACC,IRES
0.001 1 12 0 ; BETOK,ITERTY(N–R),NITMAX,NLSMX
3 1. 1. 0 IDES,FACMX,FACMN,ISWCH
700. 3500. 10. ; DLDES,DLDMX,DLDMN

A preliminary analysis had indicated that an appropriate starting length increment would be $\Delta l = 700$ which give a load change, $\Delta q_5$ of about 700 and a shortening, $\Delta p_5$ also of about 700. The solution in Figure 9.22 shows some very small increments. As indicated in the data above, a total of one hundred increments were used.

## 9.10 FURTHER WORK ON SOLUTION PROCEDURES

Inevitably, a whole range of work on solution procedures has so far been omitted from this chapter. Papers with significant reviews on the subject can be found in [W2, C4, R7, R11, R12, F1, C8, C17]. In this last section, we will now attempt to mention at least some of the work. Extensions will be given in Volume 2.

In Section 9.3, we discussed the arc-length and some related methods. A further related method, advocated by Bathe and Dvorkin [B2] uses, as a constraint, the 'constant increment of external work'. With this approach, the predictor solution would (see also (9.38)) be governed by

$$\Delta W = (\lambda + \tfrac{1}{2}\Delta\lambda_p)\mathbf{q}_{ef}^T \Delta\mathbf{p}_p = \Delta\lambda_p(\lambda + \tfrac{1}{2}\Delta\lambda_p)\mathbf{q}_{ef}^T \delta\mathbf{p}_t \tag{9.87}$$

where $\lambda$ is the load-level parameter at the end of the last increment and the work $\Delta W$ takes the place of the arc length, $\Delta l$. For subsequent iterations, the 'iterative external work' is set to zero via (see (9.23))

$$(\lambda_0 + \tfrac{1}{2}\delta\lambda)\mathbf{q}_{ef}^T \delta\mathbf{p} = (\lambda_0 + \tfrac{1}{2}\delta\lambda)(\mathbf{q}_{ef}^T \delta\bar{\mathbf{p}} + \delta\lambda \mathbf{q}_{ef}^T \delta\mathbf{p}_t) = 0 \tag{9.88}$$

where $\lambda_0$ is the 'old' (at the end of the previous iteration) value of $\lambda$. As with the 'spherical arc-length method', (9.88) leads to a quadratic equation with two roots for the load-level change, $\delta\lambda$ (compare (9.26) and (9.27)). In contrast to the arc-length method, these roots will always be real. Consequently, some workers [B5] have advocated switching to the work procedure when no real roots are obtained with the arc-length method. With the previous computer implementation of the arc-length method, we have, in these circumstances, simply cut the length increment, $\Delta l$.

Bergan [B6] has advocated an approach involving the minimisation of the out-of-balance force norm. Neither the author [C8] nor Clarke and Hancock [C4] found this method to be very successful when applied on its own. Eriksson [E1] has recommended a procedure whereby the iterative change is decomposed to include separately a component involving the lowest eigenmode (see also [J2]). Near limit points, this component is damped.

In earlier sections, we have separately discussed the arc-length methods, line searches, quasi-Newton and secant–Newton methods. Clearly these procedures can be combined and the combination of arc-length and acceleration techniques has been considered by the author [C8, C11]. A difficulty relates to the line searches which have to be enforced at a variable load level and may or may not be up hill [C16, C19]. Much further work is required.

A further set of solution procedures can be derived from so-called 'homotopy' of 'relaxation' equations whereby the equilibrium equations $\mathbf{g}(\mathbf{p}) = 0$ (here assuming the load level $\lambda$ is fixed) are embedded in an extended equation of the form

$$\mathbf{g}(\mathbf{p}, t) = \mathbf{g}(\mathbf{p}) - e^{-t}\mathbf{g}(\mathbf{p}_o) = \mathbf{0} \tag{9.89}$$

where $t$ can be considered as a 'pseudo-time' and as $t \to \infty$, one obtains the desired solution $\mathbf{g}(\mathbf{p}) = \mathbf{0}$. These methods (said to be due to Davidenko [D1]) have been much applied in both the mathematical [Al, G1, K3] and structural [P6, F2, F4] literatures. They have often been used with arc-length-type constraints and then have much in common with the previous 'continuation methods'.

A simplified variation of the method can be used to recover equilibrium from a state (related to a loading $\lambda\mathbf{q}_{ef}$) at which $\mathbf{g} \neq \mathbf{0}$ and from which standard methods fail to achieve convergence. The procedure is based on the observation that this solution is in equilibrium with a modified external loading, '$\mathbf{q}_{ef}$' $= \lambda\mathbf{q}_{ef} + \mathbf{g}$. Hence, we may restart the analysis procedure working with two load vectors following the lines of (9.45), so that

$$\text{'}\mathbf{q}_e\text{'} = \text{'}\bar{\mathbf{q}}_{ef}\text{'} + \beta\text{'}\mathbf{q}_{ef}\text{'} = (\lambda\mathbf{q}_{ef} + \mathbf{g}) - \beta\mathbf{g}. \tag{9.90}$$

In order to apply this loading '$\bar{\mathbf{q}}_{ef}$' is kept fixed while the new 'loading scalar', $\beta$, (usually written as $\lambda$) is incremented fron zero to unity in order to recover an equilibrium state related to the original intended loading ($\lambda\mathbf{q}_{ef}$). During this process, the whole range of solution strategies can be adopted, including, for example, automatic incrementation and line searches. In this basic form, the process can only be applied with elastic materials.

The 'homotopy' methods obviously have some links with a range of structural methods derived from the dynamic analogy [O1, D4, K8]. One of these is the form of dynamic relaxation as originally proposed by Otter and Day [O1, D4] which has been applied to both linar [C3] and non-linear [F13] structural problems. In this form a diagonalised mass matrix is used and, in relation to the solution of linear equations, the method can be viewed [C5] as a second-order Jacobi or Richardson process [F11].

Such a method is only one of a range of iterative methods (without a stiffness matrix) that can be used to solve both linear and non-linear equations. Similar methods include conjugate gradients [H3, P10, F8, C5] and the Lanczos method [L1, P8]. As for linear problems, a difficulty with such methods can be their slow convergence rates with badly conditioned problems (such as shells) and a range of pre-conditioning techniques have been used [K4, J1, H5, H6, C5, P17, P2, N5, C1]. These preconditioning techniques can be viewed as providing an approximate, easily factorisable stiffness marix [C5]. For non-linear structural problems, the iterative solution techniques can be embedded in an inner–outer loop whereby the outer loop is a form of Newton procedure for the non-linearities while the inner loop involves the iterative solution of the linear equations [N4, P3, P4, P5]. Because of the outer loop, the inner loop need not be solved accurately for the early outer iterations.

We have already indicated the difficulties that the arc-length method (and other methods) encounters with fiburcation points. There has been a mass of literature on the basic theory and numerical procedures [K1, K3, P5, T1–T4, A4, D6, M5, U1]. (The review by Ulrich [U1] gives 1568 papers on numerical methods for continuation and bifurcation problems!) Techniques for their implementation in finite element codes have been given in [W7, W8, A3, R10, R11, R12, K6] but, again, further work is required.

A further set of techniques that should be mentioned are the perturbation and reduced basis techniques in which various methods are used to provide a reduced number of basis vectors with which the non-linear analysis can work [C2, H1, A5, N3]. Clearly as the solution path is traced, these reduced basis vectors must be updated.

## 9.11 SPECIAL NOTATION

$a_i$ = scalar for quasi-Newton update (see (9.64))
$C_s$ = current stiffness parameter (see (9.44))
$I_d$ = desired number of iterations
$I_o$ = number of iterations for old (previous) increment
$\mathbf{K}_o$ = old fixed tangent stiffness matrix (Section 9.8)
$\mathbf{q}_{ef}$ = 'fixed' external load vector (usually to be incremented via a scalar $\lambda$)
$\bar{\mathbf{q}}_{ef}$ = fixed external load vector that will not be incremented (see (9.45))
$\mathbf{q}_i$ = internal nodal forces
$r$ = line search ratio (see (9.11))
$s$ = energy slope $= \partial\phi/\partial\eta$ (Section 9.2) or arc-length distance (Section 9.3)
$s_o = s$ at $\eta = 0$
$\mathbf{v}_i$ = vector for quasi-Newton update (see (9.63))
$\mathbf{w}_i$ = vector for quasi-Newton update (see (9.62))
$\beta$ = convergence tolerance (see (2.30))
$\beta_{ls}$ = line-search tolerance (see (9.11))
$\boldsymbol{\gamma} = \mathbf{g}_n - \mathbf{g}_o$
$\delta\mathbf{p}$ = iterative displacement change
$\delta\mathbf{p}_i^* = -\mathbf{K}_o^{-1}\mathbf{g}_i$
$\delta\mathbf{p}_t$ = tangential displacement change (see (9.23))
$\delta\bar{\mathbf{p}}$ = iterative displacement change
$\Delta l$ = incremental arc length
$\Delta\mathbf{p}$ = incremental displacement change (from last converged equilibrium state)
$\delta\lambda$ = iterative change in $\lambda$
$\Delta\lambda$ = incremental change in $\lambda$
$\lambda$ = scalar load-level parameter
$\eta$ = step length for line search
$\psi$ = scaling parameter for arc-length constraint (see (9.19))

### Subscripts

n = 'new'
o = 'old'

## 9.12 REFERENCES


[A1] Allgower, E. L., A survey of homotopy methods for smooth mappings, *Lecture Notes in Mathematics—Numerical Solution of Nonlinear Equations*, ed. E. L. Allgower *et al.*, Springer-Verlag, Berlin, pp. 2–29 (1981)

[A2] Allman, D. J., An iterative method to locate minima of a function of several variables, Royal Aircraft Estab. Tech. Report 84038 (1984).

[A3] Allman, D. J., Calculation of the stable equilibrium paths of discrete conservative systems with singular points, Royal Aircraft Estab. Tech. Report 86053, (1986); also *Computers & Structures*, **32**, 1045–1054 (1989).

[A4] Allman, D. J., On the general theory of the stability of equilibrium of discrete conservative systems, *Aeronautical Journal*, 29–35 (January 1989).

[A5] Almroth, B. O., Stern, P. & Brogan, F. A., Automated choice of global shape functions in structural analysis. *AIAA J.*, **16**, 525–528 (1978).

[B1] Bartholomew, P., A simple iterative method for the minimisation of an unconstrained function of many variables, Royal Aircraft Estab., Tech. Memo MAT/STR 1021, Farnborough, Hants. (1983).

[B2] Bathe, K.-J. & Dvorkin, E. N., On the automatic solution of nonlinear finite element equations, *Computers & Structures*, **17**, 871–879 (1983).

[B3] Bathe, K.-J. & Cimento, A. P., Some practical procedures for the solution of nonlinear finite element equations, *Comp. Meth. Appl. Mech. & Engng.*, **22**, 59–85 (1980) (BFGS).

[B4] Batoz, J. L. & Dhatt, G., Incremental displacement algorithms for non-linear problems, *Int. J. Num. Meth. Engng.*, **14**, 1262–1266 (1979). (General disp. control.)

[B5] Belleni, P. X. & Chulya, A., An improved automatic incremental algorithm for the efficient solution of nonlinear finite element equations, *Computers & Structures*, **26**, 99–110 (1987).

[B6] Bergan, P. G., Solution algorithms for non-linear structural problems, *Engineering Applications of the Finite Element Method, Computas*, 13.1–13.39 (1979); also *Computer & Structures*, **12**, 497–509 (1980).

[B7] Bergan, P. G. & Soreide, T., Solution of large displacement and instability problems using the current stiffness parameter, *Finite Elements in Non-linear Mechanics*, Tapir Press, 647–669 (1978).

[B8] Bergan, P. G., Horrigmoe, G., Krakeland, B. & Soreide, T. H., Solution techniques for non-linear finite element problems, *Int. J. Num. Meth. Engng.*, **12**, 1677–1696 (1978).

[B9] Bergan, P. G. & Soreide, T., A comparative study of different numerical techniques as applied to a nonlinear structural problem, *Comp. Meth. Appl. Mech. & Engng.*, **2**, 185–201 (1973).

[B10] Bergan, P. G. & Mollestad, E., Static and dynamic solution strategies in nonlinear analysis, *Numerical Methods for Non-linear Problems*, Vol. 2., ed. C. Taylor *et al.*, Pineridge, Swansea, pp. 3–17 (1984).

[B11] Brodlie, K. W., Gourlay, A. R. & Greenstadt, J., Rank-one and rank-two corrections to positive definite matrices expressed in product form, *J. Inst. Math. Appl.*, **11**, 73–82 (1973).

[B12] Broyden, C. G., The convergence of a double-rank minimisation 2: the new algorithm, *J. Inst. Math. Appl.*, **6**, 222–231 (1970) (BFGS).

[B13] Broyden, C. G., Quasi-Newton methods and their application to function minimisation, *Math. Comp.*, **21**, 368–381 (1967) (rank-one).

[B14] Buckley, A. G., A combined conjugate-gradient quasi–Newton minimisation algorithm, *Math. Programm.*, **15**, 200–210 (1978).

[B15] Buckley, A. & Lenir, A., QN-like variable storage conjugate gradients, *Math. Programming*, **27**, 155–175, (1983).

[C1] Carey, G. F. & Bo-Nan, J., Element-by-element preconditioned conjugate gradient algorithm for compressible flow, *Innovative Methods for Nonlinear Problems*, Pineridge, Swansea, pp. 41–49 (1984).

[C2] Carnoy, E., Postbuckling analysis of elastic structures by the finite element method, *Comp. Meth. Appl. Mech. & Engng.*, **23**, 143–174 (1980).

[C3] Cassel, A. C., Shells of revolution under arbitrary loading and the use of ficititious densities in dynamic relaxation, *Proc. Inst. Civ. Engrs.*, **45**, 65–78 (1970).

[C4] Clarke, M. J. & Hancock, G. J., A study of incremental-iterative strategies for non-linear analysis, *Int. J. Num. Meth. Engng.*, **29**, 1365–1391 (1990).

[C5] Crisfield, M. A., *Finite Elements and Solution Procedures for Structural Analysis*, Vol 1: Linear Analysis, Pineridge Press, Swansea (1986).

[C6] Crisfield, M. A. & Wills, J., Criteria for a reliable non-linear finite element system, *Reliability of Methods for Engineering Analysis*, ed. K. J. Bathe *et al.*, Pineridge press, Swansea, pp. 159–179 (1986).

[C7] Crisfield, M. A., Incremental/iterative solution procedures for non-linear structural analysis, *Numerical Methods for Non-linear Problems*, Vol. 1, ed. C. Taylor *et al.*, Pineridge Press, Swansea, pp. 261–290 (1980).

[C8] Crisfield, M. A., Solution procedures for nonlinear structural problems, *Recent Advances in Non-linear Computational Mechanics.*, ed. E. Hinton *et al.*, Pineridge, Swansea, pp. 1–40 (1982).

[C9] Crisfield, M. A., Iterative Solution Procedures for Linear and Non-linear Structural Analysis, TRRL Report LR900, Transport & Road Res. Lab., Crowthorne, Berks., UK (1979).

[C10] Crisfield, M. A., A faster modified Newton–Raphson iteration, *Comp. Meth. Appl. Meth. Appl. Mech. & Engng.*, **20**, 267–278 (1979).

[C11] Crisfield, M. A., A fast incremental/iterative solution procedure that handles 'snap-through', *Computers & Structures*, **13**, 55–62 (1981).

[C12] Crisfield, M. A. & Wills, J., Solution strategies and softening materials, *Comp. Meth. Appl. Mech. & Engng.*, **66**, 267–289 (1988).

[C13] Crisfield, M. A., Accelerating and damping the modified Newton–Raphson method, *Computers & Struct.*, **18**, 395–407 (1984).

[C14] Crisfield, M. A., Snap-through and snap-back response in concrete structures and the dangers of under-integration, *Int. J. Num. Engng.*, **22**, 751–767 )1986).

[C15] Crisfield, M. A., Overcoming limit points with material softening and strain localisation, *Numerical Methods for Nonlinear Problems*, Vol. 2, ed. C. Taylor *et al.*, *Pineridge*, Swansea, pp. 244–277 (1984).

[C16] Crisfield, M. A., Variable Step-lengths for Non-linear Structural Analysis, Transport and Road Res. Lab. Report LR1049, Crowthorne, Berks., England (1982).

[C17] Crisfield, M. A., New solution procedures for linear and nonlinear finite element analysis, *The Mathematics of Finite Elements and Applications* V, ed. J. Whiteman, Academic Press, pp. 49–81 (1985).

[C18] Crisfield, M. A. & Wills, J., Numerical analysis of a half-scale beam and slab bridge deck, *Proc. 2nd Int. Conf. on Computer Aided Analysis and Design of Concrete Structures*, ed. N. Bicanic, Pineridge, Swansea, Vol. 1, pp. 365–378 (1990).

[C19] Crisfield, M. A., An arc-length method including line searches and accelertions, *Int. J. Num. Meth. Engng.*, **19**, 1269–1289 (1983).

[C20] Crisfield, M. A., Large-deflection Elasto-plastic Buckling Analysis of Plates Using Finite Element, Transport & Road Res. Lab. Report LR593, Crowthorne, Berks., England (1973).

[C21] Crisfield, M. A., Theoretical and Experimental Behaviour of a Lightly Stiffened Box-girder Diaphragm, Transport & Road Res. Lab. Report LR961, Crowthrone, Berks., England (1980).

[C22] Crisfield, M. A., Accelerated solution techniques and concrete cracking, *Comp. Meth. in Appl. Mech. & Engng.*, **33**, 585–607 (1982).

[D1] Davidenko, D. F., On a new method of numerical solution of non-linear equations, *Dokl. Acad. Nauk. SSSR*, **88**, 601–602 (1953).

[D2] Davidon, W. C., Variable Metric Method for Minimisation, Argonne Nat. Lab. Report ANL-5990 (1959).

[D3] Davidon, W. C., Variance algorithms for minimisation, *Computer J.*, **10**, 406–410 (1968).

[D4] Day, A. S., Introduction to dynamic relaxation, *The Engineer*, **219**, 218–221 (1965).

[D5] De Borst, R., Computation of post-bifurcation and post-failure behaviour of strain-softening solids, *Computers & Structures*, **25**, 211–224 (1987).
[D6] Decker, D. W. & Keller, H. B., Path following near bifurcation, *Comm. Pure Appl. Mech.*, **34**, 149–175 (1981).
[D7] Den Heijer, C. & Rheinboldt, W. C., On steplength algorithms for a class of continuation methods, *SIAM J. Num. Analysis*, **18**, 925–948 (1981).
[D8] Dennis, J. E. & More, J., Quasi-Newton methods, motivation and theory, *SIAM Rev.*, **19**, 46–89 (1977).
[E1] Eriksson, E., Using eigenvector projections to improve convergence in non-linear finite element equilibrium iterations, **24**, 497–512 (1987).
[E2] Eriksson, A., On linear constraints for Newton–Raphson corrections and critical point searches in structural f.e. problems, *Int. J. Num. Meth.* Engng., **28**, 1317, 1334 (1989).
[E3] Eriksson, E., On some path-related measures for non-linear structural f.e. problems, *Int. J. Num. Meth. Engng.*, **26**, 1791–1803 (1988).
[F1] Felippa, C. A., Solution of nonlinear static equations, Chapter prepared for North-Holland Series on Comp. Meth. in Mechanics, Vol. on Large deflection and stability of structures, ed. K.-J. Bathe, Dept. of Mech. Engng., MIT (1986).
[F2] Felippa, C. A., Dynamic relaxation and quasi-Newton methods, *2nd Int. Conf. on Numerical Meth. for Nonlinear Problems*, Barcelona (1984).
[F3] Felippa, C. A., Procedures for computer analysis of large nonlinear structural systems, *Proc. Int. Symp. On Large Engng. Systems*, ed. A. Wexler, Pergamon Press, pp. 60–90 (1977).
[F4] Felippa, C. A., Dynamic relaxation under general incremental control, Innovative methods for Nonlinear Problems, ed. W. K. Liu *et al.*, Pineridge Press, Swansea, pp. 103–134 (1984).
[F5] Fink, J. P. & Rheinboldt, W. C., A geometric framework for the numerical study of singular points, *SIAM J. Num. Anal.*, **24**, 618–633 (1987).
[F6] Fletcher, R., A New approach to variable metric algorithms, *Computer J.*, **13**, 317–322 (1970).
[F7] Fletcher, R. Practical methods of optimisation, 2nd edition, Wiley, (1987).
[F8] Fletcher, R. & Reeves, C. M., Function minimisation by conjugate gradients, *Comp. J.*, **6**, 149–154 (1964).
[F9] Forde, B. W. R. & Sttemer, S. F., Improved arc length orthogonality methods for nonlinear finite element analysis, *Computers & Structures*, **27**, 625–630 (1987).
[F10] Fox, L. & Stanton, E., Developments in structural analysis of direct energy minimisation, *AIAA J.*, **6**, 1036–1042 (1968).
[F11] Frankel, S. P., Convergence rates of iterative treatments for partial differential equations, *Maths. Tables Aids Comput.*, **4**, 65–75 (1950).
[F12] Fried, I., Orthogonal trajectory accession to the nonlinear equilibrium curve, *Comp. Meth. Appl. Mech. & Engng.* **47**, 283–298 (1984).
[F13] Frieze, P. A., Hobbs, R. E. & Dolwing, P. J., Application of dynamic relaxation to elasto-plastic analysis of plates, *Computers & Structures*, **8**, 301–310 (1978).
[G1] Georg, K., Numerical integration of Davidenko equation, *Lecture Notes in Mathematics—Numerical Solution of Nonlinear Equations*, ed. H. B. Keller, pp. 129–161 (1981).
[G2] Geradin, M., Idelsohn, S. & Hogge, M., Computational strategies for the solution of large non-linear problems via quasi-Newton methods, *Computational Methods in Nonlinear Structural and Solid Mechanics.*, ed. A. K. Noor *et al.*, Pergamon, pp. 73–82 (1980).
[G3] Gierlinski, J. T. & Graves-Smith, T. R., A variable load iteration procedure for thin-walled structures, *Computers & Structures*, **21**, 1085–1094 (1985).
[G4] Gill, P. E. & Murray, W., Safeguarded Step-length Algorithms for Optimisations Using Descent Methods, National Physical Laboratory Report NAL 37 (1974).
[G5] Gill, P. E. & Murray, W., Quasi-Newton methods for unconstrained optimisation, *J. Inst. Math. Appl.*, **9**, 91–108 (1972).
[H1] Haftka, R. T., Mallet, R. H., Nachbar, W. Adaption of Koiter's method to finite element

analysis of snap-through buckling behaviour, *Int. J. Solids & Structs.*, **7**, 1427–1447 (1981).

[H2] Haselgrove, C. B., The solution of non-linear equations and of differential equations with two-point boundary conditions, *Computer J.*, 255–259 (1961).

[H3] Hestenes, M. & Steifel, E., Method of conjugate gradients for solving linear systems, *J. Res. Nat. Bur. Stand.*, **49**, 409–436 (1952).

[H4] Hinton, E., Abdal-Rahman, H. H. & Zienkiewicz, O. C., Computational Models for Reinforced Concrete Slab Systems, IABSE Colloq. on advanced Mechanics of Reinforced Concrete, Final Report, IABSE, Delft, pp. 303–314 (1981).

[H5] Hughes, T. J. R., Levit, I. & Winget, J., An element-by-element algorithm for problems of structural and solid mechanics, *Comp. Meth. Appl. Mech. & Engng.*, **36**, 241–254 (1983).

[H6] Hughes, T. J. R., Ferencz, R. M. & Hallquist, J. O., Large-scale vectorised implicit calculations in solid mechanics on a CRAY-XMP/48 utilising EBE preconditioned conjugate gradients, *Comp. Meth. Appl. Mech. & Engng.*, **61**, 215–248 (1987).

[I1] Irons, B. & Elsawaf, A., The conjugate-Newton algorithm for solving finite element equations, *Proc. US–German Symp. on Formulations and Algorithms in Finite Element Analysis*, ed. K. J. Bathe *et al.*, MIT, 656–672 (1977).

[J1] Jennings, A., Development of an ICCG algorithm for large sparse systems, Preconditioning Techniques in Numerical Solution of Partial Differential Equations, ed. D. J. Evans, Gordon & Breach, New York, pp. 426–438 (1983).

[J2] Jeusette, J.-P., Laschet, G., & Idelsohn, S., An effective incremental iterative method for static nonlinear structural analysis, *Comp. & Struct.*, **32**, 125–135 (1989).

[K1] Kearfott, R. B., A derivative free arc continuation method and a bifurcation technique, *Numerical Solution of Nonlinear Equations*, ed. E. L. Allgower *et al.*, Springer-Verlag, Berlin, pp. 183–198 (1981).

[K2] Keller, H. B., Global homotopies and Newton methods, Recent Advances in Numerical Analysis, ed. C. de Boor G. H. Golub, Academic Press, New York, pp. 73–94, (1978).

[K3] Keller, H. B., Numerical solution of bifurcation and nonlinear eigenvalue problems, Applications of Bifurcation Theory, ed. P. Rabinowitz, Academic Press, pp. 359–384 (1977).

[K4] Kershaw, E., The incomplete Cholesky conjugate gradient method for iterative solution of systems of linear equations, *J. Comp. Phys.*, **26**, 43–65 (1978).

[K5] Kondoh, K., & Atluri, S., Influence of local buckling on global instability: simplified, large deformation, post-buckling analyses of plane trusses, *Computers & Structures*, **21**, 613–627 (1985).

[K6] Kouhia, R. & Mikkola, M., Tracing the equilibrium path beyond simple critical points, *Int. J. Num. Meth. Engng.*, **28**, 2923–2941 (1989).

[K7] Kroplin, B., Dinkler, D. & Hillmann, J., An energy perturbation applied to nonlinear structural analysis, *Comp. Meth. Appl. Mech. & Engng.*, **52**, 885–897 (1985).

[K8] Kroplin, B. H., A viscous approach to post buckling analysis, *Engng. Struct.*, **3**, 187–189 (1981).

[K9] Kroplin, B., Dinkler, D. & Hillman, J., Global constraints in nonlinear solution strategies, *Finite Element Methods for Nonlinear Problems*, ed. P. G. Bergan *et al.*, Springer, Berlin (1986).

[K10] Kroplin, B. & Dinkler, D., Global constraints applied to static instability problems, *Numerical Methods for Nonlinear problems*, ed. C. Taylor *et al.*, Vol. 3, Pineridge, Swansea, pp. 41–60 (1986).

[L1] Lanczos, C, *Applied Analysis*, Prentice Hall, New York (1956).

[L2] Luenberger, D. G., Linear and nonlinear programming, 2nd edn, Addison-Wesley, Reading, Mass. (1984).

[M1] Matthies, H. & Strang, G., The solution of non-linear finite element equations, *Int. J. Num. Meth. Engng.*, **14**, 1613–1626 (1979).

[M2] Meek, J. L. & Loganthan, S., Large displacement analysis of space-frame structures, *Comp. Meth. Appl. Mech. & Engng.*, **72**, 57–75 (1989).

[M3] Meek, J. L. & Loganathan, S., Geometrically non-linear behaviour of space frame structures, *Computers & Structures*, **31**, 35–45 (1989).

[M3] Melhem, R. G. & Rheinboldt, W. C., A comparison of methods for determining turning points of nonlinear equations, *Computing*, **29**, 201–226 (1982).

[M4] Meek, J. L. & Tan, H. S., Geometrically nonlinear analysis of space frames by an incremental iterative technique, *Comp. Meth. Appl. Mech. & Engng.*, **47**, 261–282, (1984).

[M5] Moore, G. The numerical treatment of non-triavial bifurcation points, *Num. Funct. Anal & Optimiz.*, **2**(6), 441–472 (1980).

[N1] Nazareth, L., A relationship between the BFGS and conjugate gradient algorithms and its implications and its implications for new algorithms, *SIAM J. Num. Analysis*, **16**, 794–800 (1979).

[N2] Nocedal, J., Up-dating quasi-newton matrices with limited storage, *Math. Computation*, **35**, 773–782 (1980).

[N3] Noor, A. E. & Peters, J. M., Tracing post-limit-point paths with reduced basis techniques, *Comp. Meth. Appl. Mech. & Engng.*, **28**, 217–240 (1981).

[N4] Nour-Omid, B., Parlett, B. N. & Taylor, R. L., A Newton–Lanczos method for solution of nonlinear finite element equations, *Computers & Structures*, **16**, 241–252 (1983).

[N5] Nour-Omid, B., A preconditioned conjugate gradient method for solution of finite element equations, *Innovative Methods for Nonlinear Problems*, Pineridge, Swansea, pp. 17–40 (1984).

[O1] Otter, J. R. H. & Day, A. S., Tidal computations, *The Engineer*, **209**, 177–182 (1960).

[P1] Padovan, J. P. & Arechaga, T., Formal convergence characteristics of elliptically constrained incremental Newton–Raphson algorithms, *Int. J. Engng. Sci.*, **20**, 1077–1097 (1982).

[P2] Papadrakakis, M., Accelerating vector iteration methods, *J. Appl. Mech.*, ASME, **53**, 291–297 (1986).

[P3] Papadrakakis, M. & Gantes, C. J., Preconditioned conjugate- and secant-Newton methods for non-linear problems, *Int. J. Num. Mech. Engng.*, **28**, 1299–1316 (1989).

[P4] Papadrakakis, M., A truncated Newton–Lanczos method for overcoming limit and bifurcation points, *Int. J. Num. Meth. Engng.*, **29**, 1065–1077 (1990).

[P5] Papadrakakis, M. & Nomikos, N., Automatic non-linear solution with arc-length and Newton–Lanczos methods, *Engineering Comp.*, **7**, 1 (1980).

[P6] Park, K. C. & Rankin, C. C., A semi-implicit dynamic relaxation algorithm for static nonlinear structural analysis, *Research in Structural and Sold Mechanics*—1982, NASA Conf. Pub. 2245 (1982).

[P7] Park, K. C., A family of solution algorithms for nonlinear structural analysis based on the relaxation equations, *Int. J. Num. Meth. Engng.*, **18**, 1337–1347 (1982).

[P8] Parlett, B. N., A new look at the Lanczos algorithm for solving symmetric systems of linear equations, *Linear Algebraic Applications*, **29**, 323–346 (1980).

[P9] Pica, A. & Hinton, E., The quasi-Newton BFGS method in large deflection analysis of plates, *Num. Meth. Nonlinear Problems*, ed. C. Taylor *et al.*, Vol. 1, Pineridge, Swansea, pp. 355–366 (1980).

[P10] Polak, E. & Ribiere, Note sur la convergence de méthodes de directions conjugées, *Rev. Fr. Inform Rech. Oper.*, **16–R1**, 35–43 (1969).

[P11] Powell, G. & Simons, J., Improved iteration strategy for nonlinear structures, *Int. J. Num. Meth. Engng.*, **17**, 1455–1467 (1981).

[R1] Ramm, E., Strategies for tracing the non-linear response near limit-points, *Non-linear Finite element Analysis in structural Mechanics*, ed. W. Wunderlich, Springer-Verlag, Berlin, pp. 63–89 (1981).

[R2] Ramm, E., The Riks/Wempner approach—an extension of the displacement control method in non-linear analysis, *Non-linear Computational Mechanics*, ed. E. Hinton *et al.*, Pineridge, Swansea, pp. 63–86 (1982).

[R3] Rheinboldt, W. C., Numerical analysis of parametrised nonlinear equations, Vol. 7, Wiley (1986).

[R4] Rheinboldt, W. C., Numerical continuation methods for finite element applications, *Formulations and Computational algorithms in Finite Element Analysis*, ed. K.-J. Bathe, MIT, pp. 600–631 (1977).

[R5] Rheinboldt, W. C., Numerical methods for a class of finite dimensional bifurcation problems, *SIAM J. Num. Analysis*, **15**(1) 1–11 (1978).

[R6] Rheinboldt, W. C., Numerical analysis of continuation methods for nonlinear structural problems, *Computers and Structures*, **13**, 103–113 (1981).

[R7] Rheinboldt, W. C. & Riks, E., A Survey Solution Techniques for Nonlinear Finite Element Equations, Nat. Aerospace Lab. Report NLR 82027U, (1982). Also *State of the Art Surveys on Finite Element Techniques*, ed. A. K. Noor *et al.*, ASME, (1983).

[R8] Riks, E., An incremental approach to the solution of snapping and buckling problems, *Int. J. Solids & Structs.* **15**, 529–551 (1979).

[R9] Riks, E., The application of Newton's method to the problem of elastic stability, *J. Appl. Mech.*, **39**, 1060–6 (1972).

[R10] Riks, E., Bifurcation and Stability, a numerical approach, *Innovative methods for Nonlinear Problems*, ed. W. K. Liu *et al.*, Pineridge Press, Swansea, pp. 313–344 (1984).

[R11] Riks, E., Progress in collapse analysis, *Collapse Analysis of Structures, ASME-Pressure Vessel and Piping Conf., ASME PVP-084* (San Antonio, Texas), ed. L. H. Sobel *et al.*, New York (1984).

[R12] Riks, E., Some Computational Aspects of the Stability Analysis of Nonlinear Structures, Nat. Aero Lab. Report NLR MP 82034 U, (1981) also *Comp. Meth. Appl. Mech. & Engng.*, **47**, 219–259 (1984).

[R13] Riks, E., A unified method for the computation of critical equilibrium states of nonlinear elastic systems, *Acta Technica Scientiarum Hungaricae*, **87**, 121–141 (1978).

[R14] Riks, E. & Rankin, C. C., Bordered Equations in Continuation Methods: An Improved Solution Technique, Nat. Aero Lab. Report NLR MP 82057 U (1987).

[S1] Schmidt, W. F., Adaptive step size selection for use with the continuation method, *Int. J. Num. Meth. Engng.*, **12**, 677–694 (1978).

[S2] Schweizerhof, K. & Wriggers, P., Consistent linearization for path following methods in nonlinear f.e. analysis, *Comp. Meth. Appl. Mech. & Engng.*, **59**, 261–279 (1986).

[S3] Shanno, D. F., Conjugate gradient methods with inexact searches, *Math. O. R.*, **13** (3), 244–255 (1978).

[S4] Shanno, D. F., Conditioning of quasi-Newton methods for function minimisation, Math Computation, **24**, 322–334 (1970).

[S5] Simons, J. & Bergan, P. G., Hyperplane displacement control methods in nonlinear analysis, *Innovative methods for Nonlinear Problems*, ed. W. K. Liu *et al.*, Pineridge Press, Swansea, pp. 345–364 (1984).

[S6] Simo, J. C. Wriggers, P., Schweizerhof, K. H. & Taylor, R. L., Finite deformation postbuckling analysis involving ineasticity and contact constraints, *Innovative Methods for Nonlinear Problems*, ed. W. K. Liu *et al.*, Pineridge Press, Swansea, pp. 365–388 (1984).

[S7] Stricklin, J. A., Haisler, W. E. & Von Riesemann, W. A., Evaluation of solution procedures for nonlinear structural analysis, *AIAA J.*, **11**, 292–299 (1973).

[S8] Stricklin, J. A. & Haisler, W. E. Formulation and solution procedures for nonlinear structural analysis, *Computers and Structures*, **7**, 125–136 (1977).

[T1] Thompson, J. M. T. & Hunt, G. W., A general theory for elastic stability, Wiley, London (1973).

[T2] Thurston, G. A., Continuation of Newton's method through bifurcation points, *J. Appl. Mech.*, **36**, 425–430 (1969).

[T3] Thurston, G. A., Brogan, F. A. & Stehlin, P., Postbuckling Analysis Using a General Purpose Code, AIAA Paper 85-0719-CP, presented at the AIAA/ASME/ASCE/AHS 26th Structures, Structural Dynamics and Materials Conf., Orlando, Florida (1985).

[T4] Thurston, G. A., Newton's method: a link between continuous and discrete solutions of nonlinear problems, *Research in Nonlinear and Solid Mechanics*, compiled by H. G. McComb *et al.*, NASA, Washington (1980).

[U1] Ulrich, K., State of the Art in Numerical Methods for Continuation and Bifurcation Problems with Applications in Continuum Mechanics—a Survey and Comparative Study, Report No. 031/88, Laboratorio, Nacional de Computacao Cientifica, Rio de Janerio, Brazil (1988).

[W1] Wagner, W. & Wriggers, P., A simple method for the calculation of postcritical branches, *Engineering Computations*, **5**, 103–110 (1988).

[W2] Waszczyszyn, Z., Numerical problems of nonlinear stability analysis of elastic structures, Computers and Structures, **17**(1), 13–24 (1983).

[W3] Watson, L. T., An algorithm that is globally convergent with probability one for a class of non-linear two-point boundary value problems, *SIAM J. Num. Anal.*, **16**, 394–401 (1979).

[W4] Watson, L. T. & Holzer, M., Quadratic convergence of Crisfield's method, **17**, 69–72 (1983).

[W5] Wempner, G. A., Discrete approximations related to nonlinear theories of solids, *Int. J. Solids & Structs.*, **7**, 1581–1599 (1971).

[W6] Wolfe, M. A., *Numerical Methods for Unconstrained Optimisation—an Introduction*, Van Nostrand Reinhold, (1978).

[W7] Wriggers, P., Wagner, W. & Miehe, C., A quadratically convergent procedure for the calculation of stability points in finite element analysis, *Comp. Meth. Appl. Mech. & Engng.*, **70**, 329–347 (1988).

[W8] Wriggers, P. & Simo, J. C., A general procedure for the direct computation of turning and bifurcation points, *Int. J. Num. Meth. Engng.*, **30**, 155–176 (1990).

[W9] Wright, E. W. & Gaylord, E. H., Analysis of unbraced multistorey steel rigid frames, Proc. ASCE, *J. Struct. Div.*, **94**, 1143–1163 (1968).

# Appendix
# Lobatto rule for numerical integration

| No. of points | Position | Weighting |
|---|---|---|
| 3 | ±1 | 0.333 333 33 |
| | 0 | 1.333 333 33 |
| 4 | ±1 | 0.166 666 67 |
| | ±0.447 213 60 | 0.833 333 33 |
| | ±1 | 0.100 000 00 |
| 5 | ±0.654 653 67 | 0.544 444 44 |
| | 0 | 0.711 111 11 |
| | ±1 | 0.066 666 67 |
| 6 | ±0.765 055 32 | 0.378 474 96 |
| | ±0.285 231 52 | 0.554 858 38 |
| | ±1 | 0.047 619 04 |
| 7 | ±0.830 223 90 | 0.276 826 04 |
| | ±0.468 848 79 | 0.431 745 38 |
| | 0 | 0.487 619 04 |
| | ±1 | 0.035 714 28 |
| 8 | ±0.871 740 15 | 0.210 704 22 |
| | ±0.591 700 18 | 0.341 122 70 |
| | ±0.209 299 22 | 0.412 458 80 |
| | ±1 | 0.027 777 7778 |
| | ±0.899 757 9954 | 0.165 495 3616 |
| 9 | ±0.677 186 2795 | 0.274 538 7126 |
| | ±0.363 117 4638 | 0.346 428 5110 |
| | 0 | 0.371 519 2744 |

| No. of points | Position | Weighting |
|---|---|---|
| | $\pm 1$ | 0.022 222 2222 |
| | $\pm 0.919\,533\,9082$ | 0.133 305 9908 |
| 10 | $\pm 0.738\,773\,8651$ | 0.224 889 4320 |
| | $\pm 0.477\,924\,9498$ | 0.292 042 6836 |
| | $\pm 0.165\,278\,9577$ | 0.327 539 7612 |

# Subject index

*Index compiled by Geoffrey C. Jones.*

# Author index

*Index compiled by Geoffrey C. Jones.*